VOLUME 47

Advances in
CHROMATOGRAPHY

VOLUME 47

Advances in
CHROMATOGRAPHY

EDITORS:

ELI GRUSHKA
Hebrew University of Jerusalem
Jerusalem, Israel

NELU GRINBERG
Boehringer-Ingelheim Pharmaceutical, Inc.
Ridgefield, Connecticut, U.S.A.

CRC Press
Taylor & Francis Group
Boca Raton   London   New York

CRC Press is an imprint of the
Taylor & Francis Group, an **Informa** business

CRC Press
Taylor & Francis Group
6000 Broken Sound Parkway NW, Suite 300
Boca Raton, FL 33487-2742

First issued in paperback 2017

© 2009 by Taylor & Francis Group, LLC
CRC Press is an imprint of Taylor & Francis Group, an Informa business

No claim to original U.S. Government works

ISBN 13: 978-1-138-11342-8 (pbk)
ISBN 13: 978-1-4200-6036-2 (hbk)

Visit the Taylor & Francis Web site at
http://www.taylorandfrancis.com

and the CRC Press Web site at
http://www.crcpress.com

Contents

Contributors

Alain Berthod
Analytical Sciences Laboratory
National Center for Scientific
  Research
University of Lyon
Villeurbanne, France

Guodong Chen
Schering-Plough Research Institute
Kenilworth, New Jersey

Benjamin Chu
Department of Chemistry
Stony Brook University
Stony Brook, New York

Victor David
Department of Analytical Chemistry
Faculty of Chemistry
University of Bucharest
Bucharest, Romania

Alexandru Farca
Analytical Control Laboratory
S.C. LaborMed Pharma SA
Bucharest, Romania

Simion Gocan
Department of Analytical Chemistry
Babes-Bolyai University
Cluj-Napoca, Romania

Thomas Budde Hansen
Protein Separation and Virology
Novo Nordisk A/S
Gentofte, Denmark

Jürgen Hubbuch
Institute of Engineering
  in Life Sciences
University of Karlsruhe
Karlsruhe, Germany

Malte Kaspereit
Max Planck Institute for Dynamics
  of Complex Technical Systems
Magdeburg, Germany

Steffen Kidal
Chemistry and Purification
Novo Nordisk A/S
Bagsværd, Denmark

Janus Krarup
Protein Separation and Virology
Novo Nordisk A/S
Gentofte, Denmark

Naotaka Kuroda
Graduate School of Biomedical
  Sciences
Course of Pharmaceutical Sciences
Nagasaki University
Nagasaki, Japan

William R. LaCourse
Department of Chemistry
  and Biochemistry
University of Maryland
Baltimore, Maryland

Andrei Medvedovici
Department of Analytical Chemistry
Faculty of Chemistry
University of Bucharest
Bucharest, Romania

Urooj A. Mirza
Schering-Plough Research Institute
Kenilworth, New Jersey

Jørgen Mollerup
Department of Chemical Engineering
Technical University of Denmark
Lyngby, Denmark

Jacob Nielsen
Protein Separation and Virology
Novo Nordisk A/S
Gentofte, Denmark

Kaname Ohyama
Department of Hospital Pharmacy
Nagasaki University Hospital
  of Medicine and Dentistry
Nagasaki, Japan

Jason S. Page
Biological Sciences Division
Pacific Northwest National
  Laboratory
Richland, Washington

Birendra N. Pramanik
Schering-Plough Research Institute
Kenilworth, New Jersey

Yufeng Shen
Biological Sciences Division
Pacific Northwest National
  Laboratory
Richland, Washington

Richard D. Smith
Biological Sciences Division
Pacific Northwest National Laboratory
Richland, Washington

Arne Staby
CMC Project Planning
  and Management
Novo Nordisk A/S
Gentofte, Denmark

Fen Wan
Department of Chemistry
Stony Brook University
Stony Brook, New York

Matthias Wiendahl
Protein Separation and Virology
Novo Nordisk A/S
Gentofte, Denmark

Jun Zhang
Department of Chemistry
Stony Brook University
Stony Brook, New York

1 Macromolecules in Drug Discovery: Mass Spectrometry of Recombinant Proteins and Proteomics

Guodong Chen, Urooj A. Mirza, and Birendra N. Pramanik

CONTENTS

1.1 INTRODUCTION

There has been an enormous effort within the pharmaceutical and biopharmaceutical industries to discover active drugs that include small molecules and therapeutic proteins for the treatment of life-threatening diseases. The R&D investment by global pharmaceutical companies amounted to approximately $90 billion as reported in 2005, up roughly by 56% from 2001 [1]. These investments may have increased further over the last 3 years. The advances made in pharmaceutical and medical

1

research have significant impact in improving the treatment of common diseases such as heart disease, diabetes, and cancer in particular, as well as rare disorders such as cystic fibrosis and sickle cell anemia. Over 300 new medicines have been approved by Food and Drug Administration (FDA) in the last decade alone [2]. The drug discovery process, however, is encountering more challenges in the successful translation of emerging drug concepts into successful marketed products. Some of the marketed products were withdrawn due to adverse side effects. The overall development process has become more difficult because of expanded regulatory and marketing requirements that necessitate additional preclinical and clinical studies prior to a new drug application submission. The attrition rate of the recommended drug candidates during clinical trials is continuing to rise with about 60% of the compounds leading to the termination of drug development programs [3]. The discovery and development of a new drug costs about $1.7 billion, and it may take up to 10–12 years for the drug to reach the market [4].

The molecular weights (MWs) of small molecule pharmaceuticals generally vary from a few hundred daltons (Da) to as high as 1000 Da. The process of discovering small molecule pharmaceutical products generally involves the development of highly potent molecules that are chemically and metabolically stable, sustain good serum level, selectively bind noncovalently with target proteins, and produce the desired therapeutic response with minimal side effects [5]. Significant progress in the area of genomics and proteomics has generated new target proteins that require rapid characterization by analytical methods. In addition, the biopharmaceutical industry has marketed some 170 protein drugs and vaccines for therapeutic use in treating diseases ranging from various cancers to diabetes. These therapeutic proteins are produced in large quantities by advancement in recombinant DNA technology. Thus, proteins constitute the critical components in the discovery of both small molecule and macromolecule pharmaceuticals [6]. The structural characterization of proteins and peptides is an essential part of this process. Mass spectrometry (MS) methods are uniquely qualified to be used for the analysis of proteins and peptides, including solving a wide range of protein structural identification problems with high speed, accuracy, and sensitivity.

1.2   OVERVIEW OF MS METHODS

MS is a very powerful analytical technique for characterization of proteins. This technique is based on measurements of mass-to-charge ratios of ions in gas phase, providing the MW information as well as structural information from fragment ions [7]. The functioning of a mass spectrometer in generating a mass spectrum involves four steps: (1) the introduction of the sample, (2) the ionization of sample molecules accompanied by transfer of these ions into the gas phase, (3) the sorting of the resulting gas phase ions (mass analyzer) by their mass-to-charge ratios, and (4) the detection of the resolved ions. Through appropriate selection of ionization methods and mass analysis, mass spectrometers can be designed to perform specific analytical functions.

Two of the latest ionization techniques, electrospray ionization (ESI) [8,9] and matrix-assisted laser desorption/ionization (MALDI) [10] or soft ionization [11], have

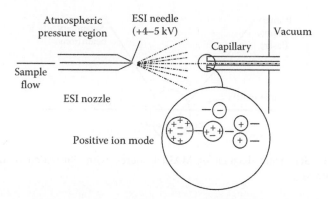

FIGURE 1.1   Schematic diagram of ESI source, illustrating positive ion formation process.

greatly expanded the role of MS in the study of proteins and peptides. The basic ESI source design consists of an ESI spray needle at a high electrical field (4–5 kV) and a thermal/pneumatic desolvation chamber (Figure 1.1). The ESI process usually involves the generation of charged microdroplets under a high electrical field and the subsequent evaporation of droplets under thermal desolvation using a drying gas (N_2). When the initially formed charged droplets become smaller droplets due to evaporation of the solvents, the surface charge density increases and the coulombic forces exceed the surface tension, with the droplets breaking into smaller droplets. Further evaporation process generates analyte ions. The formation of multiply charged ions for proteins/ peptides is one of the most important features in ESI. This allows the detection of higher MWs of proteins/peptides using a standard quadrupole mass analyzer. It also provides precise measurements of MWs of proteins/peptides via deconvolution method. A mass accuracy better than 0.01% can be achieved for proteins with masses up to 100 kDa. Because of the simplicity of the ESI source design and its operation at atmospheric pressure, ESI can easily be coupled to a high-performance liquid chromatography (HPLC) for the analysis of complex mixtures in LC (Liquid chromatography)/ MS mode. Since ESI response is directly related to the concentration of the analyte entering the ion source, the mass sensitivity can be substantially increased with a lower flow rate if the same concentration sensitivity is maintained. This has resulted in the wide use of nanospray (~nL/min) LC/MS for analysis of proteins and peptides with superior mass sensitivity at the femtomole level.

The MALDI technique has high ionization efficiency for large biomolecules including proteins/peptides, and can achieve a mass range greater than 500 kDa when coupled with a time-of-flight (TOF) mass analyzer. In this technique, the sample is mixed with a UV or IR absorbing matrix (i.e., 2,5-dihydroxybenzoic acid, sinapinic acid, α-cyano-4-hydroxycinnamic acid), which is present in large excess (5000:1 mol ratio). The resulting sample mixture is deposited on a sample target, dried, and then inserted into the mass spectrometer for direct laser irradiation/ionization (Figure 1.2). Commonly used lasers in MALDI include N_2 (337 nm), Nd-YAG (355, 266 nm) for UV, and CO_2 (3 μm) for IR. Singly charged ions are often the dominant signals in the MALDI spectrum, along with doubly, triply, or quadruply charged ions.

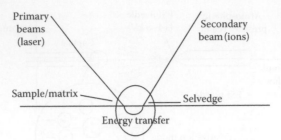

FIGURE 1.2 Schematic diagram of MALDI source with illustration of desorbed ion formation process.

The MALDI technique is an extremely sensitive method for the analysis of higher MW biomolecules at low femtomole level. The sensitivity achieved with MALDI can be ten times better than that of ESI.

Another important aspect of MS analysis is MS/MS or tandem MS experiment [7]. This technique defines the product ions (fragment ion) which are formed from a selected parent ion. Typically, the parent ions of interest are selected in the first analyzer, by their mass/charge ratios; then, these ions are fragmented by collision-induced dissociation (CID), and the resulting product ions are recorded by a second mass measurement usually involving a second mass analyzer. The main advantage of MS/MS experiments is the enhanced specificity of fragmentation which is useful when dealing with complex mixtures. There are different configurations of mass analyzers to perform MS/MS experiments. They include triple quadrupole [12], 3D and linear ion trap [13,14], TOF–TOF [15], and quadrupole-TOF [16]. The recent development of two other mass analyzers, the Orbitrap [17,18] and the Fourier transform MS (FT-MS) [19,20], provides high-resolution capabilities in multiple-step (MS^n) analysis in a single experiment. This accurate mass capability in MS/MS fashion is quite invaluable for the structure characterization of proteins and peptides.

1.3 GENERAL APPROACH FOR PROTEIN CHARACTERIZATION BY MS

1.3.1 GENERAL METHODOLOGY

The first step in protein characterization is the MW determination. Typically, a deconvoluted ESI mass spectrum is generated to give the average MW of the protein, by calculating from successive multiply charged ions in ESI-MS experiments. For example, any adjacent pair of ions (m_1 ion with n charge, m_2 ion with $n + 1$ charge, $m_1 > m_2$) can be used to calculate the number of charge n as $(m_2 - H)/(m_1 - m_2)$. Once n is known, the MW of the protein can be calculated as $m_1 n - nH$. The MW measurements of ions with a different charge are independent calculations. They can be averaged to further improve accuracy via computer algorithm. In the case of horse heart myoglobin, a single-chain globular protein of 153 amino acids contains a heme (iron-containing porphyrin) prosthetic group in the center around which the remaining

GLSDGEWQQVLNVWGKVEADIAGHGQEVLIRLFTGHPETL
EKFDKFKHLKTEAEMKASEDLKKHGTVVLTALGGILKKK
GHHEAELKPLAQSHATKHKIPIKYLEFISDAIIHVLHSKHPG
DFGADAQGAMTKALELFRNDIAAKYKELGFQG

FIGURE 1.3 Amino acid sequence of horse heart myoglobin.

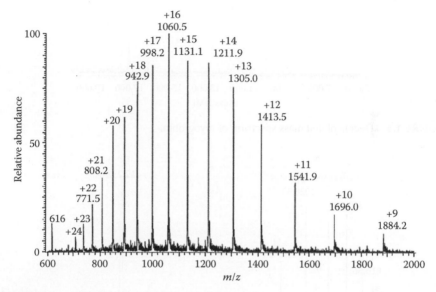

FIGURE 1.4 ESI-MS spectrum of myoglobin at pH 2.5.

apoprotein folds. Figure 1.3 illustrates its amino acid sequence information with averaged MW of 16,951.5 Da. Under acidic condition at pH 2.5, its ESI mass spectrum shows a broad charge-state distribution with charge-states ranging from +9 to +24 (Figure 1.4). If two adjacent ions at m/z 1541.9 and 1413.5 are taken for MW measurement, the number of charge n for m/z 1541.9 can be calculated to be $(1413.5 - 1)/(1541.9 - 1413.5) = 11$. Thus, the calculated MW from this pair of ions is 16,949.9 Da $(1541.9 \times 11 - 11)$. The deconvoluted spectrum (Figure 1.5) gives a measured average MW of 16,951.0 Da for this apoprotein. Under this denaturing condition, the protein is highly charged due to the higher number of sites exposed in the unfolded state (pH 2.5), and the heme ligand is no longer bound to the protein. When myoglobin is analyzed under neutral aqueous conditions (pH 7), the protein remains folded and only eight or nine basic sites are available for protonation, giving rise to a narrow charge-state distribution. The heme ligand remains bound and the mass measured is 17,567.8 Da including for both apoprotein and heme group (data not shown).

Another protein bovine cytochrome c has an averaged MW of 12,233 Da, and its ESI mass spectrum at pH 2.5 is shown in Figure 1.6 with a charge-state distribution ranging from +7 to +18. Its deconvoluted spectrum yields a measured MW of 12,232 Da. A tightly folded protein bovine ubiquitin (theoretical MW of 8565 Da)

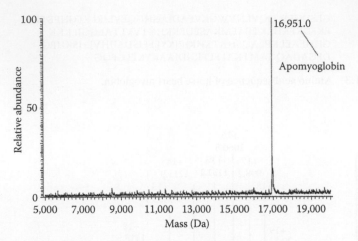

FIGURE 1.5 Deconvoluted mass spectrum of myoglobin.

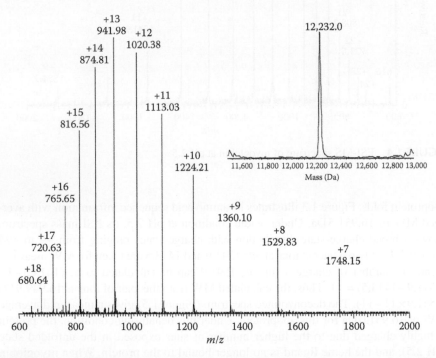

FIGURE 1.6 ESI-MS spectrum of bovine cytochrome *c* under denatured condition at pH 2.5. The insert shows the deconvoluted mass spectrum.

shows a charge-state distribution of +5 to +13 under ESI conditions (Figure 1.7) and has a measured MW of 8567 Da. For proteins with disulfide bonds, dithiothreitol (DTT) can be used to unfold the proteins. In the case of chicken egg lysozyme (theoretical MW of 14,306 Da), its ESI mass spectrum displays a charge-state distribution of +8 to +14 with a measured MW of 14,307 Da (Figure 1.8). After reduction with

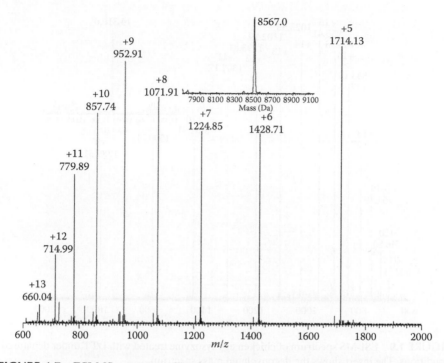

FIGURE 1.7 ESI-MS spectrum of bovine ubiquitin under denatured condition at pH 2.5. The insert shows the deconvoluted mass spectrum.

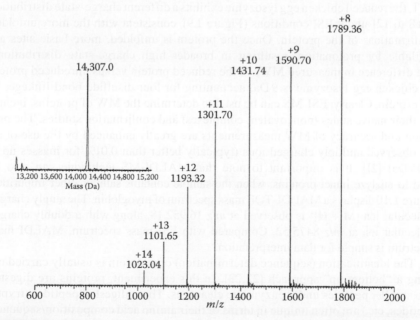

FIGURE 1.8 ESI-MS spectrum of chicken egg lysozyme under denatured conditions. The insert shows the deconvoluted mass spectrum.

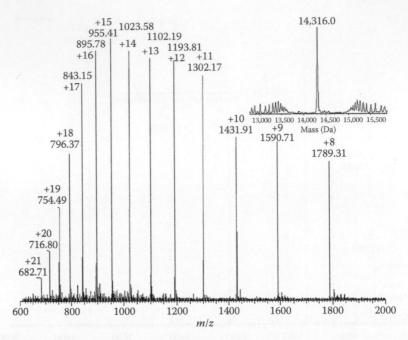

FIGURE 1.9 ESI-MS spectrum of chicken egg lysozyme treated with DTT under denatured conditions. The insert shows the deconvoluted mass spectrum.

DTT, the reduced chicken egg lysozyme exhibits a different charge-state distribution of +8 to +21 under ESI conditions (Figure 1.9), consistent with the more unfolded confirmations of the protein. Once the protein is unfolded, more basic sites are available for protonation, resulting in broader high charge-state distributions. The difference in measured MWs of the reduced protein versus unreduced protein for chicken egg lysozyme is 9 Da, accounting for four disulfide bond linkages in the protein. Clearly, ESI-MS can be used to determine the MW of proteins, including their native states (noncovalent complexes) and confirmation studies. The precision and accuracy of MW measurements are greatly enhanced by the use of all the observed multiply charged ions (typically better than 0.01% for masses up to 100 kDa) [21]. It is important to note that MALDI-MS technique can also be used to analyze intact proteins, when the sample contains salts or other impurities. Figure 1.10 displays a MALDI-TOF mass spectrum of myoglobin. The singly charged molecular ion [M + H]+ is observed at m/z 16,952.18, along with a doubly charged molecular ion at m/z 8475.52. Compared with ESI mass spectrum, MALDI mass spectrum is simple for data interpretation.

The identification (sequence determination) of a protein is usually carried out using a "bottom-up" approach [22,23]. In this experiment, proteins are digested into smaller peptides under enzymatic cleavages. These digested peptides (tryptic peptides, etc.) are often unique in terms of their amino acid composition/sequence, and separation characteristics. They can be separated/detected by LC/MS or MALDI-MS, and either compared to theoretical data for a known protein for

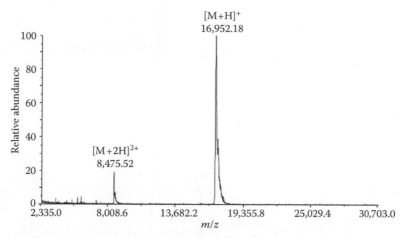

FIGURE 1.10 MALDI mass spectrum of myoglobin.

protein sequence verification, or directly searched against a genome or protein database for protein identification (peptide mass mapping). Note that an enzymatic digest of a large protein can yield fragments of incomplete digestion. For example, trypsin may not cleave at lysine–proline (K–P) bond, and R–P bonds are marginally more susceptible. Also, peptide fragments that contain two contiguous basic sites (K–K, K–R, R–R, etc.) can be observed with R or K on the N-terminal. This is a result of the poor exoprotease activity of trypsin. In the case of myoglobin, the protein was digested with trypsin at a protease-to-protein ratio of 1:25 overnight at 37°C. A total ion chromatogram of myoglobin-digested tryptic peptides is shown in Figure 1.11. The observed mass values from tryptic peptides are listed in Table 1.1. Clearly, this peptide mass mapping method can be used to confirm known protein sequences and allow structural identifications of posttranslational modifications, by comparing measured mass values with calculated mass values of predicted tryptic peptides. It is important to note here that the enzymatic digestion process can be further improved by using microwave-enhanced methodology.

1.3.2 MICROWAVE-ENHANCED REACTIONS FOR PROTEIN AND PEPTIDE CHARACTERIZATION

Microwave is a form of electromagnetic energy and it couples directly with the molecules that are heated at a high rate (about 10^{-9} s), leading to a rapid rise in temperature. A typical chemical reaction can be speeded up by as much as 1000-fold under microwave irradiation. For example, reactions that require several hours under conventional conditions can be completed in a few minutes [24–28]. Bose et al. and Pramanik et al. applied microwave-assisted Akabori reaction (with hydrazine) for rapid linear peptide and cyclic peptide analysis [29,30]. The classical Akabori reaction was devised in 1952 for the identification of C-terminus amino acids, involving the heating of a linear peptide in the presence of anhydrous hydrazine in a sealed

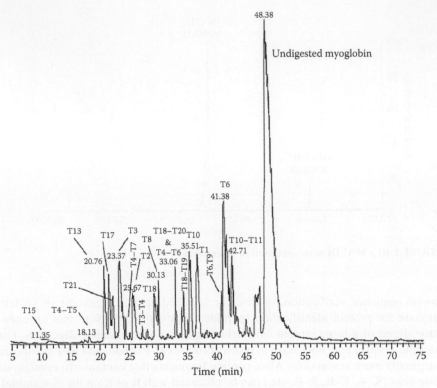

FIGURE 1.11   Total ion chromatogram of tryptic peptides of myoglobin by LC/ESI-MS.

tube for several hours [31]. The C-terminus group is liberated as free amino acid
and can be distinguished from the remaining amino acid residues that have been
converted to hydrazides. In the microwave-assisted method, the linear peptides
and hydrazine solution were exposed to a few minutes of microwave irradiation.
Then, the aliquots were analyzed by MS. In the case of a heptapeptide, H-Ala-
Pro-Arg-Leu-Arg-Phe-Tyr-OH, the initial Akabori cleavage, involving the loss of
C-terminus tyrosine from the modified heptapeptide (*m/z* 838), led to the formation
of the hexapeptide hydrazide at *m/z* 689. Two additional ions were also generated by
first-order Akabori cleavage, the tetrapeptide at *m/z* 521 and the tripeptide at *m/z* 407
at 30 min interval. In addition to C-terminus Akabori cleavage, microwave-assisted
hydrazinolysis generated sequential cleavages from the N-terminus of the modified
heptapeptide (*m/z* 838), yielding a series of ions at *m/z* 767, 670, 556, 443, and 329.
Clearly, microwave-assisted Akabori reaction can lead to rapid identification of
C-terminus amino acid in a polypeptide, including its amino acid sequence information at both C-terminus and the N-terminus. It was also found that microwave-
assisted hydrazinolysis of N-terminal substituted polypeptides followed the same
pattern as the unsubstituted peptides, while the traditional Edman degradation
approach would be unsuccessful. The presence of arginine and amino acids containing β-SH, COOH, and CONH$_2$ groups in their side chains can also be rapidly
confirmed because of their susceptibility to modifications by hydrazine.

TABLE 1.1
Tryptic Peptides of Horse Myoglobin

Code	Sequence	Expected Mass Value	Observed Peptide Mass
T1	GLSDGEWQQVLNVWGK	1817.01	T1 = 1816.9
T2	VEADIAGHGQEVLIR	1607.80	T2 = 1607.9
T3	LFTGHPETLEK	1272.44	T3 = 1272.0
T4	FDK	409.46	T3–T4 = 1663
T5	FK	294.37	T4–T5 = 685.5
T6	HIK	397.50	T4–T6 = 1062.7
T7	TEAEMK	708.81	T4–T7 = 1753.0
T8	ASEDIK	662.70	T8 = 662.30, T5–T8 = 2005
T9	K	147.20	T6–T9 = 1861.0
T10	HGTVVLTALGGILK	1379.68	T10 = 1379.0
T11	K	147.20	T10–T11 = 1507
T12	K	147.20	T12–T13 = 1983.0
T13	GHHEAELKPLAQSHATK	1855.03	T13 = 1855.0
T14	HK	284.340	T12–T14 = 2248
T15	IPIK	470.630	T15 = 470.50
T16	YIEFISDAIIHVLHSK	1886.20	T16 = 1886.0
T17	HPGDFGADAQGAMTK	1503.63	T17 = 1503.30
T18	ALELFR	748.90	T18 = 748.50
T19	NDIAAK	631.71	T18–T19 = 1361.66
T20	YK	310.37	T18–T20 = 1654.0
T21	ELGFQG	650.71	T21 = 650.40, T20–T21 = 941.50

In the case of cyclic oligopeptides, the traditional Edman degradation is not appropriate for sequencing of cyclic peptides as selective hydrolysis of peptide bonds is not easy to achieve because of the lack of free N-terminus. Other approaches using tandem MS can be difficult as the indiscriminate ring-opening pathways give a set of acylium ions of the same mass-to-charge ratio [32,33]. When microwave-assisted Akabori reaction was applied to glycine-containing cyclic peptides, microwave-assisted hydrazinolysis led to selective ring opening at the glycine residue to produce corresponding open-chain hydrazide(s) in a few minutes. The reaction mixtures were subsequently analyzed by reversed-phase (RP)-HPLC/ESI-MS and MS/MS for sequence determination. For example, a nonapeptide, cyclo(-Phe-His-Trp-Ala-Val-Gly-His-Leu-Leu-), treated with 98% hydrazine under microwave irradiation for a few minutes, generated a linear oligopeptide hydrazide (m/z 1093). The RP-HPLC/ESI-MS/MS product ion spectrum of this component (m/z 1093) gave characteristic b ions and y ions, revealing the sequence of amino acids in the cyclonoapeptide. The cleavage site was found to be at the amide bond of glycine. This is in contrast to direct MS/MS analysis of the cyclonoapeptide ions in which fragments from randomly and may not be useful for sequence determination. Another important application area in microwave technology is the use of microwave irradiation for the enzymatic digestion of proteins [34]. As discussed earlier, enzymatic cleavage to

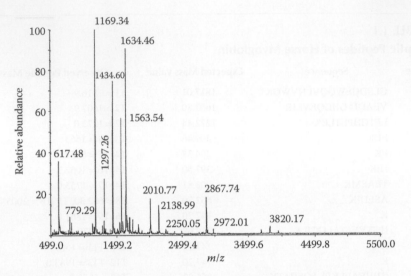

FIGURE 1.12 MALDI mass spectrum of tryptic fragments of cytochrome *c* after 10 min of microwave irradiation.

produce smaller peptide fragments of protein samples is an important step in structural characterization of proteins. Traditional enzymatic digestion method usually takes several hours, whereas microwave-assisted digestion occurs in minutes. Pramanik et al. carried out initial studies on bovine cytochrome *c*, a global protein relatively resistant to enzymatic cleavage under nondenaturing conditions. The protein was treated with trypsin at a 1:25 protease-to-protein ratio and the solution was subjected to microwave irradiation for about 10 min. The products were analyzed by MALDI-MS, and most of the expected tryptic peptides were observed in the spectrum with extensive sequence coverage (Figure 1.12). The result was similar to what was obtained using the traditional digestion approach, which took about 6 h. When horse heart myoglobin was subjected to microwave irradiation with trypsin, it showed a complete coverage of the protein in MALDI mass spectrum (Figure 1.13). Several other proteins, including bovine ubiquitin, chicken egg lysozyme, and IFN-α-2b, were also shown to exhibit the same accelerated proteolytic cleavages under microwave irradiation. Pramanik et al. also studied the action mechanism for the observed rate acceleration of the enzymatic cleavage of proteins under microwave irradiation at different microwave temperatures with different irradiation time intervals. The data suggest that the rapid increase in the reaction temperature is partially responsible for the large acceleration of digestion observed under microwave conditions [30]. Tightly folded proteins are known to require long hours for adequate proteolysis by enzymes under conventional conditions. This microwave irradiation approach can greatly enhance the proteolysis rate, improve the efficiency of protein digestion, and thus protein identification [35].

Enzymatic digested peptides can also be followed by tandem MS experiments to generate fragment ions for database searches for protein identification (sequence tagging) [36,37]. The major fragment ions in polypeptide ions are

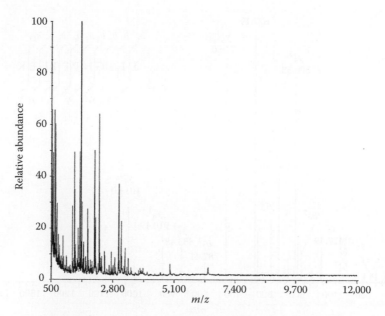

FIGURE 1.13 MALDI mass spectrum of tryptic fragments of myoglobin after 10 min of microwave irradiation.

b ions (N-terminus) and y ions (C-terminus) from cleavages of amide bonds under typical CID conditions [38,39]. Other types of fragment ions including side-chain ions can also be observed, as illustrated in Figure 1.14. These amino acid-specific fragment ions can be used to derive sequences of polypeptides. For example, a tryptic peptide T_3 (m/z 636 for +2 charge-state) of myoglobin can be dissociated

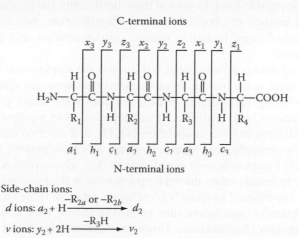

FIGURE 1.14 General fragmentation patterns of polypeptides under CID.

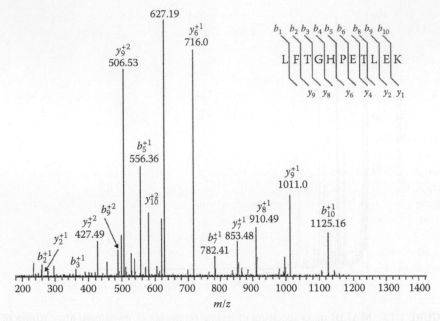

FIGURE 1.15 Product ion mass spectrum of myoglobin tryptic peptide T_3 (+2 charge-state) at m/z 636.

to give informative b and y ions, thus sequence information on T_3 (Figure 1.15). Database search based on the MS/MS information can lead to identification of the proteins. The sequence coverage from this approach varies from 5% to 70%. It is likely that posttranslational modifications can be lost during MS/MS fragmentation at the peptide level. In spite of these limitations, the bottom-up approach has become a method of choice in protein identifications because of its well-researched methodology, including mature instrumentation and excellent software development.

An emerging field in protein characterization is the employment of "top-down" method. In contrast to the bottom-up experiments, this approach directly measures intact proteins using high-resolution MS and fragments intact protein ions by MS/MS experiments [40]. Dissociation techniques used in top-down experiments include CID, electron capture dissociation [41,42], and electron transfer dissociation [43,44]. In principle, this approach covers an entire protein sequence with 100% coverage. Posttranslational modifications (i.e., glycosylation, phosphorylation) are likely to remain intact during fragmentation at the protein level. Thus, the fragment ions can be used to identify proteins by database retrieval, confirm large sections of sequences, and locate sites of modifications. This would be an ideal approach for protein identifications. However, there are significant obstacles that have to be overcome before this approach can be widely used in protein identifications. These challenges include better understanding of MS/MS mechanisms for intact proteins ions, development of advanced MS instrumentation for efficient MS/MS data acquisition, and appropriate database search algorithm for automatic data analysis.

1.4 CHARACTERIZATION OF RECOMBINANT PROTEINS AND POSTTRANSLATIONALLY MODIFIED PROTEINS

The production of recombinant therapeutic proteins by recombinant DNA techniques is an important part of biotechnology products development. The structural variations from the protein production process could affect the protein's biological activities and change the safety, potency, and stability of the therapeutic protein product. Accurate structural characterization is essential to assure the quality of the protein products. As an illustration, recombinant human granulocyte-macrophage colony-stimulating factor (rh-GM-CSF) is shown to demonstrate the role of MS in characterizing recombinant proteins. Discussions on posttranslational modifications (glycosylation and phosphorylation) would also be presented.

GM-CSF belongs to a group of interacting glycoproteins regulating the differentiation, activation, and proliferation of multiple blood cell types from progenitor stem cells [45]. It enhances the production and function of white blood cells with potential clinical applications. The mature human GM-CSF contains 127 amino acids with four cysteine residues (Figure 1.16), and has a calculated MW of 14,477.6 Da (without accounting for existing disulfide bonds) [46]. The ESI-MS experiments give a measured average MW of 14,472 Da, suggesting the presence of two disulfide bonds in the rh-GM-CSF. This was confirmed by reduction of rh-GM-CSF with β-mercaptoethanol. The charge-state distribution is shifted to higher charge-states for the reduced rh-GM-CSF, consisting of a more open form of protein structure for protonations upon reduction of disulfide bonds. There is also a 4 Da mass shift of the measured MW of reduced rh-GM-CSF (14,476 Da) from nonreduced rh-GM-CSF, indicating the presence of two disulfide bonds in the protein molecule.

To obtain the primary structural information of rh-GM-CSF, the protein was digested with either trypsin or *Staphylococcus aureus* V8 protease, followed by MS analysis of digestion mixtures. Trypsin selectively cleaves to rh-GM-CSF at the C-terminal side of arginine (R) and lysine (K), while V8 protease cleaves to the peptide bond on the C-terminal side of glutamic acid (E) residues. For tryptic digest of rh-GM-CSF, the majority of the observed ion signals could be matched with predicted tryptic peptides, with the exception of the cysteine-containing fragments T_4 (DTAAEMNETEVISEMFDLQEPTC^{54}LQTR), T_{10} (QHC^{88}PPT-PETSC^{96}ATQIITFESFK), and T_{12} (DFLLVIPFDC^{121}WEPVQE). These peptide fragments (T_4, T_{10}, T_{12}) are interconnected by disulfide bonds, as illustrated

APARSPSPSTQPWEHVNAIQEARRLLNLSRDTAAEMNETVEVI
SEMFDLQEPTC^{54}LQTRLELYKQGLRGSLTKLKGPLTMMASHYK
QHC^{88}PPTPETSC^{96}ATQIITFESFKENLKDFLLVIPFDC^{121}WEPVQE

FIGURE 1.16 Amino acid sequence of rh-GM-CSF.

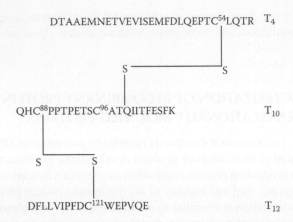

DTAAEMNETVEVISEMFDLQEPTC^{54}LQTR   T$_4$

QHC^{88}PPTPETSC^{96}ATQIITFESFK   T$_{10}$

DFLLVIPFDC^{121}WEPVQE   T$_{12}$

FIGURE 1.17   Core peptide for rh-GM-CSF.

in Figure 1.17. This disulfide-linked core peptide was detected by MS, indicating the presence of this core peptide and two disulfide bonds in rh-GM-CSF. In addition, these peptide fragments were released after treatment of the tryptic digests with dithiothreitol (reducing reagent) and MS analysis of the mixture yielded ion signals corresponding to their free sulfhydryl forms as T$_4$, T$_{10}$, and T$_{12}$, respectively, confirming the presence of two disulfide bonds in rh-GM-CSF.

The tryptic peptide mass mapping of rh-GM-CSF demonstrated the presence of two disulfide bonds and suggested two possible combinations of disulfide pairing as C54-C88/C96-C121 or C54-C96/C88-C121. However, the specific assignment of the disulfide pairing was not possible because of the absence of a tryptic cleavage site between C88 and C96 residues of T$_{10}$. The V8 protease was used to digest rh-GM-CSF and cleave the protein between each half-cystine residue at the C-terminal side of glutamic acid. The V8 protease digest of rh-GM-CSF is confirmed with most of the predicted peptides. Two disulfide-linked peptides V$_8$-SS-V$_{10}$ (PTC^{54}LQTRLE-SS-TSC^{96}ATQIITFE) and V$_{7,8}$-SS-V$_{10}$ (MFDLQE PTC^{54}LQTRLE-SS-TSC^{96}ATQIITFE) were observed, arising from incomplete cleavage at Glu(51). These MS signals disappeared upon DTT reduction reaction, thus suggesting a Cys(54)–Cys(96) disulfide bond. Furthermore, MS analysis revealed additional disulfide-linked peptides as V$_9$-SS-V$_{12,13}$ (LYKQGLRGSLTKLKGPLTMMASHYKQHC^{88}PPTPE-SS-NLKDFLLVIPFDC^{121}WEPVQE) and V$_9$-SS-V$_{11-13}$ (LYKQGLRGSLTKLKGPLTM-MASHYKQHC^{88}PPTPE-SS-SFKENLKDFLLVIPFDC^{121}WEPVQE) [34]. These data clearly established another pairing of disulfide bond between Cys(88)–Cys(121).

Two rh-GM-CSF variants (V1 and V2) were also observed during the preparation of GM-CSF. These variants were isolated from preparative HPLC and characterized using MS methods. The peptide mass mapping strategy using trypsin and V8 protease was applied for structural identification of the variants. The comparison of the trypsin and V8 protease digests of the native GM-CSF and its two variants clearly indicated that one or two methionine residues in native GM-CSF have been converted to methionine sulfoxides. In the case of V1, tryptic peptide T$_9$ had a mass increase of 16 Da, suggesting oxidation of Met(79) or Met(80). In the case of V2, both the tryptic peptide T_4 and T_9 had a mass shift of 16 Da. Therefore, V2 contains two methionine sulfoxides, one at Met(46), the other at Met(79) or Met(80). The assignment of Met(46) oxidation was supported by a mass increase of 16 Da for V8 protease-digested peptides $V_{7,8}$ and $V_{7,8}$-SS-V_{10}. Tandem MS experiments could be performed to further differentiate oxidation sites between Met(79) and Met(80). The structural assignments of V1 and V2 were further supported by MS studies of chemically modified proteins of rh-GM-CSF that have different degrees of oxidation of the four methionine residues in rh-GM-CSF.

Clearly, the MS method in combination with enzymatic digestion provides an effective approach to the characterization of GM-CSF and its variants. The MS analysis of the enzymatic digest of GM-CSF and its variants allows the determination of the MWs of the peptides resulting in the identification of modification sites, the disulfide bonding pattern, and confirmation of the cDNA-derived sequence of the protein [47].

Two important posttranslational modifications in proteins are glycosylation and phosphorylation. Carbohydrate modifications of proteins (glycosylation) are essential factors in modulating protein structures and functions [48]. There are many cases where glycan structures have been shown to have significant importance on the biological functions of a protein. Changes in levels and types of glycosylation can be associated with disease [49,50]. Glycosylation also represents the most common modification for recombinant protein products expressed in mammalian and insect cell lines, having an impact on their solubility, immunogenicity, circulatory half-life, and thermal stability. As a recent major focus in biopharmaceuticals, monoclonal antibody's biological functions and physiochemical properties often depend on the nature of the glycosylation of immunoglobulin G (IgG). Complete structural characterization of a glycoprotein requires the determination of the peptide primary sequence and the glycosylation sites, as well as the definition of the attached oligosaccharides in terms of their linear sequencing, branching, linkage, configurations, and the positional isomers [51].

In general, glycoprotein is enzymatically digested so that each glycosylation site is located within a separate peptide [52]. HPLC separation of the peptides coupled with MS precursor ion scans of sugar-specific oxonium ions, such as m/z 163 (protonated Hex), m/z 204 (protonated HexNAc), or m/z 366 (protonated Hex-HexNAc), allows the identification of glycopeptides from the mixture of peptides, for further studies [53–56]. Dissociation of a glycopeptide in MS/MS experiments can provide information on the primary sequence of the peptide, the type of sugar attached, and the amino acid residue modified by the glycosyl group [57–59]. As an example of general structural characterization of glycoproteins, MS analysis of a 150 kDa IgG glycoprotein derived from anti-interleukin 5 was illustrated [60,61]. IgG consists of two light chains and two heavy chains connected by interchain disulfide bonds to form a Y-shaped structure. The variable region of light and heavy chain contains antigen binding sites. The majority of monoclonal antibodies have only one N-linked glycosylation site in the constant region of each heavy chain. Typically the N-linked oligosaccharides of IgG consist of complex biantennary oligosaccharide structures with a core portion (reducing end) composed of two N-acetylglucosamine (GlcNAc) and three mannose (Man) residues. Variation in glycan structure occurs in the presence of a fucose residue attached to the end N-acetylglucosamine with either two (G2), one (G1), or zero (G0) terminal galactose residues and with either one or two terminal sialic acids. The profile from ESI-MS experiments illustrated that the major oligosaccharides of G0, G1, and G2 are present in the sample. The negative ion ESI-MS/MS multiple reaction monitoring experiments of PNGase F released carbohydrates were carried out, and the data confirmed the presence of the G0, G1, and G2. The linkage details within methylated glycans were further characterized by multiple-stage MS experiments using an ion trap mass spectrometer. The MS^n analysis of reduced and permethylated G0 was performed at m/z 937.5 (2+), 808.2 (2+), 678.5 (2+), 866.4 (1+), and 622.3 (1+). The MS data were consistent with the composition of $(HexNAc)_2(Fuc)_1 + (Man)_3(GlcNAc)_2$. Similarly, MS^n data on G1 show the structure to be $(Gal)_1(HexNAc)_2(Fuc)_1 + (Man)_3(GlcNAc)_2$. The use of the MS^n analysis in ion trap mass spectrometer can be effective for molecular disassembly of precursor ions of the oligosaccharides through linking of the ion fragmentation pathways and ion trees. This MS strategy for assessment of glycosylation at the molecular level is valuable for lot-to-lot evaluations of mammalian cell-derived proteins and process control monitoring.

Another important posttranslational modification in proteins is phosphorylation. Protein phosphorylation plays an essential role in intercellular communication during development, in physiological responses, and in the functioning of the nervous and immune systems [62,63]. Identification of phosphorylation sites is a critical step toward understanding the function and regulation of many protein kinases and kinase substrates. Protein kinases mediate most of the signal transduction in eukaryotic cells. By modification of the substrate activity, protein kinases also control many other cellular processes, including metabolism, transcription, cell cycle progression, apoptosis, and differentiation [64,65]. In eukaryotic cells, protein phosphorylation happens mostly on serine, threonine, and tyrosine residues (it could also happen on histidine, arginine, lysine, cysteine, glutamic acid, and aspartic acid to a much lesser extent). A complete analysis of protein phosphorylation includes identification of the phosphoproteins, the localization of the residues that are phosphorylated, and the quantitation of phosphorylation.

The general MS approach for characterization of phosphorylated proteins is based on the lability of the phosphor moiety of the phosphorylated peptides upon low-energy CID experiments. The detection of a characteristic loss allows the identification of phosphorylated peptides from an unseparated peptide mixture, or during online HPLC experiments. In the positive ion mode, a neutral loss of 98 Da (H_3PO_4 or HPO_3 and H_2O) from the phosphopeptide can be used to confirm the existence of a phosphopeptide [66,67]. The fragment ion at $m/z - 79$ (PO_3^-) is more frequently used in the negative ion mode for phosphorylation-specific precursor ion scanning [68–70]. The advantage of precursor ion scanning of $m/z - 79$ includes its applicability to all phosphopeptides with phosphorylation occurring on serine, threonine, or tyrosine. For example, in the study of the mitogen-activated kinase, MEK1 (MAP/ERK kinase), "off-pathway" phosphorylation occurs during expression of the protein [71]. MEK1 protein has 429 amino acids and its theoretical MW is 47,534 Da. The initial MS analysis on the intact protein gave measured MWs of 47,550 and 47,630 Da, suggesting the presence of one major phosphate group.

To further characterize and identify the phosphorylation site, the protein sample was digested with Glu-C or a combination of enzymes, such as Glu-C/trypsin, Glu-C/chymotrypsin, and trypsin/Asp-N (overnight at 37°C). Protein digestion by two enzymes was achieved by adding the second enzyme after the first digestion was complete. The resulting peptides were analyzed in both positive and negative ion modes. Peptide mass mapping, precursor ion scan of m/z −79 (PO_3^-), and MS/MS analysis of the enzyme cleavage products were performed using a triple-quadrupole mass spectrometer. Precursor ion scan detected a phosphorylated Glu-C fragment (m/z 3,458). However, MS/MS sequencing was not successful due to the size of this peptide. The approach of using two different enzymes was subsequently employed. Glu-C cleavage followed by chymotrypsin digestion produced a peptide (m/z 1057) that was captured by precursor ion scan. The peptide MW (MW 1056 Da) matched the same sequence region as the one obtained by the above approaches. The MS/MS data of the doubly charged ion at m/z 529.50 identified this peptide as phosphorylated MEK1 (328–336). Ser-334 was identified as the phosphorylation site, which was evident by the addition of 80 Da (HPO_3) to y_3, y_4, y_5, and y_7 as well as to b_6 and b_8 fragments. This phosphorylation site was further confirmed by using another enzyme combination with trypsin digestion followed by Asp-N. The Ser-334 residue resides in a proline-rich region of MEK1 that may have regulatory importance.

1.5 PROTEOMICS STUDIES

1.5.1 GENERAL APPROACH

Proteomics research involves the global analysis of gene expression, including identification, quantification, and characterization of proteins [72,73]. Because of the dynamic nature of cell systems, the proteome level varies with time, depending on the genome, the environment of the cell, and the cell history. Many posttranslational modifications of proteins also take place and lead to different cell functions. Analysis of such complicated systems is a challenge in proteomics research.

In order to reduce the complexity of proteins expressed in the cell as a result of different expression rates of proteins, prefractionation techniques are often used before further protein characterization [74]. Several approaches used include ion-exchange, hydrophobic interaction chromatography, and affinity chromatography to enrich low-copy-number gene products [75–77]. These prefractionation techniques provide improvements in separation resolution and increased sensitivity. After the prefractionation of complex protein samples, individual fractions can be subjected to either two-dimensional (2D) gel electrophoresis or multidimensional HPLC for further separation and analysis by MS.

The 2D gel electrophoresis has been carried out for high-resolution protein separation for a number of years [78,79]. Thousands of proteins expressed by an organism or cell can be separated using 2D gel. Gel spots of interest can be cut and digested with enzymes. The identities of proteins of interests are obtained from either peptide mass mapping or sequence tagging on digested polypeptides in combination with database search. The sequence tagging is usually performed by capillary RP-HPLC/MS and MS/MS methods. This 2D gel approach has been successful and widely used in proteomics research. One of its drawbacks is the throughput. 2D gel is a relatively slow, labor-intensive technique. It has limited abilities to resolve lower abundance proteins, membrane proteins, highly acidic or basic proteins, very large or small proteins, and hydrophobic proteins.

To address these issues in 2D gels, multidimensional HPLC/MS has been introduced and implemented. The current approach is the "shotgun" proteomics in which complex mixtures of proteins are enzymatically digested in solution to generate mixtures of peptides, and the mixtures of peptides are further separated online by 2D HPLC/MS [80,81]. The first phase of separation is often performed using strong cation-exchange LC, followed by C18 RP LC. The product ion mass spectra for individual eluting peptides are obtained online and searched against the database for protein identification. This approach can be highly automated with high throughput. However, the complexity of the sample is increased enormously because of the large number of peptides generated from each protein. In an effort to reduce the number of peptides in enzymatic digests, affinity selection strategies using affinity chromatography have been employed to selectively capture specific peptides and allow a rapid reduction in sample complexity without compromising protein identifications [82].

Another important aspect of proteomics research is the quantitation of changes in protein expressions between different states [83]. It is quite common to have 10-fold changes in protein expressions between normal state and diseased state. Traditionally, a 2D gel-matching procedure (gel imaging) can be used to compare two sets of protein mixtures run under standardized conditions from different cell states, resolving hundreds and thousands of proteins. Proteins with posttranslational modifications altering protein charges can be readily observed using 2D gel. Differential gel electrophoresis can be used to minimize problems associated with gel-to-gel reproducibility. It has a good dynamic range over four orders of magnitude. This gel-based approach is often limited by its labor-intense nature with low throughput and dependence on the expertise of the operator. Very large and very small proteins, hydrophobic proteins, and very acidic or basic species behave poorly in 2D gel. Other MS-based approaches use isotope tagging of peptides to measure the changes in expression levels between two proteomes in a single experiment. For example, various methods are developed for relative quantitation of peptides from a proteolytic digestion of complex proteins, including isotope-coded affinity tag [84], N-terminal labeling of peptides [85], and using $^{18}O/^{16}O$ labeled water [86]. One of the common issues in current chemical tagging approaches is that the dynamic range of the method may not be sufficient to cover the expression level, which may be spread over a few orders of magnitude.

A third approach in quantitative proteomics is label-free quantitation [87]. This method is based on the correlation between peptide mass spectral peak data and the abundance of the protein in the sample, using either mass spectral peak intensities of peptide ions for measuring protein amount or number of MS/MS spectra assigned to a protein as a measure of protein abundance. For peak intensity-based quantitation, at least one peptide common for a pair of samples is used for peak area calculation from extracted ion chromatogram, usually three or more peptides in common per protein needed for reproducible quantitation. A linearity of over 1000-folds can be achieved.

For spectral counting-based quantitation, at least one spectrum in either sample pair is used from MS/MS data for any peptide in a given protein, usually four or more spectra per protein required for accurate quantitation. Spectral counting is more sensitive for detecting changes in abundance, while peak intensity quantitation gives more accurate estimates of protein ratios. This label-free method is desirable if metabolic labeling is not possible, or introducing stable isotope labeling is label-intense. It is expected that new methods are to be pursued further to achieve desired quantitation results in quantitative proteomics studies.

The field of proteomics has further evolved over the past several years in terms of understanding of biological and cellular systems. A number of review articles have been published in recent years covering technological developments with emphasis on gel-free approaches, protein–protein interaction, protein quantification, and bioinformatics proteomics [88–90]. The development of bioinformatics tools has ensured that the use of software including improved statistical methodologies and filtering algorithms, be available to scientific community to further validate proteomics data. The development of novel hardware, such as Orbitrap mass spectrometer as mentioned earlier, has further assisted proteomics applications. In this system, ions are trapped and orbit around a central electrode where they induce an imaging current in the outer electrode; the frequency of the imaging current is Fourier transformed into time-domain signal producing mass spectra [17,18]. The main advantage of the Orbitrap instrument is accurate mass measurement of the trace level intact proteins including the peptide fragments assuring the accuracy of protein identification [91].

The discovery of biomarkers has been one of the major efforts in proteomics approach [88,92]. A biomarker is a measurable signal that can relate to specific biological state with particular relevance to the presence or condition of the disease. Biomarkers can be used for diagnosis and prognosis of diseases including assessing the progress of biological response or therapeutic intervention. However, despite the considerable effort, the success of finding proteomic biomarkers has been relatively disappointing.

Another area that has received considerable attention is the MALDI-MS-based imaging mass spectrometry (IMS) of proteins in tissues [93–96]. The direct analysis of tissues using IMS enables the detection of both endogenous and exogenous compounds (proteins) with molecular specificity while maintaining special orientation. These researchers have demonstrated the presence of unique protein profiles in classifying human tumor tissues and predicting patient outcomes in treatments.

1.5.2 CHARACTERIZATION OF ADENOVIRUS PROTEINS

Advances in gene therapy technology have provided new directions for combating cancers and other serious diseases by delivering therapeutic genes to target cells. The adenovirus is an icosahedral, nonenveloped, and double-stranded DNA virus that can be used as a potential vector for gene therapy [97]. The commonly used adenovirus is derived from an adenovirus serotype 5 (Ad 5) virus that has had the E1 coding sequence replaced with a 1.4 kb full-length human p53 cDNA. This 200×10^6 Da virus has at least 11 structural proteins with a wide mass range, from less than 10,000 Da to more than 100,000 Da, including hexon (II), penton base (III), peripentonal hexon-associated protein (IIIa), minor core protein (V), major core protein (VII), and other hexon-associated proteins (VI and VIII). The infectivity of the virus depends on the assembly of these structural proteins in forming a complete virion. To monitor the quality of the adenovirus and to better understand viral structure–function relationships, rapid and accurate analytical methods are needed to define the adenovirus proteins at the proteome level for its use as a therapeutic entity.

In our laboratory, we have developed MS-based assay to combine the use of multidimensional analytical techniques that include sodium dodecyl sulfate polyacrylamide gel electrophoresis (SDS-PAGE), HPLC, MALDI-MS/MS, and database search for structural characterization. It involves dissociation of intact viruses, separation of viral proteins by RP-HPLC or SDS-PAGE, enzymatic digestion of separated proteins, followed by MALDI-MS, MALDI-post source decay (PSD)-MS, and database search [98].

The adenoviral proteins were initially extracted from gel bands of SDS-PAGE for the determination of their MWs by MALDI-MS [99]. This method provided mass measurements of viral proteins with better accuracy than those obtained from SDS-PAGE (a mass accuracy of 0.1% was normally obtained for MALDI-MS). In order to gain structural information of viral proteins, each gel band was digested using trypsin, followed by peptide mass mapping and protein identification through database search. All the proteins separated by SDS-PAGE with MWs ranging from 10,000 to 100,000 Da were identified as adenoviral proteins from the SwissProt database, when searched against all taxonomies using MS-Fit algorithm. Most of the mature viral proteins II, III, IIIa, V, VI, and VII were characterized, and their presence was confirmed. In addition, precursor protein pVIII as well as the propeptide of pVIII were detected.

Another approach is to inject intact adenoviral particles onto RP-HPLC and analyze collected fractions for dissociated adenoviruses. It is well known that RP-HPLC is highly sensitive in providing faster and reproducible results even for smaller adenoviral polypeptides. Individual fractions collected from RP-HPLC were subjected to enzymatic digestion with trypsin for protein identification. For example, one of fractions collected from RP-HPLC has a measured MW of 3037 Da. This fraction was digested with trypsin and digested peptides were searched against all taxonomies using MS-Fit. It resulted in four protein identifications with low and undistinguishable molecular weight search (MOWSE) scores. The only protein correlated to an adenovirus related core protein precursor, pX, has a very low MOWSE score of 48.7. This identification could be regarded as a random hit. In addition, the search algorithm only matched four peptides with 15% sequence coverage for pX, far less than the acceptable coverage of 30%. To confirm this identification with high confidence, MALDI-PSD-MS experiments on one of the tryptic peptides at m/z 819.46 were carried out and the results are listed in Table 1.2. When searched against SwissProt database using MS-Tag program, adenoviral protein pX was unambiguously identified (Table 1.3). Clearly, information obtained from sequence tag experiments eliminated spurious protein fits and improved accuracy of identifications. It is well documented that MS-Tag search can be successful in identifying proteins,

TABLE 1.2
Fragment Ions Observed from MALDI-PSD-MS of Tryptic Peptide Ions at *m/z* 819.5 from One of Fractions Collected

Code	Observed Fragment Ions (*m/z*)
V	72.01
R	119.83
y_1-NH$_3$	158.13
y_1	174.97
GF	205.53
a_2	217.48
b_2	245.25
PGF	302.61
y_2-NH$_3$	305.62
y_2	322.53
b_3	344.42
y_3-NH$_3$	362.23
y_4-NH$_3$	459.78
y_4	476.94
y_5-NH$_3$	558.45
y_5	575.31
y_6-NH$_3$	655.78
y_6	672.83
y_7-NH$_3$	801.54

TABLE 1.3
MS-Tag Search Results from MALDI-PSD-MS of Tryptic Peptide Ions at *m/z* 819.5

Rank	Sequence	MH⁺ (Calculated)	MH⁺ (Error)	Protein MW (Da)/pI	Species	SwissProt Accession Number	Protein Name
1	(R)FPVPGFR(G)	819.4517	0.0038	8,845.7/12.88	ADE02	P14269	Late L2 MU core protein precursor (11 KD core protein) (protein X)
2	(K)VPFFPGR(G)	819.4517	0.0038	82,989.3/5.06	BACST	P14412	Peroxidase/catalase

especially in cases where unfavorable digestion conditions or modifications have made the digested peaks unidentifiable by peptide mass mapping approach.

This study of the proteome of the adenovirus type 5 vectors demonstrated an important application of MS technique in the drug discovery. The information on protein MWs, tryptic peptide mass mapping, and sequence tags of tryptic peptides derived from MALDI-MS and MALDI-PSD-MS resulted in the identification of 17 adenoviral proteins/polypeptides. The rapid and accurate identification of viral proteins in this study is significant as it provides direct evidence of the maturation stage of adenoviruses, which is closely related to viral infectivity and efficacy in gene therapy.

1.6 FUTURE PROSPECTS

Structural characterization of proteins/peptides by mass spectrometry is an important part of drug discovery process, as illustrated in this chapter. The development of recombinant therapeutic proteins and understanding of target proteins via proteomics will continue to be the driving force in defining the role of mass spectrometry in drug discovery. In addition, a relatively new field in using proteomics to obtain biomarker information would provide an opportunity for new applications in mass spectrometry. Advancements in ionization methods and mass spectrometry instrumentation can further enhance the capabilities of mass spectrometry in protein characterization. It is fully anticipated that mass spectrometry will continue to play important roles in drug discovery in the future.

ACKNOWLEDGMENT

The authors would like to thank Dr. John J. Piwinski for his support on the projects.

REFERENCES

1. Pharmaceutical Research and Manufacturers of America. 1994–2003. *New Drug Approvals* (series), Washington, DC: PhRMA.
2. Pharmaceutical Research and Manufacturers of America. 2001–2003. *New Drugs in Development* (series), Washington, DC: PhRMA.
3. Kola, I. and Landis, J. 2004. Can the pharmaceutical industry reduce attrition rates? *Nat. Rev. Drug Discov.* 3: 711–716.
4. Mullin, R. 2003. Drug development costs about $1.7 billion. *Chem. Eng. News* 81: 8–9.
5. Pramanik, B. N., Bartner, P. L., and Chen, G. 1999. The role of mass spectrometry in the drug discovery process. *Curr. Opin. Drug Discov. Devel.* 2: 401–417.
6. Lowe, J. A. III. and Jones, P. 2007. Biopharmaceuticals and the future of the pharmaceutical industry. *Curr. Opin. Drug Discov. Devel.* 10: 513–514.
7. Cooks, R. G., Chen, G., and Wong, P. 1997. Mass spectrometers. In *Encyclopedia of Applied Physics*, Ed. G. L. Trigg, pp. 289–330. New York: VCH Publishers.
8. Fenn, J. B., Mann, M., Meng, C. K., Wong, S. F., and Whitehouse, C. M. 1989. Electrospray ionization for mass spectrometry of large biomolecules. *Science* 246: 64–71.
9. Pramanik, B. N., Ganguly, A. K., and Gross, M. L. 2002. *Applied Electrospray Mass Spectrometry.* New York: Marcel Dekker.

10. Hillenkamp, F., Karas, M., Beavis, R. C., and Chait, B. T. 1991. Matrix-assisted laser desorption/ionization mass spectrometry of biopolymers. *Anal. Chem.* 63: 1193A–1203A.

11. Tanaka, K., Waki, H., Ido, Y., Akita, S., Yoshida, Y., and Yoshida, T. 1988. Protein and polymer analyses up to *m/z* 100,000 by laser ionization time-of-light mass spectrometry. *Rapid Commun. Mass Spectrom.* 2: 151–153.

12. Yost, R. A. and Enke, C. G. 1978. Selected ion fragmentation with a tandem quadrupole mass spectrometer. *J. Am. Chem. Soc.* 100: 2274–2275.

13. Cooks, R. G., Chen, G., and Weil, C. 1997. Quadrupole mass filters and quadrupole ion traps. In *Selected Topics in Mass Spectrometry in the Biomolecular Sciences*, Eds. R. M. Caprioli, A. Malorni, and G. Sindona, pp. 213–238. Dordrecht, the Netherlands: Kluwer Academic.

14. Schwartz, J. C., Senko, M. W., and Syka, J. E. P. 2002. A two-dimensional quadrupole ion trap mass spectrometer. *J. Am. Soc. Mass Spectrom.* 13: 659–669.

15. Vestal, M. L. and Campbell, J. M. 2005. Tandem time-of-flight mass spectrometry. *Methods Enzymol.* 402: 79–108.

16. Chernushevich, I. V., Loboda, A. V., and Thomson, B. A. 2001. An introduction to quadrupole-time-of-flight mass spectrometry. *J. Mass Spectrom.* 36: 849–865.

17. Hu, Q., Noll, R. J., Li, H., Makarov, A., Hardman, M., and Cooks, R. G. 2005. The orbitrap: A new mass spectrometer. *J. Mass Spectrom.* 40: 430–443.

18. Makarov, A., Denisov, E., Kholomeev, A., Balschun, W., Lange, O., Strupat, K., and Horning, S. 2006. Performance evaluation of a hybrid linear ion trap/orbitrap mass spectrometer. *Anal. Chem.* 78: 2113–2120.

19. Bogdanov, B. and Smith, R. D. 2005. Proteomics by FTICR mass spectrometry: Top down and bottom up. *Mass Spectrom. Rev.* 24: 168–200.

20. May, D., Fitzgibbon, M., Liu, Y., Holzman, T., Eng, J., Kemp, C. J., Whiteaker, J., Paulovich, A., and McIntosh, M. 2007. A platform for accurate mass and time analyses of mass spectrometry data. *J. Proteome Res.* 6: 2685–2694.

21. Smith, R. D., Loo, J. A., Edmonds, C. G., Barinaga, C. J., and Udseth, H. R. 1990. New developments in biochemical mass spectrometry: Electrospray ionization. *Anal. Chem.* 62: 882–899.

22. Henzel, W. J., Billeci, T. M., Stults, J. T., Wong, S. C., Grimley, C., and Watanable, C. 1993. Identifying proteins from two-dimensional gels by molecular mass searching of peptide fragments in protein sequence databases. *Proc. Natl. Acad. Sci. USA* 90: 5011–5015.

23. Chowdhury, S. K., Katta, V., and Chait, B. T. 1990. Electrospray ionization mass spectrometric peptide mapping: A rapid, sensitive technique for protein structure analysis. *Biochem. Biophys. Res. Commun.* 167: 686–692.

24. Gedye, R., Smith, F., Westaway, K., Ali, H., Baldisera, L., Laberge, L., and Rousell, J. 1986. The use of microwave ovens for rapid organic synthesis. *Tetrahedron Lett.* 27: 279–282.

25. Giguere, R. J., Bray, T. L., Duncan, S. C., and Majetich, G. 1986. Application of commercial microwave ovens to organic synthesis. *Tetrahedron Lett.* 27: 4945–4948.

26. Bose, A. K., Manhas, M. S., Ghosh, M., Raju, V. S., Tabei, K., and Urbanczyk, L. Z. 1990. Highly accelerated reactions in microwave oven: Synthesis of heterocycles. *Heterocycles* 30: 741–744.

27. Richter, R. C., Link, D., and Kingston, H. M. S. 2001. Microwave-enhanced chemistry. *Anal. Chem.* 73: 31A–37A.

28. Lidstrom, P., Tierney, J., Wathey, B., and Westman, J. 2001. Microwave-assisted organic synthesis. *Tetrahedron Lett.* 57: 9225–9283.

29. Bose, A. K., Ing, Y. H., Lavlinskaia, N., Sareen, C., Pramanik, B. N., Bartner, P. L., Liu, Y. -H., and Heimark, L. 2002. Microwave enhanced Akabori reaction for peptide analysis. *J. Am. Soc. Mass Spectrom.* 13: 839–850.

30. Pramanik, B. N., Ing, Y. H., Bose, A. K., Zhang, L. K., Liu, Y. H., Ganguly, S. N., and Bartner, P. L. 2003. Rapid cyclopeptide analysis by microwave enhanced Akabori reaction. *Tetrahedron Lett.* 44: 2565–2568.
31. Akabori, S., Ohno, K., and Narita, K. 1952. On the hydrazinolysis of proteins and peptides: A method for the characterization of carboxy-terminal amino acids in proteins. *Bull. Chem. Soc. Japan* 25: 214–218.
32. Ngoka, L. and Gross, M. L. 1999. Multistep tandem mass spectrometry for sequencing cyclic peptides in an ion-trap mass spectrometer. *J. Am. Soc. Mass Spectrom.* 10: 732–746.
33. Kuroda, J., Fukai, T., and Nomura, T. 2001. Collision-induced dissociation of ring-opened cyclic depsipeptides with a guanidino group by electrospray ionization/ion trap mass spectrometry. *J. Mass Spectrom.* 36: 30–37.
34. Pramanik, B. N., Mirza, U. A., Ing, Y. H., Liu, Y. -H., Bartner, P. L., Weber, P. C., and Bose, A. K. 2002. Microwave-enhanced enzyme reaction for protein mapping by mass spectrometry: A new approach to protein digestion in minutes. *Protein Sci.* 11: 2676–2687.
35. Lill, J. R., Ingle, E. S., Liu, P. S., Pham, V., and Sandoval, W. N. 2007. Microwave-assisted proteomics. *Mass Spectrom. Rev.* 26: 657–671.
36. Hunt, D. F., Yates, J. R. III, Shabanowitz, J., Winston, S., and Hauer, C. R. 1986. Protein sequencing by tandem mass spectrometry. *Proc. Natl. Acad. Sci. USA* 83: 6233–6237.
37. Yates, J. R. III. 1998. Mass spectrometry and the age of the proteome. *J. Mass Spectrom.* 33: 1–19.
38. Roepstorff, P. and Fohlman, J. 1984. Proposal for a common nomenclature for sequence ions in mass spectra of peptides. *Biomed. Mass Spectrom.* 11: 601.
39. Das, P. R. and Pramanik, B. N. 1994. Fast atom bombardment mass spectrometric characterization of peptides. In *Methods in Molecular Biology*, Vol. 36, Eds. M. W. Pennington and B. Dunn, pp. 85–106. Totowa, NJ: Humana Press.
40. Kelleher, N. L. 2004. Top-down proteomics. *Anal. Chem.* 76: 197A–203A.
41. Ge, Y., Lawhorn, B. G., ElNaggar, M., Strauss, E., Park, J. H., Begley, T. P., and McLafferty, F. W. 2002. Top down characterization of larger proteins (45 kDa) by electron capture dissociation mass spectrometry. *J. Am. Chem. Soc.* 124: 672–678.
42. Cooper, H. J., Hakansson, K., and Marshall, A. G. 2005. The role of electron capture dissociation in biomolecular analysis. *Mass Spectrom. Rev.* 24: 201–222.
43. Syka, J. E., Coon, J. J., Schroeder, M. J., Shabanowitz, J., and Hunt, D. F. 2004. Peptide and protein sequence analysis by electron transfer dissociation mass spectrometry. *Proc. Natl. Acad. Sci. USA* 101: 9528–9533.
44. Mikesh, L. M., Ueberheide, B., Chi, A., Coon, J. J., Syka, J. E., Shabanowitz, J., and Hunt, D. F. 2006. The utility of ETD mass spectrometry in proteomic analysis. *Biochim. Biophys. Acta* 1764: 1811–1822.
45. Metcalf, D., Johnson, G. R., and Burgess, A. W. 1980. Direct stimulation by purified GM-CSF of the proliferation of multipotential and erythroid precursor cells. *Blood* 55: 138–147.
46. Tsarbopoulos, A., Pramanik, B. N., Labdon, J., Reichert, P., Gitlin, G., Patel, S., Sardana, V., Nagabhushan, T. L., and Trotta, P. P. 1993. Isolation and characterization of a resistant core peptide of recombinant human granulocyte-macrophage colony-stimulating factor (GM-CSF); confirmation of the GM-CSF amino acid sequence by mass spectrometry. *Protein Sci.* 2: 1948–1958.
47. Chen, G., Pramanik, B. N., Liu, Y. -H., and Mirza, U. A. 2007. Applications of LC/MS in structure identifications of small molecules and proteins in drug discovery. *J. Mass. Spectrom.* 42: 279–287.
48. Apweiler, R., Hermjakob, H., and Sharon, N. 1999. On the frequency of protein glycosylation, as deduced from analysis of the SWISS-PROT database. *Biochim. Biophys. Acta* 1473: 4–8.

49. Dwek, M. V., Ross, H. A., and Leathem, A. J. 2001. Proteome and glycosylation mapping identifies posttranslational modifications associated with aggressive breast cancer. *Proteomics* 1: 756–762.

50. Rudd, P. M., Elliott, T., Cresswell, P., Wlison, I. A., and Dwek, R. A. 2001. Glycosylation and the immune system. *Science* 291: 2370–2376.

51. Barnes, C. A. S. and Lim, A. 2007. Applications of mass spectrometry for the structural characterization of recombinant protein pharmaceuticals. *Mass Spectrom. Rev.* 26: 370–388.

52. An, H. J., Peavy, T. R., Hedrick, J. L., and Lebrilla, C. B. 2003. Determination of *N*-glycosylation sites and site heterogeneity in glycoproteins. *Anal. Chem.* 75: 5628–5637.

53. Huddleston, M. J., Bean, M. F., and Carr, S. A. 1993. Collisional fragmentation of glycopeptides by electrospray ionization LC/MS and LC/MS/MS: Methods for selective detection of glycopeptides in protein digests. *Anal. Chem.* 65: 877–884.

54. Carr, S. A., Huddleston, M. J., and Bean, M. F. 1993. Selective identification and differentiation of *N*- and *O*-linked oligosaccharides in glycoproteins by liquid chromatography-mass spectrometry. *Protein Sci.* 2: 183–196.

55. Jedrezejewski, P. T. and Lehmann, W. D. 1997. Detection of modified peptides in enzyme digests by capillary liquid chromatography/electrospray mass spectrometry and a programmable skimmer CID acquisition routine. *Anal. Chem.* 69: 294–301.

56. Colangelo, J., Licon, V., Benen, J., Visser, J., Bergmann, C., and Orlando, R. 1999. Characterization of the *N*-linked glycosylation site of recombinant pectate lyase. *Rapid Commun. Mass Spectrom.* 13: 2382–2387.

57. Fridriksson, E. K., Beavil, A., Holowka, D., Gould, H. J., Baird, B., and McLafferty, F. W. 2000. Heterogeneous glycosylation of immunoglobulin E constructs characterization by top-down high-resolution 2-D mass spectrometry. *Biochemistry* 39: 3369–3376.

58. Tengumnuay, P., Morris, H. R., Dell, A., Panico, M., Paxton, T., and West, C. M. 1998. The cytoplasmic F-box binding protein SKP1 contains a novel pentasaccharide linked to hydroxyproline in Dictyostelium. *J. Biol. Chem.* 273: 18242–18249.

59. Kurahashi, T., Miyazaki, A., Murakami, Y., Suwan, S., Franz, T., Isobe, M., Tani, N., and Kai, H. 2002. Determination of a sugar chain and its linkage site on a glycoprotein TIME-EA4 from silkworm diapause eggs by means of LC-ESI-Q-TOF-MA and MS-MS. *Bioorg. Med. Chem.* 10: 1703–1710.

60. Liu, Y. -H., Lin, M., Ashline, D. J., Reinhold, V., Grace, M., and Pramanik, B. N. 2006. Monoclonal antibody carbohydrate structure sequencing using mass spectrometry and OSCAR: An algorithm for assigning oligosaccharide topology from MS^n data. Presented at the *54th ASMS Conference on Mass Spectrometry and Allied Topics*, Seattle, WA, May 28 to June 1.

61. Ashline, D. J., Lapadula, A. J., Liu, Y. -H., Lin, M., Grace, M., Pramanik, B. N., and Reinhold, V. N. 2007. Carbohydrate structural isomers analyzed by sequential mass spectrometry. *Anal. Chem.* 79: 3830–3842.

62. Cohen, P. 2000. The regulation of protein function by multisite phosphorylation—a 25 year update. *Trends Biochem. Sci.* 25: 596–601.

63. Graves, J. D. and Krebs, E. D. 1999. Protein phosphorylation and signal transduction. *Pharmacol. Ther.* 82: 111–121.

64. Blume-Jensen, P. and Hunter, T. 2001. Oncogenic kinase signaling. *Nature* 411: 355–365.

65. Cohen, P. 2002. Timeline: Protein kinases—the major drug targets of the twenty-first century? *Nat. Rev. Drug Discov.* 1: 309–315.

66. Covey, T., Shushan, B., Bonner, R., Schroder, W., and Hucho, F. 1991. In *Methods in Protein Sequence Analysis*, Eds. H. Jornvall, J. -O. Hoog, and A. -M. Gustavsson, p. 249. Basel: Birkhauser Verlag.

67. Chang, E. J., Archambault, V., McLachlin, D. T., Krutchinsky, A. N., and Chait, B. T. 2004. Analysis of protein phosphorylation by hypothesis-driven multistage mass spectrometry. *Anal. Chem.* 76: 4472–4483.
68. Huddleston, M. J., Annan, R. S., Bean, M. F., and Carr, S. A. 1993. Selective detection of phosphopeptides in complex mixtures by electrospray liquid chromatography/mass spectrometry. *J. Am. Soc. Mass Spectrom.* 4: 710–717.
69. Annan, R. S., Huddleston, M. J., Verma, R., Deshaies, R. J., and Carr, S. A. 2001. A multidimensional electrospray MS-based approach to phosphopeptide mapping. *Anal. Chem.* 73: 393–404.
70. Neubauer, G. and Mann, M. 1999. Mapping of phosphorylation sites of gel-isolated proteins by nanoelectrospray tandem mass spectrometry: Potentials and limitations. *Anal. Chem.* 71: 235–242.
71. Wang, S., Liu, Y. -H., Smith, C. K., and Pramanik, B. N. 2004. Mass spectrometric characterization of MEK1 phosphorylation during expression in insect cells. Presented at the *52nd ASMS Conference on Mass Spectrometry*, May 23–27, Nashville, TN.
72. Wilkins, M. R., Sanchez, J. C., Gooley, A. A., Appel, R. D., Humphery-Smith, I., Hochstrasser, D. F., and Williams, K. L. 1996. Progress with proteome projects: Why all proteins expressed by a genome should be identified and how to do it. *Biotechnol. Genet. Eng. Rev.* 13: 19–50.
73. Tyers, M. and Mann, M. 2003. From genomics to proteomics. *Nature* 422: 193–197.
74. Righetti, P. G., Castagna, A., and Herbert, B. 2001. Prefractionation techniques in proteome analysis. *Anal. Chem.* 73: 320A–326A.
75. Fountoulakis, M., Langen, H., Gray, C., and Takacs, B. 1998. Enrichment and purification of proteins of *Haemophilus influenzae* by chromatofocusing. *J. Chromatogr. A.* 806: 279–291.
76. Fountoulakis, M., Takacs, M. F., and Takacs, B. 1999. Enrichment of low-copy-number gene products by hydrophobic interaction chromatography. *J. Chromatogr. A.* 833: 157–168.
77. Fountoulakis, M. and Takacs, B. 1998. Design of protein purification pathways: Application to the proteome of *Haemophilus influenzae* using heparin chromatography. *Protein Expr. Purif.* 14: 113–119.
78. O'Farrell, P. H. 1975. High resolution two-dimensional electrophoresis of proteins. *J. Biol. Chem.* 250: 4007–4021.
79. Hamdan, M. and Righetti, P. G. 2003. Assessment of protein expression by means of 2D gel electrophoresis with and without mass spectrometry. *Mass Spectrom. Rev.* 22: 272–284.
80. McDonald, W. H. and Yates, J. R. III. 2002. Shotgun proteomics and biomarker discovery. *Dis. Markers* 18: 99–105.
81. Hancock, W. S., Wu, S. L., and Shieh, P. 2002. The challenges of developing a sound proteomics strategy. *Proteomics* 2: 352–359.
82. Geng, M., Ji, J., and Reginer, F. E. 2000. Signature-peptide approach to detecting proteins in complex mixtures. *J. Chromatogr. A.* 870: 295–313.
83. Sechi, S. 2006. *Quantitative Proteomics by Mass Spectrometry*. Totowa, NJ: Humana Press.
84. Gygi, S. P., Rist, B., Gerber, S. A., Turecek, F., Gelb, M. H., and Aebersold, R. 1999. Quantitative analysis of complex protein mixtures using isotope-coded affinity tags. *Nat. Biotechnol.* 17: 994–999.
85. Munchbach, M., Quadroni, M., Miotto, G., and James, P. 2000. Quantitation and facilitated de novo sequencing of proteins by isotopic N-terminal labeling of peptides with a fragmentation-directing moiety. *Anal. Chem.* 72: 4047–4057.
86. Stewart, I. I., Thomson, T., and Figeys, D. 2001. ^{18}O labeling: A tool for proteomics. *Rapid Commun. Mass Spectrom.* 15: 2456–2465.

87. Old, W. M., Meyer-Arendt, K., Aveline-Wolf, L., Pierce, K. G., Mendoza, A., Sevinsky, J. R., Resing, K. A., and Ahn, N. C. 2005. Comparison of label-free methods for quantifying human proteins by shotgun proteomics. *Mol. Cell. Proteomics* 4: 1487–1502.

88. Rifai, N., Gillette, M. A., and Carr, S. A. 2006. Protein biomarker discovery and validation: The long and uncertain path to clinical utility. *Nature Biotechnol.* 24: 971–983.

89. Smith, J. C., Lambert, J. -P., Elisma, F., and Figeys, D. 2007. Proteomics in 2005/2006: Developments, applications and challenges. *Anal. Chem.* 79: 4325–4344.

90. Lescuyer, P., Hochstrasser, D., and Rabilloud, T. 2007. How shall we use the proteomics toolbox for biomarker discovery? *J. Proteome Res.* 6: 3371–3376.

91. Macek, B., Waanders, L. F., Olsen, J. V., and Mann, M. 2006. Top-down protein sequencing and MS3 on a hybrid linear quadrupole ion trap-orbitrap mass spectrometer. *Mol. Cell. Proteomics* 5: 949–958.

92. Hu, S., Loo, J. A., and Wong, D. T. 2006. Human body fluid proteome analysis. *Proteomics* 6: 6326–6353.

93. Caprioli, R. M., Farmer, T. B., and Gile, J. 1997. Molecular imaging of biological samples: Localization of peptides and proteins using MALDI-TOF MS. *Anal. Chem.* 69: 4751–4760.

94. Stoeckli, M., Chaurand, P., Hallahan, D. E., and Caprioli, R. M. 2001. Imaging mass spectrometry: A new technology for the analysis of protein expression in mammalian tissues. *Nat. Med.* 7: 493–496.

95. Bunch, J., Clench, M. R., and Richards, D. S. 2004. Determination of pharmaceutical compounds in skin by imaging matrix-assisted laser desorption/ionization mass spectrometry. *Rapid Commun. Mass Spectrom.* 18: 3051–3060.

96. Li, L., Garden, R. W., and Sweedler, J. V. 2000. Single-cell MALDI: A new tool for direct peptide profiling. *Trends Biotechnol.* 18: 151–160.

97. Henry, L. J., Xia, D., Wilke, M. E., Deisenhofer, J., and Gerard, R. D. 1994. Characterization of the knob domain of the adenovirus type 5 fiber protein expressed in *Escherichia coli. J. Virol.* 68: 5239–5246.

98. Liu, Y. -H., Vellekamp, G., Chen, G., Mirza, U. A., Wylie, D., Twarowska, B., Tang, J. T., Porter, F. W., Wang, S., Nagabhushan, T. L., and Pramanik, B. N. 2003. Proteomic study of recombinant adenovirus 5 encoding human p53 by matrix-assisted laser desorption/ionization mass spectrometry in combination with database search. *Int. J. Mass Spectrom.* 226: 55–69.

99. Mirza, U. A., Liu, Y. -H., Tang, J. T., Porter, F., Bondoc, L., Chen, G., Pramanik, B. N., and Nagabhushan, T. L. 2000. Extraction and characterization of adenovirus proteins from sodium dodecylsulfide polyacrylamide gel electrophoresis by matrix-assisted laser desorption/ionization mass spectrometry. *J. Am. Soc. Mass Spectrom.* 11: 356–361.

2 Advanced Capillary Liquid Chromatography– Mass Spectrometry for Proteomics

Yufeng Shen, Jason S. Page, and Richard D. Smith

CONTENTS

2.1   INTRODUCTION

The liquid chromatography (LC)-mass spectrometry (MS) analysis of peptides has become an increasingly routine method for proteomics—the study of the entire complement of proteins, for example, expressed by a cell under a specific set of conditions at a specific time. Mixtures of peptides, such as those generated from enzymatic (e.g., trypsin) digestion of globally recovered proteins (i.e., a proteome), are typically very complex and >100,000 different molecular species may be observable using MS detection [1]. LC separations implemented prior to MS for broad protein identification have three major roles: (1) to isolate individual components or reduce complexity as much as possible, (2) to increase sensitivity by concentrating the components into narrow zones prior to MS, and (3) to eliminate or displace interfering species (e.g., salts and polymers) that may be present in proteomics samples.

A desired quality of LC separation can be achieved from the use of either multiple steps of moderate quality separations, or fewer steps of high power separations. The former approach is generally more easily accessible for very high quality separations due to the variety of commercialized LC platforms available, while the latter still often requires considerable developmental efforts (for both columns and instrumentation). In addition to proteomics data quality, other differences between these two approaches include proteomics analysis time and sample consumption (and subsequent analysis costs), as well as direct impact on potential proteomics applications that have special requirements in terms of analysis coverage, sample size, dynamic range, sensitivity, and throughput.

In this chapter, we discuss advanced LC technologies and their resultant LC-MS capabilities for various proteomics applications, using as examples the developments and applications from our laboratory.

2.2   LC COLUMNS AND THEIR CAPABILITIES FOR PROTEOMICS ANALYSIS

As columns are the central element for LC separations, we briefly discuss the different types of LC columns, including various dimensions of packed capillaries and monolithic columns for analytical proteomics applications.

2.2.1   LONG-PACKED CAPILLARY LC COLUMNS

Packed capillaries are the major LC column type and are widely used for proteomics. The separation power, quantitated by peak capacity (C_p), for separating a protein tryptic digest under mobile-phase composition, gradient reversed-phase (RP) conditions depends on the LC column length. For porous-packed capillary columns under shallow gradient conditions, this relationship can be expressed as [2]

$$C_p \approx 180 \sqrt{\frac{L}{d_p}} \tag{2.1}$$

where L and d_p are the column length (cm) and particle size (μm), respectively. Short (e.g., 10–15 cm), 5 μm porous particle-packed capillary column can generate a peak capacity of ~200, while long (e.g., 40–200 cm) capillary columns packed with

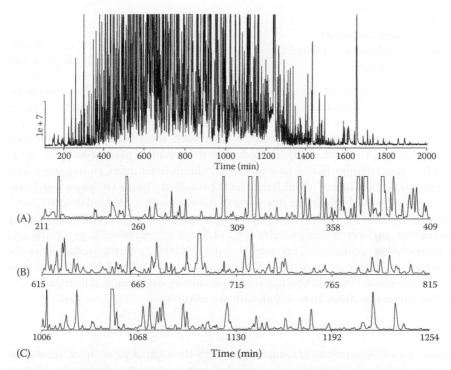

FIGURE 2.1 High-resolution LC of a cell lysate tryptic digest, using a 3 μm porous particle-packed 200 cm long capillary column operated at 20 Kpsi. A peak capacity of ~1500 is estimated with multiple single ion current chromatographic peaks (A–C) across the effective separation time window. Test sample: *S. oneidensis* tryptic digest. Detailed conditions are described in Ref. [2].

small particles (e.g., 1.4–3 μm) and manufactured [2] using a slurry packing method can generate LC peak capacities of >1000 for separating the peptides.

Figure 2.1 shows a packed capillary LC of a *Shewanella oneidensis* tryptic digest obtained using a 200 cm × 50 μm i.d. capillary column packed with 3 μm C18 porous particles. Note that the peak capacity of ~1500 obtained with this column is the highest reported to date [2], for separation of trypsin-digested peptides among various liquid-phase separations approaches. When combined with the advantages afforded by LC separations, in general—tolerance to various properties of analytes in proteomics samples (e.g., salts, neutral polymers, and charged peptides), on-column sample concentration (due to the use of gradient mobile phases), and coupling to MS (see below)—the high-peak capacities afforded by long capillary columns make LC a preferred choice for proteomics applications.

2.2.2 Narrow Bore Packed Capillary LC Columns

When LC is coupled to MS through electrospray ionization (ESI), the inner diameter of the LC column becomes an important factor as it directly affects analytical sensitivity [3]. The relationship between signal intensity and column inner diameter can be expressed as

$$I = bd_c^{-2} \qquad (2.2)$$

where

    I is the signal intensity

    d_c is the column inner diameter

    b is the constant

Figure 2.2 illustrates ESI-MS signal intensity dependence on the capillary dimension for 15–75 μm i.d. packed capillary columns that generated nanoscale flows of 20–400 nL/min.

Narrow (e.g., 15 μm i.d.) capillaries have been successfully packed for long (e.g., >80 cm) LC columns [3], and generate a separation peak capacity of up to ~1000. These columns have a ratio of ~5 for column inner diameter to particle size and produce a flow rate of ~20 nL/min when operated at a linear velocity of ~0.2 cm/s (close to the optimal value). The robust use of columns with such small inner diameter has been enabled by implementing online solid-phase extraction (SPE, see below). As smaller, uniform porous particles (e.g., ~1.5 μm) and narrower (e.g., ≤10 μm i.d.) high-resolution capillary LC columns become available, a further reduction of the flow rate to <10 nL/min is practical. However, the response of ESI-MS to <10 nL/min LC mobile-phase flow, and how much more sensitivity can be gained by reducing the mobile-phase flow from 20 to ≤10 nL/min are unknown.

2.2.3 Submicrometer Particle-Packed Capillaries Columns

While fast LC separations are required for high-throughput proteomics, increasing the speed of the LC separation unavoidably brings about a reduction in the total separation resolution achievable. The use of extremely small-sized packing particles (e.g., submicrometer) can minimize the loss of separation resolution by improving the solute mass transfer rate in the mobile phase, and subsequently the peak capacity generation rate; however, the quantitative relationship between fast gradient LC peak capacity and particle size has not been well explored.

Porous particles are attractive for proteomics due to the high column sample capacity and peak capacity [2], but availability is limited for smaller micron to submicron sizes, that is, most are ≥1.5 μm. In collaboration with the Unger group (from the Institut fuer Anaorganische Chemie und Analytische Chemie, Johannes Gutenberg-Universitaet, Germany), we have demonstrated submicrometer porous particles for fast LC separations of tryptic peptides [4,5]. Figure 2.3 shows the separations completed in 15–50 min for proteomics analysis with the use of tandem MS (i.e., MS/MS) and in 2–5 min with single-stage MS. A peak capacity of ~400 was achieved for a 50 min gradient LC separation [4], which dropped to ~100 if the separation was completed in ≤2 min [5]. Short columns (e.g., <5 cm in length) provide the best resolution for ~2 min separations, which in turn permits the use of even smaller porous particles (e.g., 0.3–0.5 μm) to further improve LC separation within a certain operating pressure range.

2.2.4 Monolithic LC Columns

Monolithic columns are characterized by low pressure drops and high column porosity (and subsequently large flow rate) [6]. The fact that fabrication of the column starts from a solution facilitates the production of narrow LC columns, with little limitation

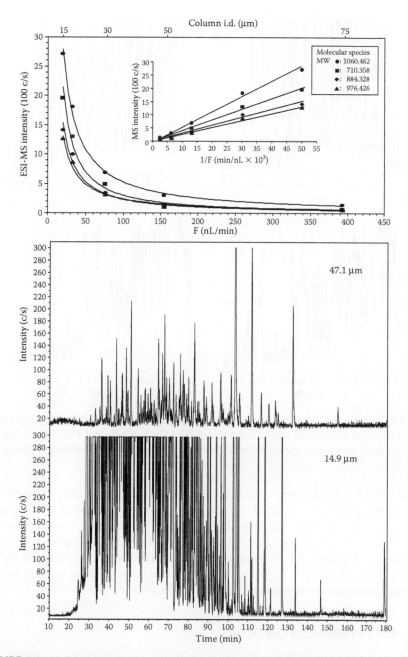

FIGURE 2.2 LC-MS sensitivity dependence on the LC column inner diameter and mobile-phase volumetric flow rate for peptides in a cell lysate tryptic digest. Packed capillary columns that were tested had inner diameters from 15 to 75 μm and generated mobile-phase flow rates from 20 to 400 nL/min. TOF MS is used for the signal detection. Two LC-MS tota-lion chromatograms show the detection sensitivity obtained from different inner diameter columns. Test sample: yeast tryptic digest. Detailed conditions are described in Ref. [3].

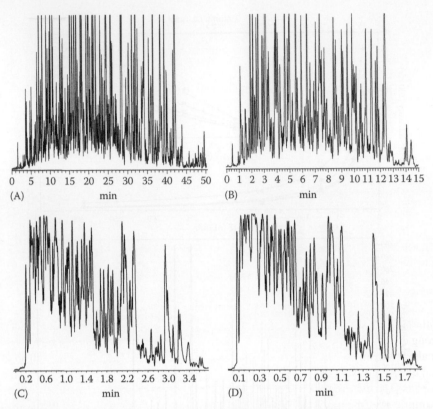

FIGURE 2.3   Fast LC separations of a cell lysate tryptic digest, using 0.8 μm porous particle-packed capillary columns in combination with LTQ MS (A and B) and TOF MS (C and D). Test sample: *S. oneidensis* tryptic digest. Detailed conditions are described in Refs. [4,5].

on the column inner diameter. A concern for proteomics applications are the limitations on the separation resolution. A peak capacity of ~400 has been reported for a long (e.g., ~80 cm) 20 μm i.d. monolithic column [7], which is equivalent to that achievable with 5 μm porous particle-packed, 30–40 cm length capillary columns according to Equation 2.2. A variation involves formation of the monolith on the capillary inner wall so as to generate a porous layer open tubular column having potential for extremely sensitive proteomics analysis [8].

2.3   ESI INTERFACE FOR LC-MS

ESI is commonly used to couple liquid-phase LC separations to gas-phase MS detection. This is due, in large part, to the ease in which ESI can generate ions from LC columns at atmospheric pressure. The electrospray process occurs when a solution, typically delivered through a capillary, is subjected to a high voltage (either positive or negative ranging from 1000 to 5000 V) which establishes an electric field between the capillary terminus (often a shaped tip of small diameter) and a counterelectrode. This electric field gradient, if large enough, will produce significant charge separation

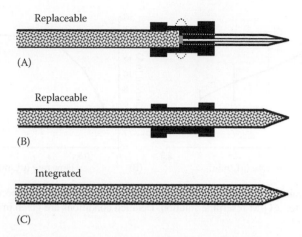

Replaceable

(A)

Replaceable

(B)

Integrated

(C)

FIGURE 2.4 Schematic of ESI emitters applied for coupling LC to MS for various proteomics analyses. (A) Replaceable ESI emitter with empty capillary, (B) replaceable ESI emitter with packed capillary, and (C) integrated ESI emitter. Detailed conditions are described in Refs. [3,4].

in the solution to form a conical meniscus, referred to as the Taylor cone [9]. The strong electric field at the apex of the Taylor cone ejects a fine liquid filament which breaks into charged droplets caused by surface wave instability [10]. These droplets then undergo solvent evaporation and columbic fission events which produce smaller droplets [11] and eventually single ions [12–14]. The stability of the electrospray can be improved, especially at lower flow rates, by tapering the capillary exit.

Figure 2.4 shows general approaches for attachment of electrospray emitters for capillary LC-MS. Figure 2.4A illustrates a short (e.g., ~1.5 cm), pretapered open tubular capillary that has an inner diameter of 1/3 or 1/2 of the capillary LC column and connected to the LC column outlet by a union. Provided the union is conductive, it can be used as the connection for electrospray high voltage. The union should have an internal bore that is at least as small as the LC column inner diameter, to reduce chromatographic peak broadening. Unions with internal bores as small as 5 μm have been used for nanoscale LC, which produced minimal peak broadening for 15 μm i.d. LC columns with separations completed in ~3 h [15]. The electrospray emitter can also be directly incorporated as part of a short prepacked capillary column (Figure 2.4B), which creates a replaceable emitter with less dead volume than an open tubular emitter. Additionally, an emitter can be created directly on the LC column (Figure 2.4C) to reduce the extracolumn dead volume for fast separations with short LC columns [16,17]. All of these emitters can also be used with monolithic materials [18] or particle/monolith mixed materials [19].

2.3.1 BENEFITS FROM NANOLITER PER MINUTE FLOW RATE ESI

ESI operating at flow rates <50 nL/min, also known as the nano-ESI regime, provides several benefits compared to traditional higher flow rate ESI [20,21]. In particular, smaller flow rates lead directly to the emission of smaller charged droplets that require less desolvation and fission events to produce gas-phase ions, which increases

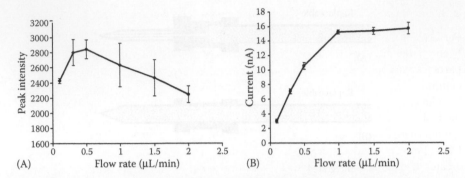

FIGURE 2.5   Peak intensity (A) and ES current (B) transmitted into the mass spectrometer versus sample flow rate for a 1 μM reserpine solution. Detailed conditions are given in Ref. [23].

ionization efficiency and therefore detection sensitivity [20,22,23]. Figure 2.5 shows the peak intensity of a common MS calibrant versus electrospray flow rate, (Figure 2.5A) along with the amount of electrospray current transmitted into the mass spectrometer for each flow rate (Figure 2.5B). The peak intensity for 1.5 μL/min was ~25,000 from ~16 nA of electrospray current transmitted into the interface. When the flow rate was reduced to 0.1 μL/min, the peak intensity was ~24,000 from only 3 nA of electrospray current. Even though the electrospray current transmitted into the mass spectrometer dropped by 80% when the flow rate was lowered to 100 nL/min, the amount of gas-phase calibrant ions in the mass spectrometer remained the same as indicated by the detector response. This observation indicates that the efficiency of ionization increased at lower flow rate. In other words, even though the amount of calibrant electrosprayed was less than a tenth at the lower flow rate, a larger portion of the calibrant was successfully ionized, which produced a peak with a similar intensity as at higher flow rate.

The electrospray current is proportional to the square root of the flow rate [22], and this relationship increases the amount of available charge per analyte as the flow rate is reduced. In turn, ionization efficiency is increased, thereby reducing the suppression effects in ESI [24,25]. In addition, the larger charge-to-analyte ratios reduce the competition between analytes, which improves quantitation. Analytes that are more surface active on the charge droplets (i.e., hydrophobic) tend to dominate the ionization process due to droplet fission events that occur on the surface and expel a majority of the charges in the droplet along with the surface-active species [23]. This leaves less surface-active species behind in the original droplet, with fewer charges and reduces their ionization efficiency. Reducing the droplet size by nano-ESI decreases the population in each individual droplet while also increasing the charge-to-analyte ratio, which increases the possibility of all species in the droplet becoming gas-phase ions. This phenomenon has been demonstrated by analyzing solutions containing a peptide and an oligosaccharide [25,26]. Even though the two species were equimolar in solution, mass spectra obtained at higher flow rates (2–4 μL/min) were dominated by the peptide peak [26]. Equal intensities of the two species were only obtained when the flow rate was reduced to <20 nL/min [26]. Nano-ESI produced a detector response that better represented the true concentrations in solution improving the quantitation.

2.3.2 Nano-ESI Coupled to Higher Flow Rate Separations

Performing LC separations using very low flow rates compatible with nano-ESI has been challenging. Its limited use is mainly due to the difficulties in packing small particles into narrow bore capillaries, and sample injection and loading onto the LC column [14]. A splitter postcolumn has been used to divert a large portion of the LC eluent to waste or a fraction collector to create a nanoliter per minute flow rate at the ESI emitter; however, this approach leaves most of the sample un-ionized [27,28]. An alternative approach has recently been demonstrated, whereby the flow from an LC column is divided into several individual electrospray emitters, each of which creates a nanoelectrospray that allows ionization of the entire sample [26]. This approach provides the benefits of nano-ESI with higher flow rate liquid separations.

An array of electrospray emitters that can be easily coupled to an LC column is pictured in Figure 2.6A. The array was made by inserting several sections of silica capillary tubing into a PEEK LC fitting and holding them in place with an epoxy. The other ends were positioned at a desired spacing by a fabricated clamp, and then ~2 cm of the protective polyimide was removed. The array of capillaries was chemically etched, using a technique that tapers the outside of the capillaries without etching the

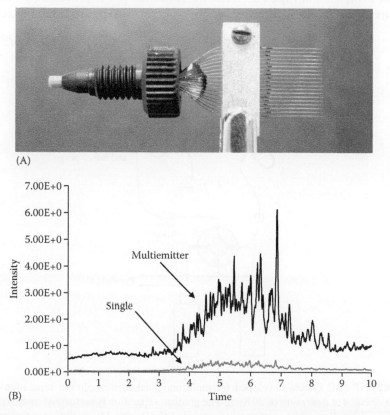

(A)

(B)

FIGURE 2.6 Picture of a 19 electrospray emitter array (A) and MS chromatograms from the analyses of a depleted human plasma sample with a traditional single electrospray emitter and a 19 electrospray emitter array (B). Detailed conditions are described in Ref. [30].

inside walls to create electrospray emitters with no internal taper, which in turn greatly reduces clogging and increases robustness [29]. The chemical etching is also self-terminating, which ensures that all the electrosprays lie in the same plane. The use of these arrays in LC-MS analysis of human plasma samples has resulted in an ~10-fold increase in sensitivity [30]. Figure 2.6B shows the MS chromatograms obtained from both a single electrospray emitter and a 19 emitter array coupled individually to the same LC column operating at $2\,\mu L/min$, with the same sample and amount injected to the column. The increase in measurement sensitivity afforded by the multiemitter, as evidenced by the more complex chromatogram, is due to the increased ionization efficiency and reduction of ion suppression effects.

2.4 AUTOMATED LC SYSTEMS WITH EXTENDED OPERATING PRESSURES

Use of small particle-packed capillary LC columns for high-resolution separations of complex proteomics samples requires an extension of LC operational pressure for transporting the mobile phase beyond 10 Kpsi (1 Kpsi = 68 atm = 69 bar = 6.9 MPa). A schematic of an automated capillary LC system [2] developed for separations at 20 Kpsi is depicted in Figure 2.7. The LC resolution power provided by this high pressure system results in a peak capacity of ~1500 (as demonstrated in Figure 2.1).

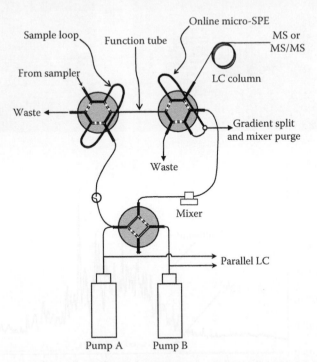

FIGURE 2.7 LC system developed for multifunctional high-resolution separations. This system operates at a pressure of 20 Kpsi. The gradient separation is performed under constant pressure, which allows the use of parallel multiple LC columns. Micro-SPE is online coupled prior to the LC column, and the function tube can be replaced according to separation and analysis requirements. Detailed conditions are described in Ref. [2].

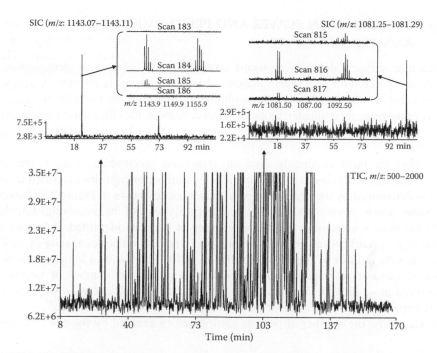

FIGURE 2.8 Efficient separation of a cell lysate tryptic digest using online micro-SPE-15 μm i.d. packed capillary LC. The loss of efficiency due to the micro-SPE is minimal, even when coupled to the 15 μm i.d. packed capillary column. Test sample: *Deinococcus radiodurans* tryptic digest. Detailed conditions are given in Ref. [11].

The automated LC system includes the following features: (1) an online coupled SPE device that precedes the LC column for sample loading, (2) the ability to operate parallel multiple capillary columns, using two LC pumps, and (3) multifunctional separations. Importantly, the online coupled SPE robustly processes rough and less-soluble proteomics samples, as well as precisely loads small-sized samples [31]. The chromatography accessory and connection modes must be carefully considered to minimize the influences imposed by the online SPE [31], and an optimal design provides very minor peak broadening even for 15 μm i.d. packed capillary columns, as shown in Figure 2.8, for a 0.25 ng proteomics sample. The use of parallel multiple capillary LC columns improves proteomics analysis throughput, by eliminating the time wasted, as one LC column is washed/re-equilibrated after each mobile-phase gradient. Implementation of the function tube shown in Figure 2.7, with ion exchange or affinity-bored column packing makes the system capable of online two-dimensional (2D)-LC or proteomics sample affinity enrichment.

A 2D-LC system with approximately the same operational pressure limit, as the automated system in Figure 2.7, has been constructed for MudPIT proteomics analysis (i.e., separations using an orthogonal combination of cation exchange and RP chromatography in a microcapillary format followed by MS/MS) [32]. Operated at a constant flow, this system also improves peptide detection compared with conventional low pressure 2D-LC.

2.5   LC SEPARATION POWER AND PROTEOMICS ANALYSIS COVERAGE

Obtaining sufficient depth of proteome coverage (or broad protein identification) is the first concern for proteomics analyses. Analysis coverage is affected by several factors that include sample size, the resolution of the separation methods, the sensitivity of the mass spectrometer, modes for data collection, the criteria used to identify proteins from collected data, and the dynamic range of protein abundances in samples.

When the most commonly used ion trap mass spectrometer (i.e., Finnegan LCQ) and data analysis criteria are applied, the number of peptides identified from MS is determined by the separation peak capacities, as shown in Figure 2.9. Direct infusion, which provides a separation peak capacity of 1, of the proteomics sample into the mass spectrometer limits the number of peptides identified. Adopting a short (e.g., 15 cm) packed column that provides a separation peak capacity of ~200 increases the number of peptides identified to ~1700 [2]. Use of a long monolithic column (separation peak capacity of ~400) increases the number of peptides identified to ~2400 [7]. The small particle-packed long LC columns (separation peak capacity of >1000) further increase the number of peptides identified to ~4500 [2]. Currently, the number of peptide identifications enabled by the use of either

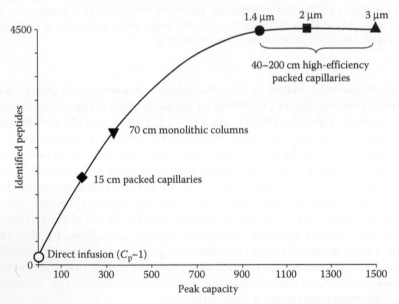

FIGURE 2.9   Separation peak capacities obtained from various types of LC columns and the number of peptide identifications enabled by these peak capacities. Manufacturers and operating conditions for various columns are described in detail in Refs. [2,7]. A conventional ion trap (i.e., Finnegan LCQ) mass spectrometer was used for the 10 h LC separations, and MS/MS peptide identifications were completed using criteria in Ref. [2]. Test sample: *S. oneidensis* tryptic digest.

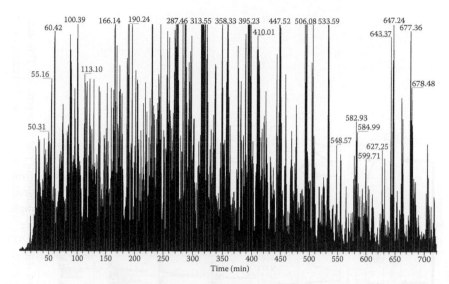

FIGURE 2.10 Single-dimensional LC-MS/MS (Finnegan LTQ) for broad identification of proteome proteins. In a single 12 h LC-MS/MS analysis, >12,000 peptides from >2,000 proteins were identified. Test sample: *S. oneidensis* tryptic digest. Detailed conditions are described in Ref. [2].

short-packed or long monolithic columns is approximately 1/3 or 2/3 of that obtained with high-resolution packed columns. Increasing the separation peak capacity from 1000 to 1500 yields a very minor gain in the number of peptides identified, which suggests that detection is now limited by the sensitivity of the mass spectrometer and that a separation with peak capacity of ~1000 should be sufficient for MS/MS (operated in data-dependent mode) detection of most of the peptides.

With the use of a faster, more sensitive linear ion trap mass spectrometer (Finnegan LTQ), the number of identified peptides can be increased ~2.5-fold compared with the conventional ion trap mass spectrometer (Finnegan LCQ). For example, a single 12 h high-resolution packed capillary LC-LTQ MS/MS run (Figure 2.10) can identify >12,000 peptides and >2,000 proteins from a proteome [2]. The analysis reproducibility (two replicates) for the number of proteins identified can be up to 99%, with ~90% proteins in common across analyses [2]. For human blood plasma, one of the most complex proteomes with a dynamic range of protein abundance that spans 10 orders of magnitude, the number of proteins identified from a single 12 h high-resolution LC-MS/MS experiment reduces to 800 [2]. In this case, the relative range of protein abundance in the sample is a key factor that determines the coverage achievable in proteomics analyses (see Section 2.6).

2.6 LC SAMPLE CAPACITY AND PROTEOMICS ANALYSIS DYNAMIC RANGE

Proteomics analysis dynamic range is determined by the detection limit of the mass spectrometer and the amount of sample available for LC separation; the use of more

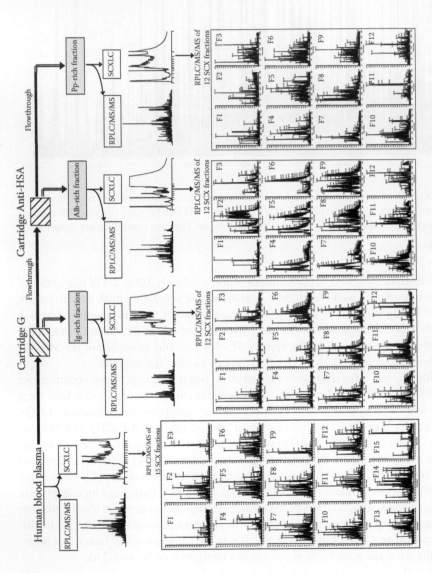

FIGURE 2.11 2D-LC-MS/MS (conventional LCQ ion trap mass spectrometer) for large dynamic range identification of proteome proteins. This strategy enabled identification of >2000 proteins with an analysis dynamic range of >8 orders in magnitude of protein abundance. Test sample: human blood plasma tryptic digest. Detailed conditions are given in Refs. [13,15].

sensitive mass spectrometers certainly improves the measurement dynamic range. For a specific mass spectrometer, the dynamic range is determined by the maximum amount of sample that can be loaded onto and separated from a certain dimension of the LC column, that is, the LC column sample capacity.

Porous particle-packed capillary columns have a much larger sample capacity than nonporous particle-packed capillary columns. The sample capacity for commonly used 50–150 μm i.d. porous-packed capillary columns is 10–100 μg (protein content) [2,32], which can support a dynamic range of protein abundance of about 4–5 orders difference in magnitude for a proteomics LC-MS/MS analysis [4,33]. Most proteins should be detectable within this dynamic range due to the LC high resolution (i.e., peak capacities of >1000) and fast data-dependant MS/MS analysis of the components (Figure 2.13 following). Advanced mass spectrometers such as linear ion trap can complete a MS/MS analysis of an individual component in ~0.3 s [4], and ~100 different components can be selected (according to their ratio of mass and charge, m/z) and analyzed from 0.5 min chromatographic peaks where the components coelute. Further enlargement of the analysis sample size (i.e., >100 μg) requires the sample to be separated into multiple fractions prior to LC-MS analysis. 2D-LC, typically with a format of strong cation exchange (SCX)-RPLC, can be implemented for this purpose, and the 2D-LC-MS results obtained for characterization of the human blood plasma proteome demonstrate a great extension of the proteomics analysis dynamic range [33,34].

Figure 2.11 depicts the strategy used for this LC- and 2D-LC-MS/MS analysis of the human blood plasma proteome. Affinity isolations are used to deplete the high abundance albumin and antibodies. Because of the potential for codepletion [34], the isolation products and side-products from each protein depletion step are examined by 2D-LC, to account for protein loss. This strategy has been demonstrated to provide 8–9 orders of magnitude of protein abundance detection range, and enable identification of human blood plasma proteins that exist at low pg-level, for example, 77 pg interleukin-12 beta chain (IL-12 p40) and ~10–30 pg fibroblast growth factor-12 [33,34]. In terms of proteomics analysis coverage, this 2D-LC strategy in combination with conventional Finnegan LCQ ion trap MS/MS led to ~2000 human plasma protein identifications [34], and >5000 proteins from human blood plasma should be realistic with more sensitive linear ion trap MS/MS.

2.7 LC SEPARATION SCALE AND PROTEOMICS ANALYSIS SENSITIVITY

Some proteomics sample sizes are very small and available in limited quantities, for example, clinical samples such as tissue biopsies. As such, ultrasensitive LC-MS has been explored for analyzing samples <50 ng [35,36], and is suitable for analysis of a limited number of cells. In an initial demonstration using high-resolution 15 μm i.d. packed capillary LC coupled to Fourier transform ion cyclotron resonance (FTICR) MS, 75-zeptomole sensitivity was achieved for identification of proteins (six peptide assignments per protein identification) and ~10-zeptomole (~6000 molecules) sensitivity for assignment of individual peptides [35]. With this level of sensitivity high abundance proteins from 0.5 pg of proteomics sample can be identified.

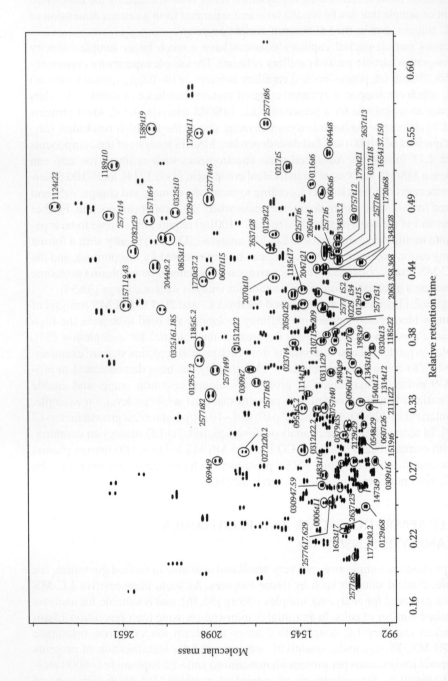

FIGURE 2.12 Ultrasensitive LC-MS proteomics analysis of 50pg proteomics sample. Approximately 150 proteins were identified from this cell lysate tryptic digest. Test sample: *D. radiodurans* tryptic digest. Detailed conditions are given in Ref. [17].

Figure 2.12 shows an example of a proteomics analysis in which ~150 proteins were identified from 50 pg of sample [36]. This analytical level of sensitivity would potentially allow the identification of proteins from 100 mammalian cells (typically 50–100 pg proteins per mammalian cell), with a dynamic range for protein abundance of ~4 orders in magnitude. The use of an online coupled micro-SPE maximized the sample loading (e.g., ~1% sample loss estimated) from a relatively large volume of solution. Note that increasing the mobile-phase flow beyond 20 nL/min, as used in this study, may result in the broadening of the chromatographic peak (e.g., average peak width of ~15 s). Additionally, enlarging the ESI emitter orifice size beyond ~2 μm i.d. may degrade the sensitivity.

In another study, the sensitivity limit for MS/MS analysis was evaluated using a 5 μm particle-packed 50 μm i.d. typical capillary LC column coupled to a linear ion trap mass spectrometer [37]. Approximately 4000 peptides and ~1000 proteins were identified from 50 ng of an *S. oneidensis* tryptic digest. This initial result highlights the potential of using narrow capillary LC columns with MS/MS to identify hundreds of proteins from <1 ng proteomics sample. Narrower (e.g., 10 μm i.d.) porous layer open tubular [8] and monolithic [38] columns have already been tested for proteomics analysis; however, the resulting protein identification sensitivity is on the same level as that obtained from a 5 μm particle-packed 50 μm i.d. capillary column. Continued efforts are needed to improve the efficiency of these columns. As such, small particle-packed narrower capillary columns are currently the best choice for ultrasensitive MS/MS proteomics analysis.

2.8 LC SEPARATION SPEED AND PROTEOMICS ANALYSIS THROUGHPUT

Analytical throughput plays an important role in most proteomics applications, as many studies require large number of measurements (e.g., replicates, time course studies). A proteomics analysis can be completed in months, weeks, days, hours, or minutes, depending on the analytical methods and requirements for analysis time, coverage, and data quality. For example, a 3–4 h LC-MS/MS analysis using LTQ mass spectrometers allows identification of ~75% of the proteins detectable at the sensitivity level of the platform [2].

Submicrometer particle-packed capillary columns have been used to speed LC separations (see description above) for optimal utilization of MS/MS. A 50 min LC-MS/MS analysis (Figure 2.3) using a linear ion trap mass spectrometer (0.3 s per MS/MS spectrum) resulted in ~4000 peptide (~1000 protein) identifications [4]. Within this 50 min time frame, the number of peptides identified increases approximately linearly as a function of the data acquisition (or analysis) time. Figure 2.13 highlights the fact that most of the acquired MS/MS spectra could be used to identify the peptides; for example, 92% of the 25 MS/MS spectra collected in just 6.5 s were utilized for peptide identification. These results reveal that the MS/MS data acquisition speed is the rate-limiting step. As such, further improvements to the separation peak capacity generation rate would have only a limited contribution toward increasing the analysis coverage. These fast separations are not as amenable to high-precision MS/MS (e.g., using a Thermo Fisher LTQ-FT and LTQ-Orbitrap mass spectrometers),

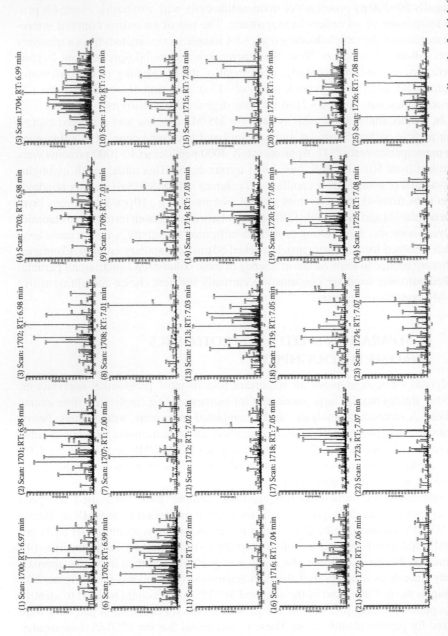

FIGURE 2.13 MS/MS spectrum generation rate from a fast LTQ mass spectrometer. In 6.5 s, 25 MS/MS spectra were collected, of which 92% were utilized for peptide identification. Test sample: *S. oneidensis* tryptic digest. Detailed conditions are described in Ref. [4].

due to the significantly longer time required to acquire high quality spectra (e.g., ~1.5 s per FT MS/MS spectrum). In this situation, long-packed capillary columns operated with shallow gradient or even 2D-LC have to be used, but at the expense of the analysis throughput.

To obtain throughput on the order of minutes (i.e., ultrahigh-throughput proteomics), single-stage MS measurements are the only option. Multiple peptides can be detected from a single MS spectrum which differs from MS/MS where one spectrum typically corresponds one peptide and only a limited number of MS/MS spectra can be acquired within such short analysis time (e.g., ~1000 MS/MS spectra in 5 min using a fast LTQ mass spectrometer). Ultrahigh-throughput peptide identification has been demonstrated using the accurate mass and time tag approach [19], and the key to this type of analysis is the ability to correlate peptide LC elution times among fast LC-MS and slow LC-MS/MS experiments [5]. Various accurate mass spectrometers, including FTICR and time-of-flight (TOF) have been evaluated, and TOF MS proved to be the best for 2–3 min proteomics analyses. Figure 2.14 shows an LC-MS 2D display of an *S. oneidensis* tryptic digest in which ~2000 peptides from >500 proteins were identified in ~2 min. Increasing the rate at which the LC peak capacity is generated (e.g., achieving >200 of peak capacity) and the rate at which a TOF MS spectrum is acquired (e.g., <0.1 s for 1 MS spectrum), while retaining a sensitivity sufficient for detecting the proteomics components, can improve the analysis coverage for ultrahigh-throughput proteomics.

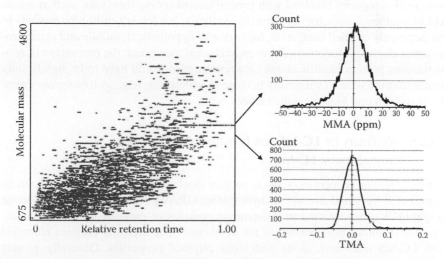

FIGURE 2.14 (See color insert following page 216.) Proteomics peptide 2D display obtained from an ultrahigh-throughput analysis using a fast LC-TOF MS platform. The mass measurement accuracy (MMA) and LC relative retention time measurement accuracy (TMA) are also shown. A 0.8 μm porous particle-packed capillary column was used for the fast LC separation, TOF MS for detection, and the accurate mass and time tag approach for protein identification. Approximately 550 proteins were identified from this ~2 min analysis. Test sample: *S. oneidensis* tryptic digest. Detailed conditions are given in Ref. [5].

2.9 OTHER PROTEOMICS ASPECTS RELATED TO LC SEPARATIONS

2.9.1 PROTEOMICS ANALYSIS REPRODUCIBILITY

The reproducibility of LC-MS-based proteomics analysis can be evaluated on the basis of a number of different criteria: the LC-MS total ion current chromatogram, the number of proteins identified, and the variety of proteins identified. The LC chromatogram can be examined for separation reproducibility, but not for proteomics analysis reproducibility (i.e., reproducibility of specific protein identifications). With correct operation, LC-MS can produce reproducible retention times and peak intensities, even for different narrow LC columns (e.g., 30 μm i.d.) and ESI emitters [3]. While comparison among the numbers of proteins (or peptides) identified from repeated LC-MS runs provides a rough evaluator of instrumental reproducibility (experimental data has shown that the variance in the number of proteins identified from repeated LC-MS can be as small as ~1% [2]), this approach provides limited insights into the reproducibility of proteins identified from the analysis.

Table 2.1 summarizes the proteomics analysis reproducibility in terms of the number of identified proteins that are obtainable from various LC-MS platforms. A similar variance of ~20% is observed for 1D-LC-MS/MS (operated with 10 h separations) and 2D-LC-MS/MS (performed with >50 fractions). The number of identified proteins varies among different categories of proteins; however, among the same category of proteins, the number of identifications is similar for the two platforms. Reproducibility was greater (>90%) for the typically more abundant proteins in categories involved with general housekeeping functions, such as amino acid biosynthesis, cofactor, and protein synthesis, but lower (<80%) for proteins in less accurately defined categories, for example, hypothetical, mobile and extrachromosomal elemental functions. From an analytical viewpoint, the percentage of non-overlapping protein identifications between samples would have to be significantly greater than the percent variability in those identifications (e.g., >>10% for our example) to discern real biological differences.

2.9.2 VERSATILITY OF LC-MS FOR IDENTIFYING PROTEOME PROTEINS FROM LC-MS

Various physical properties, such as protein isoelectric point (pI), hydrophobicity (represented by grand average of hydropathy, GRAVY value [39]), and molecular weight (MW), can be used to differentiate peptides or proteins from one another. Figure 2.15 shows distributions of the numbers of peptides and proteins identified from LC-MS platforms, along with these physical properties. Generally, protein identification based on LC-MS peptide analysis provides unbiased results compared with the proteome protein composition; however, a small discrepancy does exist. At the peptide level, both 1D- and 2D-LC provide a similar pI distribution profile for peptides (Figure 2.15A); however, fewer peptides in the MW 0.9–2 kDa and GRAVY −2.25 to 0.5 ranges are identified with the 2D-platform, most likely due to the use of SCX (Figure 2.15B and C). At the protein level, the hydrophobicity range differences

TABLE 2.1
Proteomics Analysis Reproducibility Obtainable from LC-MS Platforms

Protein Functional Categories	Predicted # Proteins	1D					2D				
		# Proteins			Percentage		# Proteins			Percentage	
		Run 1	Run 2	Overlap	Reproducibility	Coverage	50 Fractions	70 Fractions	Overlap	Reproducibility	Coverage
Amino acid biosynthesis	92	84	82	82	99	89	90	90	90	100	98
Biosynthesis of cofactors	124	80	77	72	92	58	111	108	103	94	83
Cell envelope	178	114	116	106	92	60	143	137	124	89	70
Cellular processes	277	165	158	145	90	52	226	218	201	91	73
Central intermediary metabolism	60	26	25	22	86	37	40	41	36	89	60
Disrupted reading frame	60	5	7	3	50	5	10	12	7	64	12
DNA metabolism	146	72	68	58	83	40	106	112	94	86	64
Energy metabolism	316	209	203	199	97	63	250	253	239	95	76
Fatty acid/phospholipid metabolism	65	43	44	40	92	62	55	55	52	95	80
Hypothetical	1085	142	139	104	74	10	285	296	213	73	20
Conserved hypothetical	875	325	334	284	86	32	525	528	459	87	52

(continued)

TABLE 2.1 (continued)
Proteomics Analysis Reproducibility Obtainable from LC-MS Platforms

Protein Functional Categories	Predict # Proteins	1D # Proteins Run 1	Run 2	Overlap	1D Percentage Reproducibility	Coverage	2D # Proteins 50 Fractions	70 Fractions	Overlap	2D Percentage Reproducibility	Coverage
Protein fate	194	127	123	117	94	60	158	163	150	93	77
Protein synthesis	142	109	110	104	95	73	127	132	124	96	87
Purines/pyrimidines/nucleosides/nucleotides	63	55	54	52	95	83	58	58	58	100	92
Regulatory functions	224	101	106	87	84	39	155	165	145	91	65
Signal transduction	95	53	56	49	90	52	79	82	73	91	77
Transcription	57	37	41	36	92	63	50	51	48	95	84
Transportation/binding proteins	313	159	148	136	89	43	214	199	177	86	57
Unknown	374	171	166	144	85	39	272	271	223	82	60
Total	4859	2093	2076	1851	89	38	2994	3017	2645	88	54

Notes: A Finnegan LTQ mass spectrometer was used for MS/MS experiments under conditions as described in Ref. [2]. A 40 cm × 50 μm i.d. column packed with 1.4 μm C18 particles was used for the 1D-LC [2], and a 60 cm × 150 μm i.d. column packed with 5 μm C18 particles was used for 2D-SCX/RPLC, with LC separations of 50 and 70 SCX fractions (for runs 1 and 2, respectively).

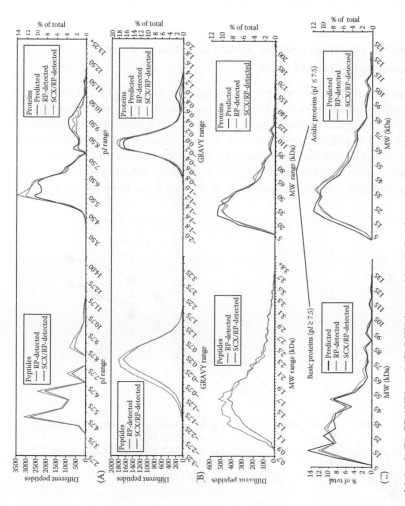

FIGURE 2.15 Ranges of (A) pI, (B) GRAVY, and (C) MW for peptides (left side) and proteins (right side) identified by LC-MS/MS. The 2D-LC-MS/MS data were obtained by fractioning the proteomics sample (500 µg of the cell lysate tryptic digest) into 10 fractions by SCX and analyzed by RPLC-MS/MS using conditions as cescribed in Table 2.1, and the 1D-LC-MS/MS was completed according to Ref. [2]. A Finnegan LTQ mass spectrometer was used for MS/MS analysis. Test sample: *S. oneidensis* tryptic digest.

of proteins identified from 1D- and 2D-LC are less apparent. The 1D-LC-MS is more powerful for detection of small acidic proteins than 2D-LC-MS platform. Proteins identified from 1D-LC-MS and 2D-LC-MS are obviously different. For example, protein overlap between the two platforms is ~80%, obviously lower than the reproducibility of each LC-MS platform (e.g., ~90%). Therefore, a comprehensive proteomics analysis using 2D-LC-MS platform should combine those obtained from 1D-LC-MS, as illustrated above, for characterization of the human plasma proteome (Figure 2.11), to minimize the possible loss of proteins.

2.9.3 LIMITING PROTEOMICS SAMPLE ANALYTICAL SIZES TO ACHIEVE QUANTITATION LC-MS DATA

A predictable but simple relationship (e.g., linearity with a slope of 1) between the proteomics sample size and the ESI-MS signal intensity provides an improved basis for proteome quantitation. As shown in Figure 2.16, such a relationship generally is observed for an LC-MS platform, but only within a limited range of proteomics sample sizes. The ESI-MS linear response at a slope of ~1 crosses 3 orders of magnitude in sample size; however, for the 50 μm i.d. capillary column and 1–2 μg-sized samples, the ESI-MS linear responses start to diverge from linearity. As a result,

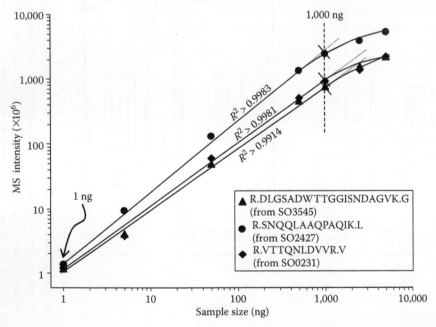

FIGURE 2.16 Proteomics sample size range for quantitative LC-MS. A 30 cm × 50 μm i.d. capillary column packed with 5 μm C18 particles was used in combination with a Finnegan LCQ to examine the three tryptic peptides that served as examples. Test sample: *S. oneidensis* tryptic digest.

divergence from linearity needs to be considered, especially for proteomics data obtained from large sample sizes used to extend the analysis dynamic range described above. The lower limit for quantitation (i.e., 1 ng sample) based on an S/N > 3 for MS single ion current peaks is greater than the peptide detection limit, as MS/MS identification can be achieved from single ion current peaks of S/N < 3, which results in an ESI-MS linear response range smaller than the proteomics analysis dynamic range described above.

2.10 FUTURE CHALLENGES

While many technological developments in LC have advanced various proteomics applications, challenges related to LC separations, including both column technology and techniques, as well as instrumentation, still need to be addressed in order to improve and extend the capabilities for LC-MS-based proteomics studies.

The LC-MS response behavior and sensitivity limits in a mobile-phase flow range of <20 nL/min that allows operation in the low nanoflow ESI regime needs to be systematically investigated using high-efficiency LC columns. The use of uniformly small (e.g., ≤1.5 μm) particles for manufacturing narrower (e.g., ≤10 μm i.d.) capillaries may be the most practical approach for this purpose. Additionally, issues of column efficiency need to be resolved so that narrow monolithic columns can be used to explore the extremes in LC-MS sensitivity. It remains to be seen whether LC-MS can achieve the level of sensitivity needed to analyze the typical size of a single mammalian cell (e.g., protein content of <100 pg), while at the same time providing a desirable dynamic range (e.g., quantify protein abundance that spans ~4 orders of magnitude). The development of submicron (e.g., 0.3–0.5 μm) porous particles is also required to improve the separation peak capacity of fast (e.g., 2–3 min) LC and subsequently, to improve the analysis coverage and dynamic range of ultrahigh-throughput proteomics.

Automated LC systems operated at 20 Kpsi are generally sufficient for identifying most of the proteins whose abundance falls within the LC-MS sensitivity limit. Increasing the pressure to 40 Kpsi would be useful for shortening the analysis time; however, difficulties associated with connecting the fused silica capillary columns to the automated switching valves would have to be resolved first. Developing a fast LC system (e.g., 1–2 min mobile-phase gradient separation) is another challenge that would require reducing the time consumed during column re-equilibrium after each mobile-phase separation. How quickly these issues will be addressed by new microfabricated devices remain uncertain.

While bottom-up (i.e., peptide level) LC-MS approaches are presently the basis for most proteomics applications, one of their biggest limitations stems from incomplete sequence coverage that limits the ability to identify protein modifications. These modifications can potentially be studied by making measurements at the intact protein level. Figure 2.17 shows an LC-MS based analysis of a yeast lysate, illustrating the MW of naturally occurring polypeptides in such samples (and the present significant bias in sensitivity toward lower MW components). In addition to the challenges associated with MS detection and identification of these components [40], the utility of LC to separate large intact proteins remains uncertain and is in need of further development.

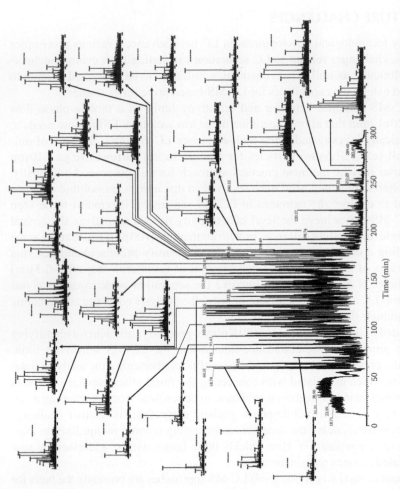

FIGURE 2.17 LC-MS of a cell lysate. Both intracellular peptides and polypeptides with MWs of 400–20,000 Da are observed, and a separation peak capacity is estimated as ~350 in the effective separation time window. Conditions: 120 cm × 100 μm i.d. column packed with 3 μm C4 particles for gradient LC separation; gradient from mobile phase A (30% acetonitrile/0.1 trifluoroacetic acid) to 70% B (90% acetonitrile/0.1 trifluoroacetic acid); Finnegan LCQ is used for detection. Test sample: yeast lysate (10 μg).

ACKNOWLEDGMENTS

Portions of this work were funded by the National Institute of Health (NIH) National Center for Research Resources (RR 018522) at the Pacific Northwest National Laboratory (PNNL), the NIH National Cancer Institute (R21 CA12619-01), the National Institute of Allergy and Infectious Diseases (NIH/DHHS through interagency agreement Y1-AI-4894-01), and the U.S. Department of Energy (DOE) Office of Biological and Environmental Research. Experimental portions of this research were performed in the Environmental Molecular Sciences Laboratory, a DOE national scientific user facility also located at PNNL, which is operated by Battelle for the DOE under Contract No. DE-AC05-76RLO 1830.

REFERENCES

1. Y. Shen, N. Tolić, R. Zhao, L. Paša-Tolić, L. Li, S.J. Berger, R. Harkewicz, G.A. Anderson, M.E. Belov, and R.D. Smith. *Anal. Chem.* 73:3011, 2001.
2. Y. Shen, R. Zhang, R.J. Moore, J.K. Kim, T.O. Metz, K.K. Hixson, R. Zhao, E.A. Livesay, H.R. Udseth, and R.D. Smith. *Anal. Chem.* 77:3090, 2005.
3. Y. Shen, R. Zhao, S.J. Berger, G.A. Anderson, N. Rodriguez, and R.D. Smith. *Anal. Chem.* 74:4235, 2002.
4. Y. Shen, R.D. Smith, K.K. Unger, D. Kumar, and D. Lubda. *Anal. Chem.* 77:6692, 2005.
5. Y. Shen, E.F. Strittmatter, R. Zhang, T.O. Metz, R.J. Moore, F. Li, H.R. Udseth, and R.D. Smith. *Anal. Chem.* 77:7763, 2005.
6. H. Kobayashi, T. Ikegami, H. Kimura, T. Hara, D. Tokuda, and N. Tanaka. *Anal. Sci.* 22:491, 2006.
7. Q. Luo, Y. Shen, K.K. Hixson, R. Zhao, F. Yang, R.J. Moore, H.M. Mottaz, and R.D. Smith. *Anal. Chem.* 77:5028, 2005.
8. G. Yue, Q. Lou, J. Zhang, S.-L. Wu, and B.L. Karger. *Anal. Chem.* 79:938, 2007.
9. G.I. Taylor. *Proc. R. Soc. London Ser. A* 280:383, 1964.
10. K. Tang and A. Gomez. *Phys. Fluids* 6:2317, 1994.
11. K. Tang and A. Gomez. *Phys. Fluids* 6:404, 1994.
12. M. Dole, L.L. Mach, R.L. Hines, R.C. Mobley, L.P. Ferguson, and M.B. Alice. *J. Chem. Phys.* 49:2240, 1968.
13. J.V. Iribarne and B.A. Thomson. *J. Chem. Phys.* 64:2287, 1974.
14. R.D. Smith, Y. Shen, and K. Tang. *Acc. Chem. Res.* 37:269, 2004.
15. Y. Shen, R. Zhao, S.J. Berger, G.A. Anderson, N. Rodriguez, and R.D. Smith. *Anal. Chem.* 74:4235, 2002.
16. Y. Shen, R.D. Smith, K.K. Unger, D. Kumar, and D. Lubda. *Anal. Chem.* 77:6692, 2005.
17. Y. Shen, E.F. Strittmatter, R. Zhang, T.O. Metz, R.J. Moore, F. Li, H.R. Udseth, and R.D. Smith. *Anal. Chem.* 77:7763, 2005.
18. T. Koerner, K. Turck, L. Brown, and R.D. Oleschuk. *Anal. Chem.* 76:6456, 2004.
19. T. Koerner, R. Xie, F. Shong, and R.D. Oleschuk. *Anal. Chem.* 79:3312, 2007.
20. M.S. Wilm and M. Mann. *Int. J. Mass Spectrom. Ion Processes* 136:167, 1994.
21. J.H. Wahl, D.R. Goodlett, H.R. Udseth, and R.D. Smith. *Electrophoresis* 14:448, 1993.
22. J. Fernandez de la Mora and I. Loscertales. *J. Fluid Mech.* 260:155, 1994.
23. J.S. Page, R.T. Kelly, K. Tang, and R.D. Smith. *J. Am. Soc. Mass Spectrom.* 18:1582, 2007.
24. K. Tang, J.S. Page, and R.D. Smith. *J. Am. Soc. Mass Spectrom.* 15:1416, 2004.
25. A. Schmidt, M. Karas, and T. Dulcks. *J. Am. Soc. Mass Spectrom.* 14:492, 2003.
26. R.T. Kelly, J.S. Page, K. Tang, and R.D. Smith. *Anal. Chem.* 79:4192, 2007.
27. M.R. Fuh and C.J. Hsieh. *J. Chromatogr. B* 736:167, 1999.
28. C.L. Andrews, C.P. Yu, E. Yang, and P. Vouros. *J. Chromatogr. A* 1053:151, 2004.

29. R.T. Kelly, J.S. Page, Q. Luo, R.J. Moore, D.J. Orton, K. Tang, and R.D. Smith. *Anal. Chem.* 78:7796, 2006.
30. R.T. Kelly, J.S. Page, R. Zhao, W.-J. Qian, H.M. Mottaz, K. Tang, and R.D. Smith. *Anal. Chem.* 80:143, 2008.
31. Y. Shen, R.J. Moore, R. Zhao, J. Blonder, D.L. Auberry, C. Masselon, L. Paša-Tolić, K.K. Hixson, K.J. Auberry, and R.D. Smith. *Anal. Chem.* 75:3596, 2003.
32. A. Motoyama, J.D. Venable, C.I. Rose, and J.R. Yates III. *Anal. Chem.* 78:5109, 2006.
33. Y. Shen, J.M. Jacobs, D.G. Camp II, R. Fang, R.J. Moore, R.D. Smith, W. Xiao, R.W. Davis, and R.G. Tompkins. *Anal. Chem.* 76:1134, 2004.
34. Y. Shen, J.K. Kim, E.F. Strittmatter, J.M. Jacobs, D.A. Camp II, R. Fang, N. Tolić, R.J. Moore, and R.D. Smith. *Proteomics* 5:4034, 2005.
35. Y. Shen, N. Tolić, C. Masselon, L. Paša-Tolić, D.G. Camp II, K.K. Hixson, R. Zhao, G.A. Anderson, and R.D. Smith. *Anal. Chem.* 76:144, 2004.
36. Y. Shen, N. Tolić, C. Masselon, L. Paša-Tolić, D.G. Camp II, M.S. Lipton, G.A. Anderson, and R.D. Smith. *Anal. Bioanal. Chem.* 378:1037, 2004.
37. R. Zhao, Y. Shen, and R.D. Smith unpublished results.
38. Q. Luo, J.S. Page, K. Tang, and R.D. Smith. *Anal. Chem.* 79:540, 2007.
39. J. Kyte and R.F. Doolittle. *J. Mol. Biol.* 157:105, 1982.
40. X. Han, M. Jin, K. Breuker, and F.W. McLafferty. *Science* 314:109–112, 2006.

3 Advances in Electrophoretic Techniques for DNA Sequencing and Oligonucleotide Analysis

Fen Wan, Jun Zhang, and Benjamin Chu

CONTENTS

3.1   INTRODUCTION

With the completion of the human genome sequence in 2003 [1], the Human Genome
Project accomplished its initial goals two years earlier than originally planned, partly
due to the rapid development of related analytical technologies. Among them, DNA
sequencing analysis by capillary electrophoresis (CE), which has played an important role, is used to determine the exact order of the bases (A, T, G, and C) by separating and identifying DNA chemical subunits with one-base resolution and high
reliability. Several breakthroughs, including the Sanger DNA sequencing developed
in 1977 by Frederick Sanger and its automation, as well as the replacement of slab
gel electrophoresis by CE, were achieved during the past two decades, leading to a
two- to three-fold enhancement in terms of efficiency and sequencing cost per base.
However, new advances are still needed to reduce the cost further and to increase the
throughput in order to satisfy the increasing demands of current biomedical research
and clinical practices. One of the primary goals of the National Human Genome
Research Institute (NHGRI) is to achieve a four to five order of magnitude decrease
in the cost of sequencing a human genome, resulting in a targeted cost of $1000 or
less for sequencing about 10 million bases, the size of a human genome [2]. If this
objective could be reached in the next 5–10 years [3], it should produce a tremendous
impact on biological research and medical applications. Nevertheless, DNA CE
remains the main reliable analytical technique at the present time.

Slab gel electrophoresis was first applied to DNA separation using a cross-linked
polyacrylamide or agarose gel as a sieving medium. However, this method has its
inherent shortcomings: gel replacement is laborious and time consuming; the separation rate is very slow since high applied electric fields cannot be used because of the
Joule heating effect. Moreover, the waste disposal and difficulty in automating the
process made it even more unfavorable. In the past 10 years, it has been gradually
replaced by CE, so as to overcome these drawbacks, especially the Joule heating
effect. CE uses a fused-silica capillary with an inner diameter (ID) of typically
25–75 μm as the separation column, allowing much higher applied voltage that
results in a substantial increase in the DNA separation rate. CE also has the advantage of much smaller sample requirements (of the order of nanogram or less), the
ability for online detection, and real-time monitoring. The overall throughput of CE
can be further increased by using capillary array electrophoresis (CAE) [4–18] and
microfabricated chips [19–28]. The application of laser-induced fluorescence (LIF)
detection on CE has greatly increased the detection sensitivity when compared with
traditional UV detection used in slab gel electrophoresis. Typically, DNA primers or
terminators are synthesized directly with a fluorescence label, such as BigDye, used
in DNA sequencing analysis.

Recently, the capillary-based technique for DNA separations seems to have
reached a choke point as it has become harder to achieve further cost reduction in
order to fulfill the growing demands from the biological and medical community.

Consequently, many new approaches such as pyrosequencing are being developed. While the outcome of these possible advances is uncertain, one can, nevertheless, envisage advances in terms of cost and throughput for future DNA sequencing analysis. For example, pyrosequencing, as one of the massively parallel sequencing systems, has demonstrated a success in reading up to 25 million bases in a 4 h run with an average read length of 100 bases [29–32], which is about 100 times faster than the current CAE system. But, at least for now, the Sanger sequencing will be widely used. It is expected to continue to be the standard method for sequencing large genomes, such as human genomes that contain complex DNA sequences, or applications that may require high accuracy [33]. In this review, we will focus on the Sanger-based sequencing analysis as well as sketches of nonelectrophoretic-based new advances of the past few years, with emphasis on the development based on sieving matrices. It should be noted that sequencing oligonucleotides (oligos), typically less than 100 bases, is often operationally different from general DNA sequencing or genome sequencing analysis. We would like to separate our discussion into two parts: DNA sequencing analysis and oligos sequencing analysis, recognizing that both analytical techniques are very important to diagnose the health of citizens in modern society.

3.2 ELECTROPHORESIS-BASED DNA SEQUENCING ANALYSIS

3.2.1 PHYSICAL MECHANISMS ASSOCIATED WITH DNA SEPARATIONS

DNA is negatively charged and will migrate in an applied electrical field, leading to a pathway for separation analysis by means of electrophoresis. However, DNA fragments with different base numbers move with essentially the same velocity in free solution because the mobility of DNA molecules is independent of molecular size because they have the same charge density. In other words, the free-flow mobility of DNA molecules is, in essence, independent of size although it is affected by temperature, DNA chain conformation, and buffer ionic strength. Hence, a sieving matrix is required to perform the separation, allowing the DNA fragments to move in different speeds based on their sizes.

To date, the Ogston model and the reptation model, as well as their modifications, are the most popular theoretical models to partially explain the separation of DNA by both, slab gel electrophoresis and CE. Figure 3.1 is a schematic diagram of DNA molecules of different sizes moving in a sieving matrix [34]. The Ogston model is useful in predicting small DNA fragments migrating in low applied electric fields with a low DNA electrophoretic mobility, which assumes that DNA molecules are moving as hard spheres in a sieving gel. However, when applied to movements of larger DNA fragments, it becomes invalid as it cannot explain why DNA with a size larger than the network pores can still pass through them and realize an effective separation. The reptation model accounts mostly for middle-sized DNA chains whereby DNA chains can reptate like a snake through the polymer network. In this model, the DNA mobility depends strongly on the DNA size, making the separation of DNA

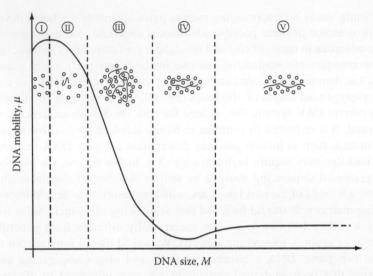

FIGURE 3.1 DNA mobility (μ) versus size (M) in a sieving matrix. Circles represent polymer/gel fibers whereas strands are DNA fragments: I, Short, rodlike DNA molecules migrate essentially as if in free flow, and mobility sometimes increases slightly with DNA size due to the transition of rod to coil; II, Ogston regime. Short random-coil fragments are separated by the porous matrix; III, Large DNA molecules must reptate snakelike to pass through the network and the mobility is a strong function of the DNA size; IV, Intermediate size regime. DNA molecules may get trapped in U-shaped conformations and the mobility ceases to be a monotonic function of the DNA size; V, Large DNA molecules orient in the direction of applied electric field and the separation becomes impossible. (From Slater, G.W., Kenward, M., McCormick, L.C., and Gauthier, M.G., *Curr. Opin. Biotechnol.*, 14, 58, 2003. With permission.)

fragments possible. However, when the DNA fragments become even larger, they will start to orient or deform, as discussed by the biased reptation model [35].

The entanglement of polymer chains in a polymer solution is different from a conventional cross-linked gel in at least one important aspect, i.e., in a polymer solution, each chain entanglement exists over a finite time period, in contrast to a covalently cross-linked polymer gel where the cross-linking points are permanent. When DNA molecules pass through an uncross-linked polymer solution, the flexible polymer chains will reptate away from the positions occupied by DNA fragments, resulting in a "constraint release" phenomenon that can deteriorate the intended separation effect. Hence, it is desirable to have the relaxation time of the entangled polymer chains to be relatively longer than the residence time of the DNA chain in the tube formed by the entangled polymer chains in order to achieve good performance. In this case, a polymer solution far above its overlap concentration (C^*, is proportional to $1/R_g^3$) is needed to form a robust network. For a given polymer solution, the relaxation time decreases with increasing polymer concentration and molecular weight. On the other hand, higher concentration implies smaller mean mesh size, which becomes a disadvantage if one wishes to separate larger DNA fragments. Thus, for a given polymer solution acting as a sieving matrix, a higher polymer molar mass is desired to sequence larger DNA fragments. Indeed, the longest read length published was carried out in a polymer matrix with ultrahigh polymer molecular weight [36]. Nevertheless, it should be noted that DNA separation could also be achieved unexpectedly in dilute polymer solution, well below its overlap concentration without the entanglement of polymer chains [37]. A transient entanglement mechanism was pointed out by Albarghouthi and Barron stating that when the DNA molecules migrate through dilute polymer solution, the DNA chains would collide with polymer chains in the vicinity and force them to be dragged in the polymer solution, leading to a difference in mobility [38]. In fact, both computational simulation [39] and experimental observation [40] have shown that the DNA molecules would undergo conformation changes when colliding with free neutral polymers.

The mechanism of DNA separation in polymer solution is more complex when compared with that in cross-linked gels. The difference between double-stranded DNA (dsDNA) and single-stranded DNA (ssDNA) has brought even more complexity. Current models have not yet been able to provide definitive and quantitative guidelines to predict what an ultimate polymer solution sieving matrix can achieve.

3.2.2 SANGER SEQUENCING METHOD

The Sanger sequencing method represents revolutionary progress in sequencing DNA and is based on the incorporation of dideoxynucleotides to terminate the growing chains [41]. Note that two methods are often used to generate fragments: random approach and primer walking. Here, we show two self-explained schematic diagrams to indicate the processes (Figures 3.2 and 3.3). Detailed strategies have been described in the review [42]. It is mentioned here to emphasize the importance of this development.

3.2.3 DNA SEQUENCING MEDIA

3.2.3.1 Polymer Solution Replacing Cross-Linked Gel as Sequencing Media

One of the early critical processes of CE is to pump the separation medium into a pretreated coated capillary. The effectiveness of the separation medium plays a central role in the performance of CE as it determines the DNA fragment migration behavior as well as the resolution [43]. Chemists have devoted considerable efforts to enhance the performance of separation medium. In the beginning stages, cross-linked chemical gels such as polyacrylamide gels were shown to be an effective approach [44]: polymer monomers and cross-linking reagents were pumped into a pretreated fused-silica capillary to allow the polymerization occurring inside the capillary to form the separation matrix. The mesh size of the matrix could be tuned by adjusting the concentration of the monomers and that of the cross-linkers. Although they were successful in separating DNA fragments, they had their own drawbacks, including instability caused by bubble formation and a relatively short column life. Moreover, the developed chemical gels often could not be used above room temperature [45]. As the separation medium was linked with the inner wall of the capillary by chemical bonding, the capillary had to be replaced when the gel lost its function.

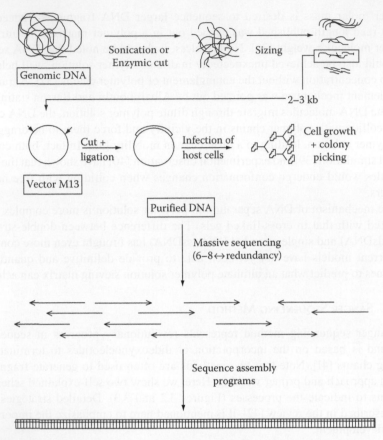

FIGURE 3.2 Random sequencing approach or shotgun. The distinct processes involve fragmentation of the DNA into 2 ± 3 kbp range; fragments are then cloned into vectors and introduced into host cells for amplification. After purification, the DNA from individual colonies is sequenced, and the results are lined up with sequence-assembly programs. (From Franca, L.T.C., Carrilho, E., and Kist, T.B.L., *Quart. Rev. Biophys.*, 35, 169, 2002. With permission.)

Polymer solution was first used as a DNA separation medium by Zhu et al. in 1989 [46]. It is a physical gel and can be replaced easily by pumping it out of the capillary. Therefore, the separation process can be carried out using a fresh "gel" solution every time in order to guarantee the reproducibility. In addition, the use of polymer solution allows for the automation in the gel injection procedure. The mechanism of separation in a physical gel is similar to that in a chemical gel as both of them allow the DNA fragments to be separated due to their migration in the mesh-like structure, except for the chain dynamics in a transient network formation in a polymer solution that could affect DNA fragment separations, especially when the DNA chain size becomes larger. According to the scaling theory, polymers in solution will start to entangle with each other when their concentration is higher than a certain value C^*, known as the overlap concentration [47].

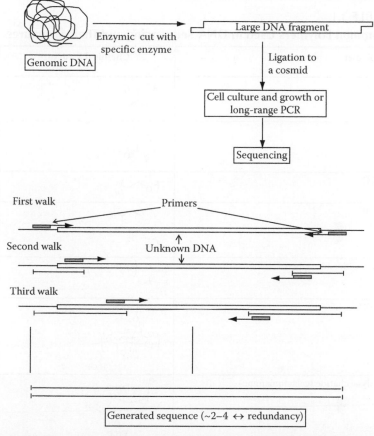

FIGURE 3.3 DNA sequencing by primer-walking strategy. Genomic DNA is cut into a large piece (~40 kbp) and inserted into a cosmid for growth. The sequencing is performed by walks, starting first from the known region of the cosmid. After the results from the first round are edited, a new priming site is located within the newly generated sequence. This procedure is repeated until the walks reach the opposite starting point. (From Franca, L.T.C., Carrilho, E., and Kist, T.B.L., *Quart. Rev. Biophys.*, 35, 169, 2002. With permission.)

When the concentration of the polymer is above the overlap concentration ($C > C^*$), the polymer chains will interpenetrate with one another to form a transient network, implying that polymer chain entanglements are not permanent. Here, the value of C^* is proportional to the molecular weight of the polymer over the cube of the radius of gyration of the polymer coils. Many hydrophilic homopolymers have been synthesized and tested as separation media, such as linear poly(acrylamide) (LPA) [48–66], poly(*N,N*-dimethylacrylamide) (PDMA) [67–75], poly(ethylene oxide) (PEO) [76–86], poly(vinyl pyrrolidone) (PVP) [87–92], cellulose and its derivatives [93–100], poly(*N*-acryloylaminopropanol) (Poly(AAP)) [101–103], polyethylene glycol with fluorocarbon tails (PEG end-capped) [104–106], and polysaccharides [107,108]. The chemical structures of these polymers are listed in Table 3.1.

TABLE 3.1

Common Polymers Used in DNA Sequencing and Their Structures

Polymer	Chemical Structure
LPA	
PDMA	
PEO	
PVP	
PEG with end-capped	$F_{2m+1}C_m\left(O\text{-}CH_2\text{-}CH_2\right)_n C_m F_{2m+1}$ $m = 6, 8$
HEC	
Polysaccharide (galactomannan)	

3.2.3.2 Electroosmosis and Capillary Coating

One of the most important issues associated with CE is the capillary coating. The use of a capillary column will bring up a phenomenon known as "electroosmosis," caused by the interactions between the silanol (–SiOH) groups hanging on the inner wall of the fused-quartz capillary and the buffer solution. The silanol groups are ionized in contact with the buffer, forming a double layer with the silanol layer being negatively charged and the outer layer positively charged. Under an applied electrical field, the cations in the outer layer will be driven toward the cathode, carrying the water of hydration with them. Thus, the whole solution will be drawn to the cathode direction as the hydrogen bonding of the water of hydration is cohesive in the bulk solution, and finally even be pumped out of the capillary. This phenomenon can be used in some applications, such as the separation of proteins by capillary zone electrophoresis [109–112]. However, for DNA separation, electroosmosis has to be suppressed in order to achieve good performance. The problem can be partially resolved by capillary coating. There are two methods addressed to overcome electroosmosis in DNA sequencing analysis. One is to pretreat the silica capillary with poly(vinyl alcohol) (PVA) [51,113] or LPA [114] covalently bonded to the inner wall of the capillary by bifunctional reagents, usually done by in situ synthesis. However, this kind of coating protocol would give rise to a series of additional problems, such as capillary fouling, coating inhomogeneity, and limited shelf life. The second approach is known as dynamic coating, which takes advantage of molecular interactions forming hydrogen bonds between the silanol groups on the inner wall and the polymer chains in the sequencing medium. The dynamic coating method is preferred as it overcomes most of the shortcomings by using the pretreatment protocol. Typically, a capillary column treated by a good dynamic coating process can run about 100 times before the performance becomes deteriorated. Then, the column can be regenerated fairly easily. However, only a few hydrophilic homopolymers have been known to show this special ability, including poly(dimethylacrylamide) (PDMA), PEO, and PVP. After carefully studying the dynamic coating behavior, Madabushi discovered that LPA had the lowest coating ability while PDMA was the best among those tested [69].

3.2.3.3 Solution Viscosity and Gel Injection

Another problem associated with DNA sequencing by CE is the separation medium viscosity. The polymer should have a high average molecular mass ($M_w \geq 2$–4 MDa) to form a relatively robust network [38,115] so as to obtain good sieving ability for long DNA chains. Unfortunately, polymer solutions with high molar mass in the sequencing buffer (usually ~7 M urea is added as the denature agent of DNA) will be of high viscosity. For instance, a 2.5% (w/v) PDMA with an average molar mass of 5.2 MDa in a sequencing buffer solution exhibits a zero-shear viscosity of 3×10^4 cP [116]. Furthermore, the corresponding LPA solution will have an even higher viscosity as LPA is more hydrophilic and exhibits larger coil size [117]. This increase in solution viscosity will cause much difficulty in injecting the sequencing medium into the capillary. As a result, a fairly high applied pressure (≥ 1000 psi) is often necessary to load the sequencing matrix into the capillary. This type of applied pressure can become one of the critical issues in designing an automated, high-throughput DNA (sequencing) analysis instrument, especially for microfabricated chips because most microchips cannot tolerate more than ~200 psi external pressure before destroying themselves and some plastic chips can only withstand ~50 psi [118].

A noticeable characteristic for hydrophilic polymers in solution is that they are non-Newtonian fluids under flow and they exhibit viscosities strongly dependent on the applied shear rate. When the shear rate reaches some critical value, the viscosity will decrease exponentially because of the disentanglement of polymer chains under flow [119,120]. This shear-thinning behavior can be utilized to help inject the high zero-shear viscosity polymer solution, such as LPA solution, into microchannels. In general, the shear-thinning behavior of polymer solutions becomes less dramatic as the polymer chains become less hydrophilic, although they also have lower zero-shear rate viscosities [117]. In addition to the shear-thinning behavior of polymer solutions, manipulation of the chemical and physical structures of polymer chains can be an effective way to change the viscosity of obtained solutions, e.g., by introducing viscosity switchable properties under thermal conditions. Thus, the polymer matrix can be easily injected (manually or automatically) into microchannels under a low solution viscosity condition, while the separation of the DNA takes place at a different temperature when the polymer solution, in a gel-like state, has a higher sieving ability in order to obtain the desired separation performance. These thermoreversible polymer solutions can be divided into two subgroups: thermothinning and thermothickening polymers. The former ones often have a lower critical solution temperature (LCST), at which the solubility-to-insolubility phase transition occurs, leading to a tremendous change in solution viscosity. Then, the gel injection can be done at a fairly low pressure when the solution has a low viscosity. For example, the random copolymer PDMA/DEA having 53% N,N-dimethylacrylamide (DMA) and 47% N,N-diethylacrylamide (DEA) showed a decrease from 2000 cP to less than 200 cP in viscosity when the solution was heated above 80°C [121]. On the other hand, thermothickening polymers have a lower viscosity at lower temperatures and will increase the solution viscosity substantially when it is being heated up to the transition temperature, sometimes called a sol-gel transition temperature. For example, for Pluronics F127, data showed that it had a transitional temperature at around 18°C and the viscosity increased from 50 cP at 5°C to 250 cP at 20°C [122].

The second type could be grafted copolymers, such as comblike grafted copolymers of poly(N-isopropylacrylamide)-g-PEO (PNIPAM-g-PEO), whose viscosity increased from 2500 cP at 31°C up to 9500 cP at 36°C, in which a transition temperature of ~36°C was detected [123].

3.2.3.4 Strategies for Making Good Sequencing Media

3.2.3.4.1 Homopolymers

LPA, which was first applied to DNA sequencing in 1993 [52], has been widely accepted as a sequencing medium due to its best sieving ability, partly because it is known as the most often used hydrophilic polymer sieving matrix. It was claimed that a higher concentration of LPA (4%) could be better for DNA fragments smaller than 450 bases, while a lower concentration (~2%) increased the resolution for the peaks corresponding to base number larger than ~450 bases [51], in agreement with theoretical prediction. Using a high molecular weight (HMW) LPA at a concentration of 2% (wt%), Karger and his coworkers achieved single-colored DNA sequencing with a read length of more than 1000 bases within 80 min at an operating temperature of 50°C and 150 V/cm applied electric voltage [51]. Similar results were obtained using an HMW LPA (~9 MDa) synthesized by inverse emulsion polymerization [120]. Other operating conditions have also been examined to increase the CE performance. For example, it was found that dimethyl sulfoxide (DMSO) could decrease the required amount of urea used as a denaturant, resulting in a reduction of the matrix viscosity. For example, a mixture of 5% DMSO and 2 M urea could keep the DNA in a denatured state, resulting in a read length of 975 bases in 40 min with 98.5% accuracy at a sequencing temperature of 70°C [124]. On the other hand, an addition of 25% (v/v) of glycerol to TTE buffer could increase the electrostatic interactions between the phosphate groups and maintain the conformation of ssDNA which would enhance the

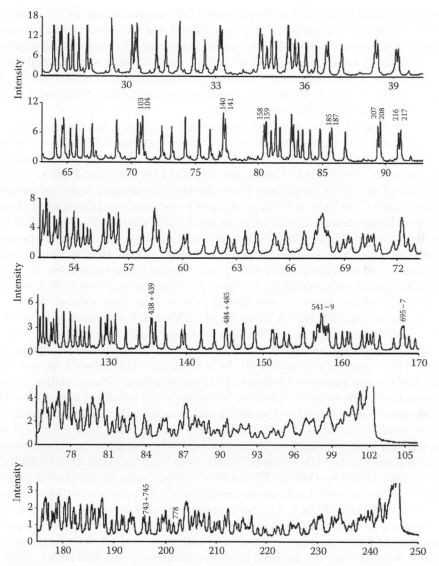

FIGURE 3.4 Comparison of separation of a C-terminated single-dye labeled DNA sequencing sample using FAM-labeled-21M13 primer on a pGEM-3Zf(+) template in 1× TTE buffer + 5 M urea with (bottom plot in each panel) and without 25% (v/v) added glycerol (top plot in each panel). The three panels show the separation of different-sized DNA fragments. Capillary: effective length, 40 cm; ID, 100 µm. DNA sample was injected electrokinetically into the column at 300 V/cm for 5 s. The separation electric field strength, 150 V/cm. (From Liang, D., Song, L., Chen, Z., and Chu, B., *J. Chromatogr. A*, 931, 163, 2001. With permission.)

sequencing performance of LPA running in a precoated capillary, though at the expense of migration time, as shown in Figure 3.4 [55].

Another modification was carried out by Barron and his coworkers who developed sparsely cross-linked "nanogels" with subcolloidal polymer structures composed of covalently linked LPA chains as a replaceable DNA sequencing matrix for capillary and microchip electrophoresis [125]. It was claimed that the existence of a sparsely cross-linked system could stabilize the entangled polymer network and bring positive effects for DNA sequencing analysis using both, conventional and chip-based electrophoresis so as to increase the CE read length. Finally, the sequencing behavior of LPA could be extensively improved by tuning up the composition of polymers, with results to be discussed in the coming sections.

Aside from LPA, an important polymer is PDMA as it has the best-known coating ability with the viscosity being much lower than that of LPA, having about the same molecular weight and concentration. Madabhushi developed a separation medium made of 6.5% of low molecular weight (LMW) (98 kDa) PDMA, with a viscosity of only 75 cP, to perform four-color DNA sequencing under an applied electric field of 160 V/cm at 40°C, resulting in a read length of 600 bases in about 2 h [69]. The separation of ssDNA using the PDMA matrix was investigated in detail by Heller [75,126–128], leading to the conclusion that the peak spacing increased with increasing PDMA concentration and molecular weight, but decreased with increasing DNA size. It was reported that sequencing analysis could be carried out successfully for up to 500 bases using 5% of PDMA with a combination of low (216 kDa) and high (1.15 MDa) molecular weights. At Stony Brook, by using a 2.5% PDMA with a M_w of 5.2 MDa, DNA sequencing up to 800 bases with a resolution of 0.5 in 96 min or 1000 bases with a resolution of 0.3 was achieved at 150 V/cm and 60 cm effective separation length at room temperature on a DNA sample prepared with FAM-labeled-21M13 forward primer on pGEM3Zf(+) and terminated with ddCTP [116]. In addition, the resolution in the sequencing electropherogram could be further improved by adding a small amount of montmorillonite clay ($2.5–5 \times 10^{-5}$ g/mL) into the polymer matrix made of LMW PDMA (~100 kDa) [74]. The authors attributed it to the structure change of the polymer matrix induced by the clay sheets, which acted as "dynamic cross-linking plates" to the PDMA chains, leading to an increase in the apparent molecular weight of PDMA. Although PDMA shows a lower viscosity and good dynamic coating ability, its main disadvantage comes from the fact that it is more hydrophobic than LPA [129]. PDMA with hydrophobic interactions could result in band shifting and peak broadening. Also, the PDMA polymer chains are more compact than that of LPA in the sequencing buffer, resulting in a network that requires more PDMA polymers forming the same mesh size when compared with LPA.

As a common hydrophilic polymer, PEO was also extensively tested as a sequencing medium. It was reported that PEO could separate ssDNA sequencing fragments up to 1000 bases in 7 h with a raw data resolution equal to 0.5 at 966 bases [76,85]. The same group also showed that by using PEO as a sieving matrix, fast separation up to 500 bases with a speed of 30 bases/min could be achieved with the aid of temperature programming [84]. While a PEO-based separation matrix could be absorbed to the fused-silica capillary wall to suppress electroosmosis, it had to be flushed by HCl after each run and the migration time was relatively long. No other additional homopolymers, to our knowledge, have been reported to sequence DNA steadily with a read length of much longer than 500 bases. Recently, cellulose derivatives, such as hydroxyethyl cellulose (HEC) [98,130,131], have drawn some attention due to their low viscosity as a nongel-sieving matrix. This low-viscosity criterion is a desirable property, especially for microchip applications.

3.2.3.4.2 Polymer Mixtures

Ideally, a good separation medium should have the qualities of high sieving ability, dynamic coating ability, and low viscosity, in order to enhance the performance of CE. Other properties, such as nontoxicity and long shelf life, are also important, but not as crucial as the above three requirements. It is difficult for a single homopolymer solution to satisfy all the three requirements since the physical properties of polymers are well defined, mainly by the chemical composition of the monomer. Generally, there are two methods that can be used to achieve the three desired features as shown in Figure 3.5 [132]: mixing of different homopolymers or copolymers to form a homogeneous solution and copolymerization of different monomer segments so that different segments are bonded together covalently.

Different molecular weights of the same polymer can be mixed together to form a separation matrix in order to achieve a desired sieving ability. As LMW polymers yield a low solution viscosity and HMW polymers work better with larger DNA fragments, mixed polymer blends of different molecular weight distributions can be tuned to optimize the DNA separation over a range of specific DNA fragment sizes. For instance, the performance of LPA sequencing matrix could be further enhanced by manipulating the molecular weight distribution, in addition to tuning operating parameters, DNA sample preparation protocols, and base-calling software. Using a mixture of 2.0% HMW LPA (9 MDa) and 0.5% LMW LPA (270 kDa) buffer solution, a significant improvement in terms of separation speed and reading accuracy was achieved under optimized conditions, resulting in a routine sequencing up to 1000 bases in less than 55 min for both standard DNA sequence fragments (M13mp18) and cloned single-stranded templates from human chromosome 17 with a base-calling accuracy between 98% and 99% [113]. The combination of 0.5% of LMW LPA increased the resolution of peaks below 100 bases without affecting that of larger

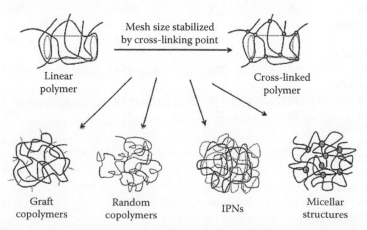

FIGURE 3.5 Schematic structure of polymer matrix. (From Chu, B. and Liang, D., *J. Chromatogr. A*, 966, 1, 2002. With permission.)

fragments, attributing to the fact that more overall concentrated polymer solution favors the separation of smaller fragments, while the larger DNA molecules were effectively separated in the meshes formed by the entanglement of longer polymer chains. Increasing the molecular weight of HMW LPA further would lead to an increase in both, the read length and the separation time simultaneously, although the viscosity will be increased as well. A comparison of migration time and read length for different compositions of LPA matrices under different temperatures and applied electrical fields are listed in Table 3.2. With further improvements of base-calling software to a resolution limit of 0.25 and modification of DNA Sanger cycle sequencing reaction to double the G-terminated fragments, a longer read length was obtained by using a sequencing solution composed of 2.0% HMW LPA of 17 MDa with 0.5% LMW LPA of 270 kDa, resulting in DNA fragments being sequenced up to 1300 bases in 2 h with an accuracy of 98.5% at 70°C and 125 V/cm [36], as shown in Figure 3.6. This is the longest read length that has been claimed in the literature.

The other approach is to mix polymers with different chemical compositions. In this particular situation, one should consider the compatibility of different polymer

TABLE 3.2
Read Length of Different Compositions of DNA Sequencing by Using LPA Mixtures at Different Temperatures and Applied Electric Fields in a Capillary with Effective Separation Length of 30 cm and ID of 75 μm [36]

LPA	Temperature (°C)	Electric Field (V/cm)	Migration Time for Base 1019 (min)	Read Length at 98.5% Accuracy (bases)
2% 17 MDa + 0.5% 270 kDa	70	125	105.4	1249
2% 10 MDa + 0.5% 270 kDa	70	125	101.0	1190
2% 17 MDa + 0.5% 270 kDa	70	125	100.0	1083
2% 10 MDa + 0.5% 270 kDa	70	125	99.5	965
2% 10 MDa + 0.5% 270 kDa	60	200	55.6	1013
2% 10 MDa	50	150	81.0	951
2% 10 MDa + 0.5% 270 kDa	70	250	44	927
2% 10 MDa + 0.5% 270 kDa	70	200	55.6	1042
2% 10 MDa + 0.5% 270 kDa	70	150	80.5	1127
2% 10 MDa + 0.5% 270 kDa	70	100	131.0	1172

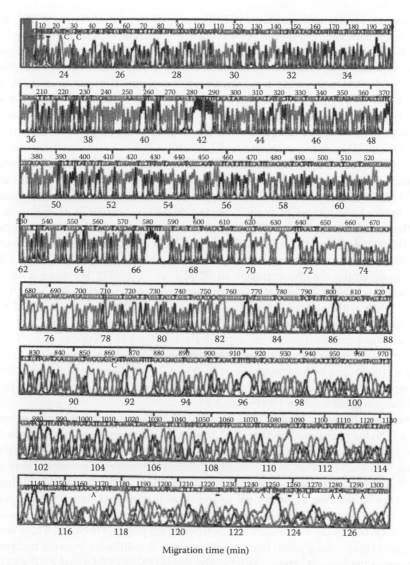

Migration time (min)

FIGURE 3.6 (See color insert following page 216.) Electropherogram with a read length of 1300 bases using the separation matrix LPA 2.0% (w/w) 17 MDa/0.5% (w/w) 270 kDa at 125 V/cm and 70°C. Samples were prepared using Universal BigDye-labeled (−21) primer cycle sequencing with AmpliTaq-FS on ssM13mp18 template. An additional G-terminated reaction was added. Conditions: Capillary: ID, 75 μm, OD, 365 μm, PVA-coated capillary with effective length 30 cm, total length 45 cm; running buffers: 50 mM Tris/50 mM TAPS/2 mM EDTA. The cathode running buffer and separation matrix also contained 7 M urea. Samples were injected at a constant electric field of 9 V/cm ($I = 0.7\,\mu A$) for 10 s and sequencing was performed at 125 V/cm. (From Zhou, H., Miller, A.W., Sosic, Z., Buchholz, B., Barron, A.E., Kotler, L., and Karger, B.L., *Anal. Chem.*, 72, 1045, 2000. With permission.)

components. For example, a mixture of PEO and hydroxypropylcellulose (HPC) were tried as a separation medium for DNA sequencing analysis by Kim and Yeung [85]. They found that the performance was very poor, compared to that of homopolymer. It may be ascribed to their poor miscibility, which is unfavorable to form a uniform robust network. On the other hand, based on the physical properties of LPA and PDMA, a mixture of LPA and PDMA was tested as the separation medium for CE at Stony Brook [133,134]. The solution was composed of PDMA at different molecular weights (8 kDa, 106 kDa, 1.1 MDa) and concentrations (0.2%, 0.5%, 1.0% (w/v)) and LPA (2.2 MDa) at a concentration of 2.5% (w/v). It was found that the incompatibility was increased with increasing molar mass and concentration of PDMA, which impaired the CE performance. By using an LMW (8 kDa) and low concentration (0.2%) of PDMA, it alleviated the incompatibility of two different homopolymers but combined the virtues of both, LPA and PDMA successfully, resulting in an excellent separation ability and a practical dynamic coating ability, yielding a one-base resolution of up to 730 bases in about 80 min at room temperature. The CE performance was further enhanced by increasing the molecular weight of LPA, while maintaining the concentration of PDMA at 0.2% [54]. By using a sieving matrix comprising of 2.25% LPA with an average molecular weight of 13 MDA and 0.2% PDMA with molecular weight of 470 kDa in 1× TTE buffer plus 7 M urea as denaturant, sequencing of BigDye Terminator v3.1 Sequencing Standard DNA samples was performed under an applied electric field of 150 V/cm at 60°C in a capillary with 50 cm effective separation length, resulting in a read length from 45 up to 1000 bases with an accuracy of 98.9% in 2 h, as shown in Figure 3.7. When the applied electric field strength was increased to 200 V/cm, sequencing up to 1000 bases was achieved in 1 h (Figure 3.8).

At Stony Brook, a kind of noncross-linked interpenetrating polymer network (quasi-IPN) has been developed for DNA sequencing analysis. The central idea is that in the quasi-IPN, polymer chains composed of two different components are intertwined with each other so as to partially suppress the polymer incompatibility and partially to take advantage of the slight incompatibility of the two polymer chains. Consequently, the two polymers try to avoid each other and thereby their chains become more extended. Then, the same mesh size can be created with fewer amounts of polymers occupying the effective space forming the polymer network. There are no permanent cross-linking points in such a network, so it is different from the traditional interpenetrating network, known as IPN. The quasi-IPN of PDMA and PVP was first synthesized and tested as a DNA separation medium at Stony Brook [43]. The results showed that the obtained matrix had combined the good separation ability of PDMA and the low viscosity of PVP with their coating ability. Furthermore, the intertwining of two different homopolymers at a molecular level successfully produced an effective separation medium.

The idea has been extended to the formation of the quasi-interpenetrating network by LPA and PDMA [135,136]. The CE performance of quasi-IPN on DNA sequencing was investigated by varying the acrylamide (AA) to DMA molar ratio, molecular weight of LPA and its size distribution [136]. On further experiments, it was concluded that the quasi-IPA made of molar ratio of AA to DMA at 10.9, i.e., roughly 11:1, showed no phase separation and PDMA at this concentration was

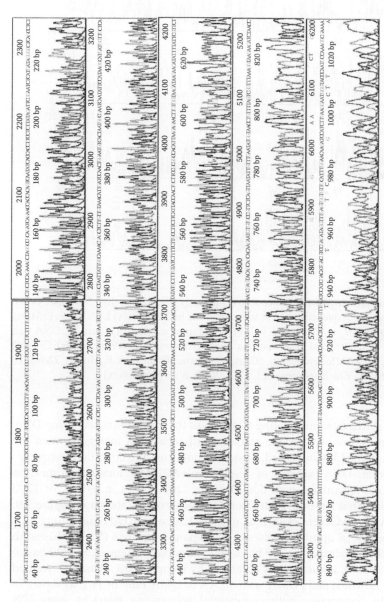

Migration time (s) and base number

FIGURE 3.7 Electropherogram obtained from a DNA sequencer provided by Professor Gorfinkel at Stony Brook University using a sequencing matrix composed of 2.25% (w/v) LPA ($M_w = 13$ MDa) + 0.9% (w/v) PDMA ($M_w = 470$ kDa) in $1\times$ TTE + 7 M urea. DNA was injected for 20 s at 41 V/cm and measurements were carried out at 150 V/cm and 60°C. A base-calling software "BASE" was used to perform the data analysis. Total capillary length, 55 cm; effective capillary length, 50 cm; ID, 50 μm. We thank Professor Vera Gorfinkel in the Department of Electrical and Computer Engineering, Stony Brook University for assistance with the DNA data analysis. (From Wan, F., He, W., Zhang, J., Ying, Q., and Chu, B., *Electrophoresis*, 27, 3712, 2006. With permission.)

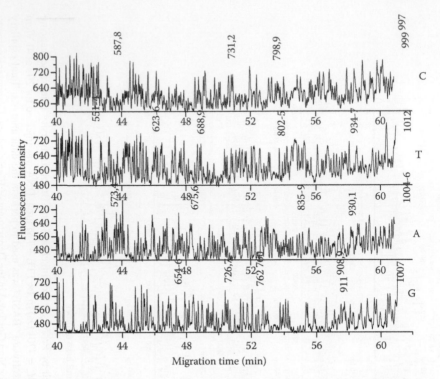

FIGURE 3.8 Sequencing of four-color DNA with only data larger than 550 bases being shown in the plot. The sequencing matrix was composed of 2.25% (w/v) LPA (M_w = 13 MDa) + 0.9% (w/v) PDMA (M_w = 470 kDa) in 1× TTE + 7 M urea, same as in Figure 3.7. DNA was injected for 20 s at 41 V/cm and measurements were carried out at 200 V/cm and 60°C. Total capillary length, 61 cm; effective length, 50 cm; ID, 50 μm. (From Wan, F., He, W., Zhang, J., Ying, Q., and Chu, B., *Electrophoresis*, 27, 3712, 2006. With permission.)

enough for effective dynamic coating to suppress the electroosmosis during DNA sequencing. In the mean time, the resolution of separation increased when the molecular weight of LPA increased. Under optimal running conditions, sequencing of one-color DNA sample of FAM-labeled-21M13 forward primer on pGEM3Zf(1) and terminated with ddCTP up to "1000 bases in 39 min, or 1200 bases in 60 min" on a lab-built instrument was achieved as shown in Figure 3.9. By using the ABI 310 Genetic Analyzer, even without optimized base-calling software, a read length of up to 700 bases of contiguous sequence (50–750 bases) in 35 min with 99.6% accuracy or 750 bases of contiguous sequence (50–800 bases) in 37 min with 98.0% accuracy was obtained, by using a 2.0% quasi-IPN formed by LPA (M_v = 7.6 MDa) and PDMA with an AA to DMA ratio of 11:1, as shown in Figure 3.10. In comparison with commercial sequencing matrices, quasi-IPN showed much better resolution than MegaBACE matrix, POP6 and POP7 for the separation of larger DNA fragments with a shorter run time.

Recently, Zhou et al. using quasi-IPN, formed by combining lower molecular weights of LPA (~3.3 MDa) with PDMA, achieved a comparable sequencing performance in a

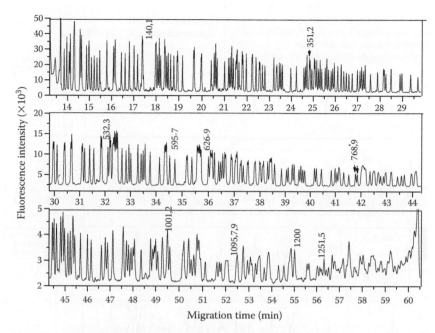

FIGURE 3.9 Single-color DNA sequencing up to 1200 bases in 60 min by using 2.0% quasi-IPN formed by LPA (M_v = 9.9 MDa) with an AA to DMA ratio of 11:1. Capillary effective length = 40 cm, ID = 75 μm; running electric field, 150 V/cm; DNA injection at 75 V/cm for 8 s; 65°C, DNA sample was prepared with FAM-labeled-21M13 forward primer on pGEM3Zf(1) and terminated with ddCTP. (From Wang, Y., Liang, D., Ying, Q., and Chu, B., *Electrophoresis*, 26, 126, 2005. With permission.)

shorter time as that obtained from the quasi-IPN comprised of higher molecular weight (~6.5 MDa) by adding gold nanoparticles, as illustrated in Figure 3.11 [137]. However, too many gold nanoparticles would decrease the resolution. The authors attributed the enhancement to the interaction between gold nanoparticles and the polymer matrix, which could form physical cross-linking points. This separation medium showed good reproducibility and long shelf life (≥8 months). Nevertheless, the presence of gold nanoparticles does add an additional component, not counting the complex nature of gold nanoparticles. It is unlikely that such an approach will be used widely. However, the experiment does illustrate the importance of chain dynamics in the polymer matrix, i.e., in order to increase the read length and resolution, the entangled polymer chains should retain the "conformation" (or entanglements) during the DNA transit time.

There are two ways to form quasi-IPNs: one is to carefully mix two HMW homopolymers above each overlap concentration; the other is to polymerize one kind of monomer in the second polymer solution, e.g., to polymerize DMA monomers to a PDMA concentration above its overlap concentration in LPA solution at the LPA concentration already in the semidilute regime. Studies at Stony Brook showed that these two methods ended up with the same results, but the second one was able to achieve a uniform separation matrix more readily, although it involved more steps in the fabrication process.

BigDye Terminator v3.0   Temperature: 60°C
Injection voltage/Time: 47.6 V/cm/15 s   Running buffer: 1× TTE+7M urea
Running voltage/cm: 200 V/cm
Capillary length: 31 cm, ID/OD = 49/361 µm

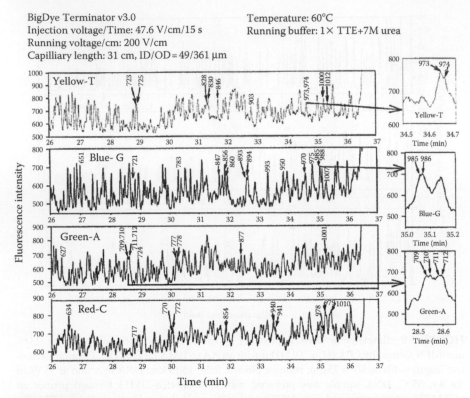

FIGURE 3.10 Last parts (base number from 600 to 1000) of four-color DNA sequencing on BigDye Terminator V 3.0 Standard DNA sample by using 2.0% quasi-IPN formed by $M_w = 7.6$ MDa LPA with an AA to DMA ratio of 11:1. Capillary effective length = 33 cm, ID = 50 µm; running voltage, 195 V/cm; DNA injection at 36 V/cm for 15 s; temperature, 60°C; Buffer, 1× TTE. (From Wang, Y., Liang, D., Ying, Q., and Chu, B., *Electrophoresis*, 26, 126, 2005. With permission.)

3.2.3.4.3 Copolymers
Compared with the mixing of different polymers, the copolymerization approach provides the possibility to adjust the qualities of the separation matrix by changing the molecular architecture and chemical composition. We can briefly divide the copolymers into three arbitrary subgroups: (1) random copolymers, such as the co-polymer of LPA and PDMA [P(AA-*co*-DMA)] [133], LPA/sugar [138,139] and PDMA/poly(*N*,*N*-diethylacrylamide) [P(DMA-*co*-DEA)] [117,121]; (2) block copolymers, such as PEO–PPO–PEO (poly(ethylene oxide)–poly(propylene oxide)–poly(ethylene oxide)) [140], and polyethylene glycol with fluorocarbon tails (PEG end-capped) [104–106]; and (3) graft copolymers, such as PNIPAM-*g*-PEO [141], LPA-*g*-PNIPAM [142], and LPA-*g*-PDMA [143,144]. The copolymers cannot only combine the desired physical properties of its components but also develop some new specific features absent in homopolymers. For instance, $PEO_{99}PPO_{69}PEO_{99}$ showed temperature-dependent viscosity properties and formed micellar structures [140,145].

In the synthesis of random copolymers for DNA sequencing, AA is usually one of the preferred monomers due to its high sieving ability of the resultant polymer

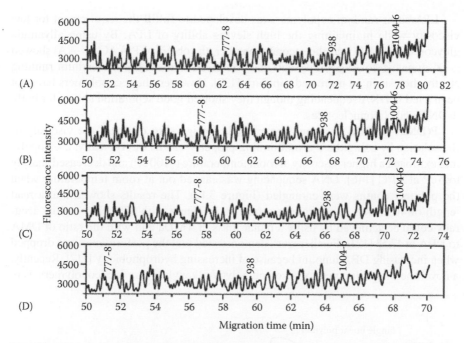

FIGURE 3.11 Last parts of electropherogram (Green-track, base A) in DNA sequencing by CE using 2.5% w/v (A) quasi-IPN, (B) quasi-IPN/GNPs-1, and (C) quasi-IPN/GNPs-2 at 50°C, and (D) quasi-IPN/GNPs-2 at 60°C. Sequencing conditions: effective/total length of bare fused-silica capillaries, 50/61 cm; ID/OD, 75/365 μm; sequencing electric field strength, 150 V/cm; DNA electrokinetic injection, 41 V/cm for 30 s; anode buffer, 1× TTE; cathode buffer, 1× TTE/7 M urea. DNA sample: BigDye Terminator v3.1 Sequencing Standard. (From Zhou, D., Wang, Y., Zhang, W., Yang, R., and Shi, R., *Electrophoresis*, 28, 1072, 2007. With permission.)

formed, while the other monomer segments have the ability to dynamically coat the capillary surface or to decrease the viscosity of the polymer solution for easier filling of the capillary or microchannel. As an example, DMA was chosen as the second monomer to obtain the poly(AA-*co*-DMA) [133]. This kind of copolymer could then possess the dynamic coating ability of PDMA, together with the high sieving ability similar to that of LPA for DNA sequencing. Different molar ratios of AA to DMA (3:1, 2:1, and 1:1) had been synthesized with the overall molecular weight being 2.2 MDa. All these copolymers had strong enough dynamic coating ability to suppress the electroosmosis, but the separation performance decreased with increasing DMA content. It has been shown that the random copolymer solution of AA and DMA with a feed ratio of 3:1 can achieve the separation performance of one-base resolution of 0.55 up to 699 bases and 0.3 up to 963 bases within 80 min at a copolymer concentration of 2.5% (w/v) under the optimized running condition for longer read length at ambient temperature. The result was even better than that of the same LPA in a precoated capillary [133]. A disadvantage of random copolymers is that at high overall copolymer molecular weight, high copolymer concentration, and comparable molar ratio of the two polymer components, microdomain inhomogeneity could occur, thereby rendering the separation matrix less effective.

LPA/sugar random copolymer was developed to fulfill the requirement for low viscosity while maintaining the high sieving ability of LPA. By using allyamide gluyconic acid (AAG) as the second monomer, the obtained P(AAG-*co*-AA) showed good dynamic coating ability and could be used as a separation medium running electrophoresis in an untreated capillary [139]. However, these copolymers have not been tried in DNA sequencing though they showed good separation ability for both, dsDNA and oligonucleotides.

P(DMA-*co*-DEA) was another example, designed to decrease the viscosity of the solution by using the thermothinning property of PDEA. Typically, 2% copolymer exhibited a higher viscosity at room temperature (~100 cP) and decreased sharply to 5 cP at 50°C [146]. DNA sequencing was carried out at room temperature when the polymer chains were entangled (Figure 3.12). The results showed that a read length of 463 bases of contiguous sequence in 78 min with 97% base-calling accuracy could be achieved by using a molar mass of 4 MDa and with the ratio of DMA to DEA at 47:53 [121]. Further work demonstrated that the performance was dropped when increasing DEA amount because of increasing hydrophobicity [117]. Recently, a novel thermogelling matrix based on poly(*N*-alkoxylacrylamide) copolymers were

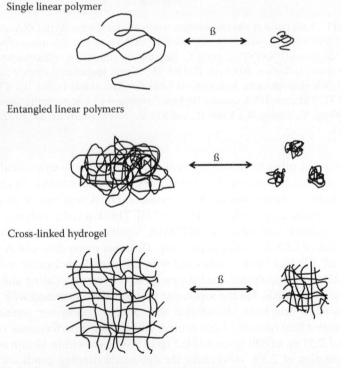

FIGURE 3.12 Schematic representation of sol-gel exhibiting polymer and hydrogel systems as they undergo a reversible volume phase transition. (From Buchholz, B.A., Doherty, E.A.S., Albarghouthi, M.N., Bogdan, F.M., Zahn, J.M., and Barron, A.E., *Anal. Chem.*, 73, 157, 2001. With permission.)

claimed as a candidate for microchannel DNA sequencing [118]. They formed entangled networks with high viscosity both, below 20°C and above 35°C. Hence, they could be easily loaded into microchannels at room temperature and form thermogels when temperature was increased beyond 35°C. Results showed that these thermogelling networks provided enhanced resolution and longer sequencing read lengths when compared to appropriate (closely related, nonthermogelling) polymer networks. In one study, the matrix containing the copolymer comprised of 90% wt/wt N-methoxyethylacrylamide and 10% wt/wt N-ethoxyethylacrylamide (M_w ~2 MDa) [P(NMEA-co-NEEA)] could be used to sequence up to 600 bases with 98.5% base-calling accuracy in 100 min. PVA-based copolymers including poly(vinyl-co-vinyl acetate) and poly(vinyl alcohol-co-1-vinyl-2-pyrrolidone) were tested at Stony Brook. It was found that they gave low resolution for ssDNA sequence though they could separate dsDNA in both, high- and low-pH regions [147].

Block copolymers often show distinguished physical properties when being dissolved in a selective solvent, which is good for one block, but poor for the other. In such a solution, the polymer can self-assemble to form micelle structures, which provide the ingredients for appropriate mesh sizes for DNA sequencing by physical cross-linking or entanglement. For example, triblock copolymers (PEO–PPO–PEO) in aqueous solution at room temperature can form a starlike micelle in a TE buffer with the PPO blocks in the core and the PEO chains forming the shell. At concentrations (~20%) appropriate for CE, the micelles were packed into a cubic structure with an aggregation number of ~60 [140], which was good for separations of small-size DNA fragments [148]. However, the mesh size was too small to effectively separate the larger DNA fragments. Also, the presence of urea in a sequencing buffer interferes with the sol-gel transition behavior of such type of block copolymers. Thus, they have not been tested as suitable separation matrices in DNA sequencing analysis.

Menchen et al. developed a set of block copolymers comprised of hydrophilic PEG end-capped with hydrophobic fluorocarbons (C_nF_{2n+1}) to test as sequencing media [104–106]. This kind of block copolymers has a critical micelle concentration (CMC) and will self-assemble into a flowerlike micelle structure, with fluorocarbons forming the core and PEG dangling outside forming the corona, as shown in Figure 3.13 [105]. With further increase in concentration, the micelles would aggregate together to form a network with PEG being the bridges. In this quasilattice state, the copolymers could be used as a separation matrix for DNA sequencing analysis. Optimum sequencing results were obtained from a 6% polymer mixture solution of C_6F_{13} end-capped and C_3F_{17} end-capped PEG 35,000 at 1:1 ratio, with a read length of 450 bases in 75 μm ID capillaries at an electric field strength of 200 V/cm.

For graft copolymers used as separation media, they often have a long chain as the backbone with some short chains attached to increase the polymer chain entanglement time in the transient polymer network. As discussed above, in a polymer network consisting of dynamically entangled polymer chains, the polymer meshes will adjust themselves to allow the larger DNA fragments to pass through by sliding the polymer chains away from each other [149], resulting in a lower resolution in the electrophoretic separation process. From another viewpoint, the relaxation time of long linear polymer chains is often not sufficiently long for effective separation of

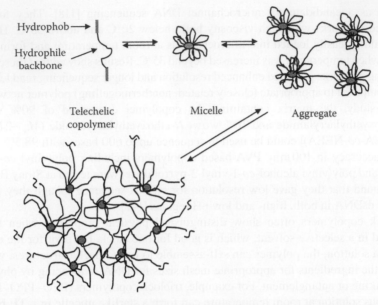

Hydrophob→
Hydrophile
backbone

Telechelic   Micelle   Aggregate
copolymer

FIGURE 3.13 Structures formed from water-soluble polymers with hydrophobic end groups. Above the CMC, polymer associates into rosette-like micelles. At higher concentrations, higher aggregates form through bridging. As aggregates become larger and more concentrated, a continuous network is generated. (From Menchen, S., Johnson, B., Winnik, M.A., and Xu, B., *Chem. Mater.*, 8, 2205, 1996. With permission.)

larger DNA fragments. If the relaxation time was shorter than that of the residence time of those larger DNA fragments, the resolution for those larger DNA fragments will again be deteriorated [150]. Graft copolymers with very short side chains on the backbone can stabilize the polymer chain entanglements in the separation matrix in order to enhance the sieving ability. A good example is the polyacrylamide grafted with poly(N-isopropylacrylamide) (LPA-g-PNIPAM) developed by Viovy and coworkers [142]. A schematic graph of this comblike graft copolymer is shown in Figure 3.14. The sequencing matrix composed of this particular type of copolymers exhibited a relatively low viscosity (~300 cP) at room temperature, allowing for easier injection, and switched to a high viscosity (~20,000 cP) at CE running temperatures (e.g., ~60°C) to form a robust network for DNA separation by aggregation of hydrophobic PNIPAM side chains. In other words, it combined the good properties of the two homopolymers including the sieving ability of the LPA and the thermal "viscosity switch" of PNIAM. As was reported, a read length of 800 bases could be achieved in less than 1 h with one-base resolution of 0.5 without complete optimization performed on a commercial ABI 310 Analyzer. At Stony Brook, on the contrary, we designed a graft copolymer with long chain PNIPAM as the backbone and short hydrophilic PEO as the branches [123,151]. In this architecture, PNIPAM-g-PEO exhibited a "coil-to-globule" transition with temperature change, i.e., at low temperature, both PNIPAM and PEO are hydrophilic and the copolymer chains take the coil conformation, while at elevated temperatures (above ~31°C), the PNIPAM becomes hydrophobic and starts to collapse, resulting in a globule structure with PNIPAM being the core and PEO

Water-soluble backbone

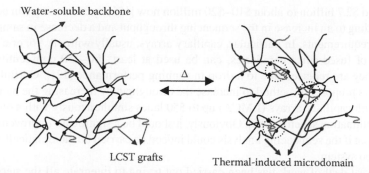

LCST grafts   Thermal-induced microdomain

FIGURE 3.14 Simplified view of mechanism for thermothickening block copolymers. By heating, the grafts undergo a microphase separation and create micelle-like aggregates which act as transient cross-links. (From Sudor, J., Barbier, V., Thirot, S., Godfrin, D., Hourdet, D., Millequant, M., Blanchard, J., and Viovy, J.-L., *Electrophoresis*, 22, 720, 2001. With permission.)

stretching out on the surface [151]. The dsDNA separation was performed at room temperature to utilize the uniform network, showing a good resolution. However, it has not been tested as a sequencing matrix, although its expected performance is not likely to yield an incremental improvement when compared with many of the existing sequencing matrices, only to demonstrate the proof of concept on thermal reversibility in network structure and solution viscosity.

Using LPA as the backbone, the Viovy group synthesized another graft copolymer by changing PNIPAM to PDMA in order to introduce the self-coating ability of PDMA [143], just like the mixture or quasi-IPN of LPA and PDMA. The sequencing performance was claimed to be comparable to long chain LPA in the absence of capillary coating and good performance was obtained even when the viscosity was as low as 200 cP.

Copolymer blends were also used as DNA separation matrices. For example, a mixture of PDEA/DMA with different ratio (one with 47%, w/w DEA/DMA, thermoresponsive; the other with 30%, w/w DEA/DMA, nonthermoresponsive) was applied to dsDNA separation, though the blend has not yet been tested as a sequencing matrix [152]. These blends were reported to be able to tune their mesh size thermally.

3.2.3.4.4 Dynamic Copolymers
Wei and Yeung applied monomeric nonionic surfactants, *n*-alkyl polyoxyethylene ethers ($C_{16}E_6$, $C_{16}E_8$, and $C_{14}E_6$) as a sequencing matrix to DNA sequencing, resulting in a read length of 600 bases on BigDye G-labeled M13 [153]. These so-called dynamic polymers have the advantage of ease of operation, solution homogeneity, low viscosity, and dynamic coating ability.

3.2.4 Capillary Array Electrophoresis

CAE has revolutionized DNA sequencing in terms of cost reduction and separation speed [154], with a reduction in the cost of sequencing a human genome from an estimated \$2.7 billion to about \$10–\$20 million now [155] by increasing run automation, leading to an increase in the sequencing throughput and a decrease in sample and reagent requirements. In particular, capillary arrays, usually with 48, 96, or a higher number of fused-silica capillaries, can be used at least 100–150 consecutive runs before they are replaced due to a drop in running performance [121]. Gorfinkel and coworkers proposed a method to increase the throughput by increasing the number of linear multicapillary arrays (LMCA) up to 550 lanes simultaneously with a dual-side laser illumination scheme [156]. Obviously, it should be a significant improvement for throughput if the sequencing analysis could indeed be carried out automatically in one integrated system.

A great deal of work has been carried out trying to integrate all the steps from DNA sample preparation to electrophoresis in a single instrument. For example, Tan and Yeung investigated an integrated online system for DNA sequencing analysis by CAE, including sample preparation, cleanup, and analysis, resulting in a read length of 370 bases with 98% accuracy [157,158].

Capillary arrays are basically the same as a single channel in terms of electrophoresis-related parameters such as matrices, running temperature, and applied electrical field. However, CAE requires a special detection system to detect the signals simultaneously for all the capillaries installed. There are two major designs currently being used in commercial instruments. One is the confocal fluorescence scanner, as applied in the MegaBACE 1000 DNA sequencing system from molecular dynamics (bought by Amersham) [159]. This method, known as the Mathies method, has the advantage of simple array design, no cross talk between capillaries and identical laser illumination for all the capillaries. However, it suffers from complicated fluorescence collection and low sensitivity. In particular, each capillary in a 96 CAE was given 1% of the time, while 99% of the DNA escaped from detection [160]. The other design uses a sheath-flow cuvette, whereby a CCD camera is used to record the image for the entire capillary set at the same time [161]. This method, as described by Kamahori and Kambara [162], is much more sensitive, whereby the fluorescence intensity is monitored continuously for each capillary with a detection time approaching 100%. However, the disadvantages come from cross talk between the images from different capillaries and unequal illumination of the capillaries. Furthermore, the sheath flow rate should be well controlled. This design has been licensed by Applied Biosystems, and applied to the ABI PRISM 3700 DNA analyzer.

The success of multiple parallel capillary in DNA sequencing analysis, in comparison with single-channel, gel-filled capillary, represents a milestone in capillary array electrophoresis. However, further improvements are needed to satisfy the new aims of National Institutes of Health (NIH)/NHGRI to achieve the goal of sequencing of human genome with a cost of about \$1000 by the year of 2014.

3.2.5 MICROCHIPS

A promising technology to increase the throughput and to decrease the cost of DNA sequencing is the use of microfluidic chips or the so-called micrototal analysis system (μTAS). It is identified as an integrated analyzing system which includes sample preparation, DNA separation, and signal detection on a single chip [163]. This new technology has the potential to increase the sequencing rate by 5–10 fold and to obtain a similar read length as in CAE instruments [3,164,165], mainly due to shorter separation channels allowed by narrower sample zone, well-controlled sample injection [166], and integration of Sanger extension reaction as well as the CE process [167]. Typically, a read length of 600–800 bases could be achieved in around 25 min on a microchip with a sequencing channel length of 20 cm [165,168], while it takes 1–2 h on CAE. Moreover, the cost for microfabricated chips could be reduced substantially, when compared with conventional CE. In particular, a set of 96-silica-fused capillary array, which is commonly used in CAE costs about $3600, based on the current price structure, while a high-quality glass chip, made from borofloat glass has a price of ~$50–$100 per 6 in. wafer, and can be manufactured at a cost of ~$400. If we assume that the glass chips have the same durability and are as effective as the capillary array, it is only ~10% of the cost for capillary arrays, even before we count on the advantage that there is no additional increase in price from a 96-channel chip to a 384-channel chip as the same labor and materials are needed to fabricate them [169]. On the other hand, plastic chips made from polymeric materials such as poly(dimethylsiloxane) (PDMS) are one or two orders of magnitude cheaper than glass chips. Hence, this novel technology, when applied widely in DNA sequencing analysis and other related fields, would bring another breakthrough in cost reduction to research in biology and medicine. However, we should remember the dynamic coating ability of polymers when the channels are changed from fused silica to polymeric materials. Then, one has to reevaluate the nature of polymers for suitable dynamic coating ability in a specific polymer substrate.

The four-color DNA sequencing on chips was first reported by Woolley and Mathies in 1995 [165]. Since then, intense research has been carried out on the development of microchip DNA sequencing including channel design, chip fabrication, separation matrices, etc., so as to create easy-to-use, high throughput, and highly automated sequencing systems. It should be noted that the critical role for the separation matrix remains the same. With a better understanding on the fundamentals of polymer network formation and transient chain dynamics, the separation matrix can be tailor-designed for specific applications and is likely to remain an essential element in new system development using microchannels and microfluidics.

3.2.5.1 Microchannel Design

Unlike separation for polymerase chain reaction (PCR) products, dsDNA, or oligonucleotides, DNA sequencing requires a long read length therefore a relatively long separation channel should be considered. It has been demonstrated that the read length significantly increased from 200 bases to ~500 bases when the separation channel was increased from 3 to 7 cm [170]. A strategy to increase the throughput with a sufficient separation distance is to densely pack more microchannels on a single device. Microchips with 16, 96, and 384 channels have been fabricated on glass wafers [171–177]. As an example, Liu et al. reported a design of automated 16-channel microchips, which routinely yielded a read length of 450 bases in 15 min using an integrated four-color confocal fluorescence detector [171]. The best result was claimed to be accomplished in less than 18 min in this integrated setup including chip loading, sample injection, and DNA separation, with a read length of 543 bases at a base-calling accuracy of ≥99%.

Another innovation was done by the Mathies group, who developed a microfabricated 96-channel microchannel plate with a four-color rotary confocal fluorescence scanner and a novel injector for uniform sieving matrix loading, as shown in Figure 3.15 [172]. It provided an effective sequencing length of 15.9 cm on a compact 150 mm diameter wafer to achieve a long read length as well as high throughput. With a BASEFINDER program, this setup produced an average read length of 430 bases with Phred Q ≥ 20 [178,179] from 95 out of 96 channels run in parallel,

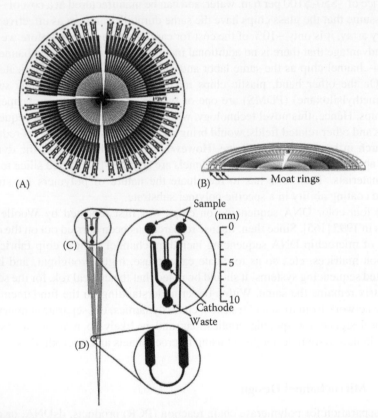

FIGURE 3.15 (A) Overall layout of 96-lane DNA sequencing microchannel plate (MCP). (B) Vertical cut-away of MCP. The concentric PMMA rings formed two electrically isolated buffer moats that lie above the drilled cathode and waste ports. (C) Expanded view of the injector. Each doublet features two sample reservoirs and common cathode and waste reservoirs. The arm from the sample to the separation channel is 85 μm wide, and the arm from the waste to the separation channel is 300 μm wide. The separation channel connecting the central anode and cathode is 200 μm wide. (D) Expanded view of the hyperturn region. The turns are symmetrically tapered with a tapering length of 100 μm, a turn channel width of 65 μm, and a radius of curvature of 250 μm. Channel widths and lengths are not drawn to scale. (From Paegel, B.M., Emrich, C.A., Wedemayer, G.J., Scherer, J.R., and Mathies, R.A., *Proc. Natl. Acad. Sci. USA*, 99, 574, 2002. With permission.)

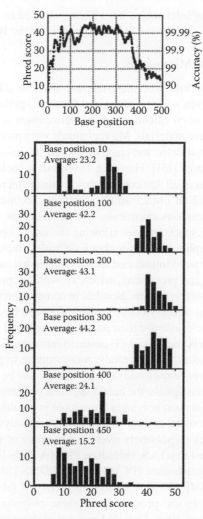

FIGURE 3.16 Average Phred score [178,179] over the 95 passed lanes as a function of base position (Upper). Miscall probabilities are listed at the right as accuracy. The distribution of Phred scores [178,179] from the 95 passed lanes is shown at six positions in the processed data (Lower). The base position and average score of each distribution is listed above the graphs. Distributions representing the beginning (10 bp), high quality (100, 200, and 300 bp), declining quality (400 bp), and end (450 bp) regions of the separation are shown. (From Paegel, B.M., Emrich, C.A., Wedemayer, G.J., Scherer, J.R., and Mathies, R.A., *Proc. Natl. Acad. Sci. USA*, 99, 574, 2002. With permission.)

as shown in Figure 3.16 [172]. However, the dense packing of these channels requires high resolution of fabrication. In addition, the separation matrix loading is another challenging problem because the viscous separation matrix is still needed to achieve better separation with a long read length as has been discussed before. That may be the reason why 384-channel-based microchips have not yet been applied to DNA sequencing, though it has been used for mutation detection. To date, the reported parallel DNA sequencing microchips have been restricted to 96 channels, although a breakthrough on such a limitation is likely to occur in the future.

3.2.5.2 Fabrication of Microchips

Generally speaking, two kinds of materials are used in μTAS: silicon/glass and polymer/plastics. μTAS was widely recognized for their applications to biomolecules in the early 1990s [180], when silicon and glass were chosen to be the substrate due to the ease in processing these materials. Microchannels were produced by the traditional photolithography method with the wet chemical etching technique being used to define the lanes and other features [181]. However, several drawbacks drove researchers to seek new substitutes. First of all, the cost to fabricate those microdevices was relatively expensive as they needed an ultraclean environment. Second, the process involved multiple steps using hazardous chemicals. Third, silicon is not transparent in the visible/UV region of the spectrum, not allowing the use of optical detection. On the other hand, glass is amorphous, which makes it difficult to etch side walls vertically.

Polymeric materials were introduced as alternatives because of their low cost in both, the material itself and processing, which allowed the production of disposable devices to avoid cross-contaminations. Besides, in comparison with silicon/glass competitors, they have several possibly better features: (1) good biocompatibility, (2) lower electroosmotic flow (~five-fold lower than that of glass or fused silica depending on the nature of separation matrix used), and (3) easier to fabricate different channel shapes. All these possibilities make them potentially more suitable to produce microchannels for DNA analysis. However, some disadvantages should also be noted, such as a higher background to fluorescence signals, the heat dissipation problem, and absorption issues. The other drawback is that most polymers are soluble in organic solvents, which limits their use in some conditions other than DNA sequencing analysis.

Many different types of polymers over a wide range of physical and chemical properties have been used in μTAS, including PMMA [182–189], cyclo-olefin copolymer (COC) [190], polycarbonate (PC) [191,192], PDMS [193–201], and thermoset polyester (TPE) [202–204]. Correspondingly, a variety of fabrication methods have been developed to enable the processing of these polymer substrates, including replication and direct fabrication. Detailed reviews of these techniques are available in Refs. [163,205,206]. Briefly, replication methods, including hot embossing, injection molding, and soft lithography, use a precise template or master to replicate many identical polymer microstructures, while direct fabrication, such as laser ablation, optical ablation, and x-ray lithography, is used to create each microdevice individually without a master. Methods are selected according to the properties of the materials and the number of identical chips needed. For example, hot embossing and injection molding are commonly used to fabricate thermoplastic materials, such as PMMA [185] and PC [207]. On the other hand, PDMS, as an elastomer, is usually built into microfluidic devices by soft lithography [197]. Thermoset polyesters were also reported to be molded by soft lithography [203].

3.2.5.3 DNA Sequencing Matrix Used in Microchips

The criteria for a sieving matrix using microchips are basically the same as that in conventional CE, except that the viscosity problem is amplified in μTAS as microchips, especially the cheaper chips that are made of plastics, cannot stand up to high external pressures (no more than ~200 psi for glass chips and ~50 psi for plastic chips) for polymer solution injection. It becomes an even more critical issue to develop new matrices suitable for chip injection as well as to yield comparable sieving abilities with long DNA fragments. Nevertheless, the same strategy, as discussed previously, can be used to seek qualified separation media, such as copolymerization to synthesize "viscosity switchable" polymers. So far, some polymers with low viscosity or viscosity switchable properties have been applied on microfluidic devices for dsDNA separation including hydroxypropylmethyl cellulose (HPMC) [208], HEC [98,209], and HPC [210]. However, the low viscosity separation media have been shown to lack the potential to achieve the required resolution for DNA sequencing analysis. At present, it remains a challenging hurdle for the replacement of CAE with μTAS in DNA sequencing analysis. Currently LPA is still widely used in microchip DNA sequencing [172,211,212] with special injection methods, as there is no other known polymer that can compete with its high sieving ability. For example, LPA was used as a separation medium in a 96-channel microplate, which was injected through the anode access port at 4100 kPa for 5 min by using a high-pressure loading system (Figure 3.15) [172]. The application of the quasi-IPN concept to the microchip format should be a worthwhile undertaking.

3.2.6 END-LABELED FREE SOLUTION ELECTROPHORESIS (ELFSE)

DNA sequencing using polymeric matrices has inherent shortcomings including loading of viscous solution to microchannels, long separation distance, and limitation in read length. These factors have prompted scientists to seek other separation methods without the use of a separation matrix. As we know, the sieving matrix is used to change the electrophoretic mobility of DNA, which otherwise is relatively size independent, in order to achieve the separation. In a given separation medium, larger DNA molecules migrated more slowly than the smaller ones and reached the detection window at a later time. Another possible way is to link an uncharged end with a fixed size (drag-tag) onto DNA molecules at different length in order to change the charge to friction ratio, resulting in different size-dependent electrophoretic mobility in free solution [213]. This method, known as end-labeled free solution electrophoresis, opens a new possible pathway for DNA sequencing as it allows DNA fragments to be sequenced in an aqueous buffer alone. It is particularly suitable for microfluidic devices because there is no need to load viscous polymer networks into microchannels.

ELFSE was first presented by Mayer et al. in 1994 [214] and a detailed theory about Gaussian polyampholytes in free-flow electrophoresis was expressed by Long et al. in 1998 [215]. From then on, studies including dynamic simulation, theoretical analysis, and experimental research have been performed to dig out its potential applications to DNA sequencing analysis or other polyelectrolyte separations [213,216–229]. It is understood that an end-labeled ssDNA molecule can adopt four possible hydrodynamic conformations, as shown in Figure 3.17 [221], which should be considered different when estimating the electrophoretic behavior in free solution. In particular, Figure 3.17A shows a simple case as the whole molecule takes the random-coil conformation. With increasing drag force, the drag-tag is separated

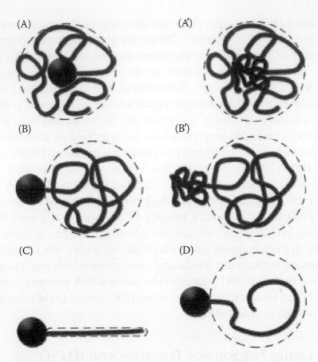

FIGURE 3.17 Schematic diagram showing various ELFSE regimes for a spherical drag-tag (A–D) and an extended polymeric drag-tag (A' + B'); the DNA is circled by dotted lines and the label is in the center (A, A') or on the left (B, B', C and D). (A' + A') The drag-tag is part of the random-coil conformation of the DNA molecule. (B + B') The drag-tag and the DNA are hydrodynamically segregated, but the random-coil conformation is not disturbed. (C) The DNA stretches out in response to the drag forces. (D) Short DNA molecules can be sterically segregated from a bulky drag-tag. (From Meagher, R.J., Won, J.-I., McCormick, L.C., Nedelcu, S., Bertrand, M.M., Bertram, J.L., Drouin, G., Barron, A.E., and Slater, G.W., *Electrophoresis*, 26, 331, 2005. With permission.)

from the undisturbed ssDNA component, leading to hydrodynamic segregation (Figure 3.17B). Further, ssDNA will be stretched out when the forces become even larger (Figure 3.17C). Figure 3.17D, on the other hand, demonstrates a short ssDNA molecule attached to a large bulky label even at a low applied electric field. Obviously, shorter DNA takes the longer migration time, inverse to the matrix electrophoresis. The resolution of separation as well as the increase in read length, so as to reach the theoretical prediction (larger than 500 bases), is based on the relative size of drag to DNA molecules as well as the applied electric fields. However, an extremely high voltage (e.g., using streptavidin label, a read length of 110 bases requires a critical field of ~8 kV/cm [221]) will bring many other problems, such as joule heating and power supply requirements. Thereby, current research is to seek a large and monodispersed drag (fixed size) to achieve a long read length for DNA sequencing. Both, natural and synthetic polymers have been tried as drag-tags, including polypeptides, polypeptoids, and PEG. However, they suffer from structural defects or polydispersity [221]. A solid-phase synthesis is able to generate monodisperse polymers, but the degree of polymerization is limited to a small number at the present time (up to 60 repeating units [230–232]). By far, the best results of ELFSE were obtained by Ren et al. [229]. Using fractionated streptavidin, a folded natural polypeptide as the drag-tag, they obtained a read length of 100 bases in 18 min within a 34 cm long noncoated capillary without any sieving matrix, as shown in Figure 3.18. Although the read length is much shorter than a typical reading of CAE, it has shown the possible applicability of such a concept for future development. Many obstacles need to be overcome before it can reach the same level of competence when compared with CAE.

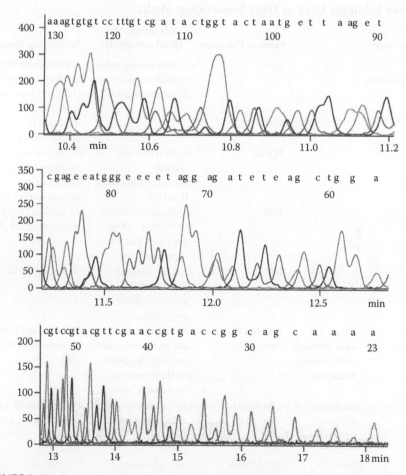

FIGURE 3.18 Separation of DNA sequencing reaction products using a purified streptavidin drag-tag and a dynamically coated capillary. The Y-axis represents the intensity of the fluorescence signal and the X-axis gives the elution time in minutes. The numbers above the peaks give the number of nucleotides (including the primer sequence). The weaker signal observed for the smaller fragments is due to their less efficient electrokinetic injection. (From Ren, H., Karger, A.E., Oaks, F., Menchen, S., Slater, G.W., and Drouin, G., *Electrophoresis*, 20, 2501, 1999. With permission.)

3.2.7 Scope of the Possible Usage of Current Matrices in DNA Sequencing and Oligo Separation

The matrices used in current electrophoretic methods for DNA sequencing analysis have been discussed above. These matrices with different compositions and structures showed different sieving behavior in the development of CE-based DNA separations, as summarized in Table 3.3. Albeit plenty of polymers have been tested as

TABLE 3.3
Polymer Solutions Used as DNA Sequencing Media

Polymer Type		Name of Polymers	Advantages(A)/ Disadvantages(D)	Sequencing Results
Linear homopolymers		LPA	(A) Best sieving ability (D) High viscosity, no dynamic coating	1000 bases with 97% accuracy in 80 min [120]; 975 bases in 40 min with 98.5% at resolution of 0.24 by adding DMSO [124]
		PDMA	(A) Best dynamic coating (D) More hydrophobic than LPA	800 bases in 96 min at a resolution of 0.5 or 1000 bases at a resolution of 0.3 [116]
		PEO	(A) Dynamic coating (D) Reflush with HCl after each run, longer sequencing time	966 bases in 7 h at a resolution of 0.5 [85]
		HEC	(A) Low viscosity (D) Short read length and inconsistent performance	570 bases in 70 min [98]
Polymer mixture	Mixture of same polymers with different molar mass	LPA HMW with LMW	(A) Good sieving ability for both small and large fragments (D) High viscosity, no dynamic coating	1300 bases in 2 h with an accuracy of 98.5% [36]
	Mixture of different polymers	LPA mixed with PDMA	(A) Good sieving ability and dynamic coating	1000 bases in 2 h with an accuracy of 98.9% or 1 h at a resolution of 0.3 [54]
	Quasi-IPN	LPA/PDMA	(A) High sieving ability and dynamic coating, less polymers to form network, suppress polymer incompatibility	750 bases of contiguous sequence in 37 min with 98.0% accuracy or 1000 bases in 39 min, or 1200 bases in 60 min [136]

TABLE 3.3 (continued)
Polymer Solutions Used as DNA Sequencing Media

Polymer Type		Name of Polymers	Advantages(A)/ Disadvantages(D)	Sequencing Results
Copolymers	Random copolymers	P(AA-*co*-DMA)	(A) Good sieving ability and dynamic coating	699 bases at a resolution of 0.55 or 963 bases in 80 min at a resolution of 0.3 [133]
		P(DMA-*co*-DEA)	(A) "Viscosity switchable" property (D) Low sieving ability for large fragments	463 bases of contiguous sequence in 78 min with 97% accuracy [121]
		P(NMEA-*co*-NEEA)	(A) "Viscosity switchable" property (D) Low sieving ability for large fragments	600 bases in 100 min with 98.5% accuracy [118]
	Block copolymers	PEG end-capped (C_nF_{2n+1})	(A) Shear-thinning low viscosity (D) Low sieving ability for large fragments	450 bases without migration time reported [104]
	Graft copolymers	LPA-*g*-PNIPAM	(A) "Viscosity switchable" property, good sieving ability	800 bases in <1 h at a resolution of 0.5 [142]
		LPA-*g*-PDMA	(A) Lower viscosity, dynamic coating, and good sieving ability	Comparable to long chain LPA [143]
Dynamic polymers		*n*-Alkyl polyoxyethylene ethers	(A) Low viscosity and dynamic coating ability	600 bases in 60 min at a resolution of 0.5 [153]
ELFSE		Free solution; Streptavidin was used as a drag-tag	(A) No loading problem, no need of coating (D) Difficulty in finding longer drag-tag	100 bases in 18 min [229]

sequencing media, LPA-based mixtures or copolymers showed the best sieving ability to date, especially for DNA fragments larger than 600 bases with relatively shorter separation time. Among them, quasi-IPN is noticeable, with good performance on separation of larger fragments with shorter run time, probably due to its unique architecture. The same characteristics can be utilized to allow the use of a shorter effective separation length in order to cut down the DNA migration time, and thereby to fulfill the need for typical shorter channels in the microchip designs. It can also combine the dynamic coating ability of the second component (e.g., PDMA, even though the polymer solution viscosity can still be a problem for quasi-IPN made of LPA and PDMA, leading to the challenge for loading the chips). However, it is a two-way issue. One can modify the loading method to solve the high polymer solution viscosity problem or change the partial composition of the quasi-IPN to decrease the viscosity. For example, by using PVP as the second component or by introducing the thermoreversible property of LPA grafted with PNIPAM. Quasi-IPN is a worthy strategy to form better sieving matrices for DNA sequencing analysis.

3.3   NON-SANGER-BASED METHODS

Alternatively, many other new approaches, which are not based on electrophoretic separation of Sanger reaction products, have been proposed. Reviews to discuss the new advances have been published [3,233], herein we provide only a brief introduction for completeness as they are beyond the range of this review. Two representative approaches are as follows: (1) sequencing by synthesis (SBS), a massively parallel sequencing method that detects sequencing data simultaneously for a larger number of DNA fragments produced in real time; and (2) nanopores sequencing, which obtains the DNA sequences by forcing a DNA chain to pass through a nanosized pore under an applied electric field.

Being a massively parallel sequencing method, SBS is characterized as the detection of a nucleotide base incorporated with the synthesis of a DNA strand with thousands of oligonucleotides simultaneously. It has a potential to sequence DNA at a rate of 1000 bases per second. Single molecular reading including zero-mode wave guide, Visigen polymerase read, molecular motors method, and exonuclease approach could be obtained by suitable label to polymerase or exonuclease, which is to place a detector next to a growing strand of DNA or to detect the fluorescence tagged nucleotides cleaved from a DNA fragment [233,234]. Besides, cycle sequencing based on ligase was reported in Ref. [235,236], being based on emulsion PCR amplification but to sequence the paired ligases. The other SBS method is base-by-base sequencing [237].

Pyrosequencing appears to be attractive as it has been applied on a commercial instrument, GS-20, owned by 454 Life Science. This method has been shown to be effective in sequencing microbial and viral genomes or other small DNA pieces. It is built on a four-enzyme real-time monitoring of DNA synthesis by biolumines-cence, which detects the light signal (bioluminescence) based on the released pyrophosphate accompanying a nucleotide. Thus, the signal can be quantitatively related to the number of bases added to the template [238]. It is good for determining unknown mutation (or resequencing), single nucleotide polymorphisms (SNPs), and tag sequence [238] due to its high sequence rate. In particular, GS20 has demonstrated success in reading up to 25 million bases on a 4 h run with an average read length of 100 bases [29–32]. The procedure can be briefly described as first attaching the ssDNA to beads, then amplifying by PCR emulsion in millions of time, finally to deposit the beads with a DNA template onto a fiber-optic wall for sequencing, as shown in Figure 3.19 [239].

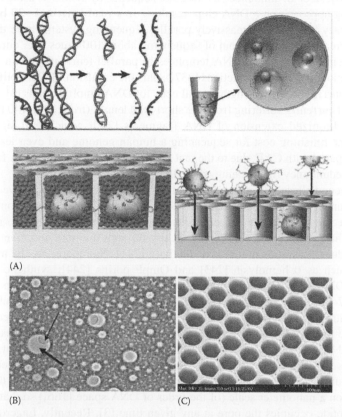

(A)

(B)   (C)

FIGURE 3.19 (See color insert following page 216.) Scheme of 454 Life Science's emulsion-based amplification of sequencing templates and placement of beads into Picotitre plates. (A) Genomic DNA is isolated, fragmented, ligated to adapters, and separated into single strands (top left). Fragments are bound to beads under conditions that favor one fragment per bead, the beads are captured in the droplets of a PCR-reaction-mixture-in-oil emulsion and PCR amplification occurs within each droplet, resulting in each bead carrying 10 million copies of a unique DNA template (top right). The emulsion is broken, the DNA strands are denatured, and beads carrying ssDNA clones are deposited into wells of a fiber-optic slide (bottom right). Smaller beads carrying immobilized enzymes required for pyrophosphate sequencing are deposited into each well (bottom left). (B) Microscope photograph of emulsion showing droplets containing a bead and empty droplets. The thin arrow points to a 28 μm bead; the thick arrow points to an approximately 100 μm droplet. (C) Scanning electron micrograph of a portion of a fiber-optic slide, showing fiber-optic cladding and wells before bead deposition. (From Margulies, M., Egholm, M., Altman, W.E., Attiya, S., Bader, J.S., Bemben, L.A., Berka, J., Braverman, M.S., Chen, Y.-J., Chen, Z., Dewell, S.B., Du, L., Fierro, J.M., Gomes, X.V., Godwin, B.C., He, W., Helgesen, S., Ho, C.H., Irzyk, G.P., Jando, S.C., Alenquer, M.L.I., Jarvie, T.P., Jirage, K.B., Kim, J.-B., Knight, J.R., Lanza, J.R., Leamon, J.H., Lefkowitz, S.M., Lei, M., Li, J., Lohman, K.L., Lu, H., Makhijani, V.B., McDade, K.E., McKenna, M.P., Myers, E.W., Nickerson, E., Nobile, J.R., Plant, R., Puc, B.P., Ronan, M.T., Roth, G.T., Sarkis, G.J., Simons, J.F., Simpson, J.W., Srinivasan, M., Tartaro, K.R., Tomasz, A., Vogt, K.A., Volkmer, G.A., Wang, S.H., Wang, Y., Weiner, M.P., Yu, P., Begley, R.F., and Rothberg, J.M., *Nature* (London, UK), 437, 376, 2005. With permission.)

Recently, Ju et al. announced a four-color sequence by synthesis using molecular engineering approach on a DNA chip, resulting in a read length of ~20 bases with high accuracy [234]. These massively parallel sequencing systems have the advantage of increasing the throughput of sequencing about 100 times over current CAE by sequencing thousands of DNA templates in parallel (currently it can achieve a parallelization of 10^5 to 10^7 reactions [237]), plus the emulsion PCR amplification to avoid the bias induced by using bacterial cells for DNA amplifications [3]. However, they are all currently suffering from the short read length (from 12 to 100 bases) due to nonsynchronized extension of DNA fragments [240], which will significantly increase the finishing cost for sequencing a human genome and even less suitable when compared with CAE, due to the fact that the more gaps in the DNA fragments, the more redundancy that is being created.

The second alternative method is to allow the polynucleotides to move through a nanopore under an applied electric field [241], which allows the separation of longer DNA molecules by the entropic effect with electric trapping [242]. The molecules with net charge are driven electrophoretically through the pore and then block the pore to induce variations in ionic conductivity [241]. Nanopores can be made from proteins such as α-hemolysin [243] and OmpF porins [244]; synthetic polymers such as PDMA [245] and polyimide [246]; and inorganic materials such as silicon oxide [247–249], silicon nitride [250–252], and multiwall carbon nanotubes [240]. The advantage of this technology comes from the fact that the polynucleotides can translocate through the nanopore very rapidly at a rate exceeding a thousand bases per second [253,254], leading to a substantial enhancement in throughput. Unfortunately, it is very difficult to obtain the signals at a single DNA base molecular level, which requires a decrease in the speed of translocation and control of the size of nanopore on a nanometer scale (of the order of DNA space [166]) so that only one polynucleotide occupies the pore at any given time [3]. Recently, Lagerqvist et al. proposed a novel idea to allow single base resolution by measuring the electrical current perpendicular to the DNA backbone when it is translocating through a pore in a solution [255]. However, all in all, this technique is still fairly far from being developed to become a robust, reliable commercial system from the proof-of-principle stage now being performed in the laboratory [256,257].

Other noticeable innovations include surface electrophoresis, the mechanical unzipping of DNA, and sequencing by hybridization. Surface electrophoresis uses a particular surface to separate DNA, which was a result of the differences in the conformation of chains at different lengths adsorbed on an attractive surface [258,259]. One of the advantages of this method is that it can separate a wide range of DNA fragment lengths from several hundreds of base pairs (bp) to the genomic size scale (megabase pairs) without substantial loss of resolution at larger DNA fragments and is suitable for DNA mapping. However, it is hard to achieve the single base resolution as required by DNA sequencing.

The mechanical zipping method is based on the presupposition that the binding strength between complementary base pairs can be detected by a force probe (i.e., atomic force microscopy) during the unzipping process. The limitation of this technique comes from the high flexibility of ssDNA chains in nature, which in turn needs a very stiff and accurate force to be used to sequence DNA [260]. Finally, sequencing by hybridization consists of two basic steps: the first one called the biochemical step is based on the complementary hybridization of an unknown ssDNA target sequence to generate fluorescence signals (sequencing spectrum); the second step, known as the combinatorial step, is to reconstruct the target sequence algorithmically from its spectrum [261–264]. Some of these new advances have been applied in sequencing SNP and resequencing or other applications which only require an average read length of less than 100 bases. However, their future in sequencing of the mammalian genome is yet to be predicted.

3.4 OLIGONUCLEOTIDE SEPARATION BY CE AND MICROCHIP ELECTROPHORESIS

Synthetic oligonucleotides (oligos) have been widely used in biochemical sciences and molecular medicine in the past few years both, as research tools and as a class of therapeutic agents against diseases, such as antisense oligodeoxynucleotides [265]. Compared to conventional pharmaceutical drugs, which bind to a target protein directly, synthetic oligonucleotides bind to the mRNA for a target protein, thereby inhibiting the protein production process. Isis' Vitravene is the first oligonucleotide drug on the market for the treatment of retinitis in AIDS patients [266]. As these drugs begin to show promise in the pharmaceutical field, advances in analysis and purification are needed to determine the quality of these chemically synthesized oligonucleotides. The rapid development of oligonucleotides as drugs presents a challenge to analytical scientists to develop matching separation methods.

Reversed phase [267] and anion-exchange high-performance liquid chromatography (HPLC) [268] have been applied to analyze and purify synthetic oligonucleotides. Although these methods can meet the needs of preparative and analytical purposes, they require a large amount of the sample. As for DNA sequencing analysis, CE combined with LIF detection has been shown to be an effective method and has been dominating the current market for the separation of oligos. In the last decade, mass spectrometry (MS) coupled with CE has gained popularity in the field of nucleic acid separation [269–271]. However, the effective length of the capillary channel needed for CE has been at least 20 cm. Microchip-based CE is one of the efficient separation platforms and has proven to be almost two orders of magnitude faster than conventional CE [272].

Minalla et al. studied the feasibility of high-resolution separation of oligonucleotides in microchannels [273]. They found that high-resolution separation of oligonucleotides was extremely challenging because it depended on sequence-specific migration, resulting in effective separation of Cy5 labeled poly(dT) in a 3 cm long channel [273], twice the length of existing commercial chip channels.

Antisense oligonucleotides are single stranded and generally 12–30 bases in length. For the separation of such small oligonucleotides, much higher concentration polymer solutions are required according to the Ogston description [274] and the scaling theory [275] in order to form smaller meshes to separate the smaller DNA molecular fragments. As a result, much higher viscosity of the solution is expected. The channels are even smaller and shorter, which brings up a new challenge for matrix design. Hereby, we would like to introduce separation media which have been tested in oligo separations and which have thermally tunable mesh sizes.

3.4.1 Oligonucleotide Separation Media

3.4.1.1 Nonthermogelling Matrices

A variety of polymer solutions, including 10% LPA [276], 14% (w/v) polyvinylpyr-rolidone (M_w = 1 MDa) [277], 12% (w/w) poly(ethylene glycol) (M_w = 20 kDa) [278], and 4% (w/v) HEC (single viscosity) [279], have been developed with some success for the separation of oligonucleotides. For example, model homo-oligomeric deoxy-nucleotides ranging in length from 12- to 24-mer were separated using a 20 mM Tris(hydroxymethyl)aminomethane-N-Tris(hydroxymethyl)methyl-3-aminopro-panesulfonic acid. Unfortunately, most sieving media at high concentrations have high viscosities, making the filling of polymer solutions into narrow bore capillary tubings very hard or impractical, especially for automated instrumentation.

Organic solvents have also been employed to enhance the solubility of polymers of interest and the selectivity [280,281]. With a matrix consisting of 5% w/v of C_{16}-derivatized 2-HEC dissolved in N-methylformamide (NMF) containing 50 mM ammonium acetate, $p(dA)_{12-18}$ and $p(dA)_{40-60}$ oligonucleotides were baseline sepa-rated [282]. Figure 3.20 shows the electropherograms of the separation of oligonucle-otides [282]. The sieving matrices consisting of hydrophobic polymers could be dissolved in organic solvents and thereby could expand the range of polymers suitable for oligonucleotide separations.

High-speed separation of antisense oligonucleotides by microchip-based CE first appeared in 1994 [283]. An ultrafast separation of oligonucleotides in less than 45 s was achieved. However, the effective separation length (3.8 cm) was relatively long when compared with commercialized systems, such as the format used in the Agilent 2100 Bioanalyzer with a separation channel of 1.5 cm. Furthermore, the siev-ing material used was a 10% in situ polymerized and viscous polyacrylamide solu-tion, making the replacement of the polymer matrix more difficult.

3.4.1.2 Thermogelling Matrices

The strategy to reduce the viscosity of the separation matrix so as to make it easier to load the "gel" into macrochips is basically the same as that for DNA sequencing analysis, except that now we pay attention to the good separation performance on much smaller DNA fragments, with base number less than 100. In particular, some triblock copolymers have attracted our attention, as they show good sieving ability for small DNA fragments though not suitable for DNA sequencing analysis. In the following sections, we discuss some of those developments.

3.4.1.2.1 F127

F127 is an amphiphilic block copolymer and has unique viscosity-adjustable proper-ties. At low temperatures, the polymer solution at concentrations above ~20% (w/v) acts as a free-flowing solution, while at elevated temperatures the solution viscosity is increased due to micelle formation by self-assembly and overlap of PEO chains in the micellar shell into a quasilattice having a face-centered cubic (FCC) structure. This copolymer was first employed as a separation medium for oligonucleotide separations at Stony Brook [145]. The oligonucleotide sizing markers were separated within a 10 cm effective length separation column in 16 min. To investigate the feasibility of

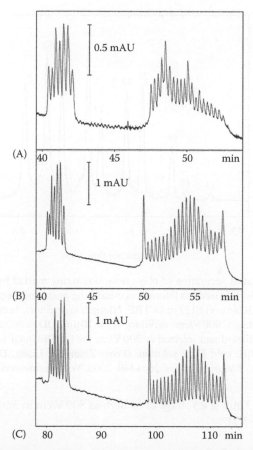

FIGURE 3.20 (A) Electropherogram showing the separation of p(dA)$_{12-18}$ and p(dA)$_{40-60}$, 1 and 2 μg/mL, respectively, using a matrix consisting of 5% w/v C$_{16}$-modified HEC in NMF. Current: 7 μA. (B) Separation of p(dA)$_{12-18}$ and p(dA)$_{40-60}$, 1 and 2 μg/mL, respectively, in 5% w/v C$_{16}$-modified HEC in NMF with 50 mM CH$_3$CO$_2$NH$_4$. Current: 10 μA. (C) Separation of p(dA)$_{12-18}$ and p(dA)$_{40-60}$, 2 and 4 μg/mL, respectively, in 5% w/v C$_{16}$-modified HEC in NMF with 50 mM CH$_3$CO$_2$NH$_4$. Current: 4–8 μA. Separations (A) and (B) were performed at room temperature and (c) at approximately 98°C. The plate number and the resolution calculated from the p(dA) 40 peak and the p(dA) 40 and 41 peaks were in (A) 440,000 and 1.4, (B) 590,000 and 2.4, and (C) 1,010,000 and 3.6, respectively. (From Sjodahl, J., Lindberg, P., and Roeraade, J., *J. Sep. Sci.*, 30, 104, 2007. With permission.)

using F127 as a sieving material in CE, a 25% (w/v) solution was chosen to test the sieving ability, resulting in a fairly good separation with all peaks, except for the 8 and 10 bases, well resolved in a 3 cm channel in less than 4.5 min. However, when the effective separation length was shortened to 1.5 cm, it was not able to separate the oligonucleotide sizing markers. Meanwhile, when the concentration was increased to 30% (w/v), the oligonucleotide sizing markers were successfully

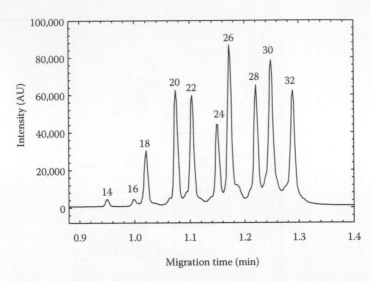

FIGURE 3.21 Ultrafast separation of oligonucleotide sizing marker by CE using F127 as separation medium and OliGreen as fluorescence-inducing agent under optimized conditions. Running conditions: 30% (w/v) F127 in 1× TBE; l (length to detector), 3 cm; L (column length), 6 cm; E (running voltage), 800 V/cm; capillary ID = 50 μm, OD = 360 μm; 1× TBE buffer, oligos were 10-fold diluted and injected at 300 V/cm for 1 s. OliGreen was 600-fold diluted and mixed with 500 μL cold F127 solution. (From Zhang, J., Liang, D., He, W., Wan, F., Ying, Q., and Chu, B., *Electrophoresis*, 26, 4449, 2005. With permission.)

separated within 1.3 min at a voltage of as high as 800 V/cm in 3 cm effective length, as shown in Figure 3.21.

3.4.1.2.2 B20-5000 ($E_{45}B_{14}E_{45}$)

$E_{45}B_{14}E_{45}$ (B20-5000) is another type of amphiphilic triblock copolymer, where E and B are oxyethylene (or ethylene oxide) and oxybutylene (or butylene oxide), respectively. The solubility of the E-block in water is much better than that of the B-block, and the temperature-dependent solubility of the B-block makes the triblock copolymer very easy to form micelles by self-assembly. As the B-block is more hydrophobic than the P-block in F127, B20-5000 has a lower critical micellar concentration (CMC) of about 6.6×10^{-4} g/mL and a smaller hydrodynamic radius in water [284]. At high concentrations, the micelles are densely packed, leading to the formation of a gel-like ordered structure. The sol-gel phase diagram of B20-5000 solution in 1× TBE buffer is shown in Figure 3.22A [285]. At polymer concentrations higher than 32% (w/v), there are two gel boundaries for each polymer concentration. At low temperatures, e.g., 10°C, and at a concentration of 32.5% (w/v), the solution is free flowing with a low viscosity of about 100 cP. Upon raising the temperature to ~30°C, the system will transform from the solution state to the gel-like state. The corresponding temperature dependence of the viscosity is shown in Figure 3.22B [285]. Further increasing the temperature to 50°C above the upper gel-to-sol transition boundary, the system returns to another low viscosity solution state. All these transitions are thermoreversible. Due to its smaller hydrophobic core, gelation occurs at a higher polymer concentration when compared with that of F127. The higher the

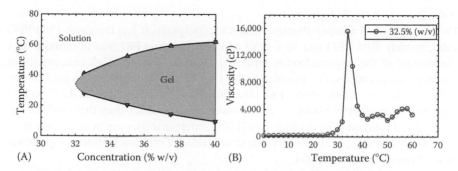

FIGURE 3.22 (A) Sol-gel phase diagram of B20-5000 ($E_{45}B_{14}E_{45}$) solution in 1× TBE buffer; (B) temperature dependence of the viscosity of 32.5% (w/v) B20-5000 solution in 1× TBE buffer. (From Zhang, J., Burger, C., and Chu, B., *Electrophoresis*, 27, 3391, 2006. With permission.)

polymer concentration, the broader the gel region in the investigated concentration ranges. Between 32% and 40% (w/v) and 10°C and 60°C, the B20-5000/1× TBE system has a body centered cubic (BCC) structure.

The application of the unique temperature-dependent viscosity-adjustable property of B20-5000 for oligonucleotide separations by microchip-based CE is demonstrated in Figure 3.23, using 35% (w/v) B20-5000 in 1× TBE solution as the separation medium. All fragments in the oligonucleotide sizing marker were well separated except that the peaks of 8, 10, and 12 bases were not detected. In another experiment, it was shown that the gel-like state was crucial to the oligonucleotide separation, by using a solution of B20-5000 at a concentration of 32.5% (w/v), which showed a better separation from 10 to 32 bases at 34°C, above the sol-gel transition point than that at 25°C, belonging to the solution state.

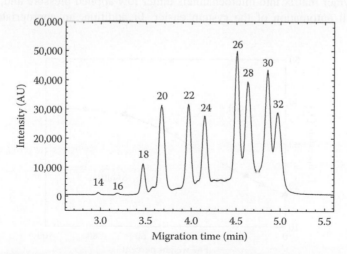

FIGURE 3.23 Electropherogram of oligonucleotide sizing marker using 35% (w/v) B20-5000 (E45B14E45) in 1× TBE buffer. Conditions: *l* (length to detector), 1.5 cm; *L* (column length), 6 cm; *E* (voltage), 200 V/cm; capillary ID = 75 μm, OD = 360 μm; electrokinetic injection at 300 V/cm for 1 s. Peak assignments are as indicated in the figure. (From Zhang, J., Burger, C., and Chu, B., *Electrophoresis*, 27, 3391, 2006. With permission.)

3.4.1.2.3 F87 ($E_{61}P_{40}E_{61}$)

F87 ($E_{61}P_{40}E_{61}$) is another Pluronic triblock copolymer. It has the same PEO/PPO ratio, namely 70% PEO and 30% PPO as F127. Pluronic F87 was introduced as a component of the oligonucleotide separation matrix [286]. At high concentrations and high temperatures, F87 micelles in the 1× TBE buffer were densely packed into a BCC structure. A 30% (w/v) F87 solution in the 1× TBE buffer at 42°C had an aggregation number of about 57, which was slightly smaller than that (~65) of the 21.2% F127 solution in its gel-like state [140]. Unfortunately, even in its gel-like state (~42°C), 30% F87 solution was not able to separate the oligonucleotide sizing marker in a 1.5 cm separation channel.

3.4.1.2.4 F87 and F127 Solution Mixtures

The advantages of using a mixture of two block copolymers instead of a single copolymer are that the DNA separation resolution is highly sensitive to the copolymer block lengths. In addition, the sol-gel phase transition temperature can be tuned by changing the weight fractions of the two constituent copolymers. It is not practical to synthesize an optimal block copolymer to meet each specification in practical applications. A series of mixed solutions were prepared using different F87/F127 weight ratios at a fixed concentration of 30% (w/v), the optimal concentration found for oligonucleotide separation [287]. The sol-gel phase diagram of 30% (w/v) F87/F127 mixture solution at different weight fractions in 1× TBE buffer is shown in Figure 3.24. It was observed that the phase transition temperature of Pluronics mixture solutions increased with increasing F87 content at the fixed concentration of 30% (w/v). The sol-gel transition temperature could be controlled over a wide range (from ~18°C to ~38°C) by varying the copolymer weight fractions. Note that all sol-gel transitions are thermoreversible. This property is helpful for filling capillary columns. In microchip electrophoresis, a polymer solution with low viscosity will allow the rapid loading of polymer matrix into microchannels under low applied pressure and, thereby, making full automation of the system easier. In addition, an appropriate sol-gel

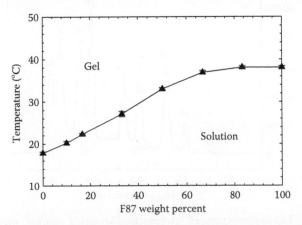

FIGURE 3.24 Sol-gel phase diagram of 30% (w/v) F87/F127 mixture solutions at different weight ratios in 1× TBE buffer. (Redrawn from data in Zhang, J., Gassmann, M., He, W.D., Wan, F., and Chu, B., *Lab Chip*, 6, 526, 2006. With permission.)

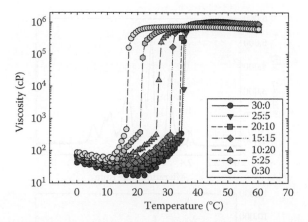

FIGURE 3.25 Temperature dependence of apparent viscosity of 30% (w/v) F87/F127 mixture solution at different weight ratios in 1× TBE buffer. (From Zhang, J., Gassmann, M., Chen, X., Burger, C., Rong, L., Ying, Q., and Chu, B., *Macromolecules* (Washington, DC), 40, 5537, 2007. With permission.)

transition temperature will also facilitate the solution filling process and help prevent the dye and oligonucleotides from being destabilized at high temperatures.

As shown in Figure 3.25 [288], the viscosity of different F87/F127 mixture solutions was less than 100 cP at low temperatures, e.g., at ~10°C. It was almost two orders of magnitude lower than those of the conventional polyacrylamide and poly(*N,N*-dimethylacrylamide) solution matrices used for DNA separations [289]. The separation of oligonucleotide sizing markers with 30% (w/v) F87/F127 matrix at a weight ratio of 1:2 in different phase regions is shown in Figure 3.26. At room temperature (~22°C), this system was a concentrated polymer solution in the solution state. The oligonucleotide separation ability of the solution in its solution state was poor, as shown in Figure 3.26A. By increasing the temperature of the 30% (w/v) F87/F127 mixture solution to about 40°C, a gel-like structure was formed, (Figure 3.26B), resulting in good separation for all fragments from 14 to 32 bases. When the temperature was elevated, the hydrophobicity of PPO was increased, the micellar number density became relatively higher, and the average distance between two neighboring micelles would then decrease. Hence, the entanglements in the overlapped micellar shells would increase, resulting in a reduction in the average mesh size which would then favor the separation of smaller DNA molecules, such as oligonucleotides.

The easy-to-tune sol-gel transition temperature of Pluronic mixture solutions could meet the requirements of many separation systems running under different operating conditions. The sieving ability of various Pluronic mixture solutions at different weight ratios was tested in their gel-like state. The 30% (w/v) F87/F127 mixture solution was found to be able to serve as an effective sieving matrix for oligonucleotide separation over a relatively broad region with F87/F127 weight ratios being varied from 5:25 to 2:1 in a 1.5 cm separation channel. After comparison, the 30% F87/F127 mixture solution with a weight ratio of 1:2 was found to have close to the best sieving ability for oligonucleotide separations. Fast separation of oligonucleotides could be achieved upon a few optimizations of running parameters (Figure 3.27),

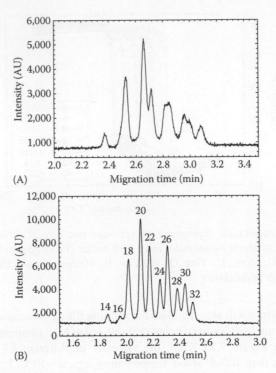

(A)

(B)

FIGURE 3.26 Electropherograms of oligonucleotide sizing marker by 30% (w/v) F87/F127 mixture solution at a weight ratio of 1:2 in 1× TBE buffer at (A) room temperature (~22°C) (solution state) and (B) 42°C (gel-like state). Conditions: *l* (length to detector), 1.5 cm; *L* (column length), 6 cm; *E* (electric field), 200 V/cm; capillary ID = 75 μm, OD = 360 μm; 1× TBE buffer, DNA electrokinetic injection at 300 V/cm for 1 s. (From Zhang, J., Gassmann, M., He, W.D., Wan, F., and Chu, B., *Lab Chip*, 6, 526, 2006. With permission.)

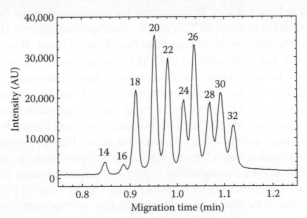

FIGURE 3.27 Fast separation of oligonucleotide sizing marker by using 30% (w/v) F87/F127 mixture solution (weight ratio of F87/F127 = 1:2) under optimized conditions. Conditions: *l* (length to detector), 1.5; *L* (column length), 6 cm; *E* (electric field), 400 V/cm; capillary ID = 75 μm, OD = 360 μm; 1× TBE buffer, DNA electrokinetic injection at 300 V/cm for 1 s. Peak assignments are as indicated in the figure. (From Zhang, J., Gassmann, M., He, W.D., Wan, F., and Chu, B., *Lab Chip*, 6, 526, 2006. With permission.)

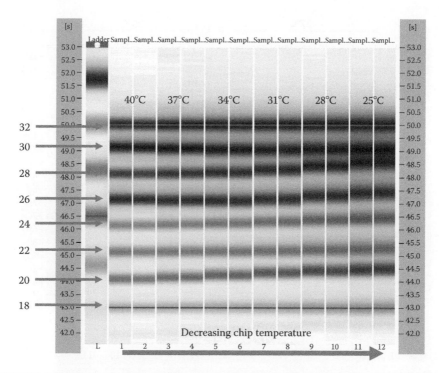

FIGURE 3.28 (See color insert following page 216.) Aligned gel-view mode of oligonucleotide separation by 30% (w/v) F87/F127 mixture solution at a weight ratio of 1:2 in 1× TBE buffer and different temperatures performed in the Agilent Bioanalyzer 2100 system with an electric field strength of 325 V/cm. (From Zhang, J., Gassmann, M., He, W.D., Wan, F., and Chu, B., *Lab Chip*, 6, 526, 2006. With permission.)

with all fragments baseline separated in 70s under an electric field of 400 V/cm. Moreover, by increasing the voltage further up to 600 V/cm, the oligonucleotide separation could be completed within 40 s at the cost of a slight degradation in resolution.

The performance for oligonucleotide separation was also tested in a commercial microchip electrophoresis instrument, known as the Agilent Bioanalyzer 2100. A 30% (w/v) F87/F127 mixture solution with a weight ratio of 1:2 was much superior in separating oligos in the size range of 8–32 bases, compared with commercial available DNA separation assay kits, such as DNA 500 assay kit. Figure 3.28 shows representative electropherograms in the aligned gel-view mode, illustrating the separation of oligonucleotide sizing markers over temperatures ranging from 25°C to 40°C at a fixed voltage of 325 V/cm. It can be seen that in the vicinity of the sol-gel transition region (25°C), the polymer mixture solution already exhibited relatively good separation ability for smaller fragments. However, the resolution for larger oligonucleotide fragments was not good, as 28-mer and 30-mer were only half separated. With temperature increased above the gel point, separation was improved, especially for the larger oligonucleotide fragments. Besides, higher temperature also caused the oligonucleotides to migrate faster, as seen in Figure 3.28.

The running voltage effect on oligonucleotide separation was also investigated on the Agilent Bioanalyzer 2100 system. The separations were performed at different

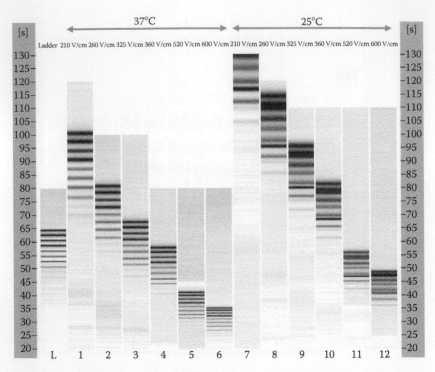

FIGURE 3.29 Unaligned gel-view mode of oligonucleotide separation at different electric field strengths (indicated) by using 30% (w/v) F87/F127 mixture solution at a weight ratio of 1:2 in 1× TBE buffer at 25°C and 37°C performed in Agilent Bioanalyzer 2100 system. (From Zhang, J., Gassmann, M., He, W.D., Wan, F., and Chu, B., *Lab Chip*, 6, 526, 2006. With permission.)

electric field strengths ranging from 210 to 600 V/cm at 25°C and 37°C, respectively, with the same separation medium, as shown in Figure 3.29 in the gel-view mode. It can be seen that all oligonucleotide fragments were baseline separated. It is noted that higher running voltages result in faster separation without the loss of high resolution at both 25°C and 37°C, although the resolution is worse at 25°C at each tested electric field. By increasing the running voltage, the resolution was improved mainly due to a decrease in diffusional band broadening. Meanwhile, the lower dielectric constant at higher polymer concentration allowed the application of higher voltages without an excessive Joule heating effect and the band broadening by the thermal gradient was effectively suppressed. In fact, oligonucleotide sizing markers were separated within 40 s at 600 V/cm. To our knowledge, it is the fastest capillary gel electrophoresis separation with high resolution in a 1.5 cm separation length reported to date for oligonucleotides in the similar size range.

3.4.1.2.5 Possible Separation Mechanism for Thermoreversible Copolymers

There are at least four qualitatively different domains in the triblock copolymer solutions introduced above, namely the condensed P or B micellar cores, the hydrated E micellar shells, the entangled E chains in the overlapping micellar shells between the closest micelles, and the water-rich interstitial gaps between micelles. Highly charged oligonucleotides should avoid hydrophobic micelle cores and migrate through hydrated PEO and interstitial domains [290]. The center-to-center distance between the nearest neighbor micelles in the BCC structure (B20-5000) was about 10.7 nm, which was much smaller than that in the FCC structure (F127) (19.8 nm). Migration of oligonucleotides through the former regions would encounter more resistance because the movements were restricted within a smaller space. Therefore, oligonucleotides could be effectively separated in a shorter separation channel in B20-5000 (1.5 cm) than F127 (3 cm).

3.4.2 OTHER SPECIFIC FEATURES IN OLIGO SEPARATION

Besides the viscosity issue, we would like to mention three particular features used in oligo separation, when compared with DNA sequencing. First, the introduction of OliGreen enhanced the sensitivity in DNA separation in comparison with conventional ethidium bromide due to its minimal fluorescence in the unbound form. Reyderman and Stavchansky successfully applied the usage of OliGreen in the determination of single-stranded oligonucleotides by capillary gel electrophoresis and quantitatively determined the short single-stranded oligonucleotides from blood plasma [291,292]. A new dye injection method was developed by mixing diluted dye solution with the polymer solution to enhance the signal-to-noise ratio (S/N). Under optimized conditions (a concentration of 10 μL 600-fold dye dilution was chosen as the amount to mix with 500 μL polymer solution), OliGreen was about six times more sensitive on average than ethidium bromide.

Secondly, the requirement of denaturant is different from DNA sequencing. As we know, urea is a commonly used denaturant in DNA sequencing analysis. The role of urea in the running buffer is to keep DNA fragments in the denatured state. Lack of denaturant in the system will result in band broadening and poor resolution of fragment peaks. However, things are different in the case of oligos, partially because the separation is completed in a much shorter time and thus the double-strand formation can be negligible. Careful investigations at Stony Brook found that the resolution obtained by the separation medium with urea was not much better than that by the medium without urea while the sensitivity was greatly decreased. Reyderman and Stavchansky conducted spectrofluorometric studies to investigate the kinetics of the fluorescent complex formation [291]. The results showed that the addition of urea drastically decreased the fluorescence of the oligo–dye complex when compared with that in 1× TBE buffer without urea (Figure 3.30), in agreement with our results. Thus, the separation medium without urea is suggested for the oligonucleotide separation.

Finally, rapid separation is inherently required by chip-based oligo separation. Due to the short separation column and low dielectric constant of the separation media, such as F127 and B20-5000 at high polymer concentrations, a higher voltage could be applied on the separation channels without external cooling [293], which could decrease diffusional band broadening and favor the small fragment separation. As discussed above, 800 V/cm could be applied for oligo size marker separations. On the contrary, the applied electrical field for DNA sequencing is somehow limited to 300 V/cm before a dramatic reduction in resolution, due to the much longer channel length.

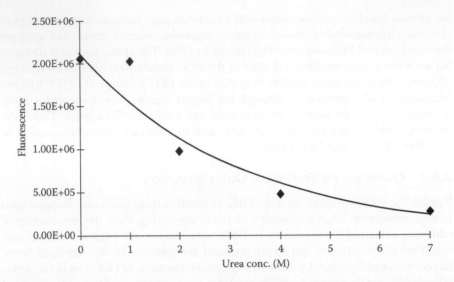

FIGURE 3.30 Effect of different urea concentrations in the Tris–borate buffer on the fluorescence of the oligo–dye complex; 1:200 dye dilution; 1.52 μg/mL oligo concentration. Complex fluorescence was corrected for the dye fluorescence in the unbound state. (From Reyderman, L. and Stavchansky, S., *J. Chromatogr. A*, 755, 271, 1996. With permission.)

3.5 CONCLUSION

The accomplishment of the human genome project was aided by the rapid development of DNA sequencing technologies. A tremendous amount of work has been performed to optimize sequencing related issues in terms of cost, read length, and throughput. In particular, sequencing matrices are one of the vital areas to be investigated extensively in order to obtain an ideal polymer matrix with high sieving ability, low viscosity, and good dynamic coating ability. In the meantime, automation, miniaturization, and integration of the whole process, including DNA isolation, sample preparation, sequencing, and analysis is another flourishing field in order to scale up DNA sequencing. Moreover, microfluidic devices have become the preferred format for the future. Though it is in the inception stage, it is potentially very promising to reduce the DNA sequencing cost further by a factor of 5–10.

At the same time, non-Sanger-based technologies are being investigated, with the potential to increase the separation rate significantly, as has been demonstrated on short DNA fragments, such as PCR products, oligonucleotides or other DNA molecules in small size (less than 100 bases). However, they are just in their proof of concept stage and suffer from a variety of problems that have to be solved before they can be used by the community in a robust and proven fashion. Under the circumstances, capillary-based electrophoresis technologies will continue to dominate the market for sequencing large genomes or other applications that require high accuracy although it may be worthwhile to combine CAE with novel nonelectrophoresis approaches to achieve the ultimate goal of NIH [2] in the future [294].

Oligonucleotide separation is another attractive application in the biomedical field. The challenge of separating oligos by CE or microchip electrophoresis comes from the fact that a much smaller mesh size is inherently needed. In other words, a higher polymer concentration will in turn increase the polymer solution viscosity, leading to difficulty in matrix loading. Several thermoreversible triblock copolymers were introduced and a fast effective separation of oligonucleotides was achieved by using a mixture of F127/F87 in a 1.5 cm long microchannel.

ACKNOWLEDGMENTS

Benjamin Chu gratefully acknowledges the support of the National Science Foundation, Polymers Program (DMR 0454887) in completing this work and the National Human Genome Research Institute (NIH 2R01HG01386-09) for the initial research.

REFERENCES

1. Collins, F.S., Morgan, M., and Patrinos, A. 2003. The human genome project: Lessons from large-scale biology. *Science* (New York) 300:286–290.
2. Collins, F.S., Green, E.D., Guttmacher, A.E., and Guyer, M.S. 2003. A vision for the future of genomics research. *Nature* (London, UK) 422:835–847.
3. Fredlake, C.P., Hert, D.G., Mardis, E.R., and Barron, A.E. 2006. What is the future of electrophoresis in large-scale genomic sequencing? *Electrophoresis* 27:3689–3702.
4. Liu, C.N., Toriello, N.M., Maboudian, R., and Mathies, R.A. 2004. High-throughput polymerase chain reaction—capillary array electrophoresis (PCR-CAE) microchip. *Special Publication—Royal Society of Chemistry* 297:297–299.
5. Olson, N.A., Khandurina, J., and Guttman, A. 2004. DNA profiling by capillary array electrophoresis with non-covalent fluorescent labeling. *Journal of Chromatography, A* 1051:155–160.
6. Emrich, C.A., Medintz, I.L., Tian, H., Berti, L., and Mathies, R.A. 2001. (Ultra)2-high throughput genetic analysis using microfabricated capillary array electrophoresis devices. *Micro Total Analysis Systems 2001, Proceedings mTAS 2001 Symposium*, 5th, Monterey, CA, Oct. 21–25, 2001:13–15.
7. Medintz, I.L., Paegel, B.M., and Mathies, R.A. 2001. Microfabricated capillary array electrophoresis DNA analysis systems. *Journal of Chromatography, A* 924:265–270.
8. Kiba, Y., Ueda, M., Abe, H., Arai, A., Nakanishi, H., Tabata, O., and Baba, Y. 1999. DNA analysis by microfabricated capillary electrophoresis device. *Nucleic Acids Symposium Series* 42:57–58.
9. Simpson, P., Woolley, A., Thorsen, T., Sensabaugh, G.F., and Mathies, R.A. 1998. High throughput DNA genotyping on capillary array electrophoresis chips. *International Congress Series* 1167:3–5.
10. Mansfield, E.S., Robertson, J.M., Vainer, M., Isenberg, A.R., Frazier, R.R., Ferguson, K., Chow, S., Harris, D.W., Barker, D.L., Gill, P.D., Budowle, B., and McCord, B.R. 1998. Analysis of multiplexed short tandem repeat (STR) systems using capillary array electrophoresis. *Electrophoresis* 19:101–107.
11. Sonehara, T., Kawazoe, H., Sakai, T., Ozawa, S., Anazawa, T., and Irie, T. 2006. Ultra-slim laminated capillary array for high-speed DNA separation. *Electrophoresis* 27:2910–2916.
12. Desmond, S.M., Shigeura, J.S., and Shigeura, J.S. 2006. Capillary array assembly for loading samples from a device, and method, 2004–887477:14 pp (Applera Corporation, USA). Application: US.

13. Kasai, S., Tsukada, K., Kita, T., and Morioka, T. 2001. Capillary array, 2001–797698:8 pp (Hitachi, Ltd., Japan). Application: US.
14. Liu, S. 2001. Microfabricated injector and capillary array assembly for high-resolution and high throughput separations of DNA for sequence analysis, 2000–US18134:41 pp (USA). Application: WO.
15. Clark, S.M. 1996. Multiplex DNA separations using bis-intercalating dyes and capillary array electrophoresis, Dissertation: 160 pp (Biochemical Methods, Berkeley, USA).
16. Madabhushi, R.S., Vainer, M., Dolnik, V., Enad, S., Barker, D.L., Harris, D.W., and Mansfield, E.S. 1997. Versatile low-viscosity sieving matrixes for nondenaturing DNA separations using capillary array electrophoresis. *Electrophoresis* 18:104–111.
17. Ueno, K. and Yeung, E.S. 1994. Simultaneous monitoring of DNA fragments separated by electrophoresis in a multiplexed array of 100 capillaries. *Analytical Chemistry* 66:1424–1431.
18. Clark, S.M. and Mathies, R.A. 1993. High-speed parallel separation of DNA restriction fragments using capillary array electrophoresis. *Analytical Biochemistry* 215:163–170.
19. Woolley, A.T., Hadley, D., Landre, P., deMello, A.J., Mathies, R.A., and Northrup, M.A. 1996. Functional integration of PCR amplification and capillary electrophoresis in a microfabricated DNA analysis device. *Analytical chemistry* 68:4081–4086.
20. Brahmasandra, S.N., Johnsona, B.N., Websterc, J.R., Burke, D.T., Mastrangeloc, C.H., and Burns, M.A. 1998. On-chip DNA band detection in microfabricated separation systems. Proceedings of SPIE—*The International Society for Optical Engineering* 3515:242–251.
21. Lee, G.-B., Chen, S.-H., Huang, G.-R., Lin, Y.-H., and Sung, W.-C. 2000. Microfabricated plastic chips by hot embossing methods and their applications for DNA separation and detection. *Proceedings of SPIE—The International Society for Optical Engineering* 4177:112–121.
22. Mueller, O., Hahnenberger, K., Dittmann, M., Yee, H., Dubrow, R., Nagle, R., and Ilsley, D. 2000. A microfluidic system for high-speed reproducible DNA sizing and quantitation. *Electrophoresis* 21:128–134.
23. Lee, G.B., Chen, S.H., Huang, G.R., Sung, W.C., and Lin, Y.H. 2001. Microfabricated plastic chips by hot embossing methods and their applications for DNA separation and detection. *Sensors and Actuators, B: Chemical* B75:142–148.
24. Lee, G.-B., Chen, S.-H., Lin, C.-S., Huang, G.-R., and Lin, Y.-H. 2001. Microfabricated electrophoresis chips on quartz substrates and their applications on DNA analysis. *Journal of the Chinese Chemical Society* (Taipei, Taiwan) 48:1123–1128.
25. Smith, E.M., Xu, H., and Ewing, A.G. 2001. DNA separations in microfabricated devices with automated capillary sample introduction. *Electrophoresis* 22:363–370.
26. Xie, W., Yang, R., Xu, J., Zhang, L., Xing, W., and Cheng, J. 2001. Microchip-based capillary electrophoresis systems. *Methods in Molecular Biology* (Totowa, NJ) 162:67–83.
27. Chen, Z. and Burns, M.A. 2005. Effect of buffer flow on DNA separation in a microfabricated electrophoresis system. *Electrophoresis* 26:4718–4728.
28. Ye, M.-Y., Yin, X.-F., and Fang, Z.-L. 2005. DNA separation with low-viscosity sieving matrix on microfabricated polycarbonate microfluidic chips. *Analytical and Bioanalytical Chemistry* 381:820–827.
29. Ronaghi, M., Uhlen, M., and Nyren, P. 1998. A sequencing method based on real-time pyrophosphate. *Science* (Washington, DC) 281:363, 365.
30. Hyman, E.D. 1988. A new method of sequencing DNA. *Analytical Biochemistry* 174:423–436.
31. Margulies, M., Egholm, M., Altman, W.E., Attiya, S., Bader, J.S., Bemben, L.A., Berka, J., Braverman, M.S., Chen, Y.-J., Chen, Z., Dewell, S.B., Du, L., Fierro, J.M., Gomes, X.V., Godwin, B.C., He, W., Helgesen, S., Ho, C.H., Irzyk, G.P., Jando, S.C., Alenquer,

M.L.I., Jarvie, T.P., Jirage, K.B., Kim, J.-B., Knight, J.R., Lanza, J.R., Leamon, J.H., Lefkowitz, S.M., Lei, M., Li, J., Lohman, K.L., Lu, H., Makhijani, V.B., McDade, K.E., McKenna, M.P., Myers, E.W., Nickerson, E., Nobile, J.R., Plant, R., Puc, B.P., Ronan, M.T., Roth, G.T., Sarkis, G.J., Simons, J.F., Simpson, J.W., Srinivasan, M., Tartaro, K.R., Tomasz, A., Vogt, K.A., Volkmer, G.A., Wang, S.H., Wang, Y., Weiner, M.P., Yu, P., Begley, R.F., and Rothberg, J.M. 2005. Genome sequencing in microfabricated high-density picolitre reactors. *Nature* (London, UK) 437:376–380.

32. Gharizadeh, B., Nordstrom, T., Ahmadian, A., Ronaghi, M., and Nyren, P. 2002. Long-read pyrosequencing using pure 2'-deoxyadenosine-5'-O'-(1-thiotriphosphate) Sp-isomer. *Analytical Biochemistry* 301:82–90.

33. Bonetta, L. 2006. Genome sequencing in the fast lane. *Nature Methods* 3:141–147.

34. Slater, G.W., Kenward, M., McCormick, L.C., and Gauthier, M.G. 2003. The theory of DNA separation by capillary electrophoresis. *Current Opinion in Biotechnology* 14:58–64.

35. Semenov, A.N., Duke, T.A.J., and Viovy, J.L. 1995. Gel electrophoresis of DNA in moderate fields: The effect of fluctuations. *Physical Review E: Statistical Physics, Plasmas, Fluids, and Related Interdisciplinary Topics* 51:1520–1537.

36. Zhou, H., Miller, A.W., Sosic, Z., Buchholz, B., Barron, A.E., Kotler, L., and Karger, B.L. 2000. DNA sequencing up to 1300 bases in two hours by capillary electrophoresis with mixed replaceable linear polyacrylamide solutions. *Analytical Chemistry* 72:1045–1052.

37. Barron, A.E., Blanch, H.W., and Soane, D.S. 1994. A transient entanglement coupling mechanism for DNA separation by capillary electrophoresis in ultradilute polymer solutions. *Electrophoresis* 15:597–615.

38. Albarghouthi, M.N. and Barron, A.E. 2000. Polymeric matrices for DNA sequencing by capillary electrophoresis. *Electrophoresis* 21:4096–4111.

39. Starkweather, M.E., Hoagland, D.A., and Muthukumar, M. 2000. Polyelectrolyte electrophoresis in a dilute solution of neutral polymers: Model studies, 33:1245–1253.

40. Todor, I. and Todorov, O.d.C.N.G.W.M.D.M. 2001. Capillary electrophoresis of RNA in dilute and semidilute polymer solutions, 22:2442–2447.

41. Viovy, J.-L. 2000. Electrophoresis of DNA and other polyelectrolytes: Physical mechanisms. *Reviews of Modern Physics* 72:813–872.

42. Franca, L.T.C., Carrilho, E., and Kist, T.B.L. 2002. A review of DNA sequencing techniques. *Quarterly Reviews of Biophysics* 35:169–200.

43. Wang, Y., Liang, D., Hao, J., Fang, D., and Chu, B. 2002. Separation of double-stranded DNA fragments by capillary electrophoresis using polyvinylpyrrolidone and poly(N,N-dimethylacrylamide) transient interpenetrating network. *Electrophoresis* 23:1460–1466.

44. Cohen, A.S., Najarian, D.R., Paulus, A., Guttman, A., Smith, J.A., and Karger, B.L. 1988. Rapid separation and purification of oligonucleotides by high-performance capillary gel electrophoresis. *Proceedings of the National Academy of Sciences of the United States of America* 85:9660–9663.

45. Carson, S., Cohen, A.S., Belenkii, A., Ruiz-Martinez, M.C., Berka, J., and Karger, B.L. 1993. DNA sequencing by capillary electrophoresis: Use of a two-laser-two-window intensified diode array detection system. *Analytical Chemistry* 65:3219–3226.

46. Zhu, M., Hansen, D.L., Burd, S., and Gannon, F. 1989. Factors affecting free zone electrophoresis and isoelectric focusing in capillary electrophoresis. *Journal of Chromatography* 480:311–319.

47. De Gennes, P.G. 1979. *Scaling Concepts in Polymer Physics*. Cornell University Press, Ithaca, NY.

48. Kleparnik, K., Mala, Z., and Bocek, P. 2001. Fast separation of DNA sequencing fragments in highly alkaline solutions of linear polyacrylamide using electrophoresis in bare silica capillaries. *Electrophoresis* 22:783–788.

49. Yan, J., Best, N., Zhang, J.Z., Ren, H., Jiang, R., Hou, J., and Dovichi, N.J. 1996. The limiting mobility of DNA sequencing fragments for both cross-linked and non-cross-linked polymers in capillary electrophoresis: DNA sequencing at 1200 V cm−1. *Electrophoresis* 17:1037–1045.

50. Kamahori, M. and Kambara, H. 1996. Characteristics of single-stranded DNA separation by capillary gel electrophoresis. *Electrophoresis* 17:1476–1484.

51. Carrilho, E., Ruiz-Martinez, M.C., Berka, J., Smirnov, I., Goetzinger, W., Miller, A.W., Brady, D., and Karger, B.L. 1996. Rapid DNA sequencing of more than 1000 bases per run by capillary electrophoresis using replaceable linear polyacrylamide solutions. *Analytical Chemistry* 68:3305–3313.

52. Ruiz-Martinez, M.C., Berka, J., Belenkii, A., Foret, F., Miller, A.W., and Karger, B.L. 1993. DNA sequencing by capillary electrophoresis with replaceable linear polyacrylamide and laser-induced fluorescence detection. *Analytical Chemistry* 65:2851–2858.

53. Wu, C., Quesada, M.A., Schneider, D.K., Farinato, R., Studier, F.W., and Chu, B. 1996. Polyacrylamide solutions for DNA sequencing by capillary electrophoresis: Mesh sizes, separation and dispersion. *Electrophoresis* 17:1103–1109.

54. Wan, F., He, W., Zhang, J., Ying, Q., and Chu, B. 2006. Scale-up development of high-performance polymer matrix for DNA sequencing analysis. *Electrophoresis* 27:3712–3723.

55. Liang, D., Song, L., Chen, Z., and Chu, B. 2001. Effect of glycerol-induced DNA conformational change on the separation of DNA fragments by capillary electrophoresis. *Journal of Chromatography, A* 931:163–173.

56. Shi, W., Zhang, Y., Chen, N., Zhang, J., Wang, L., and Zou, H. 1998. Preparation and characterization of high performance linear polyacrylamide capillary gel electrophoresis columns and study of the experimental conditions. *Fenxi Huaxue* 26:1–6.

57. Baab, B.B. and Mathies, R.A. 1997. Optimization of single-molecule fluorescent burst detection of ds-DNA: Application to capillary electrophoresis separations of 100–1000 basepair fragments. *Applied Spectroscopy* 51:1579–1584.

58. Ueda, M. and Baba, Y. 1997. Is higher concentration of linear polyacrylamide not suitable for the capillary electrophoretic separation of large DNA? *Analytical Sciences* 13:109–112.

59. Kleparnik, K., Foret, F., Berka, J., Goetzinger, W., Miller, A.W., and Karger, B.L. 1996. The use of elevated column temperature to extend DNA sequencing read lengths in capillary electrophoresis with replaceable polymer matrixes. *Electrophoresis* 17:1860–1866.

60. Barron, A.E., Sunada, W.M., and Blanch, H.W. 1996. The effects of polymer properties on DNA separations by capillary electrophoresis in uncross-linked polymer solutions. *Electrophoresis* 17:744–757.

61. Chen, N., Manabe, T., Terabe, S., Yohda, M., and Endo, I. 1994. High-resolution separation of oligonucleotides and DNA sequencing reaction products by capillary electrophoresis with linear polyacrylamide and laser-induced fluorescence detection. *Journal of Microcolumn Separations* 6:539–543.

62. Heller, C. and Viovy, J.L. 1994. Electrophoretic separation of oligonucleotides in replenishable polyacrylamide-filled capillaries. *Applied and Theoretical Electrophoresis* 4:39–41.

63. Manabe, T., Chen, N., Terabe, S., Yohda, M., and Endo, I. 1994. Effects of linear polyacrylamide concentrations and applied voltages on the separation of oligonucleotides and DNA sequencing fragments by capillary electrophoresis. *Analytical Chemistry* 66:4243–4252.

64. Pariat, Y.F., Berka, J., Heiger, D.N., Schmitt, T., Vilenchik, M., Cohen, A.S., Foret, F., and Karger, B.L. 1993. Separation of DNA fragments by capillary electrophoresis using replaceable linear polyacrylamide matrixes. *Journal of Chromatography* 652:57–66.

65. Mills, N.C. and Ilan, J. 1985. Mechanical stabilization of large pore acrylamide gels by addition of linear polyacrylamide. *Electrophoresis* 6:531–534.

66. Jeppesen, P.G.N. 1980. Separation and isolation of DNA fragments using linear polyacrylamide gradient gel electrophoresis. *Methods in Enzymology* 65:305–319.
67. Quesada, M.A. 1997. Replaceable polymers in DNA sequencing by capillary electrophoresis. *Current Opinion in Biotechnology* 8:82–93.
68. Xiong, Y., Park, S.R., and Swerdlow, H. 1998. Base stacking: pH-mediated on-column sample concentration for capillary DNA sequencing. *Analytical Chemistry* 70:3605–3611.
69. Madabhushi, R.S. 1998. Separation of 4-color DNA sequencing extension products in noncovalently coated capillaries using low viscosity polymer solutions. *Electrophoresis* 19:224–230.
70. Rosenblum, B.B., Oaks, F., Menchen, S., and Johnson, B. 1997. Improved single-strand DNA sizing accuracy in capillary electrophoresis. *Nucleic Acids Research* 25:3925–3929.
71. Zhang, J., Wang, Y., Liang, D., Ying, Q., and Chu, B. 2004. Graft copolymers of PDMA-g-PMMA: Solvent quality induced association behavior and its application to dsDNA separation. *Polymer Preprints* (American Chemical Society, Division of Polymer Chemistry) 45:816–817.
72. Liang, D. and Chu, B. 2002. Concentration gradient used in double-stranded DNA separation by capillary electrophoresis. *Electrophoresis* 23:2602–2609.
73. Chiari, M., Cretich, M., and Consonni, R. 2002. Separation of DNA fragments in hydroxylated poly(dimethylacrylamide) copolymers. *Electrophoresis* 23:536–541.
74. Liang, D., Song, L., Chen, Z., and Chu, B. 2001. Clay-enhanced DNA separation in low-molecular-weight poly(N,N-dimethylacrylamide) solution by capillary electrophoresis. *Electrophoresis* 22:1997–2003.
75. Heller, C. 1999. Separation of double-stranded and single-stranded DNA in polymer solutions. Part 1. Mobility and separation mechanism. *Electrophoresis* 20:1962–1977.
76. Fung, E.N. and Yeung, E.S. 1995. High-speed DNA sequencing by using mixed poly(ethylene oxide) solutions in uncoated capillary columns. *Analytical Chemistry* 67:1913–1919.
77. Lin, Y.-W. and Chang, H.-T. 2006. Analysis of double-stranded DNA by capillary electrophoresis using poly(ethylene oxide) in the presence of hexadecyltrimethylammonium bromide. *Journal of Chromatography, A* 1130:206–211.
78. Kuo, I.T., Chiu, T.-C., and Chang, H.-T. 2003. On-column concentration and separation of double-stranded DNA by gradient capillary electrophoresis. *Electrophoresis* 24:3339–3347.
79. Huang, M.-F., Huang, C.-C., and Chang, H.-T. 2003. Improved separation of double-stranded DNA fragments by capillary electrophoresis using poly(ethylene oxide) solution containing colloids. *Electrophoresis* 24:2896–2902.
80. Tseng, W.-L. and Chang, H.-T. 2001. A new strategy for optimizing sensitivity, speed, and resolution in capillary electrophoretic separation of DNA. *Electrophoresis* 22:763–770.
81. Tseng, W.L., Hsieh, M.M., Wang, S.J., and Chang, H.T. 2000. Effect of ionic strength, pH and polymer concentration on the separation of DNA fragments in the presence of electroosmotic flow. *Journal of Chromatography, A* 894:219–230.
82. Chen, H.-S. and Chang, H.-T. 1999. Electrophoretic separation of small DNA fragments in the presence of electroosmotic flow using poly(ethylene oxide) solutions. *Analytical Chemistry* 71:2033–2036.
83. Chen, H.-S. and Chang, H.-T. 1998. Capillary electrophoretic separation of 1 to 10 kbp sized dsDNA using poly(ethylene oxide) solutions in the presence of electroosmotic counterflow. *Electrophoresis* 19:3149–3153.
84. Fung, E., Pang, H.-M., and Yeung, E.S. 1998. Fast DNA separations using poly(ethylene oxide) in non-denaturing medium with temperature programming. *Journal of Chromatography, A* 806:157–164.

85. Kim, Y. and Yeung, E.S. 1997. Separation of DNA sequencing fragments up to 1000 bases by using poly(ethylene oxide)-filled capillary electrophoresis. *Journal of Chromatography, A* 781:315–325.

86. Chang, H.-T. and Yeung, E.S. 1995. Poly(ethyleneoxide) for high-resolution and high-speed separation of DNA by capillary electrophoresis. *Journal of Chromatography, B: Biomedical Applications* 669:113–123.

87. Wang, Q., Xu, X., and Dai, L. 2006. A new quasi-interpenetrating network formed by poly(N-acryloyl-tris(hydroxymethyl)aminomethane) and polyvinylpyrrolidone: Separation matrix for double-stranded DNA and single-stranded DNA fragments by capillary electrophoresis with UV detection. *Electrophoresis* 27:1749–1757.

88. Kim, D.-K. and Kang, S.H. 2005. On-channel base stacking in microchip capillary gel electrophoresis for high-sensitivity DNA fragment analysis. *Journal of Chromatography, A* 1064:121–127.

89. Yan, X., Hang, W., Majidi, V., Marrone, B.L., and Yoshida, T.M. 2002. Evaluation of different nucleic acid stains for sensitive double-stranded DNA analysis with capillary electrophoretic separation. *Journal of Chromatography, A* 943:275–285.

90. Pavski, V., Gao, D., and Yeung, E.S. 1998. Reusable uncoated capillary arrays for multiplexed DNA sequencing. Book of Abstracts, *216th ACS National Meeting*, Boston, August 23–27:ANYL-058.

91. Gao, Q. and Yeung, E.S. 1998. A matrix for DNA separation: Genotyping and sequencing using poly(vinylpyrrolidone) solution in uncoated capillaries. *Analytical Chemistry* 70:1382–1388.

92. Song, L., Liu, T., Liang, D., Fang, D., and Chu, B. 2001. Separation of double-stranded DNA fragments by capillary electrophoresis in interpenetrating networks of polyacrylamide and polyvinylpyrrolidone. *Electrophoresis* 22:3688–3698.

93. Jin, Y., Lin, B., and Fung, Y. 2001. Separation of deoxyribonucleic acid in glucose-hydroxy propylmethyl cellulose solution by capillary electrophoresis. *Fenxi Huaxue* 29:502–506.

94. Otim, O. 2001. The impact of urea on viscosity of hydroxethyl cellulose and observed mobility of deoxyribonucleic acids. *Biopolymers* 58:329–334.

95. Lin, W.-C., Wen, Y.-M., Hu, J.-Y., and Kuei, C.-H. 1999. Temperature effect on the separation of DNA fragments by capillary electrophoresis in hydroxyethyl cellulose solution. *Journal of Microcolumn Separations* 11:461–470.

96. Chen, H., Song, L.-G., Xiong, S.-X., and Cheng, J.-K. 1997. Separation of DNA fragments by capillary electrophoresis with non-gel sieving matrix of hydroxyethyl cellulose and sucrose. *Gaodeng Xuexiao Huaxue Xuebao* 18:1769–1773.

97. Hammond, R.W., Oana, H., Schwinefus, J.J., Bonadio, J., Levy, R.J., and Morris, M.D. 1997. Capillary electrophoresis of supercoiled and linear DNA in dilute hydroxyethyl cellulose solution. *Analytical Chemistry* 69:1192–1196.

98. Bashkin, J., Marsh, M., Barker, D., and Johnston, R. 1996. DNA sequencing by capillary electrophoresis with a hydroxyethylcellulose sieving buffer. *Applied and Theoretical Electrophoresis* 6:23–28.

99. Giovannoli, C., Anfossi, L., Tozzi, C., Giraudi, G., and Vanni, A. 2004. DNA separation by capillary electrophoresis with hydrophilic substituted celluloses as coating and sieving polymers. Application to the analysis of genetically modified meals. *Journal of Separation Science* 27:1551–1556.

100. Kang, D., Chung, D.S., Kang, S.H., and Kim, Y. 2005. Separation of DNA with hydroxypropylmethyl cellulose and poly(ethylene oxide) by capillary gel electrophoresis. *Microchemical Journal* 80:121–125.

101. Gelfi, C., Simo-Alfonso, E., Sebastiano, R., Citterio, A., and Righetti, P.G. 1996. Novel acrylamido monomers with higher hydrophilicity and improved hydrolytic stability: III. DNA separations by capillary electrophoresis in poly(N-acryloylaminopropanol). *Electrophoresis* 17:738–743.

102. Simo-Alfonso, E., Gelfi, C., Sebastiano, R., Citterio, A., and Righetti, P.G. 1996. Novel acrylamido monomers with higher hydrophilicity and improved hydrolytic stability: II. Properties of N-acryloylaminopropanol. *Electrophoresis* 17:732–737.
103. Simo-Alfonso, E., Gelfi, C., Sebastiano, R., Citterio, A., and Righetti, P.G. 1996. Novel acrylamido monomers with higher hydrophilicity and improved hydrolytic stability: I. Synthetic route and product characterization. *Electrophoresis* 17:723–731.
104. Menchen, S., Johnson, B., Winnik, M.A., and Xu, B. 1996. Flowable networks as DNA sequencing media in capillary columns. *Electrophoresis* 17:1451–1458.
105. Menchen, S., Johnson, B., Winnik, M.A., and Xu, B. 1996. Flowable networks as equilibrium DNA sequencing media in capillary columns. *Chemistry of Materials* 8:2205–2208.
106. Menchen, S., Johnson, B., Madabhushi, R., and Winnik, M. 1996. The design of separation media for DNA sequencing in capillaries. *Proceedings of SPIE—The International Society for Optical Engineering* 2680:294–303.
107. Heller, C. 1998. Finding a universal low-viscosity polymer for DNA separation. *Electrophoresis* 19:1691–1698.
108. Dolnik, V., Gurske, W.A., and Padua, A. 2001. Galactomannans as a sieving matrix in capillary electrophoresis. *Electrophoresis* 22:707–719.
109. Nashabeh, W. and El Rassi, Z. 1993. Fundamental and practical aspects of coupled capillaries for the control of electroosmotic flow in capillary zone electrophoresis of proteins. *Journal of Chromatography* 632:157–164.
110. Nashabeh, W. and Rassi, Z.E. 1992. Coupled fused silica capillaries for rapid capillary zone electrophoresis of proteins. *Journal of High Resolution Chromatography* 15:289–292.
111. Wu, C.T., Lopes, T., Patel, B., and Lee, C.S. 1992. Effect of direct control of electroosmosis on peptide and protein separations in capillary electrophoresis. *Analytical Chemistry* 64:886–891.
112. Lauer, H.H. and McManigill, D. 1986. Capillary zone electrophoresis of proteins in untreated fused silica tubing. *Analytical Chemistry* 58:166–170.
113. Salas-Solano, O., Carrilho, E., Kotler, L., Miller, A.W., Goetzinger, W., Sosic, Z., and Karger, B.L. 1998. Routine DNA sequencing of 1000 bases in less than one hour by capillary electrophoresis with replaceable linear polyacrylamide solutions. *Analytical Chemistry* 70:3996–4003.
114. Hjerten, S. 1985. High-performance electrophoresis. Elimination of electroendosmosis and solute adsorption. *Journal of Chromatography* 347:191–198.
115. Buchholz, B.A., Shi, W., and Barron, A.E. 2002. Microchannel DNA sequencing matrices with switchable viscosities. *Electrophoresis* 23:1398–1409.
116. Song, L., Liang, D., Fang, D., and Chu, B. 2001. Fast DNA sequencing up to 1000 bases by capillary electrophoresis using poly(N,N-dimethylacrylamide) as a separation medium. *Electrophoresis* 22:1987–1996.
117. Albarghouthi, M.N., Buchholz, B.A., Doherty, E.A.S., Bogdan, F.M., Zhou, H., and Barron, A.E. 2001. Impact of polymer hydrophobicity on the properties and performance of DNA sequencing matrices for capillary electrophoresis. *Electrophoresis* 22:737–747.
118. Kan, C.-W., Doherty, E.A.S., and Barron, A.E. 2003. A novel thermogelling matrix for microchannel DNA sequencing based on poly-N-alkoxyalkylacrylamide copolymers. *Electrophoresis* 24:4161–4169.
119. Bohdanecky, M. and Kovar, J. 1982. Viscosity of polymer solutions, *Polymer Science Library*, Vol. 2. Elsevier, New York.
120. Goetzinger, W., Kotler, L., Carrilho, E., Ruiz-Martinez, M.C., Salas-Solano, O., and Karger, B.L. 1998. Characterization of high molecular mass linear polyacrylamide powder prepared by emulsion polymerization as a replaceable polymer matrix for DNA sequencing by capillary electrophoresis. *Electrophoresis* 19:242–248.

121. Buchholz, B.A., Doherty, E.A.S., Albarghouthi, M.N., Bogdan, F.M., Zahn, J.M., and Barron, A.E. 2001. Microchannel DNA sequencing matrices with a thermally controlled "viscosity switch". *Analytical Chemistry* 73:157–164.

122. Wu, C., Liu, T., and Chu, B. 1998. Viscosity-adjustable block copolymer for DNA separation by capillary electrophoresis. *Electrophoresis* 19:231–241.

123. Liang, D., Song, L., Zhou, S., Zaitsev, V.S., and Chu, B. 1999. Poly(N-isopropylacrylamide)-g-poly(ethyleneoxide) for high resolution and high speed separation of DNA by capillary electrophoresis. *Electrophoresis* 20:2856–2863.

124. Kotler, L., He, H., Miller, A.W., and Karger, B.L. 2002. DNA sequencing of close to 1000 bases in 40 minutes by capillary electrophoresis using dimethyl sulfoxide and urea as denaturants in replaceable linear polyacrylamide solutions. *Electrophoresis* 23:3062–3070.

125. Doherty, E.A.S., Kan, C.-W., Paegel, B.M., Yeung, S.H.I., Cao, S., Mathies, R.A., and Barron, A.E. 2004. Sparsely cross-linked "nanogel" matrixes as fluid, mechanically stabilized polymer networks for high-throughput microchannel DNA sequencing. *Analytical Chemistry* 76:5249–5256.

126. Heller, C. 2000. Influence of electric field strength and capillary dimensions on the separation of DNA. *Electrophoresis* 21:593–602.

127. Heller, C. 1999. Separation of double-stranded and single-stranded DNA in polymer solutions. Part 2. Separation, peak width, and resolution. *Electrophoresis* 20:1978–1986.

128. Heller, C. 1998. Finding a universal low viscosity polymer for DNA separation (II). *Electrophoresis* 19:3114–3127.

129. Chiari, M., Micheletti, C., Nesi, M., Fazio, M., and Righetti, P.G. 1994. Towards new formulations for polyacrylamide matrixes: N-acryloylaminoethoxyethanol, a novel monomer combining high hydrophilicity with extreme hydrolytic stability. *Electrophoresis* 15:177–186.

130. Delnik, V. and Gurske, W.A. 1999. Capillary electrophoresis in sieving matrices. Selectivity per base, mobility slope, and inflection slope. *Electrophoresis* 20:3373–3380.

131. Liu, Y. and Kuhr, W.G. 1999. Separation of double- and single-stranded DNA restriction fragments: Capillary electrophoresis with polymer solutions under alkaline conditions. *Analytical Chemistry* 71:1668–1673.

132. Chu, B. and Liang, D. 2002. Copolymer solutions as separation media for DNA capillary electrophoresis. *Journal of Chromatography, A* 966:1–13.

133. Song, L., Liang, D., Chen, Z., Fang, D., and Chu, B. 2001. DNA sequencing by capillary electrophoresis using mixtures of polyacrylamide and poly(N,N-dimethylacrylamide). *Journal of Chromatography, A* 915:231–239.

134. Chu, B. and Liang, D. 2002. Copolymer solutions as separation media for DNA capillary electrophoresis. *Journal of Chromatography, A* 966:1–13.

135. Chu, B., Song, L., Fang, D., Liang, D., Liu, T., Wang, Y., and Ying, Q. 2005. Quasi-interpenetrating networks used as separation media, 2004–US30068:44 pp (The Research Foundation of Suny At Stony Brook, USA). Application: US.

136. Wang, Y., Liang, D., Ying, Q., and Chu, B. 2005. Quasi-interpenetrating network formed by polyacrylamide and poly(N,N-dimethylacrylamide) used in high-performance DNA sequencing analysis by capillary electrophoresis. *Electrophoresis* 26:126–136.

137. Zhou, D., Wang, Y., Zhang, W., Yang, R., and Shi, R. 2007. Novel quasi-interpenetrating network/gold nanoparticles composite matrices for DNA sequencing by CE. *Electrophoresis* 28:1072–1080.

138. Chiari, M., Damin, F., Melis, A., and Consonni, R. 1998. Separation of oligonucleotides and DNA fragments by capillary electrophoresis in dynamically and permanently coated capillaries, using a copolymer of acrylamide and beta-D-glucopyranoside as a new low viscosity matrix with high sieving capacity. *Electrophoresis* 19:3154–3159.

139. Chiari, M., Cretich, M., Riva, S., and Casali, M. 2001. Performances of new sugar-bearing poly(acrylamide) copolymers as DNA sieving matrices and capillary coatings for electrophoresis. *Electrophoresis* 22:699–706.
140. Wu, C., Liu, T., Chu, B., Schneider, D.K., and Graziano, V. 1997. Characterization of the PEO-PPO-PEO triblock copolymer and its application as a separation medium in capillary electrophoresis. *Macromolecules* 30:4574–4583.
141. Zhang, Y., Tan, H., and Yeung, E.S. 1999. Multiplexed automated DNA sequencing directly from single bacterial colonies. *Analytical Chemistry* 71:5018–5025.
142. Sudor, J., Barbier, V., Thirot, S., Godfrin, D., Hourdet, D., Millequant, M., Blanchard, J., and Viovy, J.-L. 2001. New block-copolymer thermo-associating matrices for DNA sequencing: Effect of molecular structure on rheology and resolution. *Electrophoresis* 22:720–728.
143. Barbier, V., Buchholz, B.A., Barron, A.E., and Viovy, J.-L. 2002. Comb-like copolymers as self-coating, low-viscosity and high-resolution matrices for DNA sequencing. *Electrophoresis* 23:1441–1449.
144. Viovy, J.-L. and Barbier, V. 2002. Block and comb copolymers for prevention of adsorption and/or elecroosmosis during separations, especially capillary electrophoresis of biomolecules, 2001–FR2117:66 pp (Institut Curie, Fr.; Centre National de la Recherche Scientifique (C.N.R.S.)). Application: WO.
145. Wu, C., Liu, T., and Chu, B. 1998. Viscosity-adjustable block copolymer for DNA separation by capillary electrophoresis. *Electrophoresis* 19:231–241.
146. Sassi, A.P., Barron, A., Alonso-Amigo, M.G., Hion, D.Y., Yu, J.S., Soane, D.S., and Hooper, H.H. 1996. Electrophoresis of DNA in novel thermoreversible matrixes. *Electrophoresis* 17:1460–1469.
147. Moritani, T., Yoon, K., and Chu, B. 2003. DNA capillary electrophoresis using poly(vinyl alcohol). II. Separation media. *Electrophoresis* 24:2772–2778.
148. Rill, R.L., Locke, B.R., Liu, Y., and Van Winkle, D.H. 1998. Electrophoresis in lyotropic polymer liquid crystals. *Proceedings of the National Academy of Sciences of the United States of America* 95:1534–1539.
149. Lerman, L.S. and Frisch, H.L. 1982. Why does the electrophoretic mobility of DNA in gels vary with the length of the molecule? *Biopolymers* 21:995–997.
150. Pentoney, Jr., S.L., Konrad, K.D., and Kaye, W. 1992. A single-fluor approach to DNA sequence determination using high performance capillary electrophoresis. *Electrophoresis* 13:467–474.
151. Liang, D., Zhou, S., Song, L., Zaitsev, V.S., and Chu, B. 1999. Copolymers of poly (*N*-isopropylacrylamide) densely grafted with poly(ethylene oxide) as high-performance separation matrix of DNA. *Macromolecules* 32:6326–6332.
152. Kan, C.-W., Doherty, E.A.S., Buchholz, B.A., and Barron, A.E. 2004. Thermoresponsive *N,N*-dialkylacrylamide copolymer blends as DNA sieving matrices with a thermally tunable mesh size. *Electrophoresis* 25:1007–1015.
153. Wei, W. and Yeung, E.S. 2001. DNA capillary electrophoresis in entangled dynamic polymers of surfactant molecules. *Analytical Chemistry* 73:1776–1783.
154. Marshall, E. and Pennisi, E. 1998. Hubris and the human genome. *Science* (New York) 280:994–995.
155. 2003. International Consortium Completes Human Genome Project. http://www.genome.gov/11006929.
156. Tsupryk, A., Gorbovitski, M., Kabotyanski, E.A., and Gorfinkel, V. 2006. Novel design of multicapillary arrays for high-throughput DNA sequencing. *Electrophoresis* 27:2869–2879.
157. Tan, H. and Yeung, E.S. 1997. Integrated online system for DNA sequencing by capillary electrophoresis: From template to called bases. *Analytical Chemistry* 69:664–674.
158. Tan, H. and Yeung, E.S. 1998. Automation and integration of multiplexed on-line sample preparation with capillary electrophoresis for high-throughput DNA sequencing. *Analytical Chemistry* 70:4044–4053.

159. Bashkin, J.S., Bartosiewicz, M., Roach, D., Leong, J., Barker, D., and Johnston, R. 1996. Implementation of a capillary array electrophoresis instrument. *Journal of Capillary Electrophoresis* 3:61–68.

160. Huang, X.C., Quesada, M.A., and Mathies, R.A. 1992. Capillary array electrophoresis using laser-excited confocal fluorescence detection. *Analytical Chemistry* 64:967–972.

161. Takahashi, S., Murakami, K., Anazawa, T., and Kambara, H. 1994. Multiple sheath-flow gel capillary-array electrophoresis for multicolor fluorescent DNA detection. *Analytical Chemistry* 66:1021–1026.

162. Kamahori, M. and Kambara, H. 2001. Capillary array electrophoresis analyzer. *Methods in Molecular Biology* (Totowa, NJ) 163:271–287.

163. Sun, Y. and Kwok, Y.C. 2006. Polymeric microfluidic system for DNA analysis. *Analytica Chimica Acta* 556:80–96.

164. Schmalzing, D., Koutny, L., Salas-Solano, O., Adourian, A., Matsudaira, P., and Ehrlich, D. 1999. Recent developments in DNA sequencing by capillary and microdevice electrophoresis. *Electrophoresis* 20:3066–3077.

165. Woolley, A.T. and Mathies, R.A. 1995. Ultra-high-speed DNA sequencing using capillary electrophoresis chips. *Analytical Chemistry* 67:3676–3680.

166. El-Difrawy, S.A., Srivastava, A., Gismondi, E.A., McKenna, B.K., and Enrlich, D.J. 2006. Numerical model for DNA loading in microdevices: Stacking and autogating effects. *Electrophoresis* 27:3779–3787.

167. Blazej, R.G., Kumaresan, P., and Mathies, R.A. 2006. Microfabricated bioprocessor for integrated nanoliter-scale Sanger DNA sequencing. *Proceedings of the National Academy of Sciences of the United States of America* 103:7240–7245.

168. Schmalzing, D., Adourian, A., Koutny, L., Ziaugra, L., Matsudaira, P., and Ehrlich, D. 1998. DNA sequencing on microfabricated electrophoretic devices. *Analytical Chemistry* 70:2303–2310.

169. Kan, C.-W., Fredlake, C.P., Doherty, E.A.S., and Barron, A.E. 2004. DNA sequencing and genotyping in miniaturized electrophoresis systems. *Electrophoresis* 25:3564–3588.

170. Liu, S., Shi, Y., Ja, W.W., and Mathies, R.A. 1999. Optimization of high-speed DNA sequencing on microfabricated capillary electrophoresis channels. *Analytical Chemistry* 71:566–573.

171. Liu, S., Ren, H., Gao, Q., Roach, D.J., Loder, Jr., R.T., Armstrong, T.M., Mao, Q., Blaga, I., Barker, D.L., and Jovanovich, S.B. 2000. Automated parallel DNA sequencing on multiple channel microchips. *Proceedings of the National Academy of Sciences of the United States of America* 97:5369–5374.

172. Paegel, B.M., Emrich, C.A., Wedemayer, G.J., Scherer, J.R., and Mathies, R.A. 2002. High throughput DNA sequencing with a microfabricated 96-lane capillary array electrophoresis bioprocessor. *Proceedings of the National Academy of Sciences of the United States of America* 99:574–579.

173. Paegel, B.M., Yeung, S.H.I., and Mathies, R.A. 2002. Microchip bioprocessor for integrated nanovolume sample purification and DNA sequencing. *Analytical Chemistry* 74:5092–5098.

174. Woolley, A.T., Lao, K., Glazer, A.N., and Mathies, R.A. 1998. Capillary electrophoresis chips with integrated electrochemical detection. *Analytical Chemistry* 70:684–688.

175. Woolley, A.T., Sensabaugh, G.F., and Mathies, R.A. 1997. High-speed DNA genotyping using microfabricated capillary array electrophoresis chips. *Analytical Chemistry* 69:2181–2186.

176. Woolley, A.T. and Mathies, R.A. 1994. Ultra-high-speed DNA fragment separations using microfabricated capillary array electrophoresis chips. *Proceedings of the National Academy of Sciences of the United States of America* 91:11348–11352.

177. Emrich, C.A., Tian, H., Medintz, I.L., and Mathies, R.A. 2002. Microfabricated 384-lane capillary array electrophoresis bioanalyzer for ultrahigh-throughput genetic analysis. *Analytical Chemistry* 74:5076–5083.

178. Ewing, B. and Green, P. 1998. Base-calling of automated sequencer traces using Phred. *II. Error Probabilities. Genome Research* 8:186–194.

179. Ewing, B., Hillier, L., Wendl, M.C., and Green, P. 1998. Base-calling automated sequencer traces using phred. *I. Accuracy Assessment. Genome Research* 8:175–185.

180. Ramsey, J.M. 1999. The burgeoning power of the shrinking laboratory. *Nature Biotechnology* 17:1061–1062.

181. Manz, A., Fettinger, J.C., Verpoorte, E., Luedi, H., Widmer, H.M., and Harrison, D.J. 1991. Micromachining of monocrystalline silicon and glass for chemical analysis systems. A look into next century's technology or just a fashionable craze? *TrAC, Trends in Analytical Chemistry* 10:144–149.

182. Galloway, M., Stryjewski, W., Ford, S., Llopis, S.D., Vaidya, B., and Soper, S.A. 2002. Multichannel capillary electrochromatography PMMA microdevice with integrated pulsed conductivity detector. *Micro Total Analysis Systems 2002, Proceedings of the mTAS 2002 Symposium*, 6th, Nara, Japan, Nov. 3–7, 2002 1:485–487.

183. Wabuyele, M.B., Farquar, H., Stryjewski, W.J., Hammer, R.P., Soper, S.A., Cheng, Y.-W., and Barany, F. 2003. Detection of low abundant mutations in DNA using single-molecule FRET and ligase detection reactions. *Proceedings of SPIE—The International Society for Optical Engineering* 4962:58–69.

184. Wabuyele, M.B., Farquar, H., Stryjewski, W., Hammer, R.P., Soper, S.A., Cheng, Y.-W., and Barany, F. 2003. Approaching real-time molecular diagnostics: Single-pair fluorescence resonance energy transfer (spFRET) detection for the analysis of low abundant point mutations in K-ras oncogenes. *Journal of the American Chemical Society* 125:6937–6945.

185. Wang, Y., Vaidya, B., Farquar, H.D., Stryjewski, W., Hammer, R.P., McCarley, R.L., Soper, S.A., Cheng, Y.-W., and Barany, F. 2003. Microarrays assembled in microfluidic chips fabricated from poly(methyl methacrylate) for the detection of low-abundant DNA mutations. *Analytical Chemistry* 75:1130–1140.

186. Galloway, M. and Soper, S.A. 2002. Contact conductivity detection of polymerase chain reaction products analyzed by reverse-phase ion pair microcapillary electrochromatography. *Electrophoresis* 23:3760–3768.

187. Ford, S.M., McCandless, A.B., Liu, X., and Soper, S.A. 2001. Rapid fabrication of embossing tools for the production of polymeric microfluidic devices for bioanalytical applications. *Proceedings of SPIE—The International Society for Optical Engineering* 4560:207–216.

188. Wabuyele, M.B., Ford, S.M., Stryjewski, W., Barrow, J., and Soper, S.A. 2001. Single molecule detection of double-stranded DNA in poly(methylmethacrylate) and polycarbonate microfluidic devices. *Electrophoresis* 22:3939–3948.

189. Henry, A.C., Tutt, T.J., Galloway, M., Davidson, Y.Y., McWhorter, C.S., Soper, S.A., and McCarley, R.L. 2000. Surface modification of poly(methyl methacrylate) used in the fabrication of microanalytical devices. *Analytical Chemistry* 72:5331–5337.

190. Shi, Y. and Anderson, R.C. 2003. High-resolution single-stranded DNA analysis on 4.5 cm plastic electrophoretic microchannels. *Electrophoresis* 24:3371–3377.

191. Johnson, T.J., Ross, D., and Locascio, L.E. 2002. Rapid microfluidic mixing. *Analytical Chemistry* 74:45–51.

192. Pugmire, D.L., Waddell, E.A., Haasch, R., Tarlov, M.J., and Locascio, L.E. 2002. Surface characterization of laser-ablated polymers used for microfluidics. *Analytical Chemistry* 74:871–878.

193. Sia, S.K. and Whitesides, G.M. 2003. Microfluidic devices fabricated in poly(dimethylsiloxane) for biological studies. *Electrophoresis* 24:3563–3576.

194. Gates, B.D. and Whitesides, G.M. 2003. Replication of vertical features smaller than 2 nm by soft lithography. *Journal of the American Chemical Society* 125:14986–14987.

195. Lee, J.N., Park, C., and Whitesides, G.M. 2003. Solvent compatibility of poly (dimethylsiloxane)-based microfluidic devices. *Analytical Chemistry* 75:6544–6554.

196. Wolfe, D.B., Ashcom, J.B., Hwang, J.C., Schaffer, C.B., Mazur, E., and Whitesides, G.M. 2003. Customization of poly(dimethylsiloxane) stamps by micromachining using a femtosecond-pulsed laser. *Advanced Materials* (Weinheim, Germany) 15:62–65.

197. Wu, H., Odom, T.W., Chiu, D.T., and Whitesides, G.M. 2003. Fabrication of complex three-dimensional microchannel systems in PDMS. *Journal of the American Chemical Society* 125:554–559.

198. Ng, J.M.K., Gitlin, I., Stroock, A.D., and Whitesides, G.M. 2002. Components for integrated poly(dimethylsiloxane) microfluidic systems. *Electrophoresis* 23:3461–3473.

199. Odom, T.W., Thalladi, V.R., Love, J.C., and Whitesides, G.M. 2002. Generation of 30–50 nm structures using easily fabricated, composite PDMS masks. *Journal of the American Chemical Society* 124:12112–12113.

200. McDonald, J.C. and Whitesides, G.M. 2002. Poly(dimethylsiloxane) as a material for fabricating microfluidic devices. *Accounts of Chemical Research* 35:491–499.

201. McDonald, J.C., Chabinyc, M.L., Metallo, S.J., Anderson, J.R., Stroock, A.D., and Whitesides, G.M. 2002. Prototyping of microfluidic devices in poly(dimethylsiloxane) using solid-object printing. *Analytical Chemistry* 74:1537–1545.

202. Fiorini, G.S., Yim, M., Jeffries, G.D.M., Schiro, P.G., Mutch, S.A., Lorenz, R.M., and Chiu, D.T. 2007. Fabrication improvements for thermoset polyester (TPE) microfluidic devices. *Lab on a Chip* 7:923–926.

203. Fiorini, G.S., Lorenz, R.M., Kuo, J.S., and Chiu, D.T. 2004. Rapid prototyping of thermoset polyester microfluidic devices. *Analytical Chemistry* 76:4697–4704.

204. Fiorini, G.S., Jeffries, G.D.M., Lim, D.S.W., Kuyper, C.L., and Chiu, D.T. 2003. Fabrication of thermoset polyester microfluidic devices and embossing masters using rapid prototyped polydimethylsiloxane molds. *Lab on a Chip* 3:158–163.

205. Fiorini, G.S. and Chiu, D.T. 2005. Disposable microfluidic devices: Fabrication, function, and application. *Biotechniques* 38:429–446.

206. Becker, H. and Gartner, C. 2000. Polymer microfabrication methods for microfluidic analytical applications. *Electrophoresis* 21:12–26.

207. Kellogg, G.J., Arnold, T.E., Carvalho, B.L., Duffy, D.C., and Sheppard, N.F. Jr. 2000. Centrifugal microfluidics: Applications. *Micro Total Analysis Systems 2000, Proceedings of the mTAS Symposium*, 4th, Enschede, Netherlands, May 14–18, 2000:239–242.

208. Xu, F., Jabasini, M., Liu, S., and Baba, Y. 2003. Reduced viscosity polymer matrices for microchip electrophoresis of double-stranded DNA. *Analyst* (Cambridge, UK) 128:589–592.

209. Barron, A.E., Sunada, W.M., and Blanch, H.W. 1996. Capillary electrophoresis of DNA in uncrosslinked polymer solutions: Evidence for a new mechanism of DNA separation. *Biotechnology and Bioengineering* 52:259–270.

210. Sanders, J.C., Breadmore, M.C., Kwok, Y.C., Horsman, K.M., and Landers, J.P. 2003. Hydroxypropyl cellulose as an adsorptive coating sieving matrix for DNA separations: Artificial neural network optimization for microchip analysis. *Analytical Chemistry* 75:986–994.

211. Brahmasandra, S.N., Ugaz, V.M., Burke, D.T., Mastrangelo, C.H., and Burns, M.A. 2001. Electrophoresis in microfabricated devices using photopolymerized polyacrylamide gels and electrode-defined sample injection. *Electrophoresis* 22:300–311.

212. Schmalzing, D., Tsao, N., Koutny, L., Chisholm, D., Srivastava, A., Adourian, A., Linton, L., McEwan, P., Matsudaira, P., and Ehrlich, D. 1999. Toward real-world sequencing by microdevice electrophoresis. *Genome Research* 9:853–858.

213. McCormick, L.C. and Slater, G.W. 2007. Molecular deformation and free-solution electrophoresis of DNA-uncharged polymer conjugates at high field strengths: Theoretical predictions. Part 1: Hydrodynamic Segregation. *Electrophoresis* 28:674–682.

214. Mayer, P., Slater, G.W., and Drouin, G. 1994. Theory of DNA sequencing using free-solution electrophoresis of protein-DNA complexes. *Analytical Chemistry* 66:1777–1780.

215. Long, D., Dobrynin, A.V., Rubinstein, M., and Ajdari, A. 1998. Electrophoresis of polyampholytes. *Journal of Chemical Physics* 108:1234–1244.

216. Desruisseaux, C. 2000. The electrophoretic properties of end-labeled DNA molecules in gels, polymer solutions and free-solutions: A theoretical and experimental study, Dissertation:197 pp.

217. Fabrizio, E.F., Nadim, A., and Sterling, J.D. 2003. Resolution of multiple ssDNA structures in free solution electrophoresis. *Analytical Chemistry* 75:5012–5021.

218. Won, J.-I. 2003. Design and production of protein polymer drag-tags for application in free-solution, microchannel DNA sequencing, p166 pp. (Biochemical Methods, Evanston, USA.)

219. Haynes, R.D., Meagher, R.J., Won, J.-I., Bogdan, F.M., and Barron, A.E. 2005. Comblike, monodisperse polypeptoid drag-tags for DNA separations by end-labeled free-solution electrophoresis (ELFSE). *Bioconjugate Chemistry* 16:929–938.

220. Meagher, R.J. 2005. DNA sequencing, genotyping, and analysis by end-labeled free-solution electrophoresis, p401 pp. (Biochemical Genetics, Evanston, USA.)

221. Meagher, R.J., Won, J.-I., McCormick, L.C., Nedelcu, S., Bertrand, M.M., Bertram, J.L., Drouin, G., Barron, A.E., and Slater, G.W. 2005. End-labeled free-solution electrophoresis of DNA. *Electrophoresis* 26:331–350.

222. Nedelcu, S. and Slater, G.W. 2005. Branched polymeric labels used as drag-tags in free-solution electrophoresis of ssDNA. *Electrophoresis* 26:4003–4015.

223. Won, J.-I., Meagher, R.J., and Barron, A.E. 2005. Protein polymer drag-tags for DNA separations by end-labeled free-solution electrophoresis. *Electrophoresis* 26:2138–2148.

224. Chan, Y.C., Lee, Y.-K., Wong, M., and Zohar, Y. 2006. High-throughput fabrication of sub-micron pillar arrays for free-solution DNA electrophoresis without E-beam lithography. *IEEE International Conference on Robotics and Biomimetics*, Shatin, China, July 5–9, 2005:101–104.

225. McCormick, L.C. and Slater, G.W. 2006. A theoretical study of the possible use of electroosmotic flow to extend the read length of DNA sequencing by end-labeled free solution electrophoresis. *Electrophoresis* 27:1693–1701.

226. Meagher, R.J., McCormick, L.C., Haynes, R.D., Won, J.-I., Lin, J.S., Slater, G.W., and Barron, A.E. 2006. Free-solution electrophoresis of DNA modified with drag-tags at both ends. *Electrophoresis* 27:1702–1712.

227. Won, J.-I. 2006. Recent advances in DNA sequencing by end-labeled free-solution electrophoresis (ELFSE). *Biotechnology and Bioprocess Engineering* 11:179–186.

228. Nedelcu, S., Meagher, R.J., Barron, A.E., and Slater, G.W. 2007. Electric and hydrodynamic stretching of DNA-polymer conjugates in free-solution electrophoresis. *Journal of Chemical Physics* 126:175104/175101–175104/175111.

229. Ren, H., Karger, A.E., Oaks, F., Menchen, S., Slater, G.W., and Drouin, G. 1999. Separating DNA sequencing fragments without a sieving matrix. *Electrophoresis* 20:2501–2509.

230. Vreeland, W.N., Meagher, R.J., and Barron, A.E. 2002. Multiplexed, high-throughput genotyping by single-base extension and end-labeled free-solution electrophoresis. *Analytical Chemistry* 74:4328–4333.

231. Vreeland, W.N., Slater, G.W., and Barron, A.E. 2002. Profiling solid-phase synthesis products by free-solution conjugate capillary electrophoresis. *Bioconjugate Chemistry* 13:663–670.

232. Vreeland, W.N. and Barron, A.E. 2000. Free-solution capillary electrophoresis of polypeptoid-oligonucleotide conjugates. Book of Abstracts, *219th ACS National Meeting*, San Francisco, CA, March 26–30, 2000:POLY-555.

233. Chan, E.Y. 2005. Advances in sequencing technology. *Mutation Research, Fundamental and Molecular Mechanisms of Mutagenesis* 573:13–40.

234. Ju, J., Kim, D.H., Bi, L., Meng, Q., Bai, X., Li, Z., Li, X., Marma, M.S., Shi, S., Wu, J., Edwards, J.R., Romu, A., and Turro, N.J. 2006. Four-color DNA sequencing by synthesis using cleavable fluorescent nucleotide reversible terminators. *Proceedings of the National Academy of Sciences of the United States of America* 103:19635–19640.

235. Shendure, J., Porreca, G.J., Reppas, N.B., Lin, X., McCutcheon, J.P., Rosenbaum, A.M., Wang, M.D., Zhang, K., Mitra, R.D., and Church, G.M. 2005. Accurate multiplex polony sequencing of an evolved bacterial genome. *Science* (Washington, DC) 309:1728–1732.

236. Church, G.M. 2006. Genomes for all. *Scientific American* 294:46–54.

237. Bentley, D.R. 2006. Whole-genome re-sequencing. *Current Opinion in Genetics and Development* 16:545–552.

238. Ahmadian, A., Ehn, M., and Hober, S. 2006. Pyrosequencing: History, biochemistry and future. *Clinica Chimica Acta* 363:83–94.

239. Margulies, M., Egholm, M., Altman, W.E., Attiya, S., Bader, J.S., Bemben, L.A., Berka, J., Braverman, M.S., Chen, Y.-J., Chen, Z., Dewell, S.B., de Winter, A., Drake, J., Du, L., Fierro, J.M., Forte, R., Gomes, X.V., Goodwin, B.C., He, W., Helgesen, S., Ho, C.H., Hutchinson, S., Irzyk, G.P., Jando, S.C., Alenquer, M.L.I., Jarvie, T.P., Jirage, K.B., Kim, J.-B., Knight, J.R., Lanza, J.R., Leamon, J.H., Lee, W.L., Lefkowitz, S.M., Lei, M., Li, J., Lohman, K.L., Lu, H., Makhijani, V.B., McDade, K.E., McKenna, M.P., Myers, E.W., Nickerson, E., Nobile, J.R., Plant, R., Puc, B.P., Reifler, M., Ronan, M.T., Roth, G.T., Sarkis, G.J., Simons, J.F., Simpson, J.W., Srinivasan, M., Tartaro, K.R., Tomasz, A., Vogt, K.A., Volkmer, G.A., Wang, S.H., Wang, Y., Weiner, M.P., Willoughby, D.A., Yu, P., Begley, R.F., and Rothberg, J.M. 2006. Genome sequencing in microfabricated high-density picolitre reactors. [Erratum to document cited in CA143:416895]. *Nature* (London, UK) 439:502.

240. Fan, S., Chapline, M.G., Franklin, N.R., Tombler, T.W., Cassell, A.M., and Dai, H. 1999. Self-oriented regular arrays of carbon nanotubes and their field emission properties. *Science* (Washington, DC) 283:512–514.

241. Kasianowicz, J.J., Brandin, E., Branton, D., and Deamer, D. 1996. Characterization of individual polynucleotide molecules using a membrane channel. *Proceedings of the National Academy of Sciences of the United States of America* 93:13770–13773.

242. Zeng, Y. and Harrison, D.J. 2006. Confinement effects on electromigration of long DNA molecules in an ordered cavity array. *Electrophoresis* 27:3747–3752.

243. Song, L., Hobaugh, M.R., Shustak, C., Cheley, S., Bayley, H., and Gouaux, J.E. 1996. Structure of staphylococcal α-hemolysin, a heptameric transmembrane pore. *Science* (Washington, DC) 274:1859–1866.

244. Nestorovich, E.M., Rostovtseva, T.K., and Bezrukov, S.M. 2003. Residue ionization and ion transport through OmpF channels. *Biophysical Journal* 85:3718–3729.

245. Saleh, O.A. and Sohn, L.L. 2003. An artificial nanopore for molecular sensing. *Nano Letters* 3:37–38.

246. Mara, A., Siwy, Z., Trautmann, C., Wan, J., and Kamme, F. 2004. An asymmetric polymer nanopore for single molecule detection. *Nano Letters* 4:497–501.

247. Smeets, R.M.M., Keyser, U.F., Krapf, D., Wu, M.-Y., Dekker, N.H., and Dekker, C. 2006. Salt dependence of ion transport and DNA translocation through solid-state nanopores. *Nano Letters* 6:89–95.

248. Gracheva, M.E., Xiong, A., Aksimentiev, A., Schulten, K., Timp, G., and Leburton, J.-P. 2006. Simulation of the electric response of DNA translocation through a semiconductor nanopore-capacitor. *Nanotechnology* 17:622–633.

249. Chang, H., Kosari, F., Andreadakis, G., Alam, M.A., Vasmatzis, G., and Bashir, R. 2004. DNA-mediated fluctuations in ionic current through silicon oxide nanopore channels. *Nano Letters* 4:1551–1556.

250. Li, J., Stein, D., McMullan, C., Branton, D., Aziz, M.J., and Golovchenko, J.A. 2001. Ion-beam sculpting at nanometer length scales. *Nature* (London, UK) 412:166–169.
251. Fologea, D., Uplinger, J., Thomas, B., McNabb, D.S., and Li, J. 2005. Slowing DNA translocation in a solid-state nanopore. *Nano Letters* 5:1734–1737.
252. Chen, P., Gu, J., Brandin, E., Kim, Y.-R., Wang, Q., and Branton, D. 2004. Probing single DNA molecule transport using fabricated nanopores. *Nano Letters* 4:2293–2298.
253. Deamer, D.W. and Branton, D. 2002. Characterization of nucleic acids by nanopore analysis. *Accounts of Chemical Research* 35:817–825.
254. Deamer, D.W. and Akeson, M. 2000. Nanopores and nucleic acids: Prospects for ultrarapid sequencing. *Trends in Biotechnology* 18:147–151.
255. Lagerqvist, J., Zwolak, M., and Di Ventra, M. 2006. Fast DNA sequencing via transverse electronic transport. *Nano Letters* 6:779–782.
256. Shendure, J., Mitra, R.D., Varma, C., and Church, G.M. 2004. Advanced sequencing technologies: Methods and goals. *Nature Reviews Genetics* 5:335–344.
257. Rhee, M. and Burns, M.A. 2006. Nanopore sequencing technology: Research trends and applications. *Trends in Biotechnology* 24:580–586.
258. Pernodet, N., Samuilov, V., Shin, K., Sokolov, J., Rafailovich, M.H., Gersappe, D., and Chu, B. 2000. DNA electrophoresis on a flat surface. *Physical Review Letters* 85:5651–5654.
259. Seo, Y.-S., Luo, H., Samuilov, V.A., Rafailovich, M.H., Sokolov, J., Gersappe, D., and Chu, B. 2004. DNA electrophoresis on nanopatterned surfaces. *Nano Letters* 4:659–664.
260. Voulgarakis, N.K., Redondo, A., Bishop, A.R., and Rasmussen, K.O. 2006. Sequencing DNA by dynamic force spectroscopy: Limitations and prospects. *Nano Letters* 6:1483–1486.
261. Blazewicz, J., Glover, F., Kasprzak, M., Markiewicz, W.T., Oguz, C., Rebholz-Schuhmann, D., and Swiercz, A. 2006. Dealing with repetitions in sequencing by hybridization. *Computational Biology and Chemistry* 30:313–320.
262. Leong, H.-W., Preparata, F.P., Sung, W.-K., and Willy, H. 2005. Adaptive control of hybridization noise in DNA sequencing-by-hybridization. *Journal of Bioinformatics and Computational Biology* 3:79–98.
263. Preparata, F.P. and Oliver, J.S. 2004. DNA sequencing by hybridization using semi-degenerate bases. *Journal of Computational Biology* 11:753–765.
264. Sakata, T. and Miyahara, Y. 2006. DNA sequencing based on intrinsic molecular charges. *Angewandte Chemie, International Edition* 45:2225–2228.
265. Christoffersen, R.E. 1997. Translating genomics information into therapeutics: A key role for oligonucleotides. *Nature biotechnology* 15:483–484.
266. Orr, R.M. 2001. Technology evaluation: Fomivirsen, Isis pharmaceuticals Inc/CIBA vision. *Current Opinion in Molecular Therapeutics* 3:288–294.
267. Oefner, P.J., Bonn, G.K., Huber, C.G., and Nathakarnkitkool, S. 1992. Comparative study of capillary zone electrophoresis and high-performance liquid chromatography in the analysis of oligonucleotides and DNA. *Journal of Chromatography* 625:331–340.
268. Warren, W.J and Vella, G. 1993. Analysis of synthetic oligodeoxyribonucleotides by capillary gel electrophoresis and anion-exchange HPLC. *BioTechniques* 14:598–606.
269. von Brocke, A., Freudemann, T., and Bayer, E. 2003. Performance of capillary gel electrophoretic analysis of oligonucleotides coupled on-line with electrospray mass spectrometry. *Journal of Chromatography, A* 991:129–141.
270. Willems, A.V., Deforce, D.L., Van Peteghem, C.H., and Van Bocxlaer, J.F. 2005. Development of a quality control method for the characterization of oligonucleotides by capillary zone electrophoresis-electrospray ionization-quadrupole time of flight-mass spectrometry. *Electrophoresis* 26:1412–1423.

271. Willems, A.V., Deforce, D.L., Van Peteghem, C.H., and Van Bocxlaer, J.F. 2005. Analysis of nucleic acid constituents by on-line capillary electrophoresis-mass spectrometry. *Electrophoresis* 26:1221–1253.
272. Ueda, M., Kiba, Y., Abe, H., Arai, A., Nakanishi, H., and Baba, Y. 2000. Fast separation of oligonucleotide and triplet repeat DNA on a microfabricated capillary electrophoresis device and capillary electrophoresis. *Electrophoresis* 21:176–180.
273. Minalla, A.R., Dubrow, R., and Bousse, L.J. 2001. Feasibility of high-resolution oligonucleotide separation on a microchip. *Proceedings of SPIE—The International Society for Optical Engineering* 4560:90–97.
274. Ogston, A.G. 1958. The spaces in a uniform random suspension of fibres. *Transactions of the Faraday Society* 54:1754–1757.
275. De Gennes, P.G. 1979. *Scaling Concepts in Polymer Physics*. Cornell University Press, Ithaca, NY.
276. Gelfi, C., Perego, M., Morelli, S., Nicolin, A., and Righetti, P.G. 1996. Analysis of antisense oligonucleotides by capillary electrophoresis, gel-slab electrophoresis, and HPLC: A comparison. *Antisense and Nucleic Acid Drug Development* 6:47–53.
277. Barry, J.P., Muth, J., Law, S.-J., Karger, B.L., and Vouros, P. 1996. Analysis of modified oligonucleotides by capillary electrophoresis in a polyvinylpyrrolidone matrix coupled with electrospray mass spectrometry. *Journal of Chromatography, A* 732:159–166.
278. Auriola, S., Jaaskelainen, I., Regina, M., and Urtti, A. 1996. Analysis of oligonucleotides by on-column transient capillary isotachophoresis and capillary electrophoresis in polyethylene glycol-filled columns. *Analytical Chemistry* 68:3907–3911.
279. Khan, K., Van Schepdael, A., and Hoogmartens, J. 1996. Capillary electrophoresis of oligonucleotides using a replaceable sieving buffer with low viscosity-grade hydroxyethyl cellulose. *Journal of Chromatography, A* 742:267–274.
280. Riekkola, M.L., Jussila, M., Porras, S.P., and Valko, I.E. 2000. Non-aqueous capillary electrophoresis. *Journal of Chromatography, A* 892:155–170.
281. Steiner, F. and Hassel, M. 2000. Nonaqueous capillary electrophoresis: A versatile completion of electrophoretic separation techniques. *Electrophoresis* 21:3994–4016.
282. Sjodahl, J., Lindberg, P., and Roeraade, J. 2007. Separation of oligonucleotides in N-methyl-formamide-based polymer matrices by capillary electrophoresis. *Journal of Separation Science* 30:104–109.
283. Effenhauser, C.S., Paulus, A., Manz, A., and Widmer, H.M. 1994. High-speed separation of antisense oligonucleotides on a micromachined capillary electrophoresis device. *Analytical Chemistry* 66:2949–2953.
284. Liu, T., Nace, V.M., and Chu, B. 1999. Self-assembly of mixed amphiphilic triblock copolymers in aqueous solution. *Langmuir* 15:3109–3117.
285. Zhang, J., Burger, C., and Chu, B. 2006. Nanostructured polymer matrix for oligonucleotide separation. *Electrophoresis* 27:3391–3398.
286. Zhang, J., Gassmann, M., He, W., Wan, F., and Chu, B. 2006. Reversible thermoresponsive sieving matrix for oligonucleotide separation. *Lab on a Chip* 6:526–533.
287. Zhang, J., Liang, D., He, W., Wan, F., Ying, Q., and Chu, B. 2005. Fast separation of single-stranded oligonucleotides by capillary electrophoresis using OliGreen as fluorescence inducing agent. *Electrophoresis* 26:4449–4455.
288. Zhang, J., Gassmann, M., Chen, X., Burger, C., Rong, L., Ying, Q., and Chu, B. 2007. Characterization of a reversible thermoresponsive gel and its application to oligonucleotide separation. *Macromolecules* (Washington, DC) 40:5537–5544.
289. Zhang, J., He, W.D., Liang, D.H., Fang, D.F., Chu, B., and Gassmann, M. 2006. Designing polymer matrix for microchip-based double-stranded DNA capillary electrophoresis. *Journal of Chromatography, A* 1117:219–227.
290. Rill, R.L., Locke, B.R., Liu, Y., and Winkle, D.H.V. 1998. Electrophoresis in lyotropic polymer liquid crystals. *Proc. Natl. Acad. Sci.* 95:1534–1539.

291. Reyderman, L. and Stavchansky, S. 1996. Determination of single-stranded oligode-oxynucleotides by capillary gel electrophoresis with laser induced fluorescence and on column derivatization. *Journal of Chromatography, A* 755:271–280.
292. Reyderman, L. and Stavchansky, S. 1997. Quantitative determination of short single-stranded oligonucleotides from blood plasma using capillary electrophoresis with laser-induced fluorescence. *Analytical Chemistry* 69:3218–3222.
293. Zhou, H., Miller, A.W., Sosic, Z., Buchholz, B., Barron, A.E., Kotler, L., and Karger, B.L. 2000. DNA sequencing up to 1300 bases in two hours by capillary electrophoresis with mixed replaceable linear polyacrylamide solutions. *Analytical Chemistry* 72:1045–1052.
294. Barron, A.E. 2006. DNA sequencing and genotyping. *Electrophoresis* 27:3687–3688.

4 Novel Mixed-Mode Stationary Phase for Capillary Electrochromatography

Kaname Ohyama and Naotaka Kuroda

CONTENTS

4.1 INTRODUCTION

Capillary electrochromatography (CEC), which combines the features of capillary zone electrophoresis (CZE) and high-performance liquid chromatography (HPLC), is a powerful separation technique with high efficiency, high resolution, and low consumption of mobile phase and sample [1–5]. The capillary columns for CEC, packed with several particles originally used in HPLC, are now commercially available from some manufacturers, however, these columns are still expensive and sometimes awkward to use.

The most commonly used column in CEC has been an octadecyl silica (ODS)-packed type and therefore, the focus of initial CEC research was on the separation of neutral pharmaceuticals and aromatic hydrocarbons that were well studied in reversed-phase (RP)-HPLC. But now, because of CEC's high efficiency and good compatibility with mass spectrometry, its applications dealing with biological samples are increasing and the most promising area for CEC lies in the separation of charged biomolecules. However, CEC with an ODS-packed column limits the expansion of CEC applications due to the following reasons: (1) due to a high-enough electroosmotic flow (EOF) over a broad range of pH, ODS can work only in the mobile phase with relatively high pH since the EOF decreases tremendously owing to the protonation of silanol groups on the surface of ODS in mobile phase with low pH; (2) charged analytes are not retained on the RP stationary phases enough for their separation.

From the drawbacks of ODS in CEC, a trend in CEC is directed toward the use of mixed-mode stationary phases that combine two interaction sites on the chromatographic support. Especially, CEC researchers rushed to use the ion-exchange-RP type mixed-mode stationary phases. The permanently charged functional groups in these mixed-mode stationary phases provide an increased cathodic (strong cation-exchange [SCX]–RP) or anodic (strong anion-exchange [SAX]–RP) EOF and also attract or repulse the charged analytes. The hydrophobic moiety in the stationary phases contributes to the interaction with the hydrophobic part of charged analytes. These mixed-mode stationary phases meet the demands of separating many types of charged biomolecules with a selectivity difference from RP-HPLC and RP-CEC, provided by a hybrid of electrophoretic migration and chromatographic retention involving hydrophobic and coulombic interactions in the separation mechanism. Actually, the mixed-mode stationary phases have addressed the problems in RP-CEC of charged analytes and some efforts have been devoted to the theoretical understanding of separation behavior on mixed-mode CEC [6,7]. However, the commercial selection of mixed-mode stationary phases remains small. Therefore, to exploit the full potential of CEC, advances are essential in the area of specially designed mixed-mode stationary phases which support a strong EOF over a wide range of pH and retain the charged analytes.

Capillary column formats can be divided into three types, i.e., packed column, monolithic column, and open-tubular column. During the early stage of CEC development, packed columns were mostly used, however, the frit fabrication is an art and requires experimental skill and experience in order to obtain stable columns with reproducible properties. Void and bubble formation in packed columns is a frequently encountered problem. The monolithic columns for CEC are gathering increasing attention as an alternative to the packed column format because there is no need for frits, due to their continuous structure and *in situ* preparation, thus eliminating bubble formation and yielding stable columns. For the open-tubular format, a stable surface coating needs to be created on the inner wall of a capillary to provide efficient chromatographic separation and EOF. This format also does not require the frit fabrication and avoids the technical problems experienced in packed columns. In this chapter, we review the mixed-mode stationary phases specially designed for CEC in the three column formats and their performance in CEC.

4.2 PACKED COLUMNS

The use of the capillary columns packed with SCX–RP mixed-mode stationary phases that were commercially available as packing material for HPLC was a straightforward process. Several works diverted the stationary phases containing chemically cobonding sulfonic acid and *n*-alkyl (hexyl, octyl, or octadecyl) groups on the silica surface to

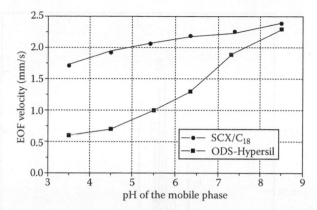

FIGURE 4.1   Dependence of EOF on the pH of eluent for an n-octadecyl bonded silica and a mixed-mode bonded silica (sulfonic acid and n-octadecyl). Capillary column: packed capillaries with Waters Spherisorb SCX/C$_{18}$, ODS-Hypersil. Conditions: mobile phase, 25 mM phosphate containing 80% v/v acetonitrile; injection, electrokinetic for 6 s at 5 kV. (Reprinted from Adam, T. and Unger, K.K., *J. Chromatogr. A*, 894, 241, 2000. With permission.)

CEC [6–8]. Adam and Unger clearly presented the boost-up of the cathodic EOF by Spherisorb SCX/C$_{18}$ and obtained high EOF even at low pH (Figure 4.1) [8].

Zhang and El Rassi pioneered the CEC column packed with specially designed mixed-mode stationary phase (octadecylsulfonated silica [ODSS], Figure 4.2) which was composed of a hydrophilic, negatively charged sublayer and a nonpolar top layer containing octadecyl ligands [9]. The preparation of ODSS was a three-layered coating process involving the hydrophilic coating of silica with a primary layer of glycidoxypropyltrimethoxysilane (sublayer). Then, a sulfonated layer was sandwiched through covalent bonding between the sublayer and an octadecyl top layer. Due to the permanently charged sulfonic acid groups, the hydrophilic nature of the sublayer, and the hydrophobic character of the top octadecyl layer, retention and selectivity of charged and relatively polar nucleosides and bases on the ODSS stationary phase were based on electrostatic, hydrophilic, and hydrophobic interactions (Figure 4.3) [10].

Applying the same idea of using fixed charge of sulfonic acid groups, Ohyama et al. newly and easily synthesized 3-(4-sulfo-1,8-naphthalimido)propyl-modified silica (SNAIP, Figure 4.4) by a single reaction of aminopropyl silica (APS) with

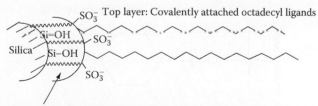

Sublayer: Covalently bound hydrophilic layer with attached sulfonic acid groups. Also shown are the residual silanols

FIGURE 4.2   Structure of the ODSS stationary phase. (Reprinted from Zhang, M. and El Rassi, Z., *Electrophoresis*, 19, 2068, 1998. With permission.)

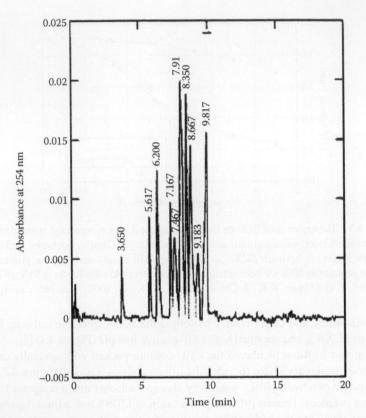

FIGURE 4.3 Separation of a mixture of purine and pyrimidine bases and their nucleosides on the ODSS capillary column. Capillary column: total length, 27 cm; effective length, 20.5 cm; 100 μm ID. Conditions: mobile phase, 4.8 mM sodium acetate (pH 4.5) containing 40% v/v acetonitrile; voltage, 20 kV; injection, electrokinetic for 2 s at 1 kV. Peaks: 1, uracil; 2, uridine; 3, thymine; 4, cytosine; 5, cytidine; 6, inosine; 7, adenine; 8, guanine; 9, adenosine; 10, guanosine. (Reprinted from Zhang, M. and El Rassi, Z., *Electrophoresis*, 20, 31, 1999. With permission.)

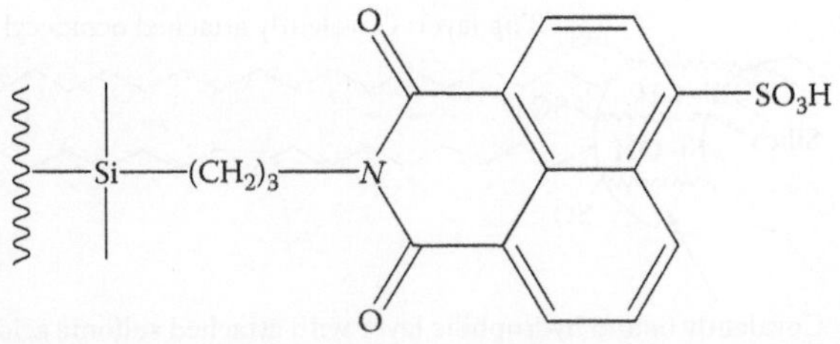

FIGURE 4.4 Structure of SNAIP stationary phase. (From Ohyama, K., Shirasawa, Y., Wada, M., Kishikawa, N., Ohba, Y., Nakashima, K., and Kuroda, N., *Electrophoresis*, 25, 3224, 2004. With permission.)

4-sulfo-1,8-naphthalic anhydride at 150°C for 5 h [11]. The unique structure of SNAIP contributed to the retention and selectivity by three interactions including hydrophobic, electrostatic, and π–π interactions. The CEC employing SNAIP was successively applied to the separations of several drugs (acidic and basic drugs) and peptides (selected and digested peptides) [11,12]. In addition, a highly polar mobile phase (up to 95% aqueous mobile phase), which usually leads to high current and bubble formation within CEC column packed with conventional RP stationary phase, can be used on SNAIP column since SNAIP is a relatively polar stationary phase. Therefore, the application scope of the SNAIP column was expanded to the separation of polar analytes (nucleosides and nucleic acid bases) with highly aqueous mobile phase [13].

Basic compounds will be eluted with serious peak tailing or cannot be eluted on SCX–RP mixed-mode stationary phases because of the strong electrostatic adsorption and so the stationary phase with positively charged functionalities (SAX–RP) can be an alternative mixed-mode phase. Anion-exchange moieties reverse the direction of EOF, however, the capillary wall in both packed and unpacked segments can contribute to the cathodic EOF. Using Spherisorb SAX column, Byrne et al. studied the contribution of capillary wall in open part of column and compared three capillary columns whose internal wall was chemically derived from amine or covered with polyvinylalcohol or normal as a reference [14]. In this experiment, they clearly indicated that the contribution of capillary wall to the anodic EOF was substantial. Scherer and Steiner developed and characterized the novel SAX–RP mixed-mode stationary phase, which was synthesized by two-step reactions: (1) reaction of activated silica with vinyltrichlorosilane resulting in a vinylsilica; (2) encapsulation of the vinylsilica via radical copolymerization in solution using octadecyl acrylate and triethylammonium methylstyrene chloride (Figure 4.5) [15]. In contrast to the description by Byrne et al., they found that the negative charges that contribute to the cathodic EOF of capillary wall in unpacked sections do not significantly influence the overall EOF

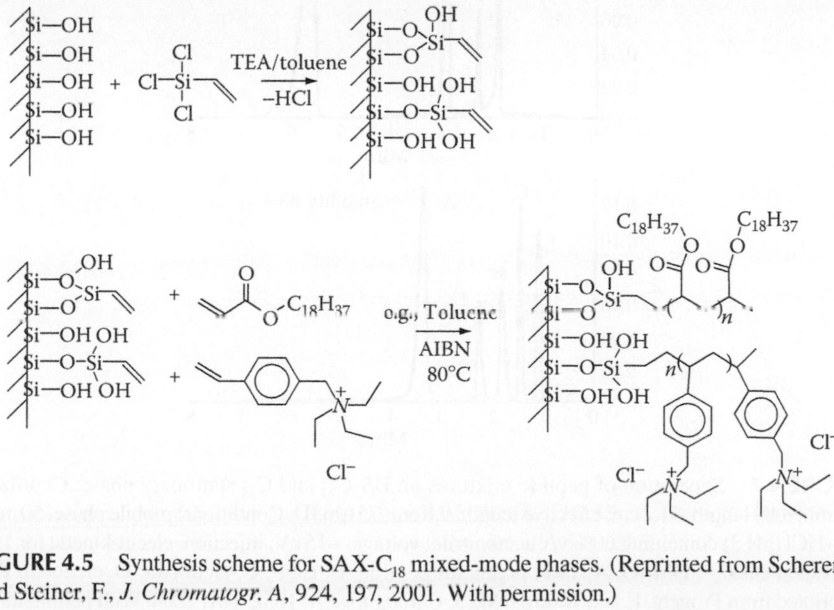

FIGURE 4.5 Synthesis scheme for SAX-C_{18} mixed-mode phases. (Reprinted from Scherer, B. and Steiner, F., *J. Chromatogr. A*, 924, 197, 2001. With permission.)

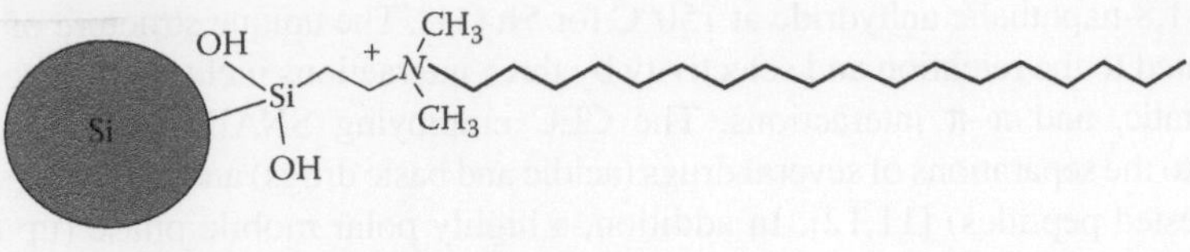

FIGURE 4.6 Structure of the Stability BS-C$_{23}$ stationary phase. (Reprinted from Progent, F. and Taverna, M., *J. Chromatogr. A*, 1052, 181, 2004. With permission.)

mobility in CEC with SAX–RP type mixed-mode column [15]. Furthermore, on a chiral stationary phase with an anion-exchange moiety, Lämmerhofer et al. showed that the EOF direction was pH dependent with considerable instability in the pH region where EOF changes from cathode to anode due to the coexistence of negatively charged residual silanol groups and positively charged amino functionalities [16]. They also found that the EOF direction depends on the type of silica and the extent of its modification [17].

Progent and Taverna introduced and evaluated a new SAX–RP type stationary phase (Stability BS-C$_{23}$) in CEC. A quaternary ammonium group is covalently bonded by a spacer at the surface of silica, which forms a relatively hydrophilic and charged sublayer, while a nonpolar upper layer of octadecyl moieties plays the hydrophobic binding layer (Figure 4.6) [18]. The stationary phase generated a consistent anodic EOF and the EOF was independent of pH over a wide range (2–12). The elution order of a synthetic peptides mixture observed on BS-C$_{23}$ clearly demonstrates a different selectivity compared with a C$_{18}$ stationary phase (Figure 4.7).

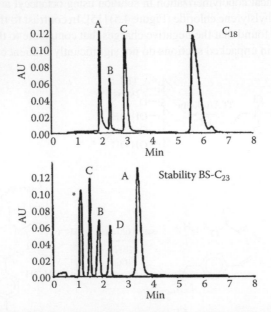

FIGURE 4.7 Separation of peptide mixtures on BS-C$_{23}$ and C$_{18}$ stationary phase. Capillary column: total length, 31.2 cm; effective length, 9.8 cm; 75 μm ID. Conditions: mobile phase, 50 mM Tris–HCl (pH 3) containing 60% v/v acetonitrile; voltage, −15 kV; injection, electrokinetic for 10 s at −5 kV. Peaks: *, DMSO; A, TRH precursor peptide; B, EGF; C, eledoisin; D, eledoisin RP. (Reprinted from Progent, F. and Taverna, M., *J. Chromatogr. A*, 1052, 181, 2004. With permission.)

(A)   (B)

(C)

FIGURE 4.8   Scanning electron micrographs of the end of a 75 µm ID continuous-bed column containing 9% sol-gel bonded large-pore ODS. (A) 1,000×, (B) 4,000×, and (C) 17,000× magnification. (Reprinted from Tang, Q., Wu, N., and Lee, M.L., *J. Micro. Sep.*, 11, 550, 1999. With permission.)

Tang and Lee proposed continuous-bed columns containing sol-gel bonded silica particles with octadecyl and propylsulfonic acid groups [19]. The columns were prepared by packing SCX-C_{18} particles into a capillary, then filling the packed capillary with a siliceous sol-gel and finally drying the column with supercritical carbon dioxide [20]. This type of column avoids problems associated with frit fabrication because of the sol-gel procedure for retaining packing materials. The structure of the sol-gel bridge between the particles is clearly visible in Figure 4.8. In the report, the proposed column showed similar retention behavior as ODS-packed column as only neutral aromatic compounds were used as model analytes, however, it will show a different selectivity for charged analytes.

Recently, Zhang and Colón prepared novel CEC columns packed with porous silica particles containing the polymeric layer produced by radical polymerization of N-isopropylacrylamide (NIPAAm) and [3-(methacryloylamino)propyl]trimethylammonium chloride (MAPTA) [21]. Figure 4.9 shows the scheme used to prepare the poly-NIPAAm-*co*-MAPTA-modified silica. The radical initiator, 4,4′-azobis (4-cyanovaleric acid), first activated APS and then, the copolymerization of NIPAAm and MAPTA on the surface of initiator-immobilized APS particles was carried out at 70°C for 5 h under N_2 atmosphere. The moiety provided by MAPTA in the polymeric layer was responsible for an anodic EOF in CEC with the stationary

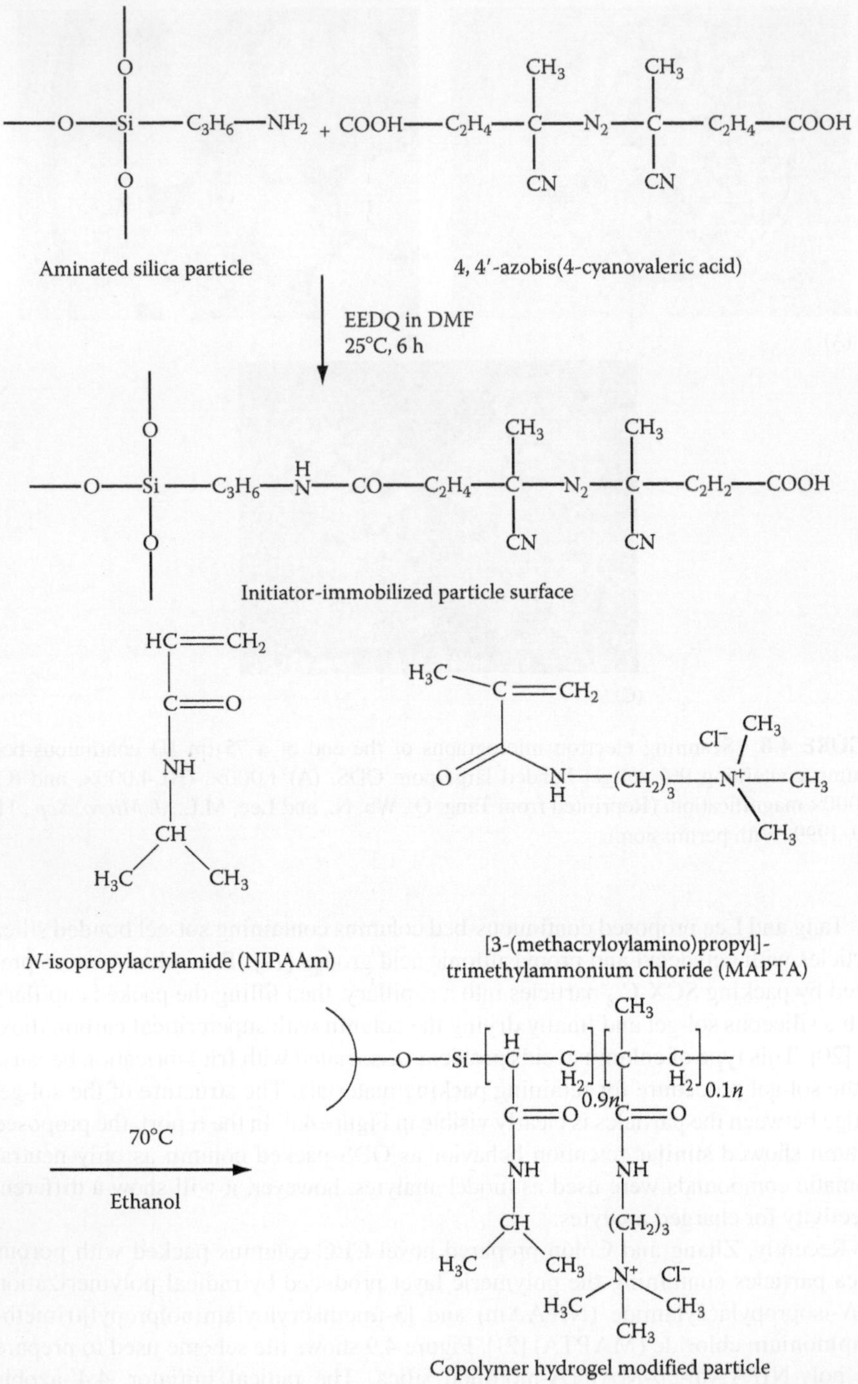

FIGURE 4.9 Scheme for modification of aminated silica surface with poly-NIPAAm-*co*-MAPTA. (Reprinted from Zhang, X. and Colón, L.A., *Electrophoresis*, 27, 1060, 2006. With permission.)

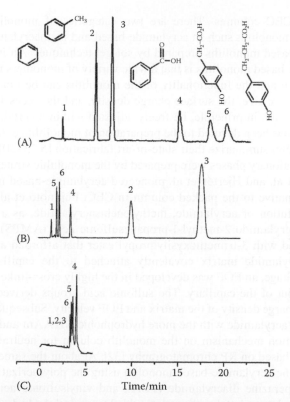

FIGURE 4.10 Separation of model compounds under different separation modes: (A) CEC, (B) HPLC, and (C) CZE. CEC and CZE conditions: mobile phase, 5 mM phosphate (pH 3.1) containing 30% v/v acetonitrile; voltage, 25 kV; injection, electrokinetic for 8 s at −5 kV. Capillary column: total length, 30 cm; effective length, 20 cm; 50 μm ID. HPLC conditions: flow rate, 1.0 mL/min. HPLC column: 5 μm C_8 commercial column, 15 cm×4.6 mm ID. Peaks: 1, DMSO; 2, benzene; 3, toluene; 4, benzoic acid; 5, *p*-hydroxyphenylacetic acid; 6, 3-(*p*-hydroxyphenyl)propionic acid. (Reprinted from Zhang, X. and Colón, L.A., *Electrophoresis*, 27, 1060, 2006. With permission.)

phase. As seen in Figure 4.10, the elution order in RP-HPLC was based on the hydrophobicity of model compounds, while the acidic compounds (4–6) eluted last in CEC, which was attributed to the mixed-mode retention mechanism (i.e., hydrophilic interaction and ion-exchange). Interestingly, the authors started the study employing the poly-NIPAAm-*co*-MAPTA stationary phase because poly-NIPAAms exhibits a thermally reversible phase transition at a lower critical solution temperature and is explored in HPLC [22]. However, in the preliminary assessment, such a behavior was not observed in CEC, which might be due to short polymeric chain and high concentration of acetonitrile in mobile phase.

4.3   MONOLITHIC COLUMNS

As described in Section 4.1, as there is no need to retain frit in CEC with monolithic columns, they are almost free from the bubble formation that distress the analyst using packed CEC columns. There are two categories of monolithic columns: polymer-based monoliths such as acrylamide-based and methacrylate-based monolith and silica-based monoliths prepared by sol-gel techniques. An interesting feature of polymer-based monoliths is that a wide variety of monomers is available for their preparation and the functionality of the monoliths can be tuned for specific demands. The pore size, the surface charge density, and the accessible chromatographic surface can, in principle, be freely adjusted within a certain range. Much research effort has been devoted to the preparation of monolithic columns for CEC and recent reviews summarize their state-of-art fabrication [5,23–30]. A number of mixed-mode stationary phases were prepared by the monolithic strategy.

Fujimoto et al. and Hjertén et al. pioneered acrylamide-based monolith as an attractive alternative to the packed column in CEC. Fujimoto et al. polymerized an aqueous solution of acrylamide, methylenebisacrylamide, as a cross-linking agent, and 2-acrylamido-2-methyl-1-propanesulfonic acid (AMPS) within capillaries pretreated with 3-trimethoxysilylpropyl ester that afforded a highly cross-linked polyacrylamide matrix covalently attached to the capillary wall [31]. By applying voltage, an EOF was developed in the highly cross-linked matrix without forcing it out of the capillary. The sulfonic acid groups derived from AMPS increased the charge density of the matrix and EOF velocity. Subsequently, the same group replaced acrylamide with the more hydrophobic NIPAAm and demonstrated that the separation mechanism on the monolith column for neutral analytes was predominantly based on RP chromatography [32]. At about the same time, Hjertén et al. realized the acrylamide-based monolith using the polymerization mixture of acrylamide, piperazine diacrylamide (PDA), and vinylsulfonic acid (VSA) with stearyl methacrylate or butylmethacrylate added to control the hydrophobicity [33]. The presence of the strongly acidic sulfonic acid groups affords a consistent EOF over a broad pH range.

In these initial reports by the two groups, the main focus was the study of RP chromatographic retention behavior on their synthesized monolithic column, using neutral and nonpolar compounds as model analytes. Enlund et al. explored the acrylamide-based monolith column for the CEC separation of charged analytes (hydrophobic amines) [34]. The hydrophobicity and charge density of the acrylamide-based monolith were systematically changed by regulating the contents of isopropyl and sulfonate groups. The best performance was obtained with a certain molar ratio for the sulfonate and isopropyl groups and resulted in efficiencies up to 200,000 N/m.

Svec and Fréchet first introduced ion-exchange functionalities into methacrylate-based monolith and presented the one-step preparation of "molded" rigid monolith column [35]. In contrast to polyacrylamide soft gels, rigid monoliths are not compressible, do not change their size on swelling, and do not require chemical anchoring to the capillary wall. The monolithic column was prepared by a simple copolymerization of mixtures of butylmethacrylate, ethylene dimethacrylate, and AMPS. A ternary porogenic solution comprising water, 1-propanol, and 1,4-butanediol was used to obtain a homogeneous polymerization mixture of hydrophobic methacrylate and hydrophilic AMPS. The composition of the porogenic solution and the amount of AMPS allowed the fine-tuning of porous properties of monoliths [36]. While the sulfonic acid groups in AMPS supported generation of the stable EOF in a wide range of pH and can also work as an ion-exchange moiety, the butyl chains provided the hydrophobic sites for RP chromatography. Although these capillary columns were employed in RP mode for CEC separation of neutral analytes in earlier studies [35–37], mixed-mode CEC separation of therapeutic peptides using same monolithic column was successfully performed by another group [38]. Furthermore, using the capillary column format as a model, the technique was expanded to *in situ* preparation of the "molded" microchip separation devices by the UV light-initiated polymerization in spite of thermally initiated polymerization [39]. On the other hand, taking advantage of its relatively higher wettability, the monolithic column proposed by Svec and Fréchet was applied to mixed-mode (SCX–RP) CEC separation of nucleosides with highly aqueous mobile phase and acidic condition [40].

In a similar manner, Wu et al. prepared the mixed-mode monolithic column by *in situ* copolymerization of 2-(sulfooxy)ethyl methacrylate (SEMA) and ethylene dimethacrylate in the presence of binary porogenic solution (1-propanol and 1,4-butanediol) [41]. A mixed-mode retention mechanism consisting of hydrophobic and electrostatic interactions was confirmed by investigating the influence of mobile phase composition on the retention of neutral compounds, aromatic anilines, and peptides. As shown in Figure 4.11, a separation of 10 peptides in acidic condition was obtained with column efficiencies up to 110,000 plates/m and asymmetry factors of less than 1.4.

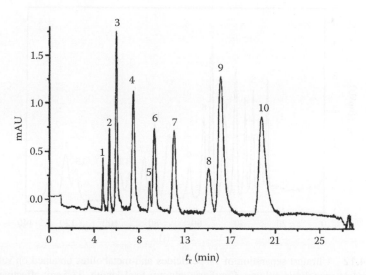

FIGURE 4.11 Separation of peptides on the SEMA-based mixed-mode monolithic column. Capillary column: total length, 30 cm; effective length, 9.5 cm; 100 μm ID. Conditions: mobile phase, 8 mM phosphate buffer (pH 3.0) containing 50% v/v acetonitrile; voltage, 5 kV; injection, electrokinetic for 2 s at 2 kV. Peaks: 1, thiourea; 2, Gly-Glu; 3, Gly-Ala; 4, Tyr-Glu; 5, Gly-Gly-Phe; 6, Ala-Val; 7, Phe-Gly; 8, Tyr-Val; 9, Trp-Gly; 10, Ala-Trp. (Reprinted from Wu, R., Zou, H., Fu, H., Jin, W., and Ye, M., *Electrophoresis*, 23, 1239, 2002. With permission.)

There are some advantages associated with polymer-based monoliths, including simple preparation and wide variation of monomers for tailor-made monolithic columns in terms of the porosity and the nature of its retentive ligand. However, in most cases, the selection of alkyl chain monomers was restricted to short alkyl chains (e.g., isopropyl, butyl, hexyl, dodecyl) because of low solubility of long alkyl ones in porogenic solvent mixtures. In this context, Bedair and El Rassi presented novel monolithic stationary phase having long alkyl chain ligands provided by the *in situ* copolymerization of pentaerythritol diacrylate monostearate (PEDAS) and AMPS in a ternary porogenic solution consisting of cyclohexanol/ethylene glycol/water [42]. The EOF velocity was easily regulated by the amount of AMPS in the polymerization solution. Using dansyl amino acids as model analytes, the mixed-mode (e.g., hydrophobic interaction as well as electrostatic interaction/repulsion) retention behavior was observed. Ultrafast separation of 17 charged and neutral pesticides and metabolites in <135 s was performed using short capillary columns of 8.5 cm (Figure 4.12).

The negatively charged groups (e.g., sulfonic acid groups) have been successfully introduced into SCX–RP mixed-mode monolith and demonstrated the suitability for the separation of neutral analytes because of a constant EOF generation over a wide range pH. However, in many cases, basic compounds were often eluted with severe peak tailing on the SCX–RP mixed-mode monolithic columns due to electrostatic

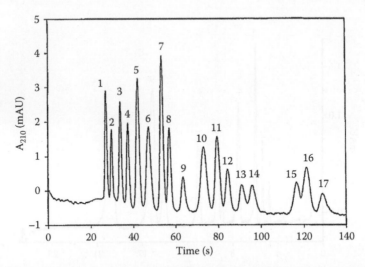

FIGURE 4.12 Ultrafast separation of 17 pesticides and metabolites obtained on sulfonated stearyl-acrylate monolithic column. Capillary column: total length, 33.5 cm; effective length, 8.5 cm; 100 μm ID. Conditions: mobile phase, 1 mM sodium phosphate (pH 6) containing 58% v/v acetonitrile; voltage, 30 kV; injection, electrokinetic for 2 s at 10 kV. Peaks: 1, oxamyl; 2, aldicarb; 3, monuron; 4, carbaryl; 5, diuron; 6, chloroxuron; 7, 1-naphthol; 8, 2,4-dichlorophenol; 9, chlorimuronethyl; 10, neburon; 11, 2,4-diisopropyl ester; 12, 2,4,5-trichlorophenol; 13, silvex; 14, dichlorprop; 15, 2,4-dibutylester; 16, 2,4,5-triisopropyl ester; 17, pentachlorophenol. (Reprinted from Bedair, M. and El Rassi, Z., *Electrophoresis*, 23, 2938, 2002. With permission.)

interaction. In this context, moderately high pH values should decrease the protonation of basic compounds and therefore, minimize their retention on the surfaces with negatively charged moieties. Based on the benefit of high pH values, Adu et al. applied the methacrylate-based mixed-mode monolith column to the separation of complex therapeutic peptides, some of which contain highly basic amino acids, at alkaline pH [38]. The monolith column was suitable for the separation at moderately high pH 9.5 because of its high stability to a wide pH range. Separations with efficiencies as high as 500,000 N/m were obtained (Figure 4.13) and the separation behavior of the therapeutic peptides could be rationalized based on their charge, molecular mass/shape, and relative hydrophobicities.

In a similar manner to packed columns, the monolithic columns containing positively charged functionalities and hydrophobic ligands are a valuable alternative to SCX–RP mixed-mode monolithic ones. On SAX–RP type mixed-mode columns, the role of the fixed positive charges is not only to generate an anodic EOF but also to reduce the retention of basic analytes by electrostatic repulsion. Compared with the columns packed with SAX–RP mixed-mode silica particles that contain silanol and amino groups, SAX–RP monolithic column prepared by organic polymer can generate a stable EOF due to a single type functionality on its matrix. Gusev et al. reported the synthetic scheme including the preparation of a monolithic matrix with

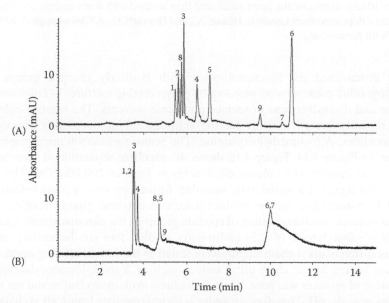

FIGURE 4.13 Separation of nine peptides under basic conditions. (A) CEC on a butyl-methacrylate-based mixed-mode monolithic column; (B) CZE on a bare fused-silica capillary. Capillary column: total length, 33.5 cm; effective length, 25 cm; 100 μm ID. Conditions: mobile phase, 50 mM sodium borate (pH 9.5)-water-acetonitrile (20:10:70, v/v/v); voltage, 10 kV; injection, electrokinetic for 2 s at 5 kV. Peaks: 1, bradykinin; 2, vasopressin; 3, LHRH; 4, substance P; 5, bradykinin fragment 1–5; 6, leucine enkephalin; 7, methionine enkephalin; 8, bombesin; 9, oxytocin. (Reprinted from Adu, J.K., Lau, S.S., Watson, D.G., Euerby, M.R., Skellern, G.G., and Tettey, J.N.A., *Electrophoresis*, 26, 3445, 2005. With permission.)

FIGURE 4.14 Scanning electron micrographs of monolithic packing in a fused-silica capillary of 75 μm ID. The specimens were first fractured and then (A) the fractured ends were cut into ca. 2 mm long pieces or (B) and (C) they were partially immersed in aqueous hydrofluoric acid for 10 min, to remove the fused-silica, and then washed with water and cut to a length of ca. 2 mm. (Reprinted from Gusev, I., Huang, X., and Horváth, C., *J. Chromatogr. A*, 855, 273, 1999. With permission.)

active groups and the functionalization with positively charged groups [43]. The monolithic columns were prepared by polymerizing mixtures of chloromethylstyrene and divinylbenzene in various porogenic solvents. The reactive chloromethyl moieties served as sites for the introduction of quaternary ammonium functionalities, *N,N*-dimethyloctylamine. The scanning electron micrographs can be seen in Figure 4.14. Figure 4.15 shows the excellent separation of three angiotensins and insulin with column efficiencies as high as 200,000 N/m. The same group also reported a methacrylic monolith formed by *in situ* polymerization of glycidyl methacrylate, methyl methacrylate, and ethylene glycol dimethacrylate, and subsequent functionalization of epoxide group on the chromatographic surface with *N*-ethylbutylamine [44]. In preliminary studies, four similar tertiary amines bearing different alkyl chain lengths were tested as the functionalizing agent. With columns having longer chain alkyl amines such as *N*-methyloctadecylamine, the separation of proteins was poor, while the column with diethylamine did not retain proteins at acidic pH. Therefore, an amine with intermediate length alkyl chain (i.e., *N*-ethylbutylamine) was used, which provided a good baseline separation of four proteins (Figure 4.16). In the case presented here solvophobic interactions were dominant and therefore, the elution order of proteins was similar to that obtained in RP chromatography. The plots of migration factors of peptides and proteins against the acetonitrile contents exhibited different trends. According to the authors, this was most likely due to the greater chromatographic retention and slower electrophoretic mobility of proteins than that of peptides in the CEC system. Furthermore,

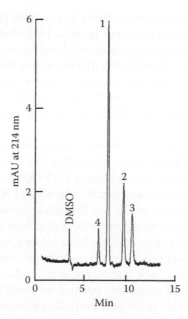

FIGURE 4.15 Separation of acidic and basic polypeptides on polystyrene-based mixed-mode monolithic column. Capillary column: total length, 31 cm; effective length, 21 cm; 75 μm ID. Conditions: mobile phase, 50 mM NaCl in 5 mM phosphate buffer with 25% acetonitrile, pH 3.0; voltage, −15 kV; injection, electrokinetic for 2 s at 5 kV. Peaks: 1, angiotensin II; 2, angiotensin I; 3, [Sar¹, Ala⁸]-angiotensin II; 4, insulin. (Reprinted from Gusev, I., Huang, X., and Horváth, C., *J. Chromatogr. A*, 855, 273, 1999. With permission.)

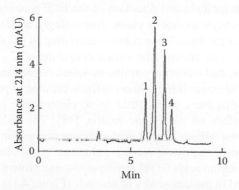

FIGURE 4.16 Separation of four proteins under isocratic elution conditions on porous methacrylic monolith column. Capillary column: total length, 39 cm; effective length, 29 cm; 50 μm ID. Conditions: mobile phase, 60 mM sodium phosphate (pH 2.5) containing 30% v/v acetonitrile; voltage, −25 kV. Peaks: 1, ribonuclease A; 2, insulin; 3, α-lactalbumin; 4, myoglobin. (Reprinted from Zhang, S., Huang, X., Zhang, J., and Horváth, C., *J. Chromatogr. A*, 887, 465, 2000. With permission.)

they evaluated the significance of temperature as a variable parameter controlling the retention behavior of both, peptides and proteins [45]. Without compromising the resolution, a twofold increase in separation speed was attained by increasing the temperature from 25°C to 55°C (Figure 4.17). The separation time could also be reduced through an increase in temperature for the separation of a tryptic digest of cytochrome c and resulted in 5 min. This is one of the first examples of employing monolith column in CEC for the separation of complex samples, especially in the context of proteomics.

The multistep preparation of monolithic column is difficult to control the amount of charged groups incorporated in the monolith and therefore, several groups had efforts to develop a single-step preparation of SAX–RP polymer-based monolithic stationary phase. The preparation and functionalization of the monolithic matrix can be combined by polymerization of the two functional monomers, one for the chromatographic interaction sites and the other for a generator of anodic EOF. Fu et al. introduced the SAX–RP mixed-mode monolithic column prepared by *in situ* polymerization of 2-(methacryloxy)ethyltrimethylammonium methyl sulfate (MEAMS) and ethylene dimethacrylate in a binary porogenic solvent (1-propanol and 1,4-butanediol) [46]. The EOF of monolithic column increased almost linearly with the increasing content of MEAMS in the polymerization mixture and therefore, the EOF is mainly determined by the amount of ionic monomer in mixtures during polymerization. As expected, the monolithic column exhibited RP-chromatographic behavior toward neutral analytes, while hydrophobic as well as electrostatic interaction/repulsion was observed for charged analytes.

Bedair and El Rassi proposed a cationic C_{17} monolithic stationary phase produced by copolymerization of PEDAS and an acrylic monomer with a quaternary amine function [47]. Four different quaternary amine acrylic monomers were tested for the maximum EOF and consequently, the best efficiency was found when using [2-(acryloyloxy)ethyl]trimethyl ammonium methyl sulfate as the quaternary amine acrylic monomer. The monolith possesses a positive zeta potential with respect to water, however, the magnitude and direction of the EOF was markedly affected by the nature of the electrolyte in mobile phase. Interestingly, anodic, zero, or cathodic EOF was observed on the monolithic column according to the electrolytes (Figure 4.18), which was due to its amphiphilic nature consisting of C_{17} chains, ester functions, hydroxyl groups, and quaternary amine moieties. At the same time, the authors applied the mixed-mode monolithic column to the separation of proteins with the use of its characteristic that there was a little or no electrostatic interaction between proteins and the surface of monolithic matrix [48]. The utility of the cationic C_{17} monolithic column allowed rapid and efficient separation of crude extracts of galactosyl transferase and cytochrome c reductase. Also, ultrafast separation of phenylthiohydration amino acids (PTH-amino acids) and proteins by short capillary column were obtained in the timescales of seconds (Figure 4.19).

Ngola et al. used novel UV-initiated acrylate-based monoliths for CEC of cationic, anionic, neutral amino acids, and peptides [49,50]. Depending on the pH of the mobile phase and the nature of the analytes, either cationic or anionic groups (quaternary amine or sulfonic acid groups) were incorporated into the monoliths for

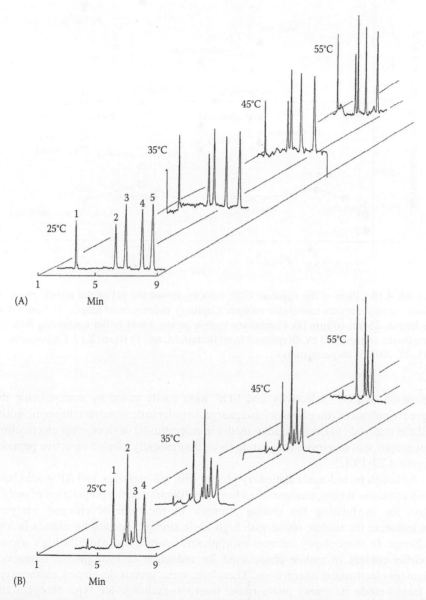

FIGURE 4.17 Effect of temperature on the separation of peptides (A) and proteins (B). Capillary column: total length, 40 cm; effective length, 30 cm; 75 μm ID. Conditions: mobile phase, 60 mM (A) or 70 mM (B) sodium phosphate (pH 2.5) containing 30% v/v acetonitrile; voltage, −30 kV; injection, electrokinetic for 2 s at −2 kV. Peaks: (A) 1, DMSO; 2, insulin; 3, angiotensin II; 4, angiotensin I; 5, [Sar^1, Ala^8]-angiotensin II. (B) 1, insulin; 2, α-lactalbumin; 3, myoglobin; 4, BSA. (Reprinted from Zhang, S., Zhang, J., and Horváth, C., *J. Chromatogr. A*, 914, 189, 2001. With permission.)

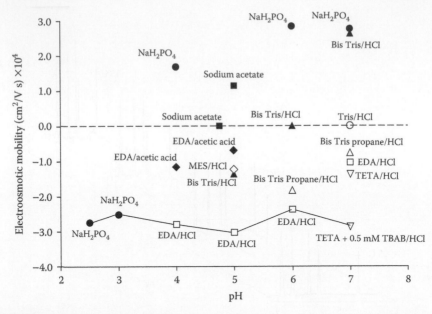

FIGURE 4.18 Plots of the apparent EOF velocity versus the pH of the mobile phase on cationic stearyl-acrylate monolithic column. Capillary column: total length, 33.5 cm; effective length, 25 cm; 100 μm ID. Conditions: mobile phase, 1 mM buffer containing 80% v/v acetonitrile; voltage, ±25 kV. (Reprinted from Bedair, M. and El Rassi, Z., *J. Chromatogr. A*, 1013, 35, 2003. With permission.)

generation of EOF. Selectivity and EOF were easily tuned by manipulating the degree of polymer hydrophobicity and charge. In order to demonstrate the applicability of these materials to the separation media in microfluidic devices, chip-electrochromatography was employed to separate three fluorescently labeled bioactive peptides (Figure 4.20) [50].

Although mixed-mode stationary phases with ion-exchange and RP modes have given attractive results, analysts often have the dilemma on the preparation of mobile phase for modulating the elution strength in the case of charged analytes. For instance, the mobile phase with high ionic strength needed for elution in ion-exchange chromatography enforces hydrophobic interactions, while the high organic modifier content in mobile phase used for reduction of hydrophobic interaction enhances electrostatic interactions. Therefore, some groups developed another type of mixed-mode stationary phase rather than ion-exchange-RP type. Hoegger and Freitag synthesized poly(*N*,*N*-dimethylacrylamide-*co*-PDA) monoliths functionalized by VSA and investigated its potential in the separation of positively charged amino acids and peptides [51,52]. In earlier report, the procedure was optimized for four types of monoliths with different hydrophobicity that was adjusted by introducing suitable comonomers, such as alkyl chain-bearing molecules, into the monolithic structure. Using a model mixture of aromatic compounds, the authors found that the retention mechanism on the monolithic column was governed neither by pure RP nor by pure normal-phase (NP) chromatography, even on monoliths having a large

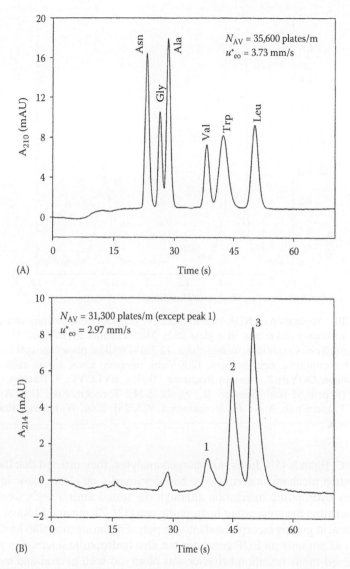

FIGURE 4.19 Ultrafast separations of PTH-amino acids (A) and proteins (B) on cationic stearyl-acrylate monolithic column. Capillary column: total length, 33.5 cm; effective length, 8.5 cm; 100 µm ID. Conditions: mobile phase, (A) 30 mM sodium phosphate (pH 2.5) containing 30% v/v acetonitrile, (B) 10 mM sodium phosphate (pH 2.5) containing 45% v/v acetonitrile; voltage, −25 kV; injection, electrokinetic for 2 s at −10 kV. Peaks: 1, concanavalin A; 2, β-lactoglobulin; 3, lysozyme. (Reprinted from Bedair, M. and El Rassi, Z., *J. Chromatogr. A*, 1013, 47, 2003. With permission.)

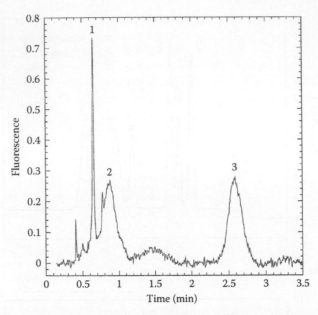

FIGURE 4.20 Separation of NDA-labeled bioactive peptides on negatively charged lauryl monolithic stationary phase cast in a glass chip. Microchannel dimensions: Deep, 25 μm; channel length, 8 cm. Conditions: mobile phase, 12.5 mM sodium phosphate (pH 7.0) containing 35% v/v acetonitrile; field strength, 1200 V/cm; injection arms, 1.0 cm each. Peaks: 1, papain inhibitor, GGYR; 2, a-casein (fragment 90–95), RYLGYL; 3, Ile-angiotensin III, RVYIHPI. (Reprinted from Shediac, R., Ngola, S.M., Throckmorton, D.J., Anex, D.S., Shepodd, T.J., and Singh, A. K., *J. Chromatogr. A*, 925, 251, 2001. With permission.)

amount of C_6 ligands [51]. Later, for charged analytes, they insisted that the mixed-mode retention mechanism seems to be based primarily on electrostatic interaction modified by hydrophilic interaction although the tested amino acids were insufficient to determine the presence of hydrophilic one [52]. Zhang and El Rassi regarded the sulfonic acid groups incorporated into the polyacrylamide monolith having dodecyl ligands as not only an EOF generator but also hydrophilic interaction sites. The RP–NP mixed-mode retention behavior was observed with neutral and moderately polar pesticides [53].

Some acrylamide monomers are polar and then, the polymerization procedures are performed in an aqueous solution, which makes the acrylamide-based monolithic stationary phases relatively hydrophilic and consequently, leads to low RP-chromatographic retention on such stationary phases. In the case of acrylamide-based monoliths in which more hydrophobic monomers are incorporated, a mixture of aqueous solution with an organic solvent is needed. As an alternative to the use of organic solvents, Wahl et al. explored host–guest complexation using derivatized cyclodextrins for the solubilization of water-insoluble monomers in aqueous medium and successfully prepared the acrylamide-based mixed-mode stationary phase consisting of methacrylamide, PDA, VSA, and an additional bisacrylamide cross-linker with octylene and dodecylene bridge [54]. Later, the technique was employed in the synthesis of methacrylate-based monolithic stationary phase consisting of different hydrophobic monomers (isobornyl, adamantyl, cyclohexyl, and phenyl methacrylate), PDA, and VSA by the same group [55]. Due to its amphiphilic nature, hydrophobic analytes were eluted with polar mobile phase according to RP retention mechanism, while polar analytes were eluted with nonpolar mobile phase according to NP retention mechanism. However, the elution of polar analytes with a polar mobile phase can be explained by the mixed-mode retention mechanism combining hydrophobic interaction with van der Waals interactions between the analytes and the stationary phase.

Up to date, most of mixed-mode monolithic stationary phases are polymer-based monoliths as they possess excellent pH stability and many monomeric precursors with different features are commercially available. An alternative to polymer-based monoliths is the silica-based monolith prepared by sol-gel chemistry. Although much less has been done compared with polymer-based monoliths, some workers presented the preparation procedure of mixed-mode silica-based monoliths. Hayes and Malik proposed a single-step procedure to prepare the silica-based monolithic stationary phase having the C_{18} ligand with a positively charged quaternary amine moiety [56]. In this method, the introduction of the C_{18} ligand was performed by addition of N-octadecyldimethyl[3-(trimethoxysilyl)propyl]ammonium chloride (C_{18}-TMS) to the sol solution that contained tetramethoxysilane (TMOS, coprecursor), phenyldimethylsilane (PheDMS, deactivation agent), and trifluoroacetic acid (catalyst). All the processes including the formation of sol-gel monolithic matrix, the introduction of the chromatographically active C_{18} ligand, and the deactivation of unreacted silanol groups were accomplished in one step (Figure 4.21). The cross-sectional view and the surface view of the monolithic stationary phase are shown in Figure 4.22. The monolithic column will exhibit the retention behavior of charged analytes based on mixed-mode interactions (i.e., anion-exchange-RP), however, this column was evaluated only in the separations of neutral analytes in the study. More recently, Ding et al. presented the single-step synthesis of a weak anion-exchange (WAX)-RP mixed-mode monolithic stationary phase that was prepared by three precursors: tetraethoxysilane, aminopropyltriethoxysilane (APTES), and octyltriethoxysilane [57]. The amino group of APTES served as a basic catalyst in the sol-gel process, an anodic EOF generator and a weak anion-exchanger. In a similar manner to mixed-mode silica materials including an anion-exchange functionality, the WAX-RP monolithic column possessed a characteristic of EOF, as shown in Figure 4.23, because the net charge on the surface of matrix was determined by both silanol and amino groups.

Allen and El Rassi introduced three different synthetic routes for the preparation of amphiphilic silica-based monoliths possessing the C_{18} ligand and positively charged groups [58]. In contrast to the single-step preparation procedures mentioned above, the preparation scheme consisted of the preparation of sol-gel silica backbone and the introduction of a hydrophobic moiety with the positive charge (Figure 4.24). The cationic C_{18}-monoliths exhibited RP-chromatographic behavior toward nonpolar analytes (alkyl benzenes), while it exhibited mixed-mode retention behavior (i.e., RP–NP) toward slightly polar analytes such as anilines and PTH-amino acids. Also, Ye et al. proposed the two-step procedure for a silica-based monolithic stationary phase chemically modified with 3-aminopropyltrimethoxysilane for pressurized CEC [59]. The amino groups on the surface of the polar stationary phase generated an anodic EOF under acidic conditions and served as a weak anion-exchanger. The anionic analytes such as nucleotides were separated by the mixed-mode retention mechanism which comprised hydrophilic interaction and WAX (Figure 4.25).

In addition to the mixed-mode CEC column with one-directional EOF (i.e., a cathodic or an anodic EOF), however, some groups applied the zwitterionic monolith column that can generate EOF with different direction depending on mobile phase pH. Although there have been some papers dealing with zwitterionic separation materials for HPLC, Fu et al. are a pioneer of the zwitterionic stationary phase

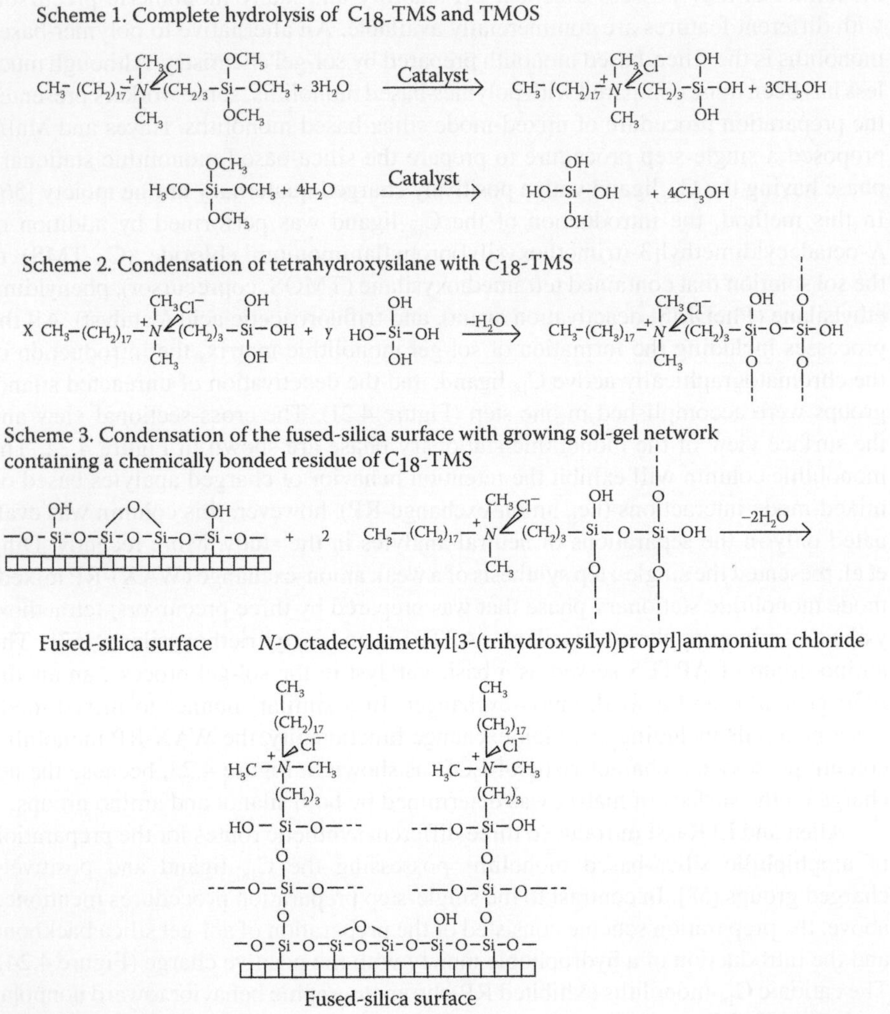

Scheme 1. Complete hydrolysis of C_{18}-TMS and TMOS

$$CH_3^- (CH_2)_{17} \overset{+}{N} \overset{CH_3Cl^-}{\underset{CH_3}{|}} (CH_2)_3 - \overset{OCH_3}{\underset{OCH_3}{\overset{|}{Si}}} - OCH_3 + 3H_2O \xrightarrow{\text{Catalyst}} CH_3^- (CH_2)_{17} \overset{+}{N} \overset{CH_3Cl^-}{\underset{CH_3}{|}} (CH_2)_3 - \overset{OH}{\underset{OH}{\overset{|}{Si}}} - OH + 3CH_3OH$$

$$H_3CO - \overset{OCH_3}{\underset{OCH_3}{\overset{|}{Si}}} - OCH_3 + 4H_2O \xrightarrow{\text{Catalyst}} HO - \overset{OH}{\underset{OH}{\overset{|}{Si}}} - OH + 4CH_3OH$$

Scheme 2. Condensation of tetrahydroxysilane with C_{18}-TMS

$$X\ CH_3-(CH_2)_{17} \overset{+}{N} \overset{CH_3Cl^-}{\underset{CH_3}{|}} (CH_2)_3 - \overset{OH}{\underset{OH}{\overset{|}{Si}}} - OH + y\ HO - \overset{OH}{\underset{OH}{\overset{|}{Si}}} - OH \xrightarrow{-H_2O} CH_3-(CH_2)_{17} \overset{+}{N} \overset{CH_3Cl^-}{\underset{CH_3}{|}} (CH_2)_3 - \overset{OH}{\underset{O}{\overset{|}{Si}}} - O - \overset{O}{\underset{O}{\overset{|}{Si}}} - OH$$

Scheme 3. Condensation of the fused-silica surface with growing sol-gel network containing a chemically bonded residue of C_{18}-TMS

$$-O-\overset{OH}{\underset{}{\overset{|}{Si}}}-O-\overset{O}{\underset{}{\overset{\diagup \diagdown}{Si}}}-O-\overset{OH}{\underset{}{\overset{|}{Si}}}-O-\overset{}{\underset{}{Si}}-O- + 2\ CH_3-(CH_2)_{17}-\overset{+}{N} \overset{CH_3Cl^-}{\underset{CH_3}{|}} (CH_2)_3 - \overset{OH}{\underset{O}{\overset{|}{Si}}} - O - \overset{O}{\underset{O}{\overset{|}{Si}}} - OH \xrightarrow{-2H_2O}$$

Fused-silica surface   *N*-Octadecyldimethyl[3-(trihydroxysilyl)propyl]ammonium chloride

Fused-silica surface

FIGURE 4.21 Synthesis scheme for silica-based monolithic stationary phase having the C_{18} ligand with a positively charged quaternary amine moiety. (Reprinted from Hayes, J.D. and Malik, A., *Anal. Chem.*, 72, 4090, 2000. With permission.)

(*continued*)

Scheme 4. Deactivation of the sol-gel ODS monolith with PheDMS

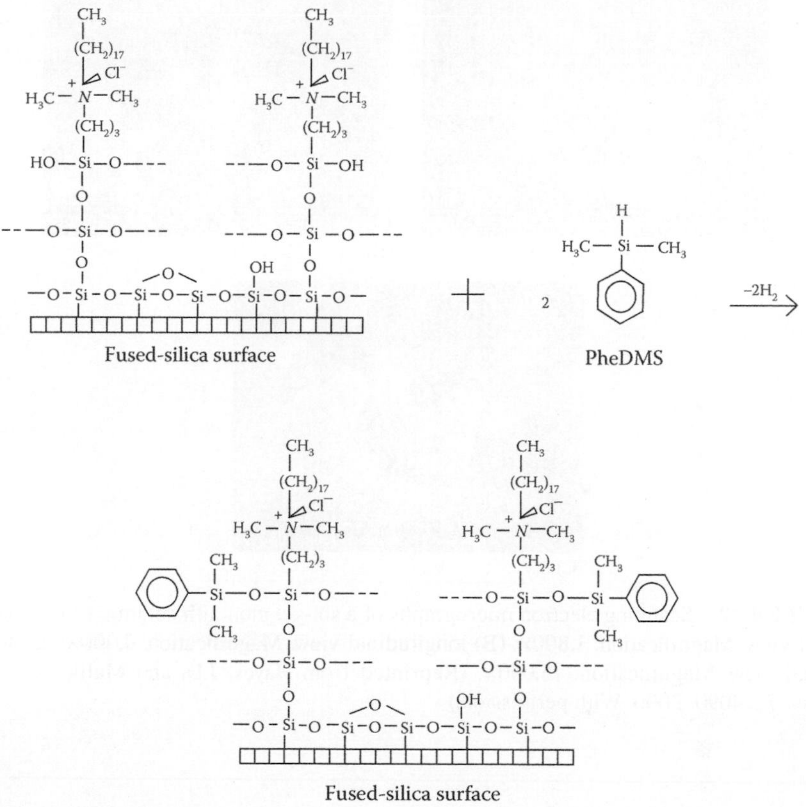

Fused-silica surface   PheDMS

Fused-silica surface

FIGURE 4.21 (continued)

designed for CEC. Most zwitterionic separation materials used in HPLC carry positively charged quaternary ammonium groups and negatively charged sulfonic acid groups and therefore, both, the positive and negative groups retain their charge over the entire pH range; i.e., the overall zwitterionic moiety maintains a zero net charge. The group pointed out that this character is a disadvantage in the case of CEC and proposed the novel zwitterionic monolithic columns carrying tertiary amine groups and acrylic acid groups which were prepared by *in situ* polymerization of butylmethacrylate, ethylene dimethacrylate, methacrylic acid, and 2-(dimethyl amino)ethyl methacrylate [60]. The EOF on the zwitterionic stationary phase can be easily controlled by adjusting the pH value of mobile phase because the monolithic column with tertiary amine groups or with acrylic acid groups resulted in a weak anion-exchanger (Figure 4.26). The proposed monolithic column exhibited the hydrophobic retention behavior toward neutral analytes. The electrostatic interaction/repulsion between the stationary phase and charged analytes can be tuned by adjusting the pH value of mobile phase. No serious peak tailing of the anilines on the stationary phase was observed with acidic and basic electrolytes and different selectivities of anilines were obtained with pH values

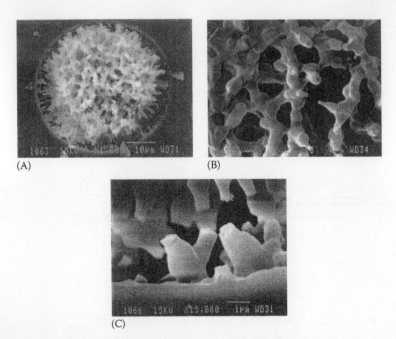

FIGURE 4.22 Scanning electron micrographs of a sol-gel monolith column. (A) cross-sectional view. Magnification, 1,800×; (B) longitudinal view. Magnification, 7,000×; (C) longitudinal view. Magnification, 15,000×. (Reprinted from Hayes, J.D. and Malik, A., *Anal. Chem.*, 72, 4090, 2000. With permission.)

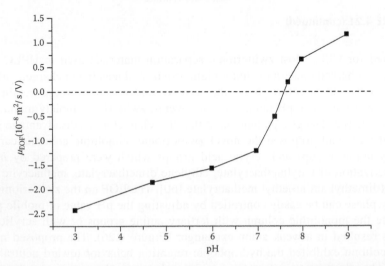

FIGURE 4.23 Plot of EOF mobility versus pH of the mobile phase on WAX-RP mixed-mode silica-based monolithic column. Conditions: mobile phase, 10 mM phosphate containing 20% v/v acetonitrile; voltage, 10 kV; injection, electrokinetic for 2 s at 3 kV. (Reprinted from Ding, G., Da, Z., Yuan, R., and Bao, J.J., *Electrophoresis*, 27, 3363, 2006. With permission.)

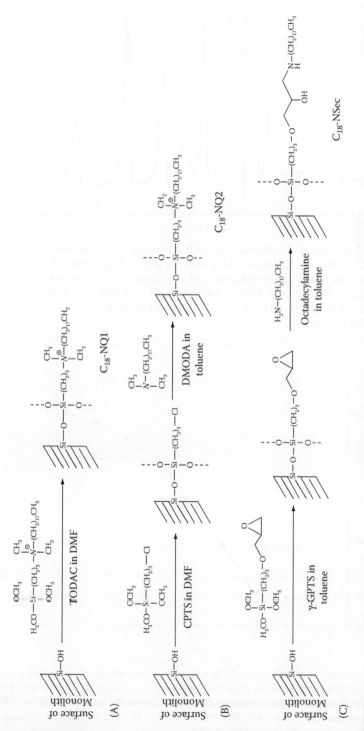

FIGURE 4.24 Schematic of three reaction pathways for amphiphilic silica-based monolithic column. (A) Reaction of [3-(trimethoxysilyl)propyl]octa-decyldimethyl ammonium chloride (TODAC) in DMF at 120°C. (B) Reaction of chloropropyltrimethoxysilane (CPTS) in DMF at 120°C followed by reaction of *N,N*-dimethyloctadecylamine (DMODA) in toluene at 70°C. (C) Reaction of (γ-glycidoxypropyl)trimethoxysilane (γ-GPTS) in toluene at 110°C followed by reaction with octadecylamine in toluene at 80°C. (Reprinted from Allen, D. and El Rassi, Z. *Analyst*, 128, 1249, 2003. With permission.)

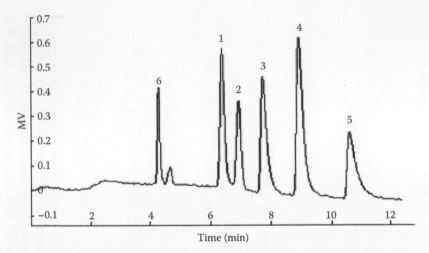

FIGURE 4.25 Separation of nucleotides on WAX-NP mixed-mode silica-based monolithic column. Capillary column: total length, 55 cm; effective length, 25 cm; 75 μm ID. Conditions: mobile phase, 40 mM triethylamine phosphate (pH 3.5) containing 60% v/v acetonitrile; voltage, 15 kV; pressure, 100 psi; flow, 0.02 mL/min. Peaks: 1, 5′-UMP; 2, 5′-AMP; 3, 5′-IMP; 4, 5′-CMP; 5, 5′-GMP. (Reprinted from Ye, F., Xie, Z., and Wong, K.-Y., *Electrophoresis*, 27, 3373, 2006. With permission.)

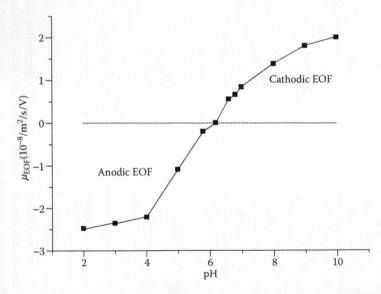

FIGURE 4.26 Effect of the pH on the EOF velocity on the zwitterionic monolith column. Capillary column: total length, 32 cm; effective length, 8.5 cm; 100 μm ID. Conditions: mobile phase, 10 mM sodium phosphate containing 40% v/v acetonitrile; voltage, 10 kV or −10 kV; injection, electrokinetic for 5 s at 5 kV. (Reprinted from Fu, H., Xie, C., Dong, J., Huang, X., and Zou, H., *Anal. Chem.*, 76, 4866, 2004. With permission.)

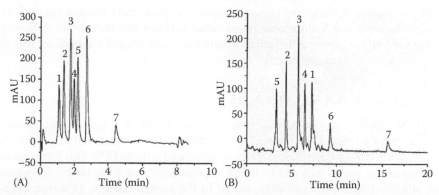

FIGURE 4.27 Separation of anilines on the zwitterionic monolith column. Conditions: mobile phase, 10 mM sodium phosphate ([A] pH 2.0 and [B] pH 8.0) containing 40% v/v acetonitrile; voltage, 20 kV. Other conditions are as specified in Figure 4.26. (Reprinted from Fu, H., Xie, C., Dong, J., Huang, X., and Zou, H., *Anal. Chem.*, 76, 4866, 2004. With permission.)

at 2.0 and 8.0 (Figure 4.27). Figure 4.28 shows typical electrochromatograms of peptides on the zwitterionic monolithic column at pH 2.0 and 8.0. Similar to anilines, the peptides were separated on the column under counterdirectional mode at pH 2.0 due to the same charge (positive) of the peptides and the stationary phase. The ionization of acrylic acid groups on the zwitterionic stationary phase worked as a weak anion-exchanger under basic mobile phase and therefore, the electrostatic interaction between the peptides and the stationary phase is not so strong as the peptides and strong cation-exchanger. Subsequently, the same group prepared the zwitterionic

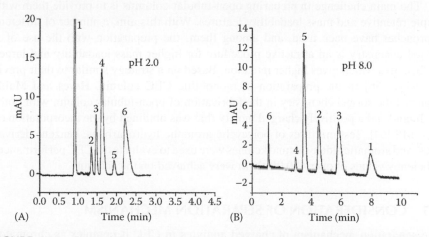

FIGURE 4.28 Separation of peptides on the zwitterionic monolith column. Conditions: mobile phase, 20 mM sodium phosphate ([A] pH 2.0 and [B] 8.0) containing 20% v/v acetonitrile; voltage, 10 kV. Other conditions are as specified in Figure 4.26. Peaks: 1, Gly-Glu; 2, Ala-Ile; 3, Gly-Phe; 4, Gly-His; 5, His-Phe; 6, Lys-Ser-Try. (Reprinted from Fu, H., Xie, C., Dong, J., Huang, X., and Zou, H., *Anal. Chem.*, 76, 4866, 2004. With permission.)

methacrylate-based stationary phase containing a chemically bonded lysine [61]. The electrostatic and hydrophobic interactions between this monolithic stationary phase and ionic analytes, and electrophoretic migration played a significant role in the separation of ionic analytes.

4.4   OPEN-TUBULAR COLUMNS

Open-tubular format represents simple column design in CEC. In this format, a stable surface-bonded coating needs to be created on the inner wall of a capillary for obtaining efficient chromatographic retention and reliable EOF. This format does not require the use of frits and therefore is practically free from bubble formation and other technical problems encountered in the packed column. However, open-tubular columns are not widely used in practical CEC. This is partially due to the slow solute diffusion in the liquid mobile phase and to the low sample capacity of the open-tubular columns.

Sawada and Jinno prepared open-tubular columns for CEC using copolymerization of N-tert-butylacrylamide with AMPS [62]. In this study, EOF was generated as a result of the incorporation of the negatively charged AMPS into the chemical structure of the polymeric coating. Huang et al. presented a functionalized polymeric porous layer grafted on the inner wall of $20\,\mu m$ ID fused-silica capillaries as a porous-layer open-tubular (PLOT) column (Figure 4.29) [63]. The porous layer was highly cross-linked and was prepared by in situ polymerization of vinylbenzyl chloride and divinylbenzene in the presence of 2-octanol as a porogen. The chloromethyl moieties on the surface of the polymeric matrix were reacted with N,N-dimethyldodecylamine to obtain C_{12} alkyl chains with positively charged sites. Separations of the four basic proteins by CEC using the column and by CZE are shown in Figure 4.30.

The main challenge in preparing open-tubular columns is to provide them with ample retentive and mass loadability features. With this aim, a number of different approaches have been used, and among them, the preparation with the use of a sol-gel chemistry is an attractive procedure for higher mass loadability and larger surface area which gives higher retention. Based on a strategy similar to their previous work [56] in the preparation of monolithic CEC column, Hayes and Malik explored the sol-gel chemistry in the fabrication of open-tubular column with both, C_{18} ligand and a positively charged moiety that was obtained by the incorporation of C_{18}-TMS [64]. Test mixtures of polycyclic aromatic hydrocarbons, benzene derivatives, and aromatic aldehydes and ketones were used to evaluate the CEC performance. Efficiency values of over $400,000\,N/m$ were achieved on the column.

4.5   CONSIDERATION OF SEPARATION MECHANISM

The separation mechanism of charged analytes in CEC is complex as chromatographic interactions, EOF, and electrophoresis contribute to the experimentally observed separation behavior. On mixed-mode CEC column, the multiple retentive interaction makes the separation mechanism more complex. The question of whether the different effects contribute in an independent or interdependent

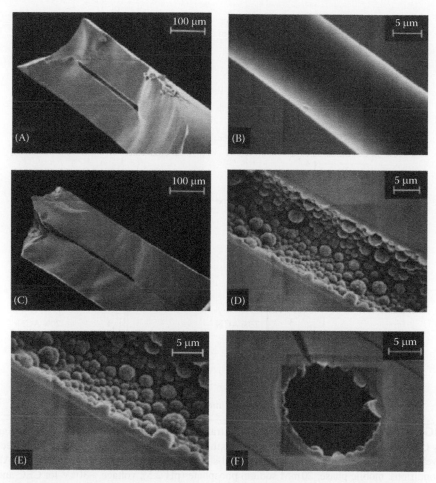

FIGURE 4.29 Scanning electron micrographs of the fused-silica capillary and a PLOT column. (A) Fractured end of fused-silica capillary of 20 μm ID; (B) enlarged lumen of the fused-silica capillary shown in (A); (C) fractured end of a PLOT column; (D) the rugulose porous layer in capillary column shown in (C); (E) the rugulose porous layer at higher magnification than in (D); (F) cross section of a PLOT column. (Reprinted from Huang, X., Zhang, J., and Horváth, C., *J. Chromatogr. A*, 858, 91, 1999. With permission.)

manner is still under discussion. Freitag and Hilbrig discussed the state of the art in the theoretical description of the individual contributions as well as models for the retention behavior [65]. The authors concluded that it is currently not possible to use the potential benefit of the presence of several contributions to the CEC separation in a controlled manner to bring about superior separations. Also, the full mathematical simulation of the factors determining the migration/separation behavior of charged analytes is not yet possible, although considerable progress has been made over the last decade [7,66–73].

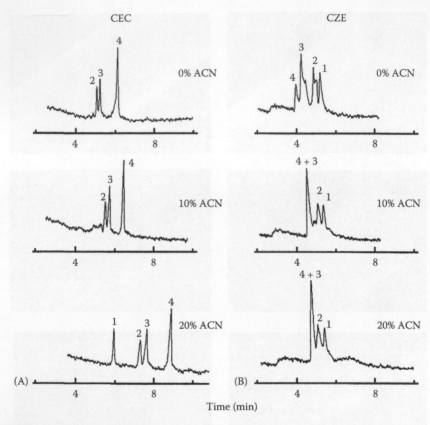

FIGURE 4.30 Effect of acetonitrile concentration on the separation of four basic proteins by CEC and CZE. (A) PLOT column (ca. 2 μm thick polymer layer); total length, 47 cm; effective length, 40 cm; 20 μm ID; (B) raw fused-silica capillary dimensions are same as (A). Conditions: mobile phase, 20 mM sodium phosphate (pH 2.5); voltage, −30 kV for CEC and 30 kV for CZE. Peaks: 1, α-chymotrypsinogen A; 2, ribonuclease A; 3, cytochrome c; 4, lysozyme. (Reprinted from Huang, X., Zhang, J., and Horváth, C., *J. Chromatogr. A*, 858, 91, 1999. With permission.)

Very recently, Al-Rimawi and Pyell investigated whether a "one-site" model or a "two-site" model is adequate for understanding the retention behavior toward charged analytes on the mixed-mode monolith column (Figure 4.31) [74]. In this study, the authors employed the methacrylate-based mixed-mode monolithic column and alkylaniline, amino acids, and peptides as model analytes. The dependence of the corrected retention factor on the concentration of the counterion ammonium and the number of methylene groups in the alkyl chain of the model analytes clearly supported the one-site model, in which the charged analytes undergo simultaneous interaction with a hydrophobic site and the ionized sulfonic acid groups. Furthermore, from the comparison of CEC of these charged analytes with electrophoretic mobility determined by CZE shows that the selectivity

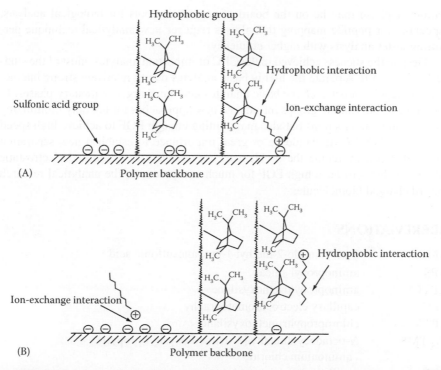

FIGURE 4.31 Schematic illustration of the interaction of a cationic solute binding on to the mixed-mode polymeric stationary phase according to (A) one-site model and (B) two-site model. (Reprinted from Al-Rimawi, F. and Pyell, U., *J. Chromatogr. A*, 1160, 326, 2007. With permission.)

observed in mixed-mode CEC is mainly attributed to chromatographic retention (hydrophobic interaction and ion-exchange), while the electrophoretic migration of these analytes plays a minor role.

4.6 CONCLUSION

Several attractive works in the CEC of charged analytes with mixed-mode type columns have been made and, at the same time, most of them provided significant progress in the design of a novel mixed-mode stationary phase suited for CEC. In initial works, the strategy for mixed-mode stationary phase was to develop ion-exchange and RP type mixed-mode stationary phase. However, the interest of some researchers shifted to another type of mixed-mode stationary phase, i.e., ion-exchange and NP, RP, and NP, and therefore, the application scope of CEC with mixed-mode column is expanding.

Although, up to date, CEC with mixed-mode stationary phase has not yet found wide acceptance as a practical separation tool in chromatographic community, one of the research trends of CEC is the development of novel type mixed-mode column and so further improvement in this area is promising. CEC with mixed-mode stationary phase may be on the board of separation tools for biological analysis, especially for peptide mapping that always requires new analytical technique permitting faster analysis with higher efficiency.

Despite the success achieved in the CEC of important analytes, most of the studies have failed to separate the highly basic analytes due to relatively strong interaction and repulsion between the basic analytes and the charged stationary phases. In this context, the leading groups introduced new approaches for not only minimizing the coulombic interactions but also maintaining enough EOF to achieve high speed separations [75–77]. It should be of great importance to establish new separation media that can overcome the problem of irreversible adsorption and electrostatic interaction but still have high EOF for much acceptance in the analytical research toward charged biomolecules.

ABBREVIATIONS

AMPS	2-acrylamido-2-methyl-1-propanesulfonic acid
APS	aminopropyl silica
APTES	aminopropyltriethoxysilane
CEC	capillary electrochromatography
CPTS	chloropropyltrimethoxysilane
C_{18}-TMS	N-octadecyldimethyl[3-(trimethoxysilyl)propyl] ammonium chloride
CZE	capillary zone electrophoresis
DMF	N,N-dimethyl formamide
DMODA	N,N-dimethyloctadecylamine
DMSO	dimethyl sulfoxide
EEDQ	1-(ethoxycarbonyl)-2-ethoxy-1,2-dihydroquinoline
EOF	electroosmotic flow
GPTS	(γ-glycidoxypropyl) trimethoxysilane
HPLC	high-performance liquid chromatography
MAPTA	(3-(methacryloylamino)propyl)trimethylammonium chloride
MEAMS	2-(methacryloxy)ethyltrimethylammonium methyl sulfate
NIPAAm	N-isopropylacrylamide
NP	normal-phase
ODS	octadecyl silica
ODSS	octadecylsulfonated silica
PDA	piperazine diacrylamide
PEDAS	pentaerythritol diacrylate monostearate
PheDMS	phenyldimethylsilane
PLOT	porous-layer open-tubular
PTH	phenylthiohydration
RP	reversed-phase
SAX	strong anion-exchange
SCX	strong cation-exchange
SEMA	2-(sulfooxy) ethyl methacrylate

SNAIP	3-(4-sulfo-1,8-naphthalimido)propyl-modified silica
TMOS	tetramethoxysilane
TODAC	[3-(trimethoxysilyl)propyl]octadecyldimethyl ammonium chloride
VSA	vinylsulfonic acid
WAX	weak anion-exchange

REFERENCES

1. Pretorius, V., Hopkins, B.J., and Shieke, J.D. Electro-osmosis: A new concept for high-speed liquid chromatography. *J. Chromatogr.*, 1974, 99, 23–30.
2. Jorgensen, J.W. and Lukas, J. High-resolution separations based on electrophoresis and electroosmosis. *J. Chromatogr.*, 1981, 218, 209–216.
3. Knox, J.H. and Grant, I.H. Miniaturization in pressure and electroendoosmotically driven liquid chromatography: Some theoretical considerations. *Chromatographia*, 1987, 24, 135–143.
4. Krull, I.S., Stevenson, R.L., Mistry, K., and Swartz, M.E. *Capillary Electrochromatography and Pressurized Flow Capillary Electrochromatography*, 2000, HNB Publishing, New York.
5. Deyl, Z. and Svec, F. *Capillary Electrochromatography*, 2001, Elsevier, Amsterdam, the Netherlands.
6. Smith, N. and Evans, M.B. Comparison of the electroosmotic flow profiles and selectivity of stationary phases used in capillary electrochromatography. *J. Chromatogr. A*, 1999, 832, 41–54.
7. Walhagen, K., Unger, K.K., and Hearn, M.T.W. Capillary electrochromatography analysis of hormonal cyclic and linear peptides. *Anal. Chem.*, 2001, 73, 4924–4936.
8. Adam, T. and Unger, K.K. Comparative study of capillary electroendosmotic chromatography and electrically assisted gradient nano-liquid chromatography for the separation of peptides. *J. Chromatogr. A*, 2000, 894, 241–251.
9. Zhang, M. and El Rassi, Z. Capillary electrochromatography with novel stationary phases. I. Preparation and characterization of octadecyl-sulfonated silica. *Electrophoresis*, 1998, 19, 2068–2072.
10. Zhang, M. and El Rassi, Z. Capillary electrochromatography with novel stationary phases: II. Studies of the retention behavior of nucleosides and bases on capillaries packed with octadecyl-sulfonated-silica microparticles. *Electrophoresis*, 1999, 20, 31–36.
11. Ohyama, K., Shirasawa, Y., Wada, M., Kishikawa, N., Ohba, Y., Nakashima, K., and Kuroda, N. Investigation of the novel mixed-mode stationary phase for capillary electrochromatography. I. Preparation and characterization of sulfonated naphthalimido-modified silyl silica gel. *J. Chromatogr. A*, 2004, 1042, 189–195.
12. Ohyama, K., Wada, M., Kishikawa, N., Ohba, Y., Nakashima, K., and Kuroda, N. Stepwise gradient of buffer concentration for capillary electrochromatography of peptides on sulfonated naphthalimido-modified silyl silica gel. *J. Chromatogr. A*, 2005, 1064, 255–259.
13. Ohyama, K., Fujimoto, E., Wada, M., Kishikawa, N., Ohba, Y., Akiyama, S., Nakashima, K., and Kuroda, N. Investigation of the novel mixed-mode stationary phase for capillary electrochromatography. Part III: Separation of nucleosides and nucleic acid bases on sulfonated naphthalimido-modified silyl silica gel. *J. Sep. Sci.*, 2005, 28, 767–773.
14. Byrne, C.D., Smith, N.W., Dearie, H.S., Moffatt, F., Wren, S.A., and Evans, K.P. Influence of the unpacked section on the chromatographic performance of duplex strong anion-exchange columns in capillary electrochromatography. *J. Chromatogr. A*, 2001, 927, 169–177.
15. Scherer, B. and Steiner, F. Application of hydrophobic anion-exchange phases in capillary electrochromatography. *J. Chromatogr. A*, 2001, 924, 197–209.

16. Lämmerhofer, M. and Linder, W. High-efficiency chiral separations of N-derivatized amino acids by packed-capillary electrochromatography with a quinine-based chiral anion-exchange type stationary phase. *J. Chromatogr. A*, 1998, 829, 115–125.

17. Lämmerhofer, M., Tobler, E., and Linder, W. Chiral anion-exchangers applied to capillary electrochromatography enantioseparation of oppositely charged chiral analytes: Investigation of stationary and mobile phase parameters. *J. Chromatogr. A*, 2000, 887, 421–437.

18. Progent, F. and Taverna, M. Retention behavior of peptides in capillary electrochromatography using an embedded ammonium in dodecacyl stationary phase. *J. Chromatogr. A*, 2004, 1052, 181–189.

19. Tang, Q. and Lee, M.L. Capillary electrochromatography using continuous-bed columns of sol-gel bonded silica particles with mixed-mode octadecyl and propylsulfonic acid functional groups. *J. Chromatogr. A*, 2000, 887, 265–275.

20. Tang, Q., Wu, N., and Lee, M.L. Continuous bed columns containing sol-gel bonded large-pore octadecylsilica for capillary electrochromatography. *J. Micro. Sep.*, 1999, 11, 550–561.

21. Zhang, X. and Colón, L.A. Evaluation of poly{-N-isopropylacrylamide-co-[3-(methacryloylamino)propyl]-trimethylammonium} as a stationary phase for capillary electrochromatography. *Electrophoresis*, 2006, 27, 1060–1068.

22. Kanazawa, H., Kashiwase, Y., Yamamoto, K., Matsushima, Y., Kikuchi, A., Sakurai, Y., and Okano, T. Temperature-responsive liquid chromatography. 2. Effects of hydrophobic groups in N-isopropylacrylamide copolymer-modified silica. *Anal. Chem.*, 1997, 69, 823–830.

23. Svec, F., Peters, E.C., Sykora, D., and Fréchet, J.M.J. Design of the monolithic polymers used in capillary electrochromatography columns. *J. Chromatogr. A*, 2000, 887, 3–29.

24. Pyell, U. Advances in column technology and instrumentation in capillary electrochromatography. *J. Chromatogr. A*, 2000, 892, 257–278.

25. Hilder, E.F., Svec, F., and Fréchet, J.M.J. Polymeric monolithic stationary phase for capillary electrochromatography. *Electrophoresis*, 2002, 23, 3934–3953.

26. Zou, H., Huang, X., Ye, M., and Luo, Q. Monolithic stationary phases for liquid chromatography and capillary electrochromatography. *J. Chromatogr. A*, 2002, 954, 5–32.

27. Allen, D. and El Rassi, Z. Silica-based monoliths for capillary electrochromatography: Methods of fabrication and their applications in analytical separations. *Electrophoresis*, 2003, 24, 3962–3976.

28. Bedair, M. and El Rassi, Z. Recent advances in polymeric monolithic stationary phases for electrochromatography in capillaries and chips. *Electrophoresis*, 2004, 25, 4110–4119.

29. Svec, F. Recent developments in the field of monolithic stationary phases for capillary electrochromatography. *J. Sep. Sci.*, 2005, 28, 729–745.

30. Eeltink, S. and Svec, F. Recent advances in the control of morphology and surface chemistry of porous polymer-based monolithic stationary phases and their applications in CEC. *Electrophoresis*, 2007, 28, 137–147.

31. Fujimoto, C., Kino, J., and Sawada, H. Capillary electrochromatography of small molecules in polyacrylamide gels with electroosmotic flow. *J. Chromatogr. A*, 1995, 716, 107–113.

32. Fujimoto, C., Fujise, Y., and Matsuzawa, E. Fritless packed columns for capillary electrochromatography: Separation of uncharged compounds on hydrophobic hydrogels. *Anal. Chem.*, 1996, 68, 2753–2757.

33. Liao, J.-L., Chen, N., Ericson, C., and Hjertén, S. Preparation of continuous beds derivatized with one-step alkyl and sulfonate groups for capillary electrochromatography. *Anal. Chem.*, 1996, 68, 3468–3472.

34. Enlund, A.M., Ericson, C., Hjertén, S., and Westerlund, D. Capillary electrochromatography of hydrophobic amines on continuous beds. *Electrophoresis*, 2001, 22, 511–517.
35. Peters, E.C., Petro, M., Svec, F., and Fréchet, J.M.J. Molded rigid polymer monoliths as separation media for capillary electrochromatography. *Anal. Chem.*, 1997, 69, 3646–3649.
36. Peters, E.C., Petro, M., Svec, F., and Fréchet, J.M.J. Molded rigid polymer monoliths as separation media for capillary electrochromatography. 1. Fine control of porous properties and surface chemistry. *Anal. Chem.*, 1998, 70, 2288–2295.
37. Peters, E.C., Petro, M., Svec, F., and Fréchet, J.M.J. Molded rigid polymer monoliths as separation media for capillary electrochromatography. 2. Effect of chromatographic conditions on the separation. *Anal. Chem.*, 1998, 70, 2296–2302.
38. Adu, J.K., Lau, S.S., Watson, D.G., Euerby, M.R., Skellern, G.G., and Tettey, J.N.A. Capillary electrochromatography of therapeutic peptides on mixed-mode butyl-methacrylate monoliths. *Electrophoresis*, 2005, 26, 3445–3451.
39. Yu, C., Svec, F., and Fréchet, J.M.J. Towards stationary phases for chromatography on a microchip: Molded porous polymer monoliths prepared in capillaries by photoinitiated *in situ* polymerization as separation media for electrochromatography. *Electrophoresis*, 2000, 21, 120–127.
40. Ping, G., Zhang, W., Zhang, L., Schmitt-Kopplin, P., Zhang, Y., and Kettrup, A. Rapid separation of nucleosides by capillary electrochromatography with a methacrylate-based monolithic stationary phase. *Chromatographia*, 2003, 57, 629–633.
41. Wu, R., Zou, H., Fu, H., Jin, W., and Ye, M. Separation of peptides on mixed mode of reversed-phase and ion-exchange capillary electrochromatography with a monolithic column. *Electrophoresis*, 2002, 23, 1239–1245.
42. Bedair, M. and El Rassi, Z. Capillary electrochromatography with monolithic stationary phases: 1. Preparation of sulfonated stearyl acrylate monoliths and their electrochromatographic characterization with neutral and charged solutes. *Electrophoresis*, 2002, 23, 2938–2948.
43. Gusev, I., Huang, X., and Horváth, C. Capillary columns with in situ formed porous monolithic packing for micro high-performance liquid chromatography and capillary electrochromatography. *J. Chromatogr. A*, 1999, 855, 273–290.
44. Zhang, S., Huang, X., Zhang, J., and Horváth, C. Capillary electrochromatography of proteins and peptides with a cationic acrylic monolith. *J. Chromatogr. A*, 2000, 887, 465–477.
45. Zhang, S., Zhang, J., and Horváth, C. Rapid separation of peptides and proteins by isocratic capillary electrochromatography at elevated temperature. *J. Chromatogr. A*, 2001, 914, 189–200.
46. Fu, H., Xie, C., Xiao, H., Dong, J., Hu, J., and Zou, H. Monolithic columns with mixed modes of reversed-phase and anion-exchange stationary phase for capillary electrochromatography. *J. Chromatogr. A*, 2004, 1044, 237–244.
47. Bedair, M. and El Rassi, Z. Capillary electrochromatography with monolithic stationary phases. II. Preparation of cationic stearyl-acrylate monoliths and their electrochromatographic characterization. *J. Chromatogr. A*, 2003, 1013, 35–45.
48. Bedair, M. and El Rassi, Z. Capillary electrochromatography with monolithic stationary phases. III. Evaluation of the electrochromatographic retention of neutral and charged solutes on cationic stearyl-acrylate monoliths and the separation of water-soluble proteins and membrane proteins. *J. Chromatogr. A*, 2003, 1013, 47–56.
49. Ngola, S.M., Fintschenko, Y., Choi, W.-Y., and Shepodd, T.J. Conduct-as-cast polymer monoliths as separation media for capillary electrochromatography. *Anal. Chem.*, 2001, 73, 849–856.
50. Shediac, R., Ngola, S.M., Throckmorton, D.J., Anex, D.S., Shepodd, T.J., and Singh, A. K. Reversed-phase electrochromatography of amino acids and peptides using porous polymer monoliths. *J. Chromatogr. A*, 2001, 925, 251–263.

51. Hoegger, D. and Freitag, R. Acrylamide-based monoliths as robust stationary phases for capillary electrochromatography. *J. Chromatogr. A*, 2001, 914, 211–222.

52. Hoegger, D. and Freitag, R. Investigation of mixed-mode monolithic stationary phases for the analysis of charged amino acids and peptides by capillary electrochromatography. *J. Chromatogr. A*, 2003, 1004, 195–208.

53. Zhang, M. and El Rassi, Z. Capillary electrochromatography with polyacrylamide monolithic stationary phases having bonded dodecyl ligands and sulfonic acid groups: Evaluation of column performance with alkyl phenyl ketones and neutral moderately polar pesticides. *Electrophoresis*, 2001, 22, 2593–2599.

54. Wahl, A., Schnell, I., and Pyell, U. Capillary electrochromatography with polymeric continuous beds synthesized via free radical polymerization in aqueous media using derivatized cyclodextrins as solubilizing agents. *J. Chromatogr. A*, 2004, 1044, 211–222.

55. Al-Rimawi, F. and Pyell, U. The use of derivatized cyclodextrins as solubilizing agents in the preparation of macroporous polymers employed as amphiphilic continuous beds in capillary electrochromatography. *J. Sep. Sci.*, 2006, 29, 2816–2826.

56. Hayes, J.D. and Malik, A. Sol-gel monolithic columns with reversed electroosmotic flow for capillary electrochromatography. *Anal. Chem.*, 2000, 72, 4090–4099.

57. Ding, G., Da, Z., Yuan, R., and Bao, J.J. Reversed-phase and weak anion-exchange mixed-mode silica-based monolithic column for capillary electrochromatography. *Electrophoresis*, 2006, 27, 3363–3372.

58. Allen, D. and El Rassi, Z. Capillary electrochromatography with monolithic-silica columns. II. Preparation of amphiphilic silica monoliths having surface-bound cationic octadecyl moieties and their chromatographic characterization and application to the separation of proteins and other neutral and charged species. *Analyst.*, 2003, 128, 1249–1256.

59. Ye, F., Xie, Z., and Wong, K.-Y. Monolithic silica columns with mixed mode of hydrophilic interaction and weak anion-exchange stationary phase for pressurized capillary electrochromatography. *Electrophoresis*, 2006, 27, 3373–3380.

60. Fu, H., Xie, C., Dong, J., Huang, X., and Zou, H. Monolithic column with zwitterionic stationary phase for capillary electrochromatography. *Anal. Chem.*, 2004, 76, 4866–4874.

61. Dong, X., Dong, J., Ou, J., Zhu, Y., and Zou, H. Capillary electrochromatography with zwitterionic stationary phase on the lysine-bonded poly(glycidyl methacrylate-*co*-ethylene dimethacrylate) monolithic capillary column. *Electrophoresis*, 2006, 27, 2518–2525.

62. Sawada, H. and Jinno, K. Preparation of capillary columns coated with linear polymer containing hydrophobic and charged groups for capillary electrochromatography. *Electrophoresis*, 1999, 20, 24–30.

63. Huang, X., Zhang, J., and Horváth, C. Capillary electrochromatography of proteins and peptides with porous-layer open-tubular columns. *J. Chromatogr. A*, 1999, 858, 91–101.

64. Hayes, J.D. and Malik, A. Sol-gel open tubular ODS columns with reversed electroosmotic flow for capillary electrochromatography. *Anal. Chem.*, 2001, 73, 987–996.

65. Freitag, R. and Hilbrig, F. Theory and practical understanding of the migration behavior of proteins and peptides in CE and related techniques. *Electrophoresis*, 2007, 28, 2125–2144.

66. Liapis, A.I. and Grimes, B.A. Modeling the velocity field of the electroosmotic flow in charged capillaries and in capillary columns packed with charged particles: Interstitial and intraparticle velocities in capillary electrochromatography systems. *J. Chromatogr. A*, 2000, 877, 181–215.

67. Wei, W., Guoan, L., and Chao, Y. Calculation of retention factors for charged solutes in capillary electrochromatography. *J. Sep. Sci.*, 2001, 24, 203–207.

68. Zhang, J., Zhang, S., and Horváth, C. Capillary electrochromatography of peptides on a column packed with tentacular weak cation-exchanger particles. *J. Chromatogr. A*, 2002, 953, 239–249.

69. Rathore, A.S. and Horváth, C. Separation parameters via virtual migration distances in high-performance liquid chromatography, capillary zone electrophoresis and electrokinetic chromatography. *J. Chromatogr. A*, 1996, 743, 231–246.

70. Rathore, A.S., Wen, E., and Horváth, C. Electroosmotic mobility and conductivity in columns for capillary electrochromatography. *Anal. Chem.*, 1999, 71, 2633–2641.

71. Xiang, R. and Horváth, C. Fundamentals of capillary electrochromatography: Migration behavior of ionized sample components. *Anal. Chem.*, 2002, 74, 767–770.

72. Liu, Z., Otsuka, K., and Terabe, S. Modeling of retention behavior in capillary electrochromatography from chromatographic and electrophoretic data. *J. Chromatogr. A*, 2002, 959, 241–253.

73. Svec, F. Csaba Horváth's contribution to the theory and practice of capillary electrochromatography. *J. Sep. Sci.*, 2004, 27, 1255–1272.

74. Al-Rimawi, F. and Pyell, U. Investigation of the ion-exchange properties of methacrylate-based mixed-mode monolithic stationary phases employed as stationary phases in capillary electrochromatography. *J. Chromatogr. A*, 2007, 1160, 326–335.

75. Li, Y., Xiang, R., Horváth, C., and Wilkins, J.A. Capillary electrochromatography of peptides on a neutral porous monolith with annular electroosmotic flow generation. *Electrophoresis*, 2004, 25, 545–553.

76. Hilder, E.F., Svec, F., and Fréchet, J.M.J. Shielded stationary phases based on porous polymer monoliths for the capillary electrochromatography of highly basic biomolecules. *Anal. Chem.*, 2004, 76, 3887–3892.

77. Okanda, F.M. and El Rassi, Z. Capillary electrochromatography with monolithic stationary phases. 4. Preparation of neutral stearyl-acrylate monoliths and their evaluation in capillary electrochromatography of neutral and charged small species as well as peptides and proteins. *Electrophoresis*, 2005, 26, 1988–1995.

5 Advanced Operating Concepts for Simulated Moving Bed Processes

Malte Kaspereit

CONTENTS

5.1 INTRODUCTION

About 10 years ago, Zhong and Guiochon stated in an earlier volume of this series with respect to simulated moving bed (SMB) chromatography that "many expect it to modify considerably the landscape of industrial separations/purifications in the near future" [1]. This contribution will not give a detailed judgment on the current state of the industrial landscape. Nonetheless, it should be emphasized that in the meantime SMB technology matured from an interesting academic engineering problem to a process that found its entry into the standard portfolio of most companies within the pharmaceutical and fine chemicals industry as well as into the syllabus of university programs for chemical and process engineering, biotechnology, etc.

The main reason for the success of SMB chromatography lies in its numerously quoted benefits with respect to process performance. In comparison to batch chromatography, the process usually facilitates drastically lower eluent consumption, higher product concentrations at high yield, and acceptable productivity.

Another important aspect that supports the more frequent utilization of SMB technology is related to the availability of useful tools for analysis and design of the process. Significant advances with respect to a detailed understanding of this process have been achieved, in particular, within the last 15 years. To some extent this is due to the fact that the SMB process is intriguing to many chemical engineers because of its similarities (as well as due to its differences) to conventional continuous separation processes like rectification. Not surprising, the first simple design method for (linear) SMB processes was based on a concept similar to the famous McCabe–Thiele diagram [2]. By now, more powerful design tools have been derived on the basis of simple models. An important driver for the recent advances is also the rapid development of computer capacities. This allowed performing optimizations on the basis of more realistic and, consequently, computationally more expensive models of SMB processes.

The field of applications of SMB chromatography has widened within the last few years. Currently, these range from rather classical hydrocarbon and sugar separations and over (still) challenging enantioseparations to ambitious biotechnological problems like purifications of proteins or plasmid DNA.

The significant research efforts that were devoted to investigations and design of this process or its application to challenging separation problems are also reflected in the number of publications related to SMB technology. This number has been rising more or less continuously since the invention of the process (see Figure 5.1).

Other indicators substantiating that SMB is a maturing technology are the numerous innovative operating concepts and structural modifications that have been

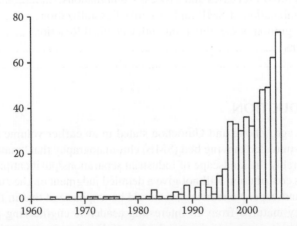

FIGURE 5.1 Number of journal articles on SMB chromatography per year since the first patent in 1961 (530 articles in total; From ISI Web of Knowledge, July 17, 2007).

suggested in order to further enhance the performance of SMB processes or to extend its range of applications to new fields like multicomponent separations or integrated reactive separations.

This chapter will review the most relevant of these advanced operating concepts. The main principles behind the different approaches as well as possible resulting benefits will be explained. However, specific limitations and disadvantages shall also be discussed.

5.2 CONVENTIONAL SMB PROCESSES

In this section, the fundamentals of classical SMB processes will be explained in order to provide an easy understanding of the more advanced operating concepts to be discussed later.

5.2.1 PRINCIPLE

SMB chromatography was developed against the background of classical continuous separation processes in the area of chemical engineering. A self-evident concept to perform such continuous separation is to apply a countercurrent between the phases involved. For example, in rectification (i.e., continuous distillation), such countercurrent is adjusted between the liquid and the gaseous phase. Only this makes possible the continuous introduction of feed and withdrawal of products.

As concerns chromatography, it turned out that a countercurrent between solid and fluid phase is too difficult to establish due to severe problems related to the transport of the solid (e.g., particle abrasion and mixing of the solid phase). It should be noted, however, that such true moving bed (TMB) chromatography remains an important theoretical concept used for process design and optimization.

The SMB concept was developed as a workaround for the mentioned solid-related problems occurring in TMB units. The classical scheme of SMB chromatography as shown in Figure 5.2 was patented by Universal Oil Products in 1961 [3] and still represents the most commonly used setup. The unit basically consists of a series connection of regular chromatographic columns. In contrast to the TMB concept, the desired countercurrent between the solid and the liquid phase is simulated here (hence the name) by periodically switching all columns in the direction opposite to the liquid flow. Alternatively, all product ports can be switched periodically in the direction of the liquid flow. The scheme contains four in- and outlet ports—two inlets for feed and desorbent and two outlets denoted as raffinate and extract. These four ports divide the unit into four zones (I–IV) that have different liquid flow rates and perform different specific tasks.

The task of the leftmost zone I is the regeneration of the solid phase. Using a rather large flow rate in this zone assures that any remaining traces of the adsorbed components are eluted and transported toward the extract port. The flow rate within zone II is adjusted such that the stronger adsorbing component B is allowed to establish a concentration wave here, while the flow is chosen high enough to ensure complete elution of the weaker adsorbing component A. Similarly, in zone III the flow is low enough to allow migration of the faster component A toward the raffinate port, while component B migrates too slow to pass this zone. Finally, in zone IV

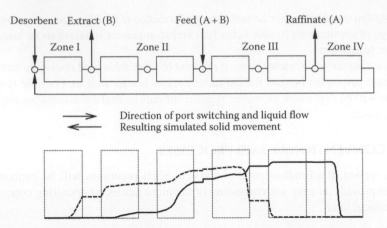

FIGURE 5.2 Classical four-zone SMB process. Top, example scheme with a 1/2/2/1 column configuration; bottom, typical internal concentration profiles at the middle of a switching interval (nonlinear adsorption isotherms). Solid line, weakly adsorbing component A; dashed line, stronger adsorbing component B.

a very low flow rate is adjusted. Therefore, the liquid phase leaving the columns in this zone should not contain any adsorbed components. With respect to these tasks just described, zones II and III are often denoted as "separation zones," while zones I and IV are frequently referred to as "regeneration zones."

Due to the periodical shifting and continuous supply or withdrawal of feed, desorbent, and products, the process reaches a periodically repetitive state after a certain number of switching intervals. This is often denoted as periodic or cyclic steady state. This situation is characterized by the fact that at the end of a switching interval the same concentration profiles are obtained throughout the unit as were at the end of the preceding interval (and will be in all successive periods). The number of switches necessary to reach this situation is difficult to predict and depends on several parameters [4]. As a rough rule of thumb, for a well-designed SMB system approximately 20 × (no. of columns) switching events are often sufficient. An important characteristic of the process is also that due to its periodic nature the concentrations at the two outlet ports vary during each interval (i.e., one observes "sawtooth profiles" for the product concentrations).

5.2.2 SOME REMARKS ON ADVANTAGES AND DISADVANTAGES

In comparison to batch chromatography, the conventional SMB process has several advantages with respect to process performance. These have been quoted numerous times. However, there are also some misconceptions and disadvantages which deserve a short discussion.

The main performance parameters for preparative chromatographic processes are the purity of the products, specific productivity (defined as mass flux of desired product per unit volume), and specific eluent consumption (amount of solvent required per mass of product obtained).

It is well known that SMB processes are superior to batch chromatography concerning their significantly lower specific eluent consumption. Consequently, they also deliver products of higher concentration. Furthermore, a reasonably high throughput (i.e., productivity) can be achieved without sacrificing the recovery yield. The latter is attributed to the fact that SMB units as such do not comprise waste streams. In addition to the above, the periodic nature of the SMB process can be regarded as an advantage (even though it makes the process more complex). In fact, this periodicity is the basis for several innovative operating concepts that enhance process performance. Examples are the asynchronous shifting of ports and the dynamic modulation of concentrations or flow rates. Such advanced operating modes will be discussed later in detail.

However, there are two more aspects often claimed to be specific advantages of SMB over batch chromatography. One is the presumably continuous nature of SMB. This is a misconception since, when considering an SMB and a reasonably designed preparative batch scheme (e.g., a "touching band separation"), one finds that both are periodic or pseudocontinuous processes. As such, both deliver products with periodically changing compositions. Furthermore, they are not very different or complex with respect to automization. Therefore, a periodically operated batch process does not differ significantly with respect to its discontinuous nature.

A second argument that was propagated for many years is that SMB processes allow for higher productivity than batch chromatography. This claim does not hold, since batch schemes can be frequently optimized such that they operate at significantly higher specific productivity (i.e., the ratio of product flux and volume of stationary phase) than the corresponding separation by SMB chromatography. To a large extent, this is due to less severe pressure drop constraints in batch systems. However, as already indicated, beyond a certain limit this is at the expense of a strongly decreasing yield.

Besides the undisputed benefits of (classical) SMB processes, there are also several disadvantages in comparison to batch chromatography. Most noteworthy is the capability of batch systems to perform multicomponent separations—often with 100% purity for each component. Obviously, since a conventional SMB unit has only two outlets, it can deliver only two pure products at the same time. Even this is possible only if the feed is a binary mixture. If the feed mixture contains more than two solutes, the classical SMB scheme can provide only the least or the strongest adsorbing component in pure form. Furthermore, the SMB process as such is more complex than batch elution. For a detailed design, so-called hybrid models (i.e., mathematical models that contain besides continuous variables, discontinuous ones also here this is due to the switching events) have to be implemented. Generally, these have to be treated using numerical methods and their solution is (still) numerically rather expensive. Other unfavorable aspects are the slightly higher investment and maintenance costs for SMB units. Moreover, batch chromatography is generally easier to monitor than an SMB process. The latter requires more than one detecting device at different positions in the unit. These detectors are often even difficult to connect (e.g., when considering a carousel unit with rotating columns). Finally, it can be difficult to apply concepts from batch chromatography that allow improved process performance (e.g., the use of gradients) to SMB processes.

However, the few specific disadvantages mentioned above are also drivers for the innovative operating concepts for SMB processes that are in the focus of this contribution.

5.2.3 Basic Design for Complete Separation

A detailed design of an SMB process requires the optimization of a rather large number of operating parameters; e.g., flow rates, feed concentration, column length and diameter, column configuration, particle size, etc. These parameters are by no means independent. Consequently, such detailed design is a challenging task. Publications covering this are rare; a practicable procedure can be found in Ref. [5]. Due to the complexity, established design procedures focus on the most important parameters, which are the four internal flow rates and the switching time of the unit.

Most SMB applications aim at the maximization of the product purity. For such cases that typically demand the complete separation of the components, useful design tools are available which allow for a direct prediction of necessary operating conditions. Methods that deserve to be mentioned are the well-known triangle theory (see, e.g., Refs. [6,7]) and the standing wave analysis [8,9]. The former considers the adsorption isotherms as the main factor influencing the behavior of an SMB unit. The latter additionally takes into account dispersive effects, but it requires a certain reconciliation using numerical models or experiments.

The triangle theory will be explained briefly since it represents the most commonly used design method. The approach is based on local equilibrium theory for a TMB process (see Section 5.2.1) and has been elaborated for most of the relevant types of adsorption isotherms [10–13]. More recently, it was further extended to the (somewhat academic) case of a generalized Langmuir adsorption isotherm [14,15].

The design of any SMB process requires the specification of the adsorption isotherms. For the sake of brevity, only here competitive mixture adsorption isotherms of the Langmuir-type will be considered. These relate the concentrations of the species on the solid surface (q_i) and in the liquid phase (c_i), respectively, in thermodynamic equilibrium by

$$q_i(\overline{c}) = \frac{a_i c_i}{1 + b_A c_A + b_B c_B} \quad i = (A, B) \tag{5.1}$$

As already mentioned, the main design parameters for a unit as shown in Figure 5.2 are the dimensionless flow rate ratios m_j for each zone j. These are defined (for a TMB unit) as quotient of the liquid zone flow rate, Q_j, and the solid flow rate, Q_S, respectively. For TMB and SMB systems, these ratios are related by

$$m_j = \frac{Q_j^{\text{TMB}} - Q_S \varepsilon_p}{Q_S (1 - \varepsilon_p)} = \frac{Q_j^{\text{SMB}} t_S - V_c (1 - \varepsilon)}{V_c (1 - \varepsilon)} \quad j = (1, ..., 4) \tag{5.2}$$

Where ε_p and ε denote the intra-particle and total porosity; and t_S and V_c the switching time and column volume, respectively.

Using equilibrium theory, analytical expressions can be derived for m_j values necessary to achieve complete separation. The solutions are valid if dispersive effects, e.g., due to axial dispersion or finite mass transfer resistances, can be neglected. For a system described by the Langmuir adsorption isotherms (Equation 5.1), optimal values for all m_j that guarantee complete separation (i.e., 100% purity— superscript "100") can be calculated explicitly [7]:

$$m_I^{100} = a_B \tag{5.3}$$

$$m_{\text{II}}^{100} = \frac{a_A}{a_B}\varpi_G \quad (5.4)$$

$$m_{\text{III}}^{100} = \frac{\varpi_G\left[\varpi_F\left(a_B - a_A\right) + a_A\left(a_A - \varpi_F\right)\right]}{a_A\left(a_B - \varpi_F\right)} \quad (5.5)$$

$$m_{\text{IV}}^{100} = \frac{1}{2}\left\{a_A m_{\text{III}} + b_A c_{A,F}\Delta m - \sqrt{\left[a_A + m_{\text{III}} + a_A c_{A,F}\Delta m\right]^2 - 4a_A m_{\text{III}}}\right\} \quad (5.6)$$

In the equations above, $c_{I,F}$ (I=A,B) denotes the feed concentrations and $\Delta m = m_{\text{III}} - m_{\text{II}}$. The index B represents the stronger adsorbing component. Furthermore, the $\varpi_{G/F}$ are the solutions of the quadratic equation

$$\left(1 + b_A c_{A,F} + b_B c_{B,F}\right)\varpi^2 - \left[a_A\left(1 + b_B c_{B,F}\right) + a_B\left(1 + b_A c_{A,F}\right)\right]\varpi + a_A a_B = 0 \quad (5.7)$$

As for the dispersive effects inherent to any real chromatographic plant, most probably flow rates calculated from the m_j^{100} above will not deliver pure products. To account for this safety factors, $\beta_j \geq 1$, are frequently introduced and that should counterbalance these deviations:

$$m_I = \begin{cases} \beta_j m_j^{100} : j = (\text{I},\text{II}) \\ m_j^{100}/\beta_j : j = (\text{III},\text{IV}) \end{cases} \quad (5.8)$$

In addition to the explicit calculation of optimum m_j-values, the design approach also allows to define a region of operating parameters within the parameter space ($m_{\text{II}}, m_{\text{III}}$) that guarantees for complete separation of the components. This region resembles a triangle (giving the method its name). An example is shown in Figure 5.3. Noteworthy,

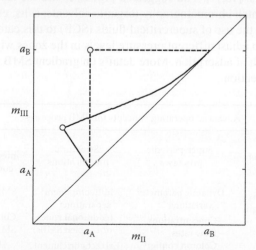

FIGURE 5.3 Regions of complete separation in the $m_{\text{II}}/m_{\text{III}}$ diagram. Dashed lines, linear adsorption isotherms; solid lines, Langmuir adsorption isotherms; points, optimum operating points. The a_i correspond to the slopes of the Langmuir isotherms (Equation 5.1) at infinite dilution (i.e., the Henry constants of the components).

the vertex of such region represents the optimal operating point in terms of productivity and solvent consumption. For details on determination of the region, the interested reader is referred to Ref. [7].

The separation triangle is a very popular tool for the basic design of SMB processes, since it gives explicit solutions and allows for a rather generic view on possible operating points. The results can also serve as good initial values in detailed design studies using more elaborated mathematical process models that account for dispersive effects (e.g., axial dispersion and mass transfer limitations).

It should be noted that equilibrium theory can also be applied to design processes with basically any desired product purity lower than 100% for extract and raffinate. A corresponding method was developed recently [16]. For a given set of desired outlet purity values, the method allows to determine optimum flow rates and product concentrations by solving a simple set of algebraic equations.

5.3　CLASSIFICATION OF ADVANCED OPERATING MODES

The conventional SMB concept described in Section 5.2.3 has been retained basically unchanged for several decades. During the last few years, however, a number of sophisticated operating modes were proposed that should either further improve the performance of the process or are intended to overcome some of its limitations. In this section, an attempt is made to classify these approaches. Figure 5.4 shows a corresponding scheme that suggests—from a process engineering perspective—four different categories.

The first category exploits the idea of manipulating the strength of adsorption in a spatially distributed manner. This basically subsumes all kinds of SMB processes incorporating gradients. Gradients are popular in batch chromatography for more than 50 years [17]. Corresponding SMB schemes have been developed more recently, Refs. [18,19]. The suggested processes use the same modifiers that are known from batch chromatography (organic solvents, salts, etc.). Note that the figure also assigns the use of supercritical fluids (SCF) to this category. This is due to the possibility to adjust different pressure levels in the zones which also lead to a distributed strength of adsorption. More details on gradient SMB processes will be given in the next section.

Advanced operating concepts for SMB processes			
Spatial manipulation of adsorption	Super-periodic processes	Structural modifications	Process integration and combination
Gradient-SMBs – Solvent strength – Temperature – Salts – SFC	Dynamic parameter variations – Concentrations – Flow rates – Column config. – Feed periods	Multicomponent separations (additional zones or coupled SMBs) Extract enrichment	SMB reactor Combination SMB/selective crystallization

FIGURE 5.4　Rough classification of advanced operating concepts for SMB processes.

The second category in Figure 5.4 is denoted as super-periodic processes. These have in common—in addition to the periodic switching of columns in classical SMB processes—further variables which are manipulated dynamically during each switching interval. This increases the degree of freedom (i.e., the number of design parameters) and therefore allows for a further optimization of the process. Examples are flow rates (Powerfeed-SMB) or feed concentrations (Modicon process) that are changed during the tact or a variable column configuration (achieved by asynchronously switching the columns, Varicol process). Selected processes belonging to this class will be discussed in Section 5.5.

The third group of processes is characterized by more or less fundamental modifications of the structure of the SMB process. Mainly, this subsumes schemes intended for multicomponent separations in order to overcome the severe limitation of SMB processes for binary separations. Such processes contain additional zones and internal recycles. Noteworthy, in contrast to the second category, they usually entail a reduced degree of freedom (e.g., due to the couplings of zones or internal recycles). Section 5.6 gives an overview on corresponding attempts.

The last category refers to processes that integrate new functionalities into the SMB process or combine SMB units with complimentary processes. Most prominent are SMB reactors, where reaction and separation are performed simultaneously in the same apparatus. Analogous to the famous concept of reactive distillation, this aims in particular at enhancing the conversion in equilibrium-limited reactions. Ideally, the products of a chemical reaction are fed into the unit and pure products are obtained at complete conversion. The main challenge is here that the feasibility depends strongly on several parameters (i.e., the stoichiometry, the order of adsorptivities, the adsorption isotherms, and the reaction rate). Also placed within this category are combined separations. An example is to purposefully couple SMB chromatography and selective crystallization in order to exploit synergisms. Integrated SMB processes and combined schemes will be discussed in Section 5.7.

5.4 SMB PROCESSES WITH GRADIENTS

The application of gradients is one of the most promising approaches to improve the performance of SMB processes. Analogous to batch chromatography, gradients are applied in order to manipulate the strength of adsorption in a purposeful way. First corresponding suggestions were made with respect to gradients of temperature [2], solvent strength [20], and pressure gradients when using SCF [21,22]. Surprisingly, detailed investigations of the most straightforward technique, solvent gradients, were performed only recently and published independently by Jensen et al. [18], and Antos and Seidel-Morgenstern [19].

The implementation of a gradient in an SMB unit is shown schematically in Figure 5.5 as the example of a gradient of solvent strength or salt concentration. In the figure, gray shadows mark the positions where modifications are established in comparison to the conventional SMB process (Figure 5.2). The rationale of the process is analogous to batch-gradient chromatography. Obviously, it is of interest to increase the solvent strength in places where fast desorption is favored. In particular, this holds for the regeneration zone I. This is attained by adding a high fraction of a

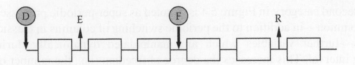

D ... Pure strong solvent (or high salt content)
F ... Feed components in weak solvent (or low salt content)

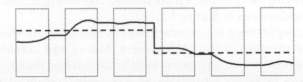

FIGURE 5.5 Principle of a solvent-gradient SMB processes (recycle stream omitted). Top, SMB setup with a strong eluent in the desorbent and weak eluent in the feed stream, respectively; bottom, resulting concentration profile of the modifier (i.e., desorption strenght throughout the unit). Dashed line, desired ideal profile; solid line, dynamic profile of a modifier with nonlinear adsorption isotherms.

strong solvent (the so-called modifier) to the desorbent stream. In contrast, adsorption should prevail in zones III and IV. Consequently, the feed should prevail and, consequently, the feed stream should contain less or none of the modifier. As a result of such gradients, significantly higher product concentrations, productivity, and lower overall eluent can be achieved. It should be noted that the same general principle as shown in the figure also holds for alternative kinds of modifiers.

Solvent-gradient SMB systems were investigated most intensively. Therefore, they will be discussed here in more detail. Suitably designed SG-SMB processes achieve remarkable performance parameters. Most notably, the specific solvent consumption can be drastically decreased and, as a consequence, to the same extent the product concentrations can be increased. Both theoretical as well as experimental studies confirmed reductions of solvent consumption by 50% [18], 60% [19], or even 90% [23], respectively. These numbers remain impressive, even when considering that in none of these investigations a detailed optimization of the two process options was performed.

There are two important aspects related to the modifier in such processes. The first is that within the SMB unit, a distinct concentration profile will be established that has to be accounted for by a proper mass balance [19]. On one hand, the modifier itself can be an (linearly or nonlinearly) adsorbing species, or it can be nonretained. On the other hand, as indicated in Figure 5.5, its resulting composition profile is not equivalent to that within a TMB process, but is determined by the periodic switching of the columns [19,26]. This problem has not been accounted for in earlier works [18].

The second aspect becomes clear when considering that the adsorption isotherm parameters for the components to be separated depend strongly on the modifier content. Therefore, expressions are required that properly describe this dependency. Usually, empirical functions have to be applied. A viable procedure was applied in

Refs. [19,25], which was based on an earlier work [26]. The modifier was assumed to adsorb according to a single-component Langmuir isotherm equation (Equation 5.1) (i.e., without any competitive effects caused by the other species). The dependency of the Langmuir parameters for the components to be separated was described according to

$$a_i(c_{mod}) = (p_{a,i} \, c_{mod})^{-r_{a,i}} \tag{5.9}$$

$$b_i(c_{mod}) = (p_{b,i} \, c_{mod})^{-r_{b,i}} \tag{5.10}$$

where c_{mod} denotes the concentration of the modifier. The p- and r-values are constants to be determined experimentally. In total, this results is eight parameters which indicate the significant experimental effort when developing a separation by gradient SMB chromatography.

Figure 5.6 exemplifies the typical relationship described by the two expressions above. The figure emphasizes the strong dependency of adsorption isotherm parameters on modifier concentration. In particular, it should be noted that in the example the separation factor $\alpha = a_2/a_1$ rises significantly with increasing c_{mod}. This has an interesting consequence. It has been demonstrated in Ref. [27] that the performance of SMB processes improves drastically with increasing separation factors. Considering further, by using a solvent gradient, the concentrations in the SMB unit also increase strongly reach, situations can arise where these concentrations the solubility limits. Since this should be regarded infeasible, in such case an isocratic SMB might be equivalent to its gradient analog. In other words, the question arises if an optimized isocratic SMB unit that operates at the solubility limit can be outperformed by an SG-SMB process or not. This aspect has not been investigated yet. However, it was touched by Abel et al. [24], who pointed out that the gradient process should at least allow for a larger separation triangle (cf. Figure 5.3) which will improve process robustness.

The design of gradient SMB chromatography is obviously more complex than an isocratic process. Different strategies were investigated. Abel et al. were able to derive design guidelines from equilibrium theory [23,24]. A different approach is to

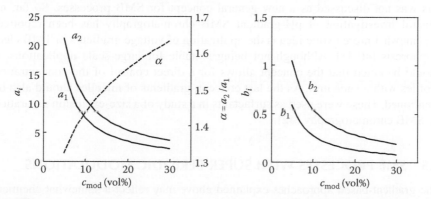

FIGURE 5.6 Dependency of Langmuir adsorption isotherm parameters on modifier concentration according to Equations 5.9 and 5.10. Left, Henry constants and separation factor $\alpha = a_2/a_1$; right, b_i values. Parameters are taken from Ref. [19].

perform parametric studies. These aim at determining a separation triangle analogous to the one shown in Figure 5.3 by scanning of the m_{II}/m_{III}-parameter space. For this purpose, a fast algorithm was applied to systems with linear isotherms [28,29]. For nonlinear processes also more detailed models were applied (e.g., Ref. [19]). More recently, advanced optimization methods have also been applied in order to determine directly the optimum operating parameters. Example is stochastic optimization [30] or a hybrid approach that combines simulated annealing and the simplex algorithm [31].

In addition to the SG-SMB processes discussed above, the use of salt gradients has also become popular [25,32–35]. These are of interest in bioseparations, where very often high salt concentrations are necessary in order to trigger desorption of strongly adsorbed proteins. Due to the relevance of this application problem, the author expects that salt gradients will be the main driving force for future developments in the field of gradient SMB processes. The principle of operation is basically the same as for SG-SMBs (i.e., a high salt concentration could be applied in the desorbent and a low salt concentration in the feed). However, an interesting aspect is that less obvious shapes of the salt profile might also be beneficial (e.g., upward vs. downward gradients), which depends to a certain extent on the objectives with respect to productivity and eluent consumption [34]. In general, however, the design procedures for salt gradient SMB processes are similar to those for solvent gradients.

Besides the use of solvent modifiers and salts, which certainly are the most important versions within this category, more gradient SMB concepts have been investigated. The use of higher temperatures in regeneration zones and lower temperatures in separation zones has been suggested already at an early stage [2]. In a theoretical study, design criteria were derived from equilibrium theory, and the process was found to be promising [36]. However, it can be expected that practical realizations could suffer severely from difficulties with respect to controlling a desired temperature profile. The temperature control must include connecting tubes and valves. Furthermore, very low switching times might be necessary to achieve uniform temperatures within the columns which will entail low productivities. As another concept, pH gradients were also suggested in Refs. [37–39], although this was not discussed as a new general concept for SMB processes. So far, no general investigation of pH-gradient SMB chromatography has been reported. A somewhat more exotic idea is the application of voltage gradients in TMB electrophoresis [40,41]. Although not being suitable for large-scale applications, it should be noted that the concept allows for a direct control of the concentration profiles within such unit. As the last concept, gradients of micelles should also be mentioned. These were used as surfactants in a study of a size-exclusion separation by SMB chromatography [42].

5.5 SMB PROCESSES WITH SUPER-PERIODIC MODULATIONS

The gradient-based approaches explained above may reflect a somewhat chemical view on the SMB process. The operating modes covered in this section might be perceived as originating rather from a "process engineering perspective." While gradient systems attempt a direct manipulation of the adsorption, the approaches to be discussed here introduce an additional degree of freedom by periodically changing different operating parameters during each switching interval.

5.5.1 MODULATION OF FLOW RATES (POWERFEED)

In conventional SMB chromatography, all volumetric flows are held constant. The idea of varying these flow rates during each switching interval was originally introduced as a patent by Kearney and Hieb [43]. Later, Kloppenburg and Gilles [44] performed optimizations of this process. More recently, the idea was investigated in more detail by Mazzotti and coworkers (see, e.g., Refs. [45,46]), who introduced the synonym Powerfeed for this operating concept. Schramm [47] applied control algorithms for the automatic optimization of the process. All authors found that when comparing Powerfeed to conventional SMB systems, significant improvements of solvent consumption and productivity can be achieved.

The principle of this process is explained in Figure 5.7. In contrast to gradient SMB systems, where modifications of the classical concept concern only feed and desorbent (cf. Figure 5.5), such modifications are applied periodically and throughout the whole unit.

Figure 5.8 shows an example of such flow rate variation. The data in the figure were adopted from Ref. [44], wherein the authors divided each switching interval into four subintervals. Piecewise constant profiles were assumed for the flow rates, which were obtained from optimization procedures.

FIGURE 5.7 Principle of Powerfeed operation. In contrast to the conventional SMB process, all flow rates can be subject to modulations within each switching interval. (Recycle stream omitted.)

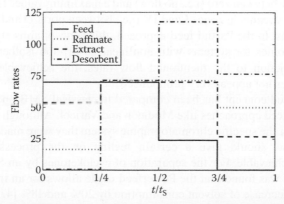

FIGURE 5.8 Example for flow rate modulations during each switching interval in a Powerfeed process (data according to optimization result in Ref. [44]).

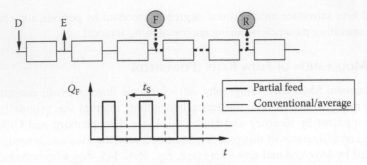

FIGURE 5.9 Partial feed concept. Top, general scheme of a corresponding setup. Dotted lines, time-dependent streams. Recycle omitted. Bottom, example for typical time profile of feed flow rate.

The figure indicates that optimal flow rate sequences cannot be designed on the basis of intuition (see, e.g., the profile of the desorbent stream). As a consequence, design procedures either rely on some optimization algorithm or on-process control concepts. Examples for successfully applied optimization methods are a simple double-layer optimization scheme [48], combinations of discretization approaches with Nonlinear programming (NLP) solvers [44,49], and genetic algorithms [45].

The Partial feed operating mode proposed by Zang and Wankat [50] has already been recognized as being a special variant of the Powerfeed concept [48]. In this operating mode, only the feed flow rate is subject to variations. More specifically, the feed is turned on only for a certain time in the middle of the switching interval. The original concept is explained in Figure 5.9. It is obviously less complex than the general Powerfeed mode described above, since only the flow rates of feed, zone III, and raffinate are time dependent. Nonetheless, in comparison to a standard SMB process, a significantly advantageous performance was obtained (reduction of solvent consumption by about 59%, productivity increase of 70%) [50].

As another specific subset of Powerfeed processes, the "Outlet Swing Strategy" might be mentioned [51]. The flow rates of raffinate and extract vary in a two-step manner, basically between zero (i.e., no flow) and a maximum value. This is achieved by manipulating streams in zones I and IV (i.e., between ports D and E and ports R and D). In contrast to the Partial feed approach, this concept aims at stretching the concentration profiles for systems with nonlinear adsorption isotherms. However, due to the restriction to the mentioned flow rates, the achievable performance improvement does not appear very pronounced.

The Powerfeed concept has been compared to classical SMB operation as well as to other advanced approaches like Modicon and Varicol. Although such comparisons hold only for the specific chromatographic system they were made for, the numbers given below should give a certain feeling on the process performance improvements achievable. For the separation of cycloketones by an SMB unit with eight columns, it was found that the Powerfeed mode allows for an increase of productivity and a decrease of solvent consumption by 20% and 18% [47], respectively. Under different chromatographic conditions for the same system, improvements were found to be 15% and 13%, respectively [52]. Zhang et al. [53] demonstrated that the benefits increase fewer columns when applied. In an elaborated optimization study, they calculated for an enantioseparation problem (racemate of 1,2,3,4-tetra-hydro-1-naphthol, Chiralpak AD, see also Ref. [54]) and obtained an increase of productivity by about 25% for a five-column unit and 45% for a four-column process, (extract purity 95%; interpolated from Figures 5.2 and 5.4 in Ref. [53]).

Despite these promising results (which have been verified also experimentally), some drawbacks of Powerfeed should be mentioned. Besides the efforts necessary to identify optimum flow rates, the changing flow rates can represent additional stress factors for the stationary phase and, depending on the practical implementation, also for the pumps of the unit [47].

5.5.2 MODULATION OF FEED CONCENTRATION (MODICON)

An alternative concept relying on dynamic parameter variation is the Modicon process [47,52,55–57]. In this operating mode, the feed concentration is varied during the switching interval. Such concentration alteration aims at a purposeful manipulation of the nonlinear concentration profiles within the unit; the concept is therefore only beneficial in the case of nonlinear adsorption isotherms.

The principle of Modicon is exemplified in Figure 5.10. As a simplification, each switching interval can again be divided into subintervals (analogous to the Power-feed mode above), in which a different feed concentration is applied. An example of a beneficial shape of such feed concentration profile is given in the middle of the figure. In the case of Langmuir adsorption isotherms (Equation 5.1), the application of a very low concentration in the beginning of each interval (e.g., by injecting only eluent) was found most useful, while the concentration should be as high as possible

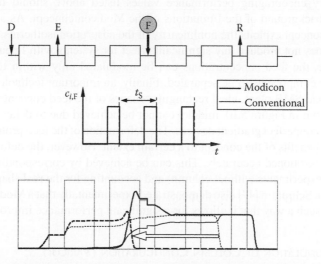

FIGURE 5.10 Principle of the Modicon process. Top, the schematic setup corresponds to a classical SMB, only the feed concentration is varied during each switching interval. Recycle stream omitted. Middle, feed concentration profile for a classical SMB processes (thin line) and Modicon (thick line). Bottom, comparison of typical internal concentration profiles for both process options.

toward the end of the interval. The result of this modulation is shown in the lower part of the figure. In the example, the same total amount of feed components is fed into the unit. In comparison to a conventional SMB process, in the Modicon process the concentration profile of the strong adsorbing component is shifted backwards (i.e., further away from the raffinate outlet; marked by the arrow). This indicates that in Modicon mode a higher feed flow rate can be used in order to reposition this front (i.e., to adjust again the same raffinate purity). Such increased flow rate entails a higher productivity of the two components and lower eluent consumption.

With respect to the design of this operating mode, it was found (as already indicated) that for adsorption isotherms of the Langmuir-type the feed mixture should be introduced as late as possible in the switching interval, with the highest possible concentration. However, in general, such a profile should be verified/optimized using a mathematical process model.

The benefits resulting from applying the Modicon concept are similar to the results given earlier for the Powerfeed mode. In comparison to the classical SMB process, improvements with respect to productivity/solvent consumption/average product concentrations were reported as 33%/22%/32% [52] or even as high as 165%/58%/151% [47]. Zhang et al. presented somewhat less enthusiastic numbers, but found very similar productivity enhancements as for the Powerfeed mode: approximately 26% for a five-column unit and 44% for a four-column process (extract purity 95%; interpolated from Figures 5.2 and 5.4 in Ref. [53]). An interesting result of the Modicon mode is also that it leads (at the same time-averaged feed concentration) to an enlargement of the separation regions (cf. Figure 5.3) which might be utilized for a more robust operation in the case of strongly nonlinear adsorption isotherms [57].

The very encouraging performance values listed above should be perceived against the background of the limitations of the Modicon concept. As already mentioned, the concept exploits the nonlinearity of the adsorption isotherms. As a consequence, it does not produce any significant effect for systems with linear isotherms. Furthermore, the feed concentrations can be modulated only within the solubility limits of the components to be separated. Finally, an important technological aspect is that in a real SMB plant, ideal rectangular shapes of the feed concentration profile (as were shown in Figure 5.10, middle) cannot be achieved due to delay times, back-mixing, and nonperfect gradient devices. The deformation of the feed profile decreases the possible benefits of the concept to a certain extent. However, the deformed profile needs to be positioned accurately. This can be achieved by corresponding measurements (e.g., experiments without column) and accounting for the real shape in model-based design. Schramm [47] also demonstrated experimentally that a Modicon process designed in such a way that it allows for a significant performance improvement.

5.5.3 Modulation of Column Configuration (Varicol)

A popular variant of the SMB process, which is also successfully marketed, is the Varicol process [54,58]. This concept is depicted in Figure 5.11. As can be seen, in this process the inlet and outlet ports are not shifted simultaneously at the end of each switching interval but asynchronously at specific times. The figure shows an example where the switching period is divided into four subintervals; at the end of

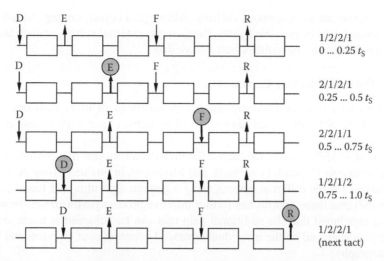

FIGURE 5.11 General principle of the Varicol process for a six-column setup. The main feature of the Varicol operating mode is the asynchronous shifting of the in- and outlet ports. In the example, the switching interval consists of four subintervals; in each of which only one of the ports is shifted. The resulting average column configuration is 1.5/1.5/1.75/1.25. (Recycle stream omitted.)

each interval, only one port is shifted (in a sequence extract, feed, desorbent, raffinate). This corresponds to a variation of the column configuration during each interval. As a consequence, one can specify time-averaged zone lengths or column configurations. For example, the switching mimic shown in the figure leads to a time-averaged column configuration of 1.5/1.5/1.75/1.25 columns per zone.

This modulation aims at the optimal exploitation of adsorbent in each zone. The most intriguing consequence of this is that Varicol allows reducing the number of necessary columns. In fact, the concept can be operated even with fewer columns than zones; e.g., with only three columns that operate all four zones. As a consequence of fewer columns, the specific productivity of Varicol processes is also higher than that of conventional SMB systems.

It should be noted that, to design a Varicol process, not only the optimum column switching policy has to be found, but also the optimum flow rates. The latter differ from those in a conventional SMB plant. This makes the design a complex optimization task, which cannot be performed on the basis of intuition only. In addition, optimum flow rates in Varicol differ from those in a conventional SMB process [54]. This makes the design a complex optimization task.

Using a genetic algorithm for this design problem, in Ref. [59] it was found that the possible benefits resulting from asynchronous switching depend strongly on purity requirements. For example, it was reported that in comparison to the equivalent regular SMB process, the productivity of Varicol is higher by 10%, 25%, and 127% for purity requirements of raffinate and extract of 90%, 95%, and 99%, respectively.

As limitations of the Varicol concept, the complexity of designing this process (see above) and its restriction to SMB systems that are based on the assembly of several multiport valves should be mentioned. The so-called carousel SMB units (which are based on a single rotary valve that holds and physically moves all columns) do not facilitate an asynchronous shifting. Although a corresponding extension of such systems should be possible by installing several external valves, no such attempt has been published to the author's best knowledge.

5.5.4 FURTHER (HYBRID) CONCEPTS

An obvious idea is to exploit simultaneous modulations of more than one of the operational variables discussed above (i.e., flow rates, concentration, and column configuration).

Zhang et al. [53] investigated such different hybrid concepts that combine two or all of the concepts (Varicol, Powerfeed, and Modicon), in one unit. They performed optimizations for the different options using a genetic algorithm and found (as can be expected) that these approaches further improve process performance. However, they also concluded that the additional gain that can be achieved is much smaller than the gain achieved by the individual advanced concepts over conventional SMB chromatography.

A significant problem related to any kind of these approaches is to find optimum parameters, because the complexity of the design problem increases drastically by such combination; it is practically impossible to design them based on engineering experience or intuition. Furthermore, it is very time-consuming to study in detail all possible process options using mathematical process models. In this context, a very promising approach has been suggested by Kawajiri and Biegler, who performed an optimization-based synthesis of the optimal process, for example, by combining Powerfeed and Varicol. The main idea is to first create the so-called superstructure that inherits all possible options (even such options that might at first appear detrimental) [60]. This is followed by an optimization that identifies the optimal subprocess and the optimal parameters [49,61,62]. This approach should be extensible to also include further parameters (e.g., gradients, multicomponent problems, chemical reactions, etc.).

As the last aspect of such hybrid schemes, it should be considered that—despite the advantages that are theoretically possible—they will most certainly suffer from a decreased robustness of operation (i.e., the sensitivity of the process against small disturbances will be high). This will be more severe if more parameters are varied. Usually, such issues are addressed only by researchers who study control schemes for SMB processes. However, with respect to the hybrid schemes above, this problem has not yet been investigated sufficiently.

5.6 SMB PROCESSES WITH STRUCTURAL MODIFICATIONS

This section will explain operating concepts that entail significant structural modifications of the classical four-zone SMB scheme.

5.6.1 MULTICOMPONENT SEPARATIONS

The inherent restriction of conventional SMB systems to binary separation has been a motivation for development of sophisticated SMB processes for multi-component problems. A large number of suggestions have been published, in particular for ternary separations. Here, only the fundamental principles will be explained.

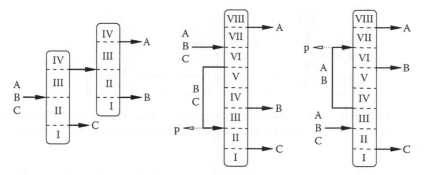

FIGURE 5.12 Examples for the general SMB-based approaches capable of performing ternary separations. Left, coupling of several SMB units. Middle, eight-zone SMB system performing the split A/BC. Right, eight-zone process with a split AB/C. Necessary purge streams are marked by "p."

Figure 5.12 shows some example processes that clarify the possible approaches for ternary separations. The first and most obvious is to couple several SMB units (left; see, e.g., Ref. [63]). An alternative to this is also to combine SMB and batch chromatography [64]. The problem of such series connections (and the explosion of the number of possibilities with increasing number of components) is well known from the design of distillation trains. The other main idea with respect to SMB is the introduction of further zones (see, e.g., Ref. [65] for system with nine zones). In the middle of the figure, an eight-zone SMB system is shown that performs a complete separation (a sharp split) between the least adsorbing compound A and the mixture B+C; the latter is then recycled to a lower part of the unit. The opposite scenario is shown in the right, where the split is performed between the two heavy components B and C. Besides the options shown in the figure, there is also the possibility to use different adsorbents packed into alternating columns [66].

There are several significant challenges related to multicomponent SMB separations. Certainly an issue is that, usually the feed of the second unit (in SMB cascades) and the recycle in multizone systems are strongly diluted. This limits the overall performance. Furthermore, in multizone SMB systems, the volume of diluted recycle will usually be too large to be reintroduced completely. As indicated in Figure 5.12, purge streams are necessary to overcome this problem [67]. Another relevant aspect related to both approaches is the reduced degree of freedom resulting from the coupling. In case of directly coupled processes (i.e., SMB cascades), one design parameter for the connected unit is lost. For multizone systems, the switching time is equal throughout all zones which represents a serious limitation.

As can be seen from Figure 5.12, there is a vast range of possible processes, in particular if more than three components should be separated. A detailed discussion of all options is beyond the scope of this work; the interested reader is instead referred to several reviews. A comprehensive and detailed overview was given by Chin and Wang [68], who also addressed hardware issues (e.g., valve design and system setup). Further reviews were given, e.g., by Kurup et al. [69] and Keßler and Seidel-Morgenstern [67].

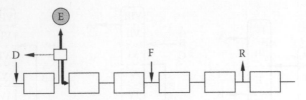

FIGURE 5.13 Scheme of an SMB process with an enriched extract stream. The output of zone I is withdrawn completely, enriched (e.g., by evaporation), and partially recycled into zone II.

5.6.2 SMB Process with Extract Enrichment

An interesting process first described in a patent [70] is an SMB system combined with a recycle of enriched extract (EE). Following the vogue of appointing a name to new SMB concepts, in two current investigations the process was called M3C-SMB [71] and EE-SMB [72].

Figure 5.13 explains the concept. The main idea is to withdraw completely the liquid output of zone I and to enhance its concentrations (e.g., by an evaporation step). A part of this liquid is kept as product and the remainder is recycled into zone II. For systems with nonlinear and favorable adsorption isotherms (e.g., Langmuir isotherms [Equation 5.1]), the now high concentration of the stronger adsorbing component entering zone II causes a displacement of the less adsorbing component. The concentration front of this weak species is moved to the right (i.e., toward the feed port). Therefore, the extract purity increases or, alternatively, the throughput can be enhanced. Paredes et al. [72] predicted a possible productivity enhancement of 150% for a model system.

As has been demonstrated in Refs. [71,72], the process can be designed by using equilibrium theory. An important aspect of both design procedures is that the authors require the concentration of the recycle to exceed a certain minimum limit. This limit is given by the so-called watershed point. The intention of this is to convert the dispersive waves of the components in zone II (as shown in the bottom of Figure 5.2) into shock waves. However, as indicated in Ref. [72], a positive effect has already been obtained by concentration enrichments below the watershed point.

It should be noted that for the design of the concept several constraints have to be considered. The main question is whether the watershed point can be reached or not. On one hand, usually rather high concentrations are necessary for this; on the other hand, these must lie within the solubility range. To conclude, for a given chromatographic system, the following parameters must fit together: (1) the mass stream obtained from zone I, (2) the solubility of the compounds, and (3) the thermodynamics (i.e., adsorption isotherms).

The reader might have noticed that the process requires an additional unit operation (e.g., an evaporator). Therefore, it might well be classified as one of the process combination concepts that will be discussed in Section 5.7. However, the main aspect considered here is the recycle with its highly concentrated stream rather than the unit used for the enrichment.

5.7   INTEGRATED SMB PROCESSES AND PROCESS COMBINATIONS

The SMB process as such is nowadays quite well understood. Therefore, it is not surprising that current work focuses also on its assembly within its typical plant environment. It is desired to identify and exploit synergisms that may arise when combining SMB chromatography with other unit operations. This section will discuss the two main possibilities. The first is to couple SMB units to other processes (here: to crystallization). The second is to integrate a chemical reaction into the system (SMB reactors).

5.7.1   Combination of SMB and Selective Crystallization

A well-known fact in engineering is that costs of a separation decrease significantly if purity requirements can be lowered. A very cost-intensive application for SMB chromatography is the separation of enantiomers. This is the background for the suggested process combination in Figure 5.14.

In this scheme, the SMB is coupled to one or two crystallizers. These can deliver pure solid enantiomers if their feed streams exceed a certain minimum purity (the eutectic composition) with respect to the desired enantiomer. Therefore, the SMB unit is operated only at limited purity which allows for several possible benefits (i.e., significantly less or a cheaper chiral stationary phase, higher throughput, etc.).

The idea of such coupling was suggested by Lim et al. [73]; its full potential was recognized first by Blehaut and Nicoud [74]. A more detailed discussion was presented by Lorenz et al. [75]. As indicated above, the potential of the combination stems from the improved performance of SMB chromatography when lowering the

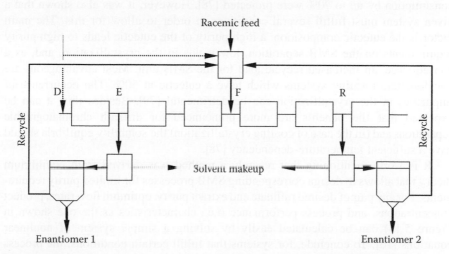

FIGURE 5.14   Schematic setup of a combined process of SMB chromatography and enantioselective crystallization. The SMB process delivers only partially separated products. Pure enantiomers are obtained from crystallization. Unresolved mother liquor is recycled.

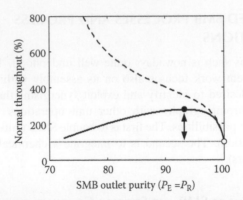

FIGURE 5.15 Performance of a stand-alone SMB process (dashed line) and the process combination with crystallization (solid line). Open symbol, throughput of the stand-alone SMB process; filled symbol, process combination. Note that both options deliver the target enantiomers with 100% purity.

purity requirements. This relationship is exemplified in Figure 5.15, where a purity-throughput characteristic of an SMB process is shown (dashed line). It can be seen already that a slight decrease of the purity below (typically required) 100% allows for a strongly increased throughput. From such a characteristic, (which can be determined *a priori* from, e.g., model-based calculations) the throughput of the combined scheme can be calculated using simple algebraic relations [76].

The concept was investigated in detail for different enantiomeric [76–79] and diastereomeric systems [79]. It was demonstrated that under favorable conditions the performance of the process combination can be significantly higher than that of a stand-alone separation by SMB chromatography. Under somewhat idealized assumptions, possible increase of productivity by up to 200% and a decrease of solvent consumption by up to 70% were projected [78]. However, it was also shown that a given system must fulfill several conditions in order to allow for this. The main factor is the eutectic composition; a high purity of the eutectic leads to high purity requirements on the SMB separation, low yields in the crystallization, and, as a consequence, an increased recycle load for the SMB unit. Most advantageous are conglomerate forming systems which have a eutectic at 50%. The occurrence of impurities in the crystallization can be detrimental [79]. Besides this, it can be expected that the benefits are more pronounced for difficult chromatographic separations and (in the case of cooling crystallization) the solubility equilibria should have a sufficient temperature-dependency [78].

It is worth mentioning that recently a method was derived from equilibrium theory that allows to design corresponding SMB processes for limited purity requirements. For any pair of desired raffinate and extract purity, optimum flow rates, product concentrations, and process performance (i.e., characteristics as the one shown in Figure 5.15) can be calculated easily by solving a simple system of nonlinear equations [16]. To conclude, for systems that fulfill certain conditions, the process combination of SMB chromatography and selective crystallization can be promising enough to consider the additional efforts necessary to establish the somewhat complex process scheme in Figure 5.14.

5.7.2 INTEGRATION OR COMBINATION WITH CHEMICAL REACTIONS

As the last important field of current research, the combination of SMB chromatography and chemical reactions will be considered. In particular, the idea to perform an equilibrium-limited chemical reaction *within* an SMB unit is intriguing for many engineers (which is certainly inspired by other successful examples like reactive distillation).

The investigations performed in this direction are manifold; again, only the main ideas can be summarized here. Figure 5.16 demonstrates the three different options that are used to deal with equilibrium-limited chemical reactions. The classical concept (top) is the reactor–separator setup, where an SMB process is connected to some chemical reactor to separate its output; unconverted reactants are recycled. This was suggested, e.g., for application in sucrose inversion (i.e., sucrose = fructose + glucose) and dextran production [80], and was reviewed more recently also in the context of other biotransformation reactions [81]. Another option, which is an extension of this concept, is the Hashimoto process (middle) [82]. This was developed for isomerization problems. Here, the reaction is functionally distributed within the unit by including several reactors in certain zones. It should be mentioned that so

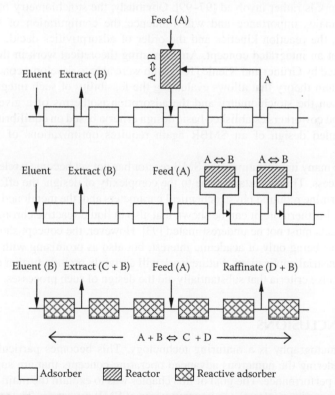

FIGURE 5.16 Examples for the general possibilities for combining SMB processes with chemical reactions. Top, classical reactor–separator system with recycle (e.g., for isomerization problems). Middle, the partially integrated Hashimoto process (isomerization). Bottom, fully integrated SMB reactor using a reactive adsorbent for a reaction of type $A + B \leftrightarrow C + D$.

far this concept was applied only for certain limiting cases. It was investigated for systems with anti-Langmurian adsorption behavior and the stronger adsorbing component as target product; e.g., for fructose–glucose isomerization using enzymes as catalysts and ion exchange resins as adsorbent [82,83].

The last concept in the figure is the SMB reactor (SMBR, bottom). In an SMBR, a reactive adsorbent is packed into columns usually arranged as a classical four-zone scheme (bottom). This process is useful for reactions with two products, i.e., for reactions with a stoichiometry of the form $A \leftrightarrow C+D$ or $A+B \leftrightarrow C+D$, respectively. In the example shown, the reactants are introduced through the feed port (A) and the desorbent (B). Ideally, the products C and D are obtained in pure form at the two outlets. The stoichiometry $A \leftrightarrow C+D$ holds, e.g., for the already mentioned sucrose inversion and was investigated, e.g., in Refs. [84–86]. Esterification and transesterification reactions, for which SMB reactors have been studied most frequently, follow the scheme $A+B \leftrightarrow C+D$; corresponding studies were reported in, e.g., Refs. [87–93]. There are many more aspects that have been investigated with respect to SMB reactors. The reader is kindly referred to several publications that give a corresponding overview [94–96].

The design of SMB reactors is a challenging task; the corresponding mathematical framework is rather involved [97–99]. Obviously, the stoichiometry of the reaction is of major importance and will influence the configuration of the setup. In addition, the reaction kinetics and the order of adsorptivities decide about the feasibility of an integrated concept. An interesting theoretical work in this context was published by Grüner and Kienle [100], who were able to develop a concept based on equilibrium theory that allows evaluating the feasibility of such integrated process based on the stoichiometry and the adsorption isotherms of a given system. Mazzotti and coworkers established basic design criteria based on equilibrium theory [102]. Detailed design of an SMBR again requires optimizations of numerical models.

Despite many investigations, the SMB reactor has not yet become a relevant commercial success. This is certainly due to the complexity of design, the efforts necessary to determine required physicochemical parameters and, the mentioned feasibility restrictions. Furthermore, it can be shown that also in liquid reactive chromatography thermal effects must not be underestimated [93]. However, the concept should not be considered as being only of academic interest, but also as promising with respect to possible industrial applications. Future work will certainly provide further qualitative and quantitative criteria that substantially aid the design of such processes.

5.8 CONCLUSIONS

SMB chromatography is a maturing technology. This becomes particularly clear when considering the numerous advanced operating concepts that are suggested to improve its performance. The goal of this chapter was to explain the main principles that allow to realize performance improvements of SMB processes. The toeholds for this are the direct manipulation of adsorption strength (i.e., gradients), dynamic variation of operational variables, modifications of process structure, and purposeful combination with or integration of additional functionalities.

Not all proposed approaches could be covered in detail, and there are constantly evolving more process schemes with rising complexity. Several schemes combine characteristics of both SMB and batch chromatography (e.g., the so-called Japan Organo, or JO process [103]). Some concepts even exploit almost all of the above main principles. For example, a partially countercurrent concept was developed that employs multiple columns, asynchronous switching, and gradients for multicomponent separations of biomolecules [104]. Also, the chromatographic single-column processes that try to mimic SMB behavior (i.e., steady-state recycling [105,106] or the so-called single-column SMB analogs [107,108]) are not considered here. They represent interesting alternatives to SMB units, in particular, when it comes to investment costs.

When considering the large number of innovative operating modes that are available for SMB chromatography, one must keep in mind that the design of any advanced scheme becomes more and more tedious with increasing number of manipulated variables. Thus, a particular challenge for the future is to support practitioners with respect to two important problems: to choose the right process configuration for a given separation problem (which is still reasonably simple enough to be implemented in practice), and to find the optimal design parameters for this process. In light of this background, it appears of high interest to devise corresponding criteria and guidelines that support the choice of the operating mode as well as to develop further tools and methods for process design.

REFERENCES

1. G. Zhong and G. Guiochon, in *Advances in Chromatography*, Vol. 39, P.R. Brown and E. Grushka (Eds.), Marcel Dekker, New York, p. 352, 1998.
2. D.M. Ruthven and C.B. Ching, *Chem. Eng. Sci.*, 44:1011, 1989.
3. D.B. Broughton and C.G. Gerhold, U.S. Patent 2.985.589, 1961.
4. H. Kniep, PhD thesis, Otto-von-Guericke Universität Magdeburg. 30, 1998.
5. F. Charton and R.-M. Nicoud, *J. Chromatogr. A*, 702:97, 1995.
6. G. Storti, M. Mazzotti, M. Morbidelli, and S. Carra, *AIChE J.*, 39:471, 1993.
7. M. Mazzotti, G. Storti, and M. Morbidelli, *J. Chromatogr. A*, 769:3, 1997.
8. Z. Ma and N.-H.L. Wang, *AIChE J.*, 43:2488, 1997.
9. T. Mallmann, B.D. Burris, Z. Ma, and N.-H.L. Wang, *AIChE J.*, 44:2628, 1998.
10. M. Mazzotti, G. Storti, and M. Morbidelli, *AIChE J.*, 40:1825, 1994.
11. M. Mazzotti, G. Storti, and M. Morbidelli, *AIChE J.*, 42:2784, 1996.
12. M. Mazzotti, G. Storti, and M. Morbidelli, *AIChE J.*, 43:64, 1997.
13. C. Migliorini, M. Mazzotti, and M. Morbidelli, *AIChE J.*, 46:1384, 2000.
14. M. Mazzotti, *Ind. Eng. Chem. Res.*, 45:6311, 2006.
15. M. Mazzotti, *J. Chromatogr. A*, 1126:311, 2006.
16. M. Kaspereit, A. Seidel-Morgenstern, and A. Kienle, *J. Chromatogr. A*, 1162:2, 2007.
17. R.S. Alm, R.J.P. Williams, and A. Tiselius, *Acta Chem. Scand.*, 6:826, 1952.
18. T.B. Jensen, T.G.P. Reijns, H.A.H. Billiet, and L.A.M. van der Wielen, *J. Chromatogr. A*, 873:149, 2000.
19. D. Antos and A. Seidel-Morgenstern, *Chem. Eng. Sci.*, 56:6667, 2001.
20. R.-M. Nicoud, M. Perrut, and G. Hotier, U.S. Patent 5.422.007, 1995.
21. J.Y. Clavier, R.-M. Nicoud, and M. Perrut, in *High Pressure Chemical Engineering*, P.R. von Rohr and C. Trepp (Eds.), Elsevier Science, London, pp. 429–434, 1995.
22. M. Mazzotti, G. Storti, and M. Morbidelli, *J. Chromatogr. A*, 786:309, 1997.

23. S. Abel, M. Mazzotti, and M. Morbidelli, *J. Chromatogr. A*, 944:23, 2002.
24. S. Abel, M. Mazzotti, and M. Morbidelli, *J. Chromatogr. A*, 1026:47, 2004.
25. L. Gueorguieva, L.F. Vallejo, U. Rinas, and A. Seidel-Morgenstern, *J. Chromatogr. A*, 1135:142, 2006.
26. E. Soczewinski, *Anal. Chem.*, 41:179, 1969.
27. M. Kaspereit, P. Jandera, M. Skavrada, and A. Seidel-Morgenstern, *J. Chromatogr. A*, 944:249, 2002.
28. D. Beltscheva, P. Hugo, and A. Seidel-Morgenstern, *J. Chromatogr. A*, 989:31, 2003.
29. D. Antos and A. Seidel-Morgenstern, *J. Chromatogr. A*, 944:77, 2002.
30. G. Ziomek, M. Kaspereit, J. Jezowski, A. Seidel-Morgenstern, and D. Antos, *J. Chromatogr. A*, 1070:111, 2005.
31. K. Kaczmarski and D. Antos, *Acta Chromatogr.*, 17:20, 2006.
32. J. Houwing, H.A.H. Billiet, and L.A.M. van der Wielen, *J. Chromatogr. A*, 944:189, 2002.
33. J. Houwing, S.H. van Hateren, H.A.H. Billiet, and L.A.M. van der Wielen, *J. Chromatogr. A*, 952:85, 2002.
34. J. Houwing, T.B. Jensen, S.H. van Hateren, H.A.H. Billiet, and L.A.M. van der Wielen, *AIChE J.*, 49:665, 2003.
35. P. Li, G.H. Xiu, and A.E. Rodrigues, *AIChE J.*, 53:2419, 2007.
36. C. Migliorini, M. Wendlinger, M. Mazzotti, and M. Morbidelli, *Ind. Eng. Chem. Res.*, 40:2606, 2001.
37. N. Gottschlich, S. Weidgen, and V. Kasche, *J. Chromatogr. A*, 719:267, 1996.
38. N. Gottschlich and V. Kasche, *J. Chromatogr. A*, 765:201, 1997.
39. J. Thommes, J.P. Pieracci, M. Bisschops, A.M. Sonnenfeld, L. Conley, and M. Pennings, U.S. Patent 7.220.356, 2007.
40. B.M. Thome and C.F. Ivory, *Electrophoresis*, 28:1477, 2007.
41. B.M. Thome and C.F. Ivory, *J. Chromatogr. A*, 1129:119, 2006.
42. D.A. Horneman, M. Ottens, J.T.F. Keurentjes, and L.A.M. van der Wielen, *J. Chromatogr. A*, 1113:130, 2006.
43. M. Kearney and K. Hieb, U.S. Patent 5.102.553, 1992.
44. E. Kloppenburg and E. Gilles, *Chem. Eng. Technol.*, 22:813, 1999.
45. Z. Zhang, M. Mazzotti, and M. Morbidelli, *J. Chromatogr. A*, 1006:87, 2003.
46. Z. Zhang, M. Morbidelli, and M. Mazzotti, *AIChE J.*, 50:625, 2004.
47. H. Schramm, PhD thesis, Otto-von-Guericke Universität Magdeburg, Shaker, Aachen (Germany), 2005.
48. Y. Zang and P.C. Wankat, *Ind. Eng. Chem. Res.*, 42:4840, 2003.
49. Y. Kawajiri and L.T. Biegler, *AIChE J.*, 52:1343, 2006.
50. Y. Zang and P.C. Wankat, *Ind. Eng. Chem. Res.*, 41:2504, 2002.
51. P.S. Gomes and A.E. Rodrigues, *Sep. Sci. Technol.*, 42:223, 2007.
52. H. Schramm, A. Kienle, M. Kaspereit, and A. Seidel-Morgenstern, *Chem. Eng. Sci.*, 58:5217, 2003.
53. Z. Zhang, M. Mazzotti, and M. Morbidelli, *Korean J. Chem. Eng.*, 21:454, 2004.
54. O. Ludemann-Hombourger, R.-M. Nicoud, and M. Bailly, *Sep. Sci. Technol.*, 35:1829, 2000.
55. H. Schramm, M. Kaspereit, A. Kienle, and A. Seidel-Morgenstern, *Chem. Eng. Technol.*, 25:1151, 2002.
56. H. Schramm, S. Grüner, and A. Kienle, *J. Chromatogr. A*, 1006:3, 2003.
57. H. Schramm, M. Kaspereit, A. Kienle, and A. Seidel-Morgenstern, *J. Chromatogr. A*, 1006:77, 2003.
58. O. Ludemann-Hombourger, G. Pigorini, R.-M. Nicoud, D. Ross, and G. Terfloth, *J. Chromatogr. A*, 947:59, 2002.
59. Z. Zhang, K. Hidajat, A. Ray, and M. Morbidelli, *AIChE J.*, 48:2800, 2002.
60. Y. Kawajiri and L.T. Biegler, *Ind. Eng. Chem. Res.*, 45:8503, 2006.
61. Y. Kawajiri and L.T. Biegler, *AIChE J.*, 52:1343, 2006.
62. Y. Kawajiri and L.T. Biegler, *J. Chromatogr. A*, 1133:226, 2006.

63. P.C. Wankat, *Ind. Eng. Chem. Res.*, 40:6185, 2001.
64. G. Ziomek, D. Antos, L. Tobiska, and A. Seidel-Morgenstern, *J. Chromatogr. A*, 1116:179, 2006.
65. R. Wooley, Z. Ma, and N.-H.L. Wang, *Ind. Eng. Chem. Res.*, 37:3699, 1998.
66. K. Hashimoto, Y. Shirai, and S. Adachi, *J. Chem. Eng. Jpn.*, 26:52, 1993.
67. L.C. Keßler and A. Seidel-Morgenstern, *J. Chromatogr. A*, 1126:323, 2006.
68. C.Y. Chin and N.-H.L. Wang, *Sep. Purif. Rev.*, 33:77, 2004.
69. A.S. Kurup, K. Hidajat, and A.K. Ray, *Ind. Eng. Chem. Res.*, 45:6251, 2006.
70. M. Bailly, R.-M. Nicoud, A. Philippe, and O. Ludemann-Hombourger, Patent Application WO/2004/039468, 2004.
71. S. Abdelmoumen, L. Muhr, M. Bailly, and O. Ludemann-Hombourger, *Sep. Sci. Technol.*, 41:2639, 2006.
72. G. Paredes, H.K. Rhee, and M. Mazzotti, *Ind. Eng. Chem. Res.*, 45:6289, 2006.
73. B. Lim, C. Ching, R.B.H. Tan, and S. Ng, *Chem. Eng. Sci.*, 50:2289, 1995.
74. J. Blehaut and R.-M. Nicoud, *Analusis Mag.*, 26:M60, 1998.
75. H. Lorenz, P. Sheehan, and A. Seidel-Morgenstern, *J. Chromatogr. A*, 908:201, 2001.
76. M. Kaspereit, K. Gedicke, V. Zahn, A.W. Mahoney, and A. Seidel-Morgenstern, *J. Chromatogr. A*, 1092:55, 2005.
77. M. Amanullah and M. Mazzotti, *J. Chromatogr. A*, 1107:36, 2006.
78. M. Kaspereit, PhD thesis, Otto-von-Guericke Universität Magdeburg. Shaker, Aachen (Germany), 2006.
79. K. Gedicke, M. Kaspereit, W. Beckmann, U. Budde, H. Lorenz, and A. Seidel-Morgenstern, *Chem. Eng. Res. Des.*, 85:928, 2007.
80. P.E. Barker, G. Ganetsos, J. Ajongwei, and A. Akintoye, *Chem. Eng. J. Bioch. Eng.*, 50:B23, 1992.
81. M. Bechtold, S. Makart, M. Heinemann, and S. Panke, *J. Biotechnol.*, 124:146, 2006.
82. K. Hashimoto, S. Adachi, H. Noujima, and Y. Ueda, *Biotechnol. Bioeng.*, 23:2371, 1983.
83. T. Borren and H. Schmidt-Traub, *Chem. Ing. Tech.*, 76:805, 2004.
84. J. Fricke, M. Meurer, J. Dreisörner, and H. Schmidt-Traub, *Chem. Eng. Sci.*, 54:1487, 2004.
85. M. Meurer, U. Altenhöner, J. Strube, and H. Schmidt-Traub, *J. Chromatogr. A*, 769:71, 1997.
86. D. Azevedo and A.E. Rodrigues, *Chem. Eng. J.*, 82:95, 2001.
87. M. Kawase, T. Suzuki, K. Inoue, K. Yoshimoto, and K. Hashimoto, *Chem. Eng. Sci.*, 51:2971, 1996.
88. G. Ströhlein, Y. Assuncao, N. Dube, A. Bardow, M. Mazzotti, and M. Morbidelli, *Chem. Eng. Sci.*, 61:5296, 2006.
89. J.P. Meissner and G. Carta, *Ind. Eng. Chem. Res.*, 41:4722, 2002.
90. C. Migliorini, M. Fillinger, M. Mazzotti, and M. Morbidelli, *Chem. Eng. Sci.*, 54:2475, 1999.
91. V.M.T.M. Silva and A.E. Rodrigues, *AIChE J.*, 51:2752, 2005.
92. W.F. Yu, K. Hidajat, and A.K. Ray, *Chem. Eng. J.*, 112:57, 2005.
93. T. Sainio, M. Kaspereit, A. Kienle, and A. Seidel-Morgenstern, *Chem. Eng. Sci.*, 62:5674, 2007.
94. F. Lode, M. Houmard, C. Migliorini, M. Mazzotti, and M. Morbidelli, *Chem. Eng. Sci.*, 56:269, 2001.
95. M. Mazzotti, A. Kruglov, B. Neri, D. Gelosa, and M. Morbidelli, *Chem. Eng. Sci.*, 51:1827, 1996.
96. G. Ströhlein, F. Lode, M. Mazzotti, and M. Morbidelli, *Chem. Eng. Sci.*, 59:4951, 2004.
97. B. Cho, R. Aris, and R. Carr, *Proc. R. Soc. Lond. A*, 383:147, 1982.
98. H.-K. Rhee, R. Aris, and N.R. Amundson, *First-Order Partial Differential Equations. Vol. I—Theory and Applications of Single Equations*, Dover Publications, Mineola, NY, 2001.

99. H.-K. Rhee, R. Aris, and N.R. Amundson, *First-Order Partial Differential Equations. Vol. II—Theory and Applications of Hyperbolic Systems of Quasilinear Equations*, Dover Publications, Mineola, NY, 2001.
100. S. Grüner and A. Kienle, *Chem. Eng. Sci.*, 59:901, 2004.
101. G. Ströhlein, M. Mazzotti, and M. Morbidelli, *Chem. Eng. Sci.*, 60:1525, 2005.
102. Z.Y. Zhang, K. Hidajat, and A.K. Ray, *Ind. Eng. Chem. Res.*, 41:3213, 2002.
103. V. Mata and A.E. Rodrigues, *J. Chromatogr. A*, 939:23, 2001.
104. G. Ströhlein, L. Aumann, M. Mazzotti, and M. Morbidelli, *J. Chromatogr. A*, 1126:338, 2006.
105. C.M. Grill, *J. Chromatogr. A*, 796:101, 1998.
106. T.Q. Yan and C. Orihuela, *J. Chromatogr. A*, 1156:220, 2007.
107. N. Abunasser and P.C. Wankat, *Sep. Sci. Technol.*, 40:3239, 2005.
108. R.C.R. Rodrigues, J.M.M. Araújo, and J.P.B. Mota, *J. Chromatogr. A*, 1162:14, 2007.

6 Advances in Resins for Ion-Exchange Chromatography

*Arne Staby, Jacob Nielsen, Janus Krarup, Matthias
Wiendahl, Thomas Budde Hansen, Steffen Kidal,
Jürgen Hubbuch, and Jørgen Mollerup*

CONTENTS

6.1 INTRODUCTION

Liquid chromatography is a dynamic technique for separation of molecules, both
analytically and in preparative scale, using liquid as the mobile phase. For separation
of proteins and peptides, liquid chromatography is used in various modes depending
on the nature of ligands immobilized onto the stationary phase. Ligands on ion-
exchange resins are charged in a broad pH range. If the resin ligand is positively
charged the mode of operation is called anion-exchange chromatography, while
cation-exchange chromatography is applied with negatively charged ligands, and

ion-exchange chromatography is basically used to separate proteins based on their overall or local difference in charge. In analytical liquid chromatography, loading of the chromatographic column is very low to obtain the best separation of the distinct molecules present in the sample, and various elution gradients and modifiers are applied to achieve maximum resolution. In preparative liquid chromatography, high column loading is applied to get high productivities and good process economy, and a separation is developed to obtain optimal purity and yield of the target protein or peptide. Gradients and modifiers are applied in the simplest way possible to obtain the desired purity and yield. The nature of gradients in ion-exchange chromatography is linked to the mode of operation, e.g., salt or pH and occasionally organic solvent, displacers, or other modifiers to increase selectivity in bind-and-elute or flow-through modes of operation. These operational modes may be combined with each other and with displacement effects, and the transition between the two modes of operation is not well defined.

Ion-exchange chromatography is an important and integrated part of biopharma-ceutical downstream processing due to the fairly mild conditions at which elution takes place, and because the native structure and activity of the protein of interest usually are preserved in this aqueous environment. During purification process development, selection of ion-exchange resins is based on knowledge and a number of scientific and economical considerations. Considerations include comparison of characteristics and performance of resins for the specific purification application. Resin suppliers put an extensive effort in the attempt to increase overall performance of their products, and significant gains in selectivity, capacity, physical stability, etc., have been obtained over the years. For purification of therapeutic antibodies, there is a tendency to apply generic purification processes including selection of ion-exchange resins to minimize development and manufacturing cost; however, it is still neces-sary for the biopharmaceutical industry to compare the performance of various preferred resins both for purification of therapeutic antibodies and especially for other recombinant proteins and peptides.

Although industrial comparison of resins is usually considered to be classified information by the companies, an increasing number of papers have recently been presented at international conferences, and several papers evaluating cation-exchangers have been published by commercial suppliers, and academic and non-commercial institutions comparing various chromatographic parameters including: dynamic [2,5–7,9,15,16,19,22,28,30–33,94], static [2,3,6,11,16,27,31–33,93], and ionic [2–6,11,22,24,29,30,94] capacities, binding strength [4,10,15,17–20,22,25,26,31–33, 93,94], elution dependence on pH [8,10,14,15,17,18,31–33], efficiency [6,9,10,12, 14,17,22,24,30–33], resolution [2,3,5–7,11,14,15,17,19–23,93], adsorption isotherms [6,12,15,18,27,30], pressure drop [5,6,11,14,17], compressibility [3,5,6,14], protein recovery [3,5,6,14,17,94], operating flowrate [3,5,6,9], cost [5,23], chemical stability [5,11,14,17,94], base matrix chemistry [2,4,11,13,14,17,23], pore size distribution [1,4,9,17,30], and others. The test proteins used are typically lysozyme, chy-motrypsinogen, cytochrome *c*, IgG, and others. Similarly, a number of papers evaluating anion-exchangers have been published by commercial suppliers, and academic and noncommercial institutions comparing various chromatographic parameters including: dynamic [2,5,7,29,36,37,39,42,44,46–48,51,52,55,59,66–71], static [2,34,37,44,45,54,59,66,68–71], and ionic [2,5,29,34,37,41,46,47,50,51,55–58] capacities, titration curves [46,55,57,69–70], binding strength [35,38,49,50–53,62,64, 69–71], elution dependence on pH [8,46,49,52,61,62,69–71], efficiency [7,12,35,36, 40,43,47,66,69–70], resolution [2,5,7,8,34,36,37,42,43,46,49,56,62–64,71], adsorption isotherms [40,42,45,47,48,54,58,60], pressure drop and compressibility [5,34,37,46, 57,58,60,66,67], protein recovery [5,34,47,51,57,61,67], operating flow rate [5,34,36, 37,39,42,44,46–49,57,59,60,63,66], cost [5,36,37], chemical stability [5,37,41,46], spacer arm [2,41,43], base matrix chemistry [2,41,43,46,57], pore size distribution [50,58,65,66,68], and others. The test proteins used are typically serum albumins and others. The number of resins, compared in these papers, is two to four; however, the papers given in Table A.1 present evaluation of more than four ion-exchangers. The paper by Lohrmann et al. [72] also contains a comparison of various parameters for six weak cation-exchangers with lysozyme and aprotinin; however, the names of the resins used are not stated, making results of this chapter difficult to compare with others. All resins presented in this paper are listed in Table A.2.

Traditional screening techniques have been employed, in the above-mentioned papers, to gain knowledge of ion-exchangers, and this approach has been a cornerstone in the purification process development in many companies; however, this current methodology is rather time- and material-consuming. New approaches, including high-throughput screening (HTS)/robot techniques [71,73–75,98], have been employed lately to increase the speed of process development within the biopharmaceutical industry. These HTS techniques will probably be state of the art in future purification process development, either as stand-alone techniques or in conjunction with mathematical modeling, when the accuracy of the measurements with 96-well HTS format or miniature columns has increased.

We have chosen to assess most of the advances in resins for ion-exchange chromatography purification, in this chapter, from comparison studies performed by our group [31–33,69–71] with new unpublished studies, due to differences in experimental setup by the various authors. The difference in experimental conditions between various authors makes a direct comparison very difficult; however, we would report general trends observed by the various authors. The ion-exchangers cover a broad range of commercially available base matrix chemistries including agarose, agarose/ dextran, acrylic and vinylic polymers, ceramics, and polystyrene-divinylbenzene. The selected test proteins cover a broad range of isoelectric points and molecular weights. They include both common test proteins such as bovine serum albumin (BSA), lysozyme, and myoglobin, and additionally proteins and peptides obtained at Novo Nordisk and Novozymes (Anti-FVII mAb [IgG] [76] and aprotinin [77], heparin-binding protein [HBP] [78], human growth hormone [hGH], insulin precursor [79], and lipolase [80]). Properties of proteins and peptides used in the studies are given in Table 6.1.

A number of parameters impact the resin performance in general, and we have chosen to focus on few of these parameters, i.e., parameters that may easily be compared by a standard experimental chromatography setup. However, examples of parameters that require direct comparison between resins and conditions such as removal of specific impurities will also be presented as well as verification studies of obtained results. The exact experimental setup including raw materials, equipment,

TABLE 6.1
Properties of Test Proteins

Protein	pI	Molecular Weight (kDa)
Anti-FVII mAb (IgG)	~6–7	150
Aprotinin	~10.5	6
BSA	5.0–5.2	69
HBP	~9	28
hGH	~5	22
Insulin precursor	5.3	6
Lipolase	~4.3	35
Lysozyme	~11	14
Myoglobin	7–8	18

data handling, and theoretical considerations are given elsewhere [31–33,69–71], as well as discussions on any limitations of the experimental setup.

6.2 PARAMETERS IMPACTING RESIN PERFORMANCE

The chromatographic resins assessed in this chapter, may be used in a variety of applications including capture, intermediate purification, and final purification or polishing in a downstream process. The different purification modes are not clearly defined and may depend on whether the origin is a recombinant or a natural/wild-type expression system, presence of reactants, cell debris, fats, degree of previous purification including filtration, centrifugation, etc. In general, resins used for capture are characterized by having a fairly large particle size and a high binding capacity in order to concentrate the target protein, remove water, operate at a reasonably high flow rate, and avoid clogging of the column by fermentation products. Another characteristic is a fairly large particle size distribution resulting in a lower price of these bulk resins. In a special mode of chromatographic capture of proteins from fermentation broth, expanded bed adsorption, resin particles should be dense in order to apply high flow rates in a fluidized state allowing cell debris, etc., to pass through the column. Intermediate purification is employed for further removal of host cell proteins (HCP) and easily removed product-related impurities, while the resins for polishing should have a high selectivity and are characterized by having a smaller particle size and a narrow particle size distribution with high resolution, optionally for high-pressure operation. Resins for polishing are usually the most expensive due to their ability to provide sharper peaks and high resolution between the target protein and its related impurities.

Ion-exchange resins are characterized by their specific ligands and their charge. Positively charged ligands define an anion-exchanger, while negatively charged ligands define a cation-exchanger. Ion-exchangers are traditionally further split up into two categories, strong and weak ion-exchangers based on their apparent pH operation window. Strong anion-exchange ligands are mostly quaternary aminomethyl or quaternary aminoethyl (Q or QAE) types, with pH working ranges of ~ 3–13, and the weak anion-exchange ligands are often diethyl aminoethyl (D or DEAE) types with a pH working range of 3–9. Strong and weak cation-exchange ligands usually contain sulfonic acid and carboxylic acid groups, respectively, with pH working ranges of ~ 3–13 and 6–13, and they are frequently referred to as "S"- and "C"-types. Most suppliers manufacture one of each type of ligand, i.e., four different ion-exchangers having one or more base matrix chemistries and with one or more particle sizes; however, some suppliers provide several ligand chemistries of both strong and weak ion-exchangers, e.g., Tosoh Bioscience. Recently, new types of the so-called mixed-mode ligands have merged combining the charged and hydrophobic groups to form the ligand, giving these multimodal resins a higher degree of salt tolerance during operation, e.g., Capto MMC and Capto adhere from GE Healthcare.

There are many ways of conducting ion-exchange chromatography. Managing ion-exchange chromatography may include the use of counterions for elution (e.g., chloride ions for anion-exchange), pH changes for elution (typically to the opposite pH range of the proteins pI), in flow-through mode (binding of impurities, but not or only weakly the target protein), and again others using competition for elution (e.g., calcium ions against positive counterions on the anion-exchange surface or alcohols or temperature to change protein conformation to obtain elution). All these modes of operation are ion-exchange chromatography methods because an ion-exchange resin is used, and it is the stationary phase that defines the mode of chromatography, not the means of elution.

Ion-exchange chromatography is universal, thus if a concept is working, e.g., for anion-exchange in the high pH range, the person skilled in the art of ion-exchange chromatography would also expect the similar concept to work for cation-exchange in the low pH range and vice versa. Thus, it can be stated, that a finding, concept, use, or discovery, proven effective for one mode of ion-exchange chromatography will also be successful for the other ion-exchange chromatography mode.

6.2.1 pH

Process analytical technologies are increasingly gaining foothold in the biopharmaceutical industry; however, traditional process challenge and validation issues are still of great importance during process development and manufacturing. Among parameters to validate in ion-exchange chromatography, pH is a key factor which would be controlled within predetermined ranges due to protein stability, potential lack of binding or separation, and other issues. Awareness of resin capabilities to maintain or change the pH during operation is thus an essential knowledge in process development, and such knowledge may be obtained from the data of protein retention as a function of pH in linear salt gradient elution.

Experimental results of retention at linear gradient elution as a function of pH for various proteins are presented in Figures 6.1 and 6.2. Figure 6.1 presents data of five proteins on an anion-exchange resin (MacroPrep 25Q) to be considered for polishing operation, while Figure 6.2 presents data of four proteins on a cation-exchange resin (SP Sepharose BB) usually used for capture processes. In Figure 6.1, the isoelectric points of BSA and lipolase are below the experimental pH range of 6–9 as

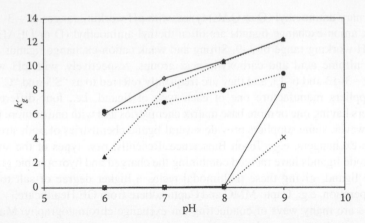

FIGURE 6.1 pH dependence plot (k_g'–pH) of various proteins on MacroPrep 25Q anion-exchanger. The pH dependence was determined by eluting a 20 μL pulse of a 1 g/L protein solution by a 20 CV linear gradient from 0 to 1 M NaCl in 25 mM Tris + 25 mM Bis-Tris propane buffer. Column dimensions were 10 × 0.46 or 10 × 0.5 cm. Symbols: ◇, IgG; □, aprotinin; ▲, BSA; ●, lipolase; ×, myoglobin.

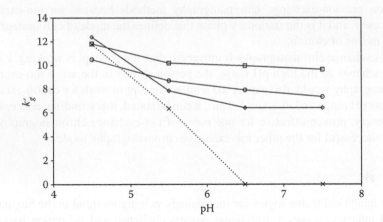

FIGURE 6.2 pH dependence plot (k_g'–pH) of various proteins on SP Sepharose BB cation-exchanger. The pH dependence was determined by eluting a 20 μL pulse of a 1 g/L protein solution by a 20 CV linear gradient from 0 to 1 M NaCl in 16.7 mM MES + 16.7 mM HEPES + 16.7 mM sodium acetate buffer. Column dimensions were 10 × 0.46 or 10 × 0.5 cm. Symbols: ◇, IgG; □, aprotinin; ○, lysozyme; ×, myoglobin.

given in Table 6.1, and the observed pH dependence displays the expected correlation of increasing retention with increasing pH. A similar trend is observed for IgG having an isoelectric point of ~6.5. The pI of myoglobin is within the experimental pH range, and the results show the expected correlation of no binding at pH 6–8, but some retention at pH 9. A similar trend is obtained for aprotinin which has a pI above the experimental range, thus a rather high retention is observed at pH 9, and though the overall charge of the protein is positive at pH 9, the retention experienced could be due to local areas with many negative charges. This finding is made for several anion- and cation-exchangers [31–33,69–71], and it is analogous to previous discoveries by Kopaciewicz et al. [81] who found retention of a number of proteins up to one pH unit below the pI of the proteins on a Mono Q resin.

For protein retention on SP Sepharose BB in Figure 6.2, equivalent findings are obtained with the expected trend of decreasing retention with increasing pH. For lysozyme and aprotinin with isoelectric points far above the experimental pH range, a very small decrease in retention is observed. For IgG and myoglobin with pI in the experimental range, more dramatic changes in retention are observed with increasing pH, especially for myoglobin. Again, IgG displays considerable binding at pH above its pI for cation-exchange chromatography, thus corresponding to the finding of aprotinin on MacroPrep 25Q, while myoglobin demonstrates the more traditional perception of protein binding as a function of pH.

Experimental results of retention in linear gradient elution as a function of pH may also be presented as retention data of a single protein on different resins. Figures 6.3 and 6.4 present the data measured with lysozyme on cation-exchange resins suitable for capture processes, and with lipolase on anion-exchange resins for polishing steps, respectively. Figure 6.3 shows that the expected degree of decreasing retention with increasing pH is fairly modest for most resins with lysozyme, and the general trend is a parallel retention pattern with no crossover between the resins. However, at pH 4.5 for CM Hyper Z, the increase in retention is pronounced almost twofold from pH 7.5 to pH 4.5. The pK_A of the carboxylic acid group of the weak cation-exchangers is ~4.5. The recommendation of most suppliers is to use weak cation-exchangers above pH 6 to obtain full dissociation of and binding to the carboxylic acid group on these resins; however, the highest

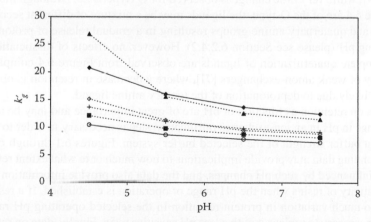

FIGURE 6.3 pH dependence plot (k_g'–pH) of lysozyme on various cation-exchangers for capture processing. The pH dependence was determined by eluting a 20 µL pulse of a 1 g/L protein solution by a 20 CV linear gradient from 0 to 1 M NaCl in 16.7 mM MES + 16.7 mM HEPES + 16.7 mM sodium acetate buffer. Column dimensions were 10 × 0.46 or 10 × 0.5 cm. Symbols: +, SP Toyopearl 650c; o, SP Sepharose BB; ■, CM Cellufine C-500; ◇, CM Toyopearl 650m; ♦, SP Toyopearl 550c; ▲, CM HyperZ.

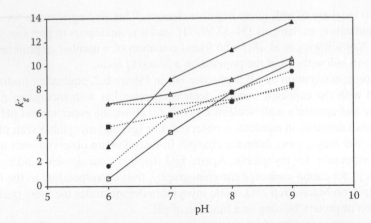

FIGURE 6.4 pH dependence plot (k_g'–pH) of lipolase on various anion-exchangers for polishing processes. The pH dependence was determined by eluting a 20 μL pulse of a 1 g/L protein solution in 20 CV linear gradient from 0 to 1 M NaCl in 25 mM Tris + 25 mM Bis-Tris propane buffer. Column dimensions were 10 × 0.46 or 10 × 0.5 cm. Symbols: +, Q HyperD 20; •, Source 30Q; △, Poros 50 HQ; ▫, TSKGel Q-5PW-HR; ▲, UNO Q-1; ■, Q Sepharose HP.

binding was still obtained at the lowest pH for all weak cation-exchange resins, and no dissociation effects of ligands were observed at these linear conditions.

The data on anion-exchange resins in Figure 6.4 also display the expected trend of increasing retention of lipolase with increasing pH; however, the trend displays a more complex retention pattern compared to cation-exchange resins with multiple cases of crossover. Highest degree of change is obtained for TSKGel Q-5PW-HR and UNO Q-1, while very little change is observed for Q HyperD 20. Although the resins in Figure 6.4 are of the Q-type, the ligands may be a mixture of different secondary, ternary, and quaternary amine groups resulting in a gradual release of protons with increasing pH (please see Section 6.2.4.2). However, no effects of dissociation due to incomplete quaternization of ligands are observed from Figure 6.4 compared to the study of weak anion-exchangers [71], where a decrease in retention is observed at pH 9, likely due to deprotonation of the ternary amine ligand.

Data on retention as function of pH are of great importance and may be applied if a change in pH in an ion-exchange purification step is necessary in order to obtain sufficient buffer strength of the selected buffer system. Figures 6.1 through 6.4 and corresponding data may provide implications to how much or to what extent retention will be influenced by such pH changes, and the data also provide information on the pH sensitivity of resins when the pH range of operation is established. If a resin displays too much variation in protein retention in the selected operating pH range, it may be necessary to replace it with a less pH sensitive resin. Finally, data on retention as function of pH are useful for development of step gradient programs for elution by a change of pH, to avoid salt elution and for planning of flow-through mode operation. Based on the experience gained through papers [31–33,69–71], we have in Table 6.2 ranked the influence of pH on retention for anion- and cation-exchangers. The grading of course depends on the proteins applied in the investigation, but Table 6.2 gives a

TABLE 6.2

Approximate pH Dependence Order of Anion- and Cation-Exchange Resins Based on Our General Experience and Gradient Retention Measurements with Lipolase and IgG and Aprotinin and Lysozyme, Respectively

pH Sensitivity of Retention	Anion-Exchangers	Cation-Exchangers
Lower		Fractogel SE HICAP (M)
	Q HyperD 20	Heparin Sepharose FF
	Q Sepharose FF	Heparin Ceramic HyperD M
	Poros 50 D	Fractogel EMD SO$_3^-$ (M)
	Fractogel EMD DEAE (M)	CM Sepharose FF
	DEAE Sepharose FF	Heparin Toyopearl 650 m
	Poros QE/M	CM Cellufine C-500
	Toyopearl DEAE 650 (M)	SP Sepharose BB
	DEAE Ceramic HyperD 20	SP Sepharose FF
	MacroPrep 25Q	SP Toyopearl 650 m
	MacroPrep DEAE Support	TSKGel SP-5PW-HR20
	Q Sepharose HP	SP Toyopearl 650 c
	Poros 50 HQ	SP Sepharose XL
Higher	Q Zirconia	CM Toyopearl 650 m
	Q Sepharose XL	Source 30 S
	Fractogel EMD TMAE 650 s	MacroPrep CM
	Express-Ion Q	MacroPrep 25 S
	Toyopearl QAE 550 c	Fractogel EMD COO⁻ (M)
	Separon HemaBio 1000Q	Poros 50 HS
	Source 30Q	SP Toyopearl 550 c
	TSKGel Q-5PW-HR	S Ceramic HyperD 20
	UNO Q-1	CM Hyper Z
	Toyopearl SuperQ 650 c	MacroPrep High S
		CM Ceramic HyperD F

Note: Resins with the lowest pH sensitivity of retention are at the top of the table.

fairly good indication of the pH sensitivity of the retention based on our general experience and the gradient retention measurements with lipolase and IgG, and aprotinin and lysozyme, respectively. Furthermore, cation-exchangers generally bind proteins stronger, thus allowing for higher salt concentrations than anion-exchangers wherefore cation-exchange may be preferred for capture processes.

Comparing with Table 6.2, similar pH dependence patterns for anion-exchangers were obtained by Andersson et al. [52]. They studied various Amersham Biosciences resins (now: GE Healthcare) in salt gradient mode in the pH range 6–10, with numerous proteins, including BSA and pepsinogen. The resins were Q Sepharose FF, Q Sepharose XL, DEAE Sepharose FF, ANX Sepharose FF (low sub), and ANX Sepharose FF (high sub), and in addition the affinity chromatography resins Amino

Sepharose FF and Benzamidine Sepharose FF employed, in this study, as weak anion-exchange resins. Their ranking of retention as a function of pH for Q Sepharose FF, Q Sepharose XL, and DEAE Sepharose FF for BSA was similar to the studies in Table 6.2, that is, in the lower end—typically less than doubled retention from 6 to 10, and the same trend was achieved for the other resins in their study. For the weak anion-exchangers, a decreased retention was observed at pH 9 and especially at pH 10, possibly due to deprotonization of the ligand amino groups as observed in Ref. [71]. Retention as a function of pH for pepsinogen displayed a more constant or even negative trend over the entire pH range for weak anion-exchangers, explained by the acidic nature of the test protein [25].

6.2.2 BINDING STRENGTH

The most common mode of ion-exchange purification processes is the binding and elution of the target component and impurities through an alteration of the salt concentration and thus the conductivity of the mobile phase. Conductivity of the sample for loading is often known, and though, e.g., a diafiltration step may be included in the process, it is generally advantageous to avoid additional handling and conditioning steps. A key aspect of resin selection during purification process development is thus to be acquainted with the aptness of resins for a specific purpose. Knowledge of retention as a function of salt concentration or conductivity, also called binding strength, is such a key aspect in resin selection. This information is usually measured as isocratic retention or partition coefficient data as a function of salt concentration or conductivity.

Experimental binding strength data are obtained by isocratic runs at linear conditions, and characterized by plots of retention as a function of conductivity shown below. Stronger binding marks the need for more salt for elution to occur and thus a larger total ionic strength, I_{total}. Figures 6.5 and 6.6 illustrate binding strength plots of selected anion-exchangers suitable for capture and intermediate purification processes using test proteins with low isoelectric point, IgG and lipolase, respectively. For IgG in Figure 6.5, high binding strength is obtained for QAE Toyopearl 550c, Q Sepharose XL, and Fractogel EMD DEAE (M). Resins with the weakest binding for IgG are DEAE Toyopearl 650 m and Express-Ion Q. A similar pattern is shown for lipolase in Figure 6.6, i.e., high binding strength for QAE Toyopearl 550c and Q Sepharose XL, and weak binding for DEAE Toyopearl 650 m. However, for lipolase a fairly weak binding is obtained on Fractogel EMD DEAE (M), while intermediate binding strength is obtained on Express-Ion Q.

Conductivity data for anion-exchangers typically represent salt concentrations in the range of 0–350 mM NaCl with the current test proteins. High and low binding strength for QAE Toyopearl 550c and DEAE Toyopearl 650 m, respectively, are in good agreement with general results obtained for the 550 and 650 series from Toyopearl, both on anion-exchangers [69–71] and on corresponding cation-exchangers [31–33]. Fractogel resins with the same base matrix chemistry as Toyopearl resins, may display both strong and weak binding strength as discussed above, but in most cases they display binding strengths in the lower end both for anion- and cation-exchangers.

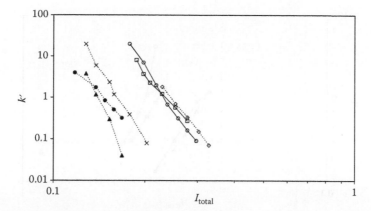

FIGURE 6.5 Binding strength plots ($k'-I_{total}$) of IgG on various anion-exchangers for capture or intermediate purification. Binding strength was determined by eluting a 20 μL pulse of 1 g/L protein solution in 25 mM Tris + 25 mM Bis-Tris propane buffer pH 8 at various isocratic NaCl concentrations. Column dimensions were 10 × 0.46 or 10 × 0.5 cm. Symbols: ×, Q Zirconia; o, Q Sepharose XL; ◇, QAE Toyopearl 550c; •, DEAE Toyopearl 650 m; ▫, Fractogel EMD DEAE 650M; ▲, Express-Ion Q.

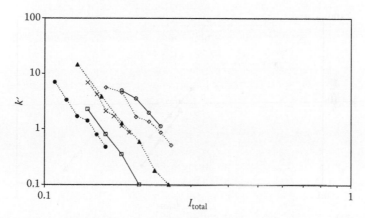

FIGURE 6.6 Binding strength plots ($k'-I_{total}$) of lipolase on various anion-exchangers for capture or intermediate purification. Binding strength was determined by eluting a 20 μL pulse of 1 g/L protein solution in 25 mM Tris + 25 mM Bis-Tris propane buffer pH 8 at various isocratic NaCl concentrations. Column dimensions were 10 × 0.46 or 10 × 0.5 cm. Symbols: ×, Q Zirconia; o, Q Sepharose XL; ◇, QAE Toyopearl 550c; •, DEAE Toyopearl 650 m; ▫, Fractogel EMD DEAE 650M; ▲, Express-Ion Q.

Data on binding strength for selected strong cation-exchangers for polishing and intermediate purification processes are presented in Figures 6.7 and 6.8, for lysozyme and aprotinin, respectively. Both test proteins have high isoelectric point. The strongest binding for the lysozyme data shown in Figure 6.7 is achieved on MacroPrep 25 S resin, and the weakest binding is obtained on Source 30 S and

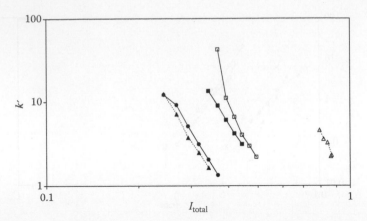

FIGURE 6.7 Binding strength plots (k'–I_{total}) of lysozyme on various cation-exchangers for final purification. Binding strength was determined by eluting a 20 μL pulse of 1 g/L protein solution in 16.7 mM MES + 16.7 mM HEPES + 16.7 mM sodium acetate buffer pH 5.5 at various isocratic NaCl concentrations. Column dimensions were 10 × 0.46 or 10 × 0.5 cm. Symbols: △, MacroPrep 25 S; ■, S Ceramic HyperD 20; □, Poros 50 HS; ▲, Source 30 S; ●, TSKGel SP-5PW-HR20.

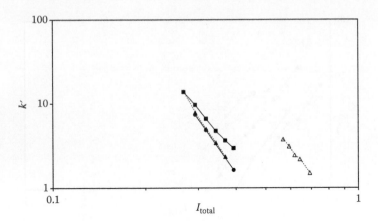

FIGURE 6.8 Binding strength plots (k'–I_{total}) of aprotinin on various cation-exchangers for final purification. Binding strength was determined by eluting a 20 μL pulse of 1 g/L protein solution in 16.7 mM MES + 16.7 mM HEPES + 16.7 mM sodium acetate buffer pH 5.5 at various isocratic NaCl concentrations. Column dimensions were 10 × 0.46 or 10 × 0.5 cm. Symbols: △, MacroPrep 25 S; ■, S Ceramic HyperD 20; ▲, Source 30 S; ●, TSKGel SP-5PW-HR20.

TSKGel SP-5PW-HR20. Corresponding results are obtained for aprotinin in Figure 6.8, where strongest binding is achieved on MacroPrep 25 S and weakest binding is obtained on Source 30 S and TSKGel SP-5PW-HR20.

Conductivity data for cation-exchangers typically represent salt concentrations in the range of 0–875 mM NaCl [31–33] for the test proteins employed. Very high binding strength for MacroPrep 25 S is in good agreement with general results obtained for MacroPrep cation-exchange resins [33] and on the corresponding anion-exchangers

[69,71]. Poros resins tend to be in the higher binding strength area, while Source resins have a propensity to be in the medium to lower binding strength area. Ceramic HyperD resins appear in the intermediate to high binding strength area.

From Figures 6.5 through 6.8, it is evident that cation-exchangers generally bind stronger than anion-exchangers at equal salt concentration, and although some scatter occurs, data generally display a parallel trend in retention as a function of conductivity. Thus, an approximate general binding strength order of ion-exchange resins may be set up, and experience and observations published in Refs. [31–33,69–71] are summarized in Table 6.3. The grading is to some extent dependent on the proteins

---

TABLE 6.3

Ranking of the Approximate Binding Strength of Anion- and Cation-Exchange Resins Based on Our General Experience and Isocratic Retention Measurements with Lipolase and IgG, and Aprotinin and Lysozyme, Respectively

Binding Strength	Anion-Exchangers	Cation-Exchangers
High		Capto MMC[a,b]
	Poros QE/M	MacroPrep 25 S
	MacroPrep DEAE Support	MacroPrep High S
	MacroPrep 25Q	MacroPrep CM
	Toyopearl QAE 550c	SP Toyopearl 550c
	Q Sepharose XL	CM Ceramic HyperD F
	Q Sepharose HP	Poros 50HS
	Q Sepharose FF	S Ceramic HyperD 20
	Capto Q[a]	SP Toyopearl GigaCap 650 m[a]
	Poros 50 HQ	Fractogel EMD SO$_3^-$ (M)
Medium	Q HyperD 20	SP Sepharose FF
	Q Zirconia	SP Sepharose XL
	UNO Q-1	SP Sepharose BB
	Fractogel EMD DEAE (M)	Capto S[a]
	Source 30Q	CM Hyper Z
	Poros 50 D	MacroCap SP
	Express-Ion Q	TSKGel SP-5PW-HR20
	DEAE Ceramic HyperD 20	Source 30 S
Low	Fractogel EMD TMAE 650s	CM Toyopearl 650 m
	DEAE Sepharose FF	SP Toyopearl 650 m
	Toyopearl SuperQ 650c	CM Sepharose FF
	Separon HemaBio 1000Q	Fractogel SE HICAP (M)
	TSKGel Q-5PW-HR	SP Toyopearl 650c
	Toyopearl DEAE 650 (M)	Fractogel EMD COO$^-$ (M)
	Q-Cellthru Bigbeads Plus	CM Cellufine C-500

Note: Highest binding strength is at the top of the table.

[a] Approximate ranking based on static HTS binding capacity strength measurements.

[b] Capto MMC has medium capacity at low conductivity, but very high capacity at high conductivity.

applied in the investigation and could possibly also depend on the counterion used for elution. The ranking is based on our general experience and results obtained with lipolase and IgG, and aprotinin and lysozyme, respectively. We have attempted to group the resins into three groups of having strong, medium, and weak binding strengths, respectively. Table 6.3 presents a large amount of data, and various trends may be found as discussed below. Sepharose resins tend to have a higher binding strength as anion-exchangers, but a medium binding strength as cation-exchangers. Strong ion-exchangers (Q- and S-types) are as a majority represented in the strong binding strength region (and vice versa); however, this is likely due to the ligand density of each specific resin. Data of pH sensitivity of retention may also provide information on binding strength, but there is no direct link between general variation of retention as a function of pH and binding strength. Isocratic elution ion-exchange chromatography is indeed an equilibrium process. In fact, a high enough volume of an eluent with a given conductivity will cause elution; however, these relations are exponentially increasing. Thus, it may end up with very high volumes for elution to occur, but elution will in principle always happen.

The literature provides a multitude of data on binding strength obtained in isocratic or gradient mode. Lenhoff and coworkers have performed numerous studies [4,25,35] of binding-strength-related measurements and resin comparison in general, as shown in Table A.1. In the thesis of Bai [35], anion-exchange resins were studied and data were presented as log k' as a function of NaCl concentration with α-lactalbumin and pepsinogen. With α-lactalbumin the binding strength ranking is Fractogel EMD DEAE (M), Fractogel EMD DMAE (M), Q Sepharose FF, SuperQ Toyopearl 650 m, DEAE Sepharose FF, Fractogel EMD TMAE (M), and Toyopearl DEAE 650 (M). For pepsinogen, binding strength order of SuperQ Toyopearl 650 m and Fractogel EMD TMAE (M) was interchanged, but otherwise as given for α-lactalbumin. Six of the seven anion-exchange resins are mentioned in Table 6.3, and exactly the same elution order was found compared to pepsinogen, except for the binding strength order of Fractogel EMD DEAE (M) and Q Sepharose FF which was the opposite. Papers of DePhillips and Lenhoff [4,25] present plots of log k' as a function of NaCl concentration for cation-exchangers with lysozyme, α-chymotrypsinogen, cytochrome c, FGF-1, and FGF-2. The binding strength order of lysozyme was SP Toyopearl 550c, SP Spherodex M, Fractogel EMD SO$_3^-$(M), SP Sepharose FF, SP Toyopearl 650 m, CM Toyopearl 650 m, CM Spherodex M, Fractogel EMD COO$^-$ (M), and CM Sepharose FF. Seven of the nine cation-exchange resins are mentioned in Table 6.3, and the same elution order was found except that the order of SP Toyopearl 650 m and CM Toyopearl 650 m and of Fractogel EMD COO$^-$ (M) and CM Sepharose FF was opposite. However, all four resins are still to be regarded as resins with fairly low binding strength. Compared to the other test proteins in Refs. [4,25], changes in the binding strength order of specific resins occur. Retention measurements of FGF-1 and FGF-2 on Amicon's Cellufine Sulfate were part of the study. Still, SP Toyopearl 550c, SP Spherodex M, Fractogel EMD SO$_3^-$(M), and Cellufine Sulfate should be regarded as resins with a high binding strength, SP Sepharose FF as a resin with medium binding strength, and the others as resins with a low binding strength.

Binding strength measurements using the current methodology are associated with thorough and time-consuming experimentation on standard small-scale process development equipment. As stated previously, new approaches have been employed lately to increase the speed of process development within the biopharmaceutical industry, including robot-based HTS platforms. These techniques will probably be state of the art in future purification process development. The current status of HTS techniques is that they work excellently in a 96-well static format for measurement of static capacities and adsorption isotherms; however, there is still some room for improvement in applications in dynamic mode operation. Though binding strength measurements are usually performed in dynamic mode, we have tried to set up experiments to provide the same and possibly additional information on binding-strength related issues from static mode measurements. We will call these results binding capacity strength, as they provide information on binding strength linked to binding capacity. The data are plotted as solid phase static capacity as a function of solution salt concentration at an equilibrium protein concentration of 1 g/L.

Figures 6.9 and 6.10 show results of such binding capacity strength measurements, where new resins not previously tested are compared to well-known resins.

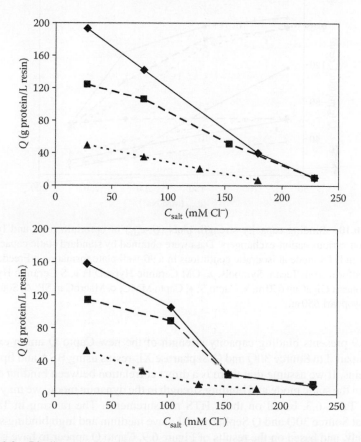

FIGURE 6.9 Binding capacity strength plots (Q–c_{salt}) of BSA (top) and lipolase (bottom) on various anion-exchangers. Data were obtained by standard static capacity measurements in HTS mode at isocratic conditions in a 96-well plate format on a Freedom EVO robotic platform from Tecan. Symbols: ◆, Q Sepharose XL; ■, Capto Q; ▲, Source 30Q.

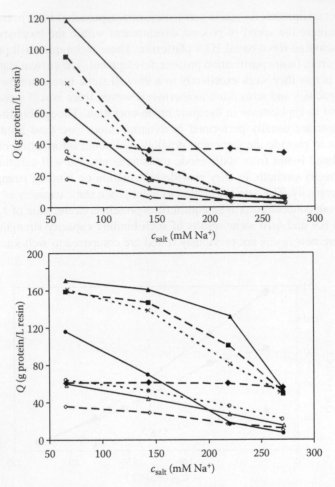

FIGURE 6.10 Binding capacity strength plots (Q–c_{salt}) of aprotinin (top) and lysozyme (bottom) on various cation-exchangers. Data were obtained by standard static capacity measurements in HTS mode at isocratic conditions in a 96-well plate format on a Freedom EVO robotic platform from Tecan. Symbols: ▲, CM Ceramic HyperD F; ■, S Ceramic HyperD F; ×, SP Toyopearl GigaCap 650 m; ●, Capto S; ◆, Capto MMC; ○, MacroCap SP; △, Source 30 S; ◇, CM Toyopearl 650 m.

Figure 6.9 presents binding capacity strength of the new Capto Q anion-exchange resin compared to Source 30Q and Q Sepharose XL resins using BSA and lipolase as test proteins. If we assume that there is a direct correlation between binding capacity strength in the static mode and binding strength in the dynamic mode, we may include resins in Table 6.3, based on these HTS measurements. The ranking in Table 6.3 shows that Source 30Q and Q Sepharose XL have medium and high binding strength, respectively, and based on the results of Figure 6.9, Capto Q appear to have a binding capacity strength and thus a binding strength between these two resins, although very close to Q Sepharose XL. We have thus included Capto Q in Table 6.3 as a high binding strength resin.

Figure 6.10 shows binding capacity strength data of the new SP Toyopearl GigaCap 650 m, Capto S, Capto MMC, and MacroCap SP cation-exchange resins compared to CM Ceramic HyperD F, S Ceramic HyperD F, Source 30 S, and CM Toyopearl 650 m resins, using aprotinin and lysozyme as test proteins. Comparing the sequence of the well-known resins in Figure 6.10 with the binding strength sequence of Table 6.3, we experience exactly the same sequence using the same test proteins indicating that the assumption of similarity between binding capacity strength and binding strength measures is reasonable. It is further noticed that the data in Figure 6.10 (top) obtained with aprotinin display a similar trend for all resins, except Capto MMC, which has been developed as special high salt-tolerant resin and thus displays an almost equal capacity over the entire experimental salt concentration range. When lysozyme is used as a test protein, Figure 6.10 (bottom), the capacity of Capto S decreases more dramatically with increasing salt concentration indicating a higher salt sensitivity than the other resins. Based on the results shown in Figure 6.10, the four new resins have been included in Table 6.3: Capto MMC and SP Toyopearl GigaCap 650 m have been ranked as high binding strength resins although the capacity of Capto MMC is in the medium range at low salt concentration compared to other resins, and Capto S and MacroCap SP have been marked as medium binding strength resins.

Binding strength measurements are of utmost importance for resin selection and optimization of ion-exchange processes. If the target protein binds weakly due to a high conductivity loading solution or due to the presence of various additives, modifiers, etc., the choice of resins for further testing should be among those at the top of Table 6.3, to secure sufficient binding. In case of a target protein binding strongly to all resins, a resin represented in the bottom of Table 6.3 would probably be selected for further testing to minimize the salt consumption for elution. In a flow-through mode operation where the target protein passes or only weakly binds to the resin while retaining the impurities, resins in the bottom of Table 6.3 may also apply; however, it also depends on the capacity of that resin for binding of impurities. Applying binding capacity strength evaluation might be a very relevant and possibly better alternative to normal binding strength measurements at linear conditions for selection of resins for testing in process development, as binding strength is evaluated with capacity. However, dynamic limitations may not be sufficiently elucidated by this methodology, and experimental testing is necessary anyway to evaluate separation abilities and specificity of the given resins.

6.2.3 Capacity

Productivity and throughput are of utmost importance to the industry for development, implementation, and daily operation of large-scale purification processes. Process economy has played a secondary role in traditional biopharmaceutical biotech industry where productions of a few kilogram of material per year could supply the world market, e.g., interferons, coagulation factors, growth factors, etc. Here focus would be on increasing fermentation expression levels of cells derived from recombinant technology. For commodity products such as insulin and more recently monoclonal antibodies, however, process economy has played a major role for many years wherefore changes and improvements to existing facilities and regulatory approved processes have occurred on a regular basis. A key feature in improving the productivity is by increasing the amount of material to be processed in each run and thus the amount of material to be bound on a given column, also called column capacity.

The resin capacity for different proteins is often specified by suppliers based on different modes of measurement (dynamic or static), experimental conditions (pH, salt/conductivity, protein concentration), reference (capacity per milliliter wet resin or g dry resin), etc. making it very difficult to compare the resins based on tabular values from the vendors. The static capacity is usually referred to as the maximum protein-binding capacity at given solvent and protein concentration conditions, which varies significantly for specific proteins, while dynamic capacity is a measure of the highest obtainable capacity during normal operation. The dynamic capacity concept is regularly accompanied by a breakthrough fraction of 2%, 5%, 10%, 50%, or another number referring to the amount of target material in the flow-through during measurement or loading. A 5% or 10% breakthrough capacity, $Q_{5\%}$ or $Q_{10\%}$, is a typical industry standard for comparison and evaluation of accessible capacity in manufacturing mode. Some suppliers provide information on small ion capacity (ligand density), i.e., the amount of salt (counterions), acid or base necessary to titrate the ligands, as a help of potential capacity; however, ligand density along with pore volume, pore structure, base matrix material, etc., determines the true protein capacity.

Tables 6.4 and 6.5 present dynamic and static capacities for BSA and lysozyme, respectively, with small ion capacity obtained from suppliers, if available, and an approximate grading of resins based on dynamic capacities from previous experience [31–33,69–71] is given. Determination of the dynamic capacity of anion-exchangers presented in Table 6.4 was performed by frontal analysis experiments at ~25% of the maximum recommended flow-rate/pressure. Actual flow-rates are given elsewhere [31–33,69–71]. Static capacities of Table 6.4 were obtained by standard measurements of solution UV absorbance change in a beaker of >100 mL, after one day of incubation with resin at slow agitation. Similar methods were applied to obtain data for the cation-exchangers in Table 6.5. Tables 6.4 and 6.5 represent an enormous amount of data, and although small differences in capacity are achieved between resins, and their true capacity order may be slightly different depending on conditions, various trends are found and discussed below. Suppliers offering both weak and strong ion-exchangers typically provide resins with both high and somewhat lower capacity. MacroPrep resins appear to be in the lower to medium capacity range; however, this may be compensated for by high selectivity for explicit applications. Heparin resins should not be used for cation-exchange only, as they generally have low capacity if no affinity effect is obtained. Sepharose XL resins appear to have the highest capacity for both anion- and cation-exchangers, and in general, it may be stipulated that if a specific cation-exchanger has a high capacity, the corresponding anion-exchanger is also most likely to have a high capacity and vice versa. In a few cases, a peculiar result of higher dynamic than static capacity was obtained, e.g., for CM Sepharose FF and Source 30 S, but these results may be due to aggregation/dimerization of BSA or similar nonlinear effects.

The dynamic capacity, $Q_{10\%}$, provides information on the useful capacity, and when compared to the static capacity it gives a measure of the flow-rate dependence

TABLE 6.4
Binding Capacity of Anion-Exchange Resins for 1 mg/mL Solution of BSA

Resin	Ionic Capacity (μmol/mL)	Dynamic Capacity $Q_{10\%}$ (mg/mL)	Static Capacity (mg/mL)
Q-Cellthru Bigbeads Plus	—	<0.1	<1
MacroPrep DEAE Support	175 ± 75	16	37
Separon HemaBio 1000Q	~100/g	18	25
Toyopearl DEAE 650 (M)	80–120	20	23
MacroPrep 25Q	220 ± 40	21	—
UNO Q-1	—	24	—
Fractogel EMD TMAE 650 s	—	28	—
Toyopearl QAE 550 c	280–380	33	68
Poros QE/M	—	38	—
Source 30Q	—	38	—
TSKGel Q-5PW-HR	120–180	42	—
Fractogel EMD DEAE (M)	—	42	67
Poros 50 HQ	—	45	73
Express-Ion Q	1 meq/dry g	50	—
Q Sepharose FF	180–250	54	—
DEAE Sepharose FF	110–160	59	77
Q Sepharose HP	140–200	64	78
Toyopearl SuperQ 650 c	200–300	66	98
Q Zirconia	—	76	—
DEAE Ceramic HyperD 20	≥200	76	82
Poros 50 D	—	77	110
Capto Q[a]	160–220		124
Q HyperD 20	>250	85	—
Q Sepharose XL	180–260	176	244

Note: Ionic capacity obtained from suppliers. Dynamic capacity data obtained at low flow-rate.
[a] Approximate order based on static HTS binding capacity measurements.

of the uptake rate. Excluding a few results, the fraction utilized at $Q_{10\%}$ ranges from ~ 50% to 90% of the static capacity. Soft resins and resins causing hindered diffusion are more subjective to flow-rate changes [69]. The ionic capacities presented in Tables 6.4 and 6.5 demonstrate large difference for the assorted resins. Although some data are missing, ionic capacity is in the range of 60–250 μmol/mL for cation-exchangers, and in a slightly higher range of 80–300 μmol/mL for anion-exchangers. As shown in the two tables, there is no direct link between ionic capacity and protein capacity; however, an optimal ionic capacity exists for a pair of a given ligand and a given base matrix with a specific protein, e.g., Toyopearl DEAE and BSA [82].

Reasonable agreement was obtained between results of Tables 6.4 and 6.5 and supplier data for BSA and lysozyme; however, any difference observed was generally in favor of the suppliers. Discrepancy in capacity data may be due to a different experimental setup, most pronounced for Fractogel EMD resins. Dennis et al. [34]

TABLE 6.5

Binding Capacity of Cation-Exchange Resins for 1 mg/mL Solution of Lysozyme

Resin	Ionic Capacity (μmol/mL)	Dynamic Capacity $Q_{10\%}$ (mg/mL)	Static Capacity (mg/mL)
Heparin Sepharose FF	~28	4	—
Heparin Toyopearl 650 m	~35	17	29
MacroPrep CM	210 ± 40	18	50
Heparin Ceramic HyperD M	~35–70	20	28
TSKGel SP-5PW-HR20	60–120	28	41
SP Toyopearl 650 c	120–170	29	41
SP Toyopearl 650 m[a]	130–170	30	45
MacroPrep High S	160 ± 40	40	—
CM Toyopearl 650 m	80–120	41	52
Fractogel EMD COO⁻ (M)	—	41	50
Poros 50 HS	—	43	63
Capto MMC[b]	70–90	—	64
MacroCap SP[b]	100–130	—	66
Fractogel SE HICAP (M)	—	51	—
MacroPrep 25 S	110 ± 30	53	—
Fractogel EMD SO₃⁻ (M)	—	74	—
CM Hyper Z	—	76	95
Source 30 S	—	79	76
CM Cellufine C-500	—	91	106
SP Sepharose BB	180–250	92	138
S Ceramic HyperD 20	≥150	98	—
SP Sepharose FF	180–250	104	—
Capto S[b]	110–140	—	118
Toyopearl SP 550 c	120–180	107	139
CM Sepharose FF	90–130	136	93
Toyopearl GigaCap S-650 m[b]	150 ± 50	—	159
CM Ceramic HyperD F[a]	≥250	126	261
SP Sepharose XL	180–250	188	337

Note: Ionic capacity obtained from suppliers. Dynamic capacity data obtained at low flow-rate.

[a] Approximate order based on high flow-rate dynamic binding capacity measurements.

[b] Approximate order based on static HTS binding capacity measurements.

measured static capacity for BSA and lysozyme for anion- and cation-exchangers, respectively, for 70 different commercial ion-exchange resins. In comparison to Tables 6.4 and 6.5, different resin capacity values were obtained by Dennis et al. [34] due to very different experimental conditions; however, the capacity orders of resins between the two studies are in good agreement. Dennis et al. [34] also presented experimental ionic capacity data, and similar to our findings, there are no direct links between ionic capacity, static, and dynamic protein capacity. Many other authors present dynamic and static protein capacity measurements [6,13,16,18,19,46,48], and in general good agreement was obtained with Tables 6.4 and 6.5.

As for binding strength measurements, the current methodology for measurement of protein capacities is associated with time- and material-consuming experimentation. Thus, applying a more standardized scaled-down approach with HTS/robot techniques to increase the speed of process development is obvious especially for static binding capacity measurement. These techniques work perfectly in a 96-well static format for measurement of static capacities and adsorption isotherms [110]. Determination of adsorption isotherms is in fact a more fair method for evaluation of protein-binding capacity as different protein solution concentrations are taken into account.

Figures 6.11 through 6.13 present results of adsorption isotherm determination of new resins compared to well-known resins with no salt addition, i.e., at counterion concentrations resulting from the applied buffer and pH adjustment. The resulting

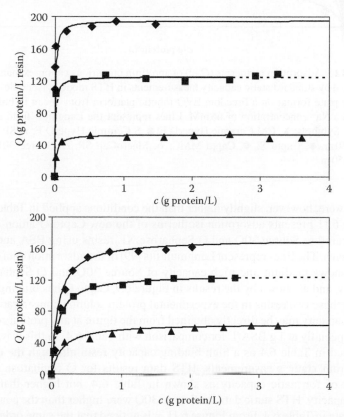

FIGURE 6.11 Adsorption isotherms ($Q-c$) of BSA (top) and lipolase (bottom) on various anion-exchangers. Data were obtained by standard static capacity measurements in HTS mode at isocratic conditions in a 96-well plate format on a Freedom EVO robotic platform from Tecan in buffer corresponding to a Cl⁻ concentration of 28 mM. Lines represent the Langmuir fit to adsorption data results. Symbols: ♦, Q Sepharose XL; ■, Capto Q; ▲, Source 30Q.

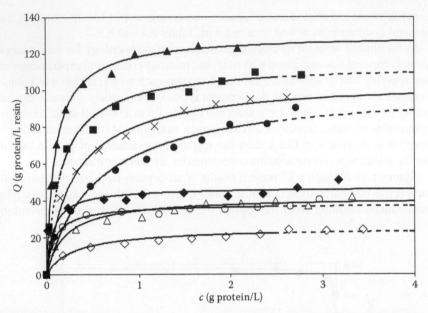

FIGURE 6.12 Adsorption isotherms (Q–c) of aprotinin on various cation-exchangers. Data were obtained by standard static capacity measurements in HTS mode at isocratic conditions in a 96-well plate format on a Freedom EVO robotic platform from Tecan in buffer corresponding to a Na$^+$ concentration of 66 mM. Lines represent the Langmuir fit to adsorption data results. Symbols: ▲, CM Ceramic HyperD F; ■, S Ceramic HyperD F; ×, SP Toyopearl GigaCap 650 m; ●, Capto S; ◆, Capto MMC; ○, MacroCap SP; △, Source 30 S; ◇, CM Toyopearl 650 m.

conditions were, however, slightly higher than the conditions applied in Tables 6.4 and 6.5. Figure 6.11 presents adsorption isotherms of the new Capto Q anion-exchange resin compared to Source 30Q and Q Sepharose XL resins using BSA and lipolase as test proteins. The lines represent Langmuir fits [99] of the adsorption data obtained. Table 6.4 shows medium and high capacity of Source 30Q and Q Sepharose XL, respectively, and as shown by the results in Figure 6.9, Capto Q has binding capacity in between these two resins in the experimental protein solution concentration range. The static capacity may be directly obtained from the figure at any protein concentration and especially at 1 g BSA/L for comparison with Table 6.4. In this way, Capto Q was included in Table 6.4 as a high binding capacity resin although the data were obtained from static measurements. HTS data results for Q Sepharose XL were below that of for static capacity as shown in Table 6.4, but higher than that for dynamic capacity. HTS static data for Source 30Q were higher than the conventional dynamic data in Table 6.4. From Figure 6.11, it is noticed that the same order of binding capacity between the three resins was obtained for BSA and lipolase, and a good representation of data was obtained by the Langmuir fit.

Figures 6.12 and 6.13 present adsorption isotherms of the new SP Toyopearl GigaCap 650 m, Capto S, Capto MMC, and MacroCap SP cation-exchange resins compared to previously investigated CM Ceramic HyperD F, S Ceramic HyperD F,

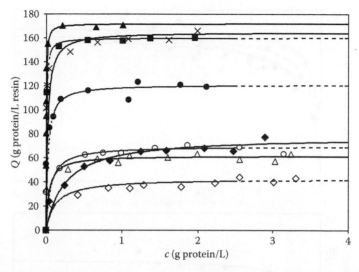

FIGURE 6.13 Adsorption isotherms $(Q-c)$ of lysozyme on various cation-exchangers. Data were obtained by standard static capacity measurements in HTS mode at isocratic conditions in a 96-well plate format on a Freedom EVO robotic platform from Tecan in buffer corresponding to a Na^+ concentration of 66 mM. Lines represent the Langmuir fit to adsorption data results. Symbols: ▲, CM Ceramic HyperD F; ■, S Ceramic HyperD F; ×, SP Toyopearl GigaCap 650 m; ●, Capto S; ◆, Capto MMC; ○, MacroCap SP; △, Source 30 S; ◇, CM Toyopearl 650 m.

Source 30 S, and CM Toyopearl 650 m resins using aprotinin and lysozyme as test proteins, respectively. As for the anion-exchangers, the lines are Langmuir fit of the adsorption data. Table 6.5 shows high capacity of CM Ceramic HyperD F and S Ceramic HyperD F, medium capacity of Source 30 S, and lower capacity of CM Toyopearl 650 m. There is a good agreement between the order of the capacities of these resins and the order of the adsorption isotherms in Figures 6.12 and 6.13. The sequence of the adsorption isotherms of the new resins in Figures 6.12 and 6.13 is SP Toyopearl GigaCap 650 m > Capto S > Capto MMC > MacroCap SP, and compared to previously tested resins: S Ceramic HyperD F > SP Toyopearl GigaCap 650 m and MacroCap SP ≈ Source 30 S. It is further noticed that the trends for resins in Figures 6.12 and 6.13 are basically parallel as expected, except for Capto MMC and partly for SP Toyopearl GigaCap 650 m with lysozyme, and for Capto S with aprotinin. Whether this trend is a true feature of resins or a result of lack of accuracy of HTS techniques at very low protein concentrations is not known. Based on the results of Figures 6.12 and 6.13, the four new resins have been included in Table 6.5. As for anion-exchangers, the adsorption data measured by HTS for these cation-exchangers were generally lower than the static capacity data reported in Table 6.5.

Figure 6.14 presents adsorption isotherms of Capto MMC and MacroCap SP for aprotinin at the same salt concentrations. These adsorption isotherms give a good representation of binding capacity as a function of protein solution concentration at various salt concentrations, and thus of the susceptibility of binding capacity to salt concentration in ion-exchange chromatography. The Capto MMC resin developed

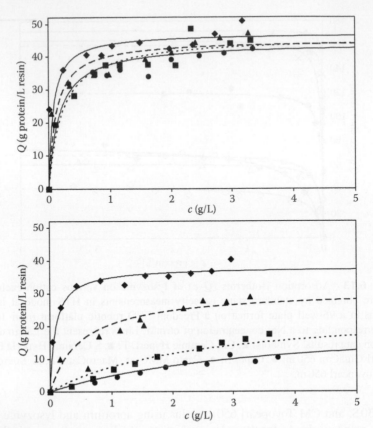

FIGURE 6.14 Adsorption isotherms (Q–c) of aprotinin on Capto MMC (top) and Macro-Cap SP (bottom) cation-exchangers. Data were obtained by standard static capacity measurements in HTS mode at isocratic conditions in a 96-well plate format on a Freedom EVO robotic platform from Tecan in buffer at various salt concentrations. Lines represent the SMA fit to adsorption data results. Symbols: ◆, 66 mM Na⁺; ▲, 143 mM Na⁺; ■, 220 mM Na⁺; ●, 271 mM Na⁺.

for high salt concentration operation thus demonstrates a very low susceptibility to salt concentration as all data points and steric mass action (SMA) fits are very close to each other, while MacroCap SP with more traditional ion-exchange attributes illustrates larger dependence of binding capacity on salt concentration. The use of MacroCap resin for purification of aprotinin would be unlikely in bind-and-elute mode with the current capacity values.

Binding capacities of industrial mixtures such as fermentation broth, give a more realistic picture of what to expect in the progress of process development [84]. However, binding capacity for pure proteins is still important as it allows for ranking of resins as shown in Tables 6.4 and 6.5 without interference from specific fermentation broth components, and resins tested are intended for different purposes as previously stated. In Section 6.4, examples of binding capacities of proteins in true fermentation

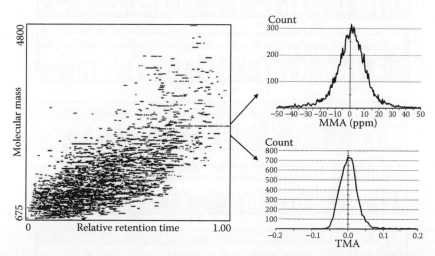

FIGURE 2.14 Proteomics peptide 2D display obtained from an ultrahigh-throughput analysis using a fast LC-TOF MS platform. The mass measurement accuracy (MMA) and LC relative retention time measurement accuracy (TMA) are also shown. A 0.8 μm porous particle-packed capillary column was used for the fast LC separation, TOF MS for detection, and the accurate mass and time tag approach for protein identification. Approximately 550 proteins were identified from this ~2 min analysis. Test sample: *S. oneidensis* tryptic digest. Detailed conditions are given in Ref. [5].

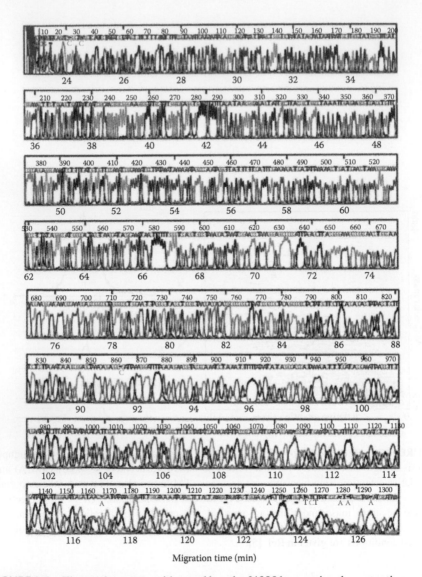

Migration time (min)

FIGURE 3.6 Electropherogram with a read length of 1300 bases using the separation matrix LPA 2.0% (w/w) 17 MDa/0.5% (w/w) 270 kDa at 125 V/cm and 70°C. Samples were prepared using Universal BigDye-labeled (–21) primer cycle sequencing with AmpliTaq-FS on ssM13mp18 template. An additional G-terminated reaction was added. Conditions: Capillary: ID, 75 μm, OD, 365 μm, PVA-coated capillary with effective length 30 cm, total length 45 cm; running buffers: 50 mM Tris/50 mM TAPS/2 mM EDTA. The cathode running buffer and separation matrix also contained 7 M urea. Samples were injected at a constant electric field of 9 V/cm (*I* = 0.7 μA) for 10 s and sequencing was performed at 125 V/cm. (From Zhou, H., Miller, A.W., Sosic, Z., Buchholz, B., Barron, A.E., Kotler, L., and Karger, B.L., *Anal. Chem.*, 72, 1045, 2000. With permission.)

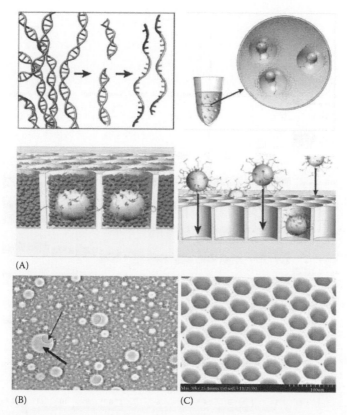

(A)

(B)   (C)

FIGURE 3.19 Scheme of 454 Life Science's emulsion-based amplification of sequencing templates and placement of beads into Picotitre plates. (A) Genomic DNA is isolated, fragmented, ligated to adapters, and separated into single strands (top left). Fragments are bound to beads under conditions that favor one fragment per bead, the beads are captured in the droplets of a PCR-reaction-mixture-in-oil emulsion and PCR amplification occurs within each droplet, resulting in each bead carrying 10 million copies of a unique DNA template (top right). The emulsion is broken, the DNA strands are denatured, and beads carrying ssDNA clones are deposited into wells of a fiber-optic slide (bottom right). Smaller beads carrying immobilized enzymes required for pyrophosphate sequencing are deposited into each well (bottom left). (B) Microscope photograph of emulsion showing droplets containing a bead and empty droplets. The thin arrow points to a 28 μm bead; the thick arrow points to an approximately 100 μm droplet. (C) Scanning electron micrograph of a portion of a fiber-optic slide, showing fiber-optic cladding and wells before bead deposition. (From Margulies, M., Egholm, M., Altman, W.E., Attiya, S., Bader, J.S., Bemben, L.A., Berka, J., Braverman, M.S., Chen, Y.-J., Chen, Z., Dewell, S.B., Du, L., Fierro, J.M., Gomes, X.V., Godwin, B.C., He, W., Helgesen, S., Ho, C.H., Irzyk, G.P., Jando, S.C., Alenquer, M.L.I., Jarvie, T.P., Jirage, K.B., Kim, J.-B., Knight, J.R., Lanza, J.R., Leamon, J.H., Lefkowitz, S.M., Lei, M., Li, J., Lohman, K.L., Lu, H., Makhijani, V.B., McDade, K.E., McKenna, M.P., Myers, E.W., Nickerson, E., Nobile, J.R., Plant, R., Puc, B.P., Ronan, M.T., Roth, G.T., Sarkis, G.J., Simons, J.F., Simpson, J.W., Srinivasan, M., Tartaro, K.R., Tomasz, A., Vogt, K.A., Volkmer, G.A., Wang, S.H., Wang, Y., Weiner, M.P., Yu, P., Begley, R.F., and Rothberg, J.M., *Nature* (London, UK), 437, 376, 2005. With permission.)

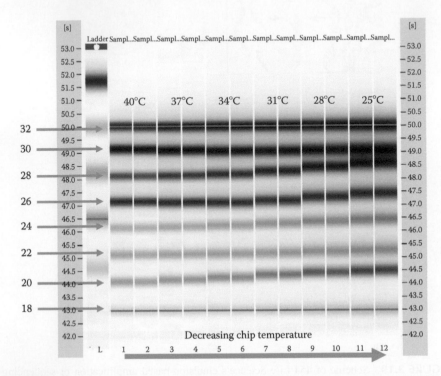

FIGURE 3.28 Aligned gel-view mode of oligonucleotide separation by 30% (w/v) F87/F127 mixture solution at a weight ratio of 1:2 in 1×TBE buffer and different temperatures performed in the Agilent Bioanalyzer 2100 system with an electric field strength of 325 V/cm. (From Zhang, J., Gassmann, M., He, W.D., Wan, F., and Chu, B., *Lab Chip*, 6, 526, 2006. With permission.)

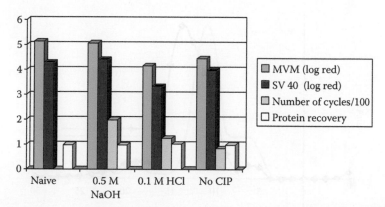

FIGURE 6.19 Evaluation of MVM and SV40 viral clearance of various CIP methods upon repetitive use of Q Sepharose FF. (Data from Norling, L., Lute, S., Emery, R., Khuu, W., Voisard, M., Xu, Y., Chen, Q., Blank, G., and Brorson, K., *J. Chromatogr. A*, 1069, 79, 2005.)

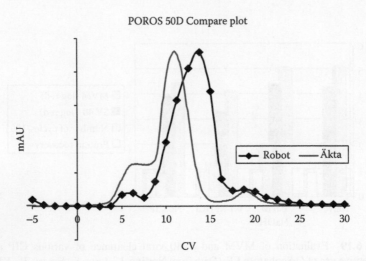

FIGURE 6.22 Comparison of HTS (Robot) and traditional chromatography system (Äkta) setup. Separation type is salt gradient elution of insulin-related molecule (main peak) from related impurities on Poros 50 D anion-exchanger.

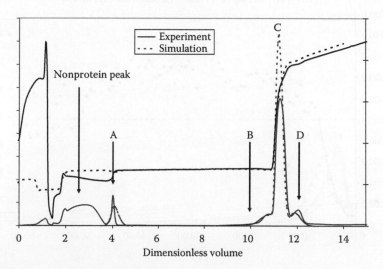

FIGURE 6.23 Comparison of simulation and pilot plant ion-exchange separation of four-component mixture of related molecules.

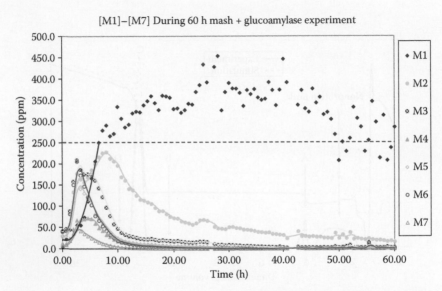

FIGURE 7.15 Concentration of M1–M7 over a 60 h liquefied corn mash plus glucoamylase experiment monitored by online MD-HPAEC-PED. Data are fit using a moving average of period = 2 h. The limit of linearity for glucose is 250 ppm, points outside linear range not included in trend line.

broth compared to pure protein capacity are discussed. Binding capacity data are very important to the biopharmaceutical industry. If process economy is a large constraint or in cases where two or more resins perform equally well during process development with respect to parameters such as resolution, flow-rate, price, etc., the resin with the highest protein capacity will be chosen for the manufacturing process to increase the productivity.

6.2.4 Particle Structure

An economic focus area of utmost importance in the biopharmaceutical industry is to decrease the number of purification steps and thus minimizing full manufacturing cost (FMC). Hereby manual labor and the risk associated with an overly complex process are reduced, and yields will be increased. To optimize purification steps in a given process, selectivity and separation ability are key measures. HETP and van Deemter plots [85] at binding or nonbinding conditions are often used to describe flow dependence of separation abilities, but testing for the specific application is always necessary to find the optimal resin. Optimal selection decision is influenced by the physical appearance of resin particle size, ligand structure, pore structure, and thus base matrix chemistry. The physical appearance of resins may vary as shown in the scanning electron microscopy (SEM) pictures in Figure 6.15, and several forms have been available over the years, but the most common shape available today is spherical, porous particles in a variety of sizes.

6.2.4.1 Particle Size

The spherical particles are usually produced by chemical synthesis depending on the base matrix material, and although exceptions occur, particle size distribution is often achieved and controlled by subsequent sieving. Particle size distribution was measured by our group using coulter counting of new resins [32,33,69–71], and the results were verified with SEM pictures. Tables 6.6 and 6.7 show the results of the mean particle size (50%, v/v) of anion- and cation-exchangers, respectively.

Generally, mean particle size in Tables 6.6 and 6.7 is in agreement with supplier data for most resins; however, much smaller particle diameters were found for Sepharose resins and Q-Cellthru Bigbeads Plus. The difference in results between these studies and supplier data for Sepharose resins is discussed elsewhere [31]. The SEM particle size verification methodology is encumbered with uncertainty due to measurement of a very small and possibly not representative sample, but it gives an indication of the validity of size distribution results obtained by coulter counting. Shrinkage of resins due to dehydration during SEM imaging may occur; however, a fairly good agreement of the mean particle size is obtained in Table 6.7 by the two independent methods, except for Sepharose resins. Results by Nash and Chase [6] on mean particle size for Source 30 S is in very good agreement with results given in Table 6.7, while good agreement with the supplier data is obtained for SP Sepharose FF. Particle size distribution for Heparin Ceramic HyperD M and Poros 50 HS are in agreement with general results obtained for corresponding resins by Weaver and Carta [16], thus the methodology of Tables 6.6 and 6.7 appears to be reliable.

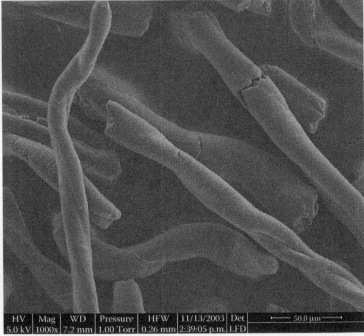

FIGURE 6.15 SEM images of ion-exchange resins at different magnification. Scale appears in the separate figures. Resins are (top) CM HyperD F; (bottom) Express-Ion C.

TABLE 6.6
Comparison of Particle Size for Various Anion-Exchange Resins Determined by Coulter Counting and as Given by Suppliers

Resin	Particle Size (μm), Supplier Data	Mean Particle Size (μm), Coulter Counting
Separon HemaBio 1000Q	10 (8–12)	11
TSKGel Q-5PW-HR	20 (15–25)	22
Poros QE/M	20	22
Q HyperD 20	20	16
MacroPrep 25Q	25	20
Source 30Q	30	24
Fractogel EMD TMAE 650s	30 (20–40)	25
Q Sepharose HP	34 (24–44)	18
MacroPrep DEAE Support	50	54
Poros 50 D	50	49
Poros 50 HQ	50	37
DEAE Ceramic HyperD 20	40–90	59
Fractogel EMD DEAE (M)	40–90	60
Toyopearl DEAE 650 (M)	40–90	59
Q Zirconia	76	35
DEAE Sepharose FF	90 (45–165)	67
Q Sepharose FF	90 (45–165)	42
Q Sepharose XL	90 (45–165)	24
Express-Ion Q	60–130	47
Toyopearl QAE 550c	100 (50–150)	20
Toyopearl SuperQ 650c	100 (50–150)	78
Q-Cellthru Bigbeads Plus	400 (300–500)	—
UNO Q-1	—[a]	—[a]

[a] Monolithic column.

Many suppliers provide resins with different numerical particle size depending on use, i.e., for capture, intermediate purification, or polishing. Figure 6.16 presents the particle size distribution measurements of four beads from two different suppliers, which are derived from the same base matrix chemistry made by Tosoh Bioscience, i.e., TSKGel Q-5PW-HR, Fractogel EMD TMAE 650s, SP Toyopearl 650m, and SP Toyopearl 650c. An extra coarse quality also exists for some chemistries. Figure 6.16 not only illustrates the fairly normal particle size distribution of resins, but also that the distribution becomes broader as the particle size increases. This feature adds to the explanation on why more narrow peaks may be achieved from smaller particles, where both the mean size and the distribution play a significant role.

Data on particle size distribution are necessary for selection of column filters for industrial chromatography columns. If resin particles are smaller than stated or have a different shape than expected, they may clog up the filter leading to increased

TABLE 6.7

Comparison of Particle Size for Various Cation-Exchange Resins Determined by Coulter Counting, SEM, and as Given by Suppliers

Resin	Particle Size (μm), Supplier Data	Mean Particle Size (μm), Coulter Counting	Mean Particle Size (μm), SEM
S Ceramic HyperD 20	20	—	21 (14–27)
TSKGel SP-5PW-HR20	20 (15–25)	22	21 (16–26)
MacroPrep 25 S	25	20	—
Source 30 S	30	27	32 (30–32)
CM Ceramic HyperD F	50	44	49 (25–73)
MacroPrep CM	50	62	44 (25–62)
MacroPrep High S	50	61	62 (39–84)
Poros 50 HS	50	44	43 (19–67)
S Ceramic HyperD F	50	57	59 (42–75)
Fractogel EMD COO$^-$ (M)	40–90	67	—
Fractogel EMD SO$_3^-$ (M)	40–90	57	48 (33–62)
Fractogel SE HICAP (M)	40–90	59	40 (25–55)
CM Toyopearl 650 m	65 (40–90)	62	54 (38–69)
Heparin Toyopearl 650 m	65 (40–90)	62	—
SP Toyopearl 650 m	65 (40–90)	62	54 (38–69)
CM Hyper Z	75	67	—
Heparin Ceramic HyperD M	80	75	79 (62–95)
CM Cellufine C-500	53–125	73	—
CM Sepharose FF	90 (45–165)	57	43 (24–62)
Heparin Sepharose FF	90 (45–165)	67	52 (30–73)
SP Sepharose FF	90 (45–165)	66	56 (30–82)
SP Sepharose XL	90 (45–165)	66	60 (32–87)
Toyopearl SP 550 c	100 (50–150)	80	91 (46–135)
Toyopearl SP 650 c	100 (50–150)	80	100 (58–142)
SP Sepharose BB	200 (100–300)	99	131 (65–196)

column back-pressure and the risk of damaging the column and the chromatographic resin. It is also important to know the methodology used by suppliers to measure particle size distribution of resins in the selection of column filters if differences or problems arise. Taking into consideration the resolution of various resins, knowledge of resin particle size is also essential. Comparison should be performed on particles of same size and distribution, and if not available, this should be taken into account upon selection of a resin for implementation in an industrial process. Finally, particle size distribution and SEM pictures may be used to determine the damage and lifetime of resins by the amount of lines generated by repeated use and clean-in-place (CIP).

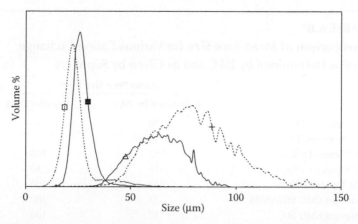

FIGURE 6.16 Particle size distribution (%, v/v) of ion-exchange resins in 0.9% NaCl measured by coulter counting. Symbols: □, TSKGel Q-5PW-HR; ■, Fractogel EMD TMAE 650s; △, SP Toyopearl 650m; +, SP Toyopearl 650c. Differences in curve area are due to different data collection frequency.

6.2.4.2 Pore Size, Base Matrix Chemistry, and Ligand Structure

While particle size has a huge influence on diffusion distance in the resin particle, the general speed of diffusion is determined by the pore size and structure. The pore size of particles should be large enough for proteins to basically diffuse unhindered into the particle, but not too large to avoid loss of capacity. This phenomenon is a key parameter for separation to occur in size-exclusion chromatography and gel permeation chromatography, but it is also very important for ion-exchange chromatography. Several methods for pore size determination exist, including mercury intrusion, nitrogen or gas adsorption, and microscopy; however, the most representative method applied for protein and polypeptide separation resins appears to be inverse size-exclusion chromatography (ISEC) [1,68,86,87]. Lenhoff and coworkers [1,68] have been very active in the pore size measurement area lately, and Yao and Lenhoff [65] recently published a thorough review on pore size determination and effects on chromatographic resin performance.

Table 6.8 presents the comparison data on mean pore size for a number of cation-exchangers obtained by ISEC [1]. Some discrepancy was obtained between ISEC data and supplier data in general, possibly due to different measurement methods. There is apparently a large difference in pore size between the strong cation-exchange Toyopearl and Fractogel EMD resins though they originate from the same base material. This observation may be due to a high tentacle structure ligand capacity in the Fractogel EMD resin. Furthermore as shown in Table 6.8, pore size depends not only on base matrix materials but also on the medium in which measurement is performed. From the data by DePhillips and Lenhoff [1] acrylic-based materials tend to increase pore size as a function of salt concentration, whereas silica-based material seems to decrease. This phenomenon may have a major influence in large-scale operation during elution and CIP with large variations in resin volume, pore accessibility, etc. Comparison of Toyopearl 550 and 650 series presents a more open structure of the

TABLE 6.8

Comparison of Mean Pore Size for Various Cation-Exchange
Resins Determined by ISEC and as Given by Suppliers

	Mean Pore Size (nm)	
Resin	Calculated by ISEC	Supplier Data
SP Sepharose FF	50	—
CM Sepharose FF	55	—
SP Toyopearl 650 m	53	100
SP Toyopearl 550 c	18	30
CM Toyopearl 650 m	148	100
Fractogel EMD SO_3^- 650 M	33	100
Fractogel EMD SO_3^- 650 M (1 M NaCl)	59	100
Fractogel EMD COO$^-$ 650 M	161	100
SP Spherodex M	69	100
SP Spherodex M (1 M NaCl)	43	100
CM Spherodex M	21	100

Source: Data from DePhillips, P. and Lenhoff, A.M., *J. Chromatogr. A*, 883, 39, 2000.

Toyopearl 650 m resins, which is in agreement with the findings of Yao et al. [88] by electron tomography and in general with the purpose of these two Toyopearl resin categories.

Several papers and reviews exist on description of base matrix chemistry, and a recent paper by Jungbauer [89] gives an excellent overview. Commercially available base matrix chemistries include cellulose, dextran, agarose, acrylamide, polystyrene-divinylbenzene, methacrylate, and various ceramic and composite materials. Selection of base matrix chemistry for the specific, commercial purification task depends not only on separation ability of pore structure and ligand of the material, but also on applied operating pressure and CIP methods. Thus selection is usually influenced by available equipment and the combination of particle characteristics (particle size and chemistry, pore size, and ligands).

Ion-exchange ligands are basically divided into four categories: strong and weak, anion- and cation-exchange ligands as previously stated, and the weak ion-exchange ligands may be affected by the pH operating range as discussed in Section 6.2.1. Some strong anion-exchange resins may also have a degree of "weak" character due to problems with the supply of pure quaternary amino ion raw material for ligands. To what extent the degree of quarternization is missing on strong anion-exchangers may be found from the comparison of titration curves of the resins [69,70]. Figure 6.17 presents titration curves of a number of strong anion-exchange resins, and the titration curves of a blank titration and DEAE Sepharose FF for comparison. A "true" strong anion-exchange resin would have a titration curve corresponding to the blank titration, while a weak or weaker anion-exchange resin would have a titration curve tending more to that of DEAE Sepharose FF. As shown in Figure 6.17, Q Zirconia

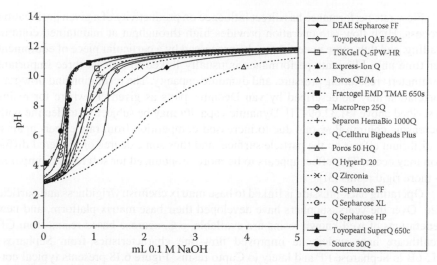

FIGURE 6.17 Titration curves of 2 mL anion-exchange resins in 1 M KCl from pH 3 to 12 with 0.1 M NaOH. A blank titration is given by a solid curve.

and QAE Toyopearl 550c display trends toward the titration curve of DEAE Sepharose FF indicating the presence of tertiary amino groups, while Poros QE/M seems to contain a lot of tertiary amino groups with the titration curve placed well below that of DEAE Sepharose FF. The titration curves of resins like Q Sepharose FF and Express-Ion Q follow a blank titration curve indicating that they are truly strong anion-exchange resins.

Various modifications such as grafting methods [82], may affect ligands and pore structure. This method or similar method has been used to create or modify attributes of resins such as Ceramic HyperD, Fractogel EMD, and Sepharose XL resins, providing resins with unique mass transport properties, multiple ionic exchange features, and/or increased capacity and salt concentration independent binding. The best-kept secret of resin suppliers is, however, the chemistry of their ligand spacers, e.g., the chemistry used to covalently bind the ionic groups to the base matrix wall. As for ligand density, an optimal spacer length generally exists balancing increased spacer length to avoid wall effects with decreased spacer length to decrease the hindered pore diffusion.

6.2.5 OTHER GENERAL PARAMETERS

Apart from parameters discussed above and the ability to perform the desired separation and provide sufficient resolution, many other parameters influence the choice of resins for preparative application in biopharmaceutical industry. Other important parameters include operating back-pressure, flow-rate, large-scale applicability in general, resin lifetime, ease of cleaning the resin, compatibility with solvents, salts, additives, and extreme pH, temperature stability, protein recovery, resin lot-to-lot consistency, safe supply of resin in appropriate amounts, and column packing stability, i.e., the right combination of resin, frits, and column material (glass/steel).

Flow-rate performance has direct influence on productivity of a given purification process. High flow-rate operation provides high throughput at maintained column loading, and thus more material may be processed by a particular piece of equipment per time unit. Flow-rate performance is usually linked to at least three important parameters: resolution, pressure, and dynamic capacity. The resolution and flow-rate correlation may be described by van Deemter plots as given elsewhere for resins described above [31,32,69–71]. Dynamic capacity may be subjective to decline with increasing column flow-rate due to increased competition from rate effects against equilibrium effects on the particle surface, and thus some degree of hindered diffusion may occur. This effect appears to be more pronounced for soft resins compared to more rigid resins [31].

Operating back-pressure is linked to base matrix chemistry/rigidness and particle size. Over the years, suppliers have developed their base matrix platform, and new generations of well-known resins are available, e.g., agarose-based resins from GE Healthcare have undergone improved flow-rate characteristics from Sepharose CL-6B to Sepharose FF and lately to Capto resins. Figure 6.18 presents typical correlation plots of column back-pressure as a function of flow-rate for four cation-exchange resins based on data from Nash and Chase [6]. Smaller particles offer not only better resolution but also higher pressure drop which may affect operating flow-rate and/or put extra demand on equipment, as illustrated by Source 15 S and Source 30 S. The approximate pressure limit for agarose-based Sepharose FF resins is 5 bar depending on column wall support, while higher back-pressure and thus resolution may be achieved with polystyrene-based and smaller particle size Source and Poros materials. High back-pressure may lead to irreversible deformation or breaking stress of resin particles; however, water-based solvents and surface modification may increase physical stability of resin particles as described by Müller et al. [90].

The ion-exchanger of choice is applicable for large-scale operation, consequently very small particles below, e.g., 10 μm or very soft resin material should be avoided.

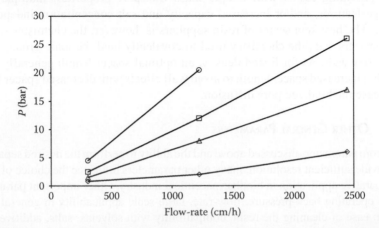

FIGURE 6.18 Plot of column back-pressure, P, across 10 cm bed versus flow-rate for selected cation-exchangers. (Data from Nash, D.C. and Chase, H.A. *J. Chromatogr. A*, 807, 185, 1998.) Resins are: ◇, SP Sepharose FF; △, Source 30 S; □, Poros 20 SP; and o, Source 15 S.

Column packing stability, i.e., the right combination of resin, frits, and column material (glass/steel) is also important. Some resin materials such as methacrylate may expand or shrink during use due to the different salts, solvents, and pH applied, and packing of such material should be done in a column fit for the purpose. Maintaining column packing stability may cause some large-scale problems if new resins are to be applied in an existing facility with little flexibility on choice of columns. If more rigid and/or small particle size resins are used, the column may have to withstand medium to high pressure, and steel columns may be applied. However, glass or at least transparent columns are preferred, because channeling, particulate matter, or other problems may be detected in due time. The column material should of course be compatible with solvents and salts applied. Therefore, chloride is often substituted with noncorrosive anions like acetate for steel columns and tubing; however, acetate has a lower elution strength which must be taken into account [71].

Ion-exchangers for industrial use must endure many different physical and chemical conditions, and these parameters are frequently tested. Data on case of cleaning and CIP are the expected standard information to be provided by suppliers with new ion-exchangers, where the benchmark today is sufficient volumes of 0.1–1 M NaOH at room temperature. These data are usually provided as resin lifetime of standard purification cycles with NaOH CIP and hundreds of cycles are expected for competitive ion-exchangers. Due to this fact, silica-based ion-exchangers are rarely employed in new industrial processes. Figure 6.19 presents data from Norling et al. [67] comparing various CIP agents for protein recovery and repetitive removal of vira on Q Sepharose FF. As shown in the figure, all agents (salt, low pH, and high pH) provide acceptable protein recovery ~100%, but only NaOH presents unchanged viral removal after ~200 cycles.

Ion-exchangers must also be compatible with various solvents like alcohols [31], salts [33,53], additives such as urea [91], extreme pH for CIP, and purification of proteins with extreme isoelectric point, elevated or reduced temperature [92], or a

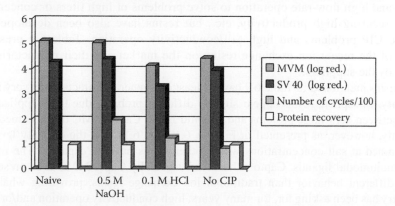

FIGURE 6.19 (See color insert following page 216.) Evaluation of MVM and SV40 viral clearance of various CIP methods upon repetitive use of Q Sepharose FF. (Data from Norling, L., Lute, S., Emery, R., Khuu, W., Voisard, M., Xu, Y., Chen, Q., Blank, G., and Brorson, K., *J. Chromatogr. A*, 1069, 79, 2005.)

combination of several of these parameters. Finally, a very important parameter for evaluation is the protein recovery of ion-exchange resins. Some ion-exchange resins or the combination of resin and separation conditions may have a tendency to induce protein or peptide aggregation in various forms, including gelation, dimerization, oligomerization, polymerization, fibrillation, etc., on the column or just after elution, resulting in decreased yield of the specific purification step. Handling of protein recovery issues is usually dealt with by resin substitution or fast change of postcolumn conditions, e.g., pH adjustment after elution.

Manufacture of recombinant, biopharmaceutical proteins requires safe supply of consistent chromatographic resin material in appropriate amounts, due to fairly rigid boundaries and parameter ranges of Good Manufacturing Practice (GMP) processes [79]. Testing of lot-to-lot consistency is a common task for industry typically evaluating three different batches of resins. Evaluations performed by industry are rarely published, but Mazza and Cramer [22] published a similar study on lot-to-lot consistency of Source 15 S, and they found very good agreement between the different batches testing various parameters including efficiency, ionic capacity, retention time in isocratic mode, breakthrough curves, and displacement chromatography effects. Suppliers are also constantly being audited by customers from the biopharmaceutical industry for evaluation of their quality systems, supply policies, capacity, response times, backup possibilities, drug master files of resins, etc. Thus, parameters other than just the ones which may be tested in a laboratory play an important role in the selection of ion-exchangers for industrial use.

6.3   NEW RESINS

Chromatographic ion-exchangers available today are in general capable of solving the purification problems of the biopharmaceutical industry; however, some improvements would still be beneficial. Several new resins on the market have been targeted against mAb purification with the necessary characteristics of large pores and high flow-rate operation to solve problems of high titers or concentrations, securing high productivity, etc., but resins have also been developed to handle CIP problems and high salt/conductivity operation. Table 6.9 presents some of the recent ion-exchange resins on the market and their characteristics given by the suppliers.

Resins mentioned in Table 6.9 have at least one general characteristic: very high capacity. Unosphere resins appear slightly different probably due to its application area between very large proteins and plasmid DNA. Capto MMC displays medium capacity, however, as presented in Figures 6.10 and 6.14, and the capacity level is maintained at salt concentrations where other resins have no capacity. The resins with multimodal ligands, Capto adhere and especially Capto MMC thus present a truly different behavior than traditional ion-exchange resins, providing what the industry has been asking for, for many years: high conductivity operation and/or ion-exchange-like operation without prior buffer exchange. The drawbacks may be lower initial capacity and limited pH operation range [51,94]. The ligand chemistry reminds of the mixed-mode, hydrophobic charge induction principle developed some years

TABLE 6.9
Characteristics of Recent, Commercial Ion-Exchangers, Supplier Data

Resin	Supplier	Type	Particle Size (μm)	Capacity (g/L)	Characteristic
Fractoprep TMAE	Merck	Strong AIE	30–150	>100	CIP stability, etc.
Fractoprep DEAE	Merck	Weak AIC	30–150	>100	CIP stability, etc.
Fractoprep SO₃	Merck	Strong CIE	30–150	>100	CIP stability, etc.
Toyopearl GigaCap Q-650M	Tosoh	Strong AIE	75	162	High flow-rate, etc.
Toyopearl GigaCap S-650M	Tosoh	Strong CIE	40–90	IgG: 150	High flow-rate, etc.
Toyopearl Megacap II SP-550EC	Tosoh	Strong CIE	100–300	Insulin: ~120	Capture, etc.
Capto DEAE	GE HC	Weak AIE	90	>90	High flow-rate, etc.
Capto MMCᵃ	GE HC	Multimodal CIE	75	>45	High conductivity operation
Capto Q	GE HC	Strong AIE	90	>100	High flow-rate, etc.
Capto S	GE HC	Strong CIE	90	>120	High flow-rate, etc.
MacroCap SP	GE HC	Strong CIE	50	—	Pegylated proteins
Capto adhereᵃ	GE HC	Multimodal AIE	75	—	mAb flow-through mode
Unosphere Q	Bio-Rad	Strong AIE	120	—	Plasmid DNA
Unosphere S	Bio-Rad	Strong CIE	80	IgG: 13	mAb Capture

Note: Capacity is with BSA for anion-exchangers and lysozyme for cation-exchangers unless otherwise specified.

ᵃ Multimodal resins may be characterized as mixed-mode resins between ion-exchange and hydrophobic interaction.

ago as an attempt to substitute Protein A resins for mAb purification capture, and where elution was accomplished by pH decrease [95]; however, additional chemical features and a true ion exchange ligand part are added to these Capto resin ligands. Although Capto adhere is to be used as a high salt-tolerant anion-exchanger mainly for mAb aggregate removal in flow-through mode, a truly high conductivity operation anion-exchanger optional for capture processing is not available, and the biopharmaceutical industry would appreciate such a product from suppliers.

Another characteristic of resins presented in Table 6.9 is the fairly large particle size and particle size distribution of recently provided resins. The current focus on mAb purification is reflected in the new resin developments by large pores for good diffusion and high flow-rate/productivity and less on selectivity differences and peak sharpness when aggregates and other related impurities are the primary targets of removal. An area not yet covered by any suppliers is monodisperse large particle size ion-exchange resins optionally for simultaneously increased productivity and degree of purification already at the capture step. The monodisperse or narrow particle size distribution would allow for improved flow-rate operation with lower back-pressure or smaller particles for capture without the risk of column clogging. The biopharmaceutical industry would welcome new ion-exchange products with these features. An additional tool to increase productivity is simulated moving bed (SMB)-inspired systems for semicontinuous chromatography purification operation. Ion-exchange resins with monodisperse nature may pack better, operate more robust, and thus secure optimal performance of SMB-related operation.

The third characteristic of recent ion-exchange resins presented in Table 6.9 is their potential ability to improve FMC of the biopharmaceutical industry. Fractoprep resins are developed to provide enhanced physical and mechanical stability while maintaining high capacity and protein recovery, thus productivity and in-use time (number of cycles) should be improved. Toyopearl GigaCap resins and traditional Capto resins have also been developed to obtain enhanced physical stability and capacity thereby increasing productivity, and from the limited data in Tables 6.4 and 6.5, their developments seem to have been accomplished. Further studies with flow-rate as variable parameter would no doubt confirm this. The high focus on productivity have turned attention in the direction of alternative, nonchromatographic or semichromatographic methods, and parts of the biopharmaceutical industry have long desired to implement adsorptive membranes to increase throughput; however, concerns of poor capacity and selectivity have resulted in limited progress. Still, attempts to use membranes instead of flow-through mode anion-exchangers for mAb aggregate removal seem promising. Several companies provide adsorptive membranes including the Mustang Q and S membrane-based filters and units from Pall. Another opportunity may be in the area between membranes and stacked monoliths in the future as described by Jungbauer [89] or in monolithic materials [97]. Membrane processes have been the cornerstone in purification of industrial enzymes for many years, and similar applications are likely to be implemented to some extent in the biopharmaceutical industry as they provide cheaper and more productive processes.

Resins presented in Table 6.9 are all from well-esteemed suppliers, and there is a high barrier of entry for new suppliers as it requires a certain company size, drug master files on resins, safe and adequate supply to industry, no lot-to-lot variation, etc., that is a large investment on top of new and improved products compared to existing suppliers. Thus, new suppliers with an excellent product may never be able to penetrate the market, and they probably need a partnership with one or several from industry during early development. Industry may be stuck with choices of resins early on, thus it is important for suppliers to have their material included in pilot productions for early clinical trials and wait 5–10 years before the real money appears. Even more beneficial for suppliers is to have their resins included in any purification process platforms of industry, e.g., for mAb purification. Then certain income is secured for a long period.

6.4 APPLICATIONS

Testing of model systems is always convenient as conditions are always well defined, and sufficient amounts of test material are usually available. However, verification of findings from model systems on other proteins and true applications, and screening and comparison directly on the purification task in question are equally important.

Figure 6.20 presents results of binding strength tests with a hGH sample in salt gradient operation mode on DEAE Ceramic HyperD 20 for four elution salts, namely NaH_2PO_4, sodium acetate, sodium sulfate, and NaCl. The hGH sample comprised hGH and a spiked amount of the four amino acid extended precursor MEAE-hGH. DEAE Ceramic HyperD 20 intended for polishing presents excellent separation efficiency for most salts at the given conditions, but the binding strength order as a function of the salt type in Figure 6.20 is rather complicated. Elution order of chloride and acetate is as expected according to the association to the Hofmeister series: with the same linear gradient slope proteins elute faster with chloride than acetate, because chloride has increased elution strength compared to acetate. This attribute is occasionally exploited in process development, where chloride may be substituted with acetate if binding is difficult to achieve and as previously stated due to the corrosive nature of chloride. Dihydrogen phosphate and sulfate are expected to have lower elution strength than acetate according to the series; however, hGH is eluted faster for both ion types even compared to chloride, and no baseline separation was achieved with sulfate although similar elution strength was obtained for dihydrogen phosphate and sulfate. A simple explanation for the elution behavior with sulfate is the divalent nature of this ion resulting in a general higher ionic strength at the same molar concentration. Overall, various effects including elution strength seem to be in play such as salt type-depending interaction with protein surfaces. MES used as low-concentration buffer component, which is of sulfate-type and negatively charged,

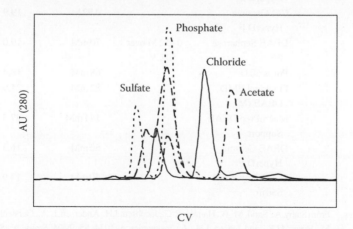

FIGURE 6.20 Separation of MEAE-hGH and hGH using different elution salts. Separation was determined by applying a 250 µL sample of 1 g/L protein solution in 50 CV linear gradient from 0 to 0.2 M sodium salt in 10 mM MES buffer, pH 6.0 through a 10 × 0.5 cm column.

was not assumed to interfere with the results. Tugcu et al. [53] also performed retention studies as a function of sodium salt type for numerous proteins on Source 15Q and Q Sepharose HP, and the same retention order was found, i.e., chloride elution occurred later than sulfate elution.

Table 6.10 presents binding strength studies in salt gradient elution mode of various ion-exchangers with three different proteins: insulin precursor, HBP, and hGH.

TABLE 6.10

Binding Strength Verification Studies in Gradient Mode of Pure Test Proteins on Various Ion-Exchangers

Protein	Resin	Gradient (NaCl)	c_{salt} at t_R	t_R (CV)
Insulin precursor	SP Sepharose BB	0–1 mol/kg over 20 CV	221 mmol/kg	8.6
	SP Toyopearl 550 c		253 mmol/kg	9.6
	SP Toyopearl 650 c		157 mmol/kg	5.7
HBP	SP Sepharose FF	0–1.5 M over 30 CV	1.19 M	25.4
	CM Sepharose FF		0.90 M	19.6
	SP Toyopearl 650 m		0.79 M	17.3
	CM Toyopearl 650 m		0.90 M	19.6
	S Ceramic HyperD 20		0.54 M	12.4
	CM Ceramic HyperD F		0.92 M	19.9
hGH	DEAE Sepharose FF	0–0.2 M over 50 CV	70 mM	19.0
	Poros 50 D		68 mM	18.4
	Fractogel EMD DEAE (M)		82 mM	21.9
	MacroPrep DEAE Support		144 mM	37.5
	DEAE Ceramic HyperD F		60 mM	16.3
	DEAE Toyopearl 650 m		50 mM	13.9

Sources: From Staby, A., Sand, M.-B., Hansen, R.G., Jacobsen, J.H., Andersen, L.A., Gerstenberg, M., Bruus, U.K., and Jensen, I.H., *J. Chromatogr. A*, 1034, 85, 2004; Staby, A., Sand, M.-B., Hansen, R.G., Jacobsen, J.H., Andersen, L.A., Gerstenberg, M., Bruus, U.K., and Jensen, I.H., *J. Chromatogr. A*, 1069, 65, 2005; Staby, A., Jensen, R.H., Bensch, M., Hubbuch, J., Dünweber, D.L., Krarup, J., Nielsen, J., Lund, M., Kidal, S., Hansen, T.B., and Jensen, I.H., *J. Chromatogr. A*, 1164, 82, 2007.

Based on the results of model proteins in Table 6.3, the following order of resins for binding strength should be assumed for insulin precursor: Toyopearl SP 550c > SP Sepharose BB > Toyopearl SP 650c. As shown in Table 6.10, the order of resins for pure insulin precursor is as expected indicating that isocratic data obtained for model proteins will give a good suggestion to relative elution in gradient mode of different resins. For HBP, the most likely order of resins for binding strength based on model results in Table 6.3 should be CM Ceramic HyperD F > S Ceramic HyperD 20 > SP Sepharose FF > CM Toyopearl 650 m ≥ SP Toyopearl 650 m ≥ CM Sepharose FF, although the order of the three latter resins should be close or equal. The binding strength order of HBP in Table 6.10 is, however, SP Sepharose FF > CM Ceramic HyperD F > CM Toyopearl 650 m = CM Sepharose FF > SP Toyopearl 650 m > S Ceramic HyperD 20. The most surprising outcome is the decreased retention of Ceramic HyperD resins in the light of the hypothesis, especially the fairly low retention of S Ceramic HyperD 20. Whether this is due to the experimental setup or special features of HBP in conjunction with ceramic HyperD resins is not known. Still, the elution order of the other resins is as expected. For hGH, the expected elution order based on Table 6.3 would be MacroPrep DEAE Support > Fractogel EMD DEAE (M) > Poros 50 D > DEAE Ceramic HyperD F > DEAE Sepharose FF > DEAE Toyopearl 650 m. In Table 6.10, only DEAE Sepharose FF has changed position in the elution order with higher retention than DEAE Ceramic HyperD F. Otherwise, the elution order was as expected and the data in Table 6.3 are in general verified with some degree of uncertainty as previously stated.

Figure 6.21 presents a verification investigation of applicability of model results in Table 6.5 on dynamic capacity measurements for cation-exchange resins with aprotinin. General conditions for the breakthrough experiments were comparison of sorption capacities with insulin precursor in yeast fermentation broth and in pure state at similar settings, i.e., at the same pH, residence time, solution conductivity, aprotinin concentration, scale, buffer, and temperature. Figure 6.21 shows the breakthrough curves for Sepharose FF, Toyopearl 650 m, Ceramic HyperD F, MacroPrep, and Fractogel EMD M strong and weak cation-exchange resins. According to Table 6.5 where data were obtained for lysozyme, the sequence of binding capacity of pure aprotinin solution would be CM Ceramic HyperD F > CM Sepharose FF > SP Sepharose FF > S Ceramic HyperD F > Fractogel EMD SO_3^- M > Fractogel EMD COO⁻ M > CM Toyopearl 650 m > MacroPrep High S > SP Toyopearl 650 m > MacroPrep CM. Figure 6.21 displays the following approximate dynamic capacity order: S Ceramic HyperD F > CM Ceramic HyperD F ≥ SP Sepharose FF > CM Sepharose FF > Fractogel EMD SO_3^- M > CM Toyopearl 650 m > SP Toyopearl 650 m > MacroPrep High S > Fractogel EMD COO⁻ M ≥ MacroPrep CM at 5% breakthrough. Although some breakthrough curves do not have a steep increase indicating high flow rates (or poor column packing), the order seems reliable and fairly close to expectations from Table 6.5: high capacity of Ceramic HyperD and Sepharose FF resins and lower capacity for MacroPrep resins. If no other parameters than dynamic capacity were to be considered, CM Ceramic HyperD F, S Ceramic HyperD F, SP Sepharose FF, CM Sepharose FF, and possibly Fractogel EMD SO_3^- resins would be selected for further testing in process-development due to their high pure component capacity. For aprotinin in feedstock

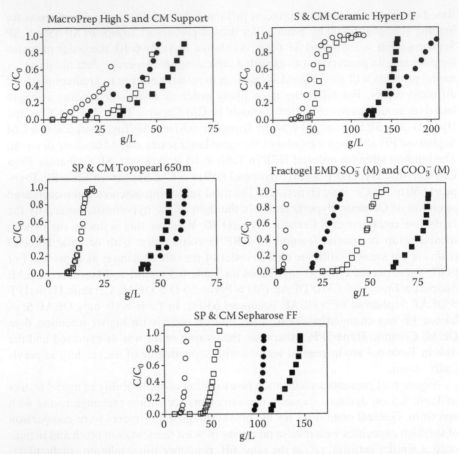

FIGURE 6.21 Breakthrough curves for aprotinin in pure state and in fermentation broth on various corresponding weak and strong cation-exchangers. Flow rates are at ~ 50% of maximum recommended flow-rate. Symbols: ■, pure state on strong cation-exchanger; ●, pure state on weak cation-exchanger; □, fermentation broth on strong cation-exchanger; ○, fermentation broth on weak cation-exchanger.

a different result is achieved, and of course, a general decrease in aprotinin capacity is obtained due to competitive binding from HCP and medium components. The five selected resins still do not have the highest capacities, and their order has changed to S Ceramic HyperD F > SP Sepharose FF > CM Ceramic HyperD F ≥ Fractogel EMD SO$_3^-$ > CM Sepharose FF, and MacroPrep High S actually appears to have higher dynamic capacity than CM Sepharose FF for this application. Based on these data, Fractogel EMD SO$_3^-$ might be the best choice of the five resins, because it binds less HCP and utilizes ~ 60% of its pure component dynamic capacity at 5% breakthrough for aprotinin binding; however following that argument, the best of all ten resins might be MacroPrep High S, which was omitted in the first place, since it has almost the same capacity for aprotinin in feedstock and utilizes more than 70% of its pure component capacity for aprotinin binding. Thus, the most pure product with respect to HCP and medium components may appear using the MacroPrep High S if the different resins have similar selectivities, and/or aprotinin is eluted from the column by a step gradient. Other parameters including column life time, process economy constraints, and analytical results are not considered here, but pure component data including binding capacity data will in general provide a good basis for resin selection. However, these data may not tell the complete and true story.

With the appearance of HTS and robot techniques, intensive testing of column materials and various conditions has become more attractive even if material for early process development is scarce. In previous sections, use of HTS in the static mode resulted in various reliable outcomes, i.e., Figures 6.9 through 6.14; however, the optimal advantage of HTS would be to perform screening in dynamic mode. Figure 6.22 presents a comparison of the same anion-exchange separation on insulin-related molecules performed by HTS and conventional chromatography on the same column material. The column was a 200 μL Atoll column packed with Poros 50 D for HTS and a 1 mL column for conventional chromatography. As shown in Figure 6.22, a fair agreement is obtained between the two experimental setups detecting impurity peaks on both sides of the main peak. Retention agreement in column volume (CV) is also fair; however, there is a distinct degree of peak broadening in both methods as should be expected from this degree of scaling down (column wall effects, etc.), but most pronounced for the HTS setup. With the current status of dynamic HTS for ion-exchange chromatography development, it is to be used for initial screening and comparison as a stand-alone tool and not for final optimization of methods.

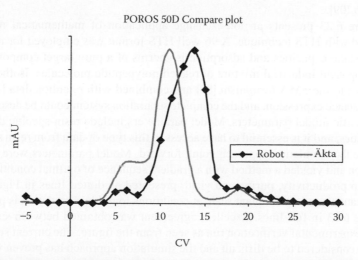

FIGURE 6.22 (See color insert following page 216.) Comparison of HTS (Robot) and traditional chromatography system (Äkta) setup. Separation type is salt gradient elution of insulin-related molecule (main peak) from related impurities on Poros 50 D anion-exchanger.

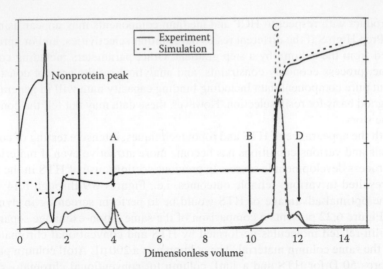

FIGURE 6.23 (See color insert following page 216.) Comparison of simulation and pilot plant ion-exchange separation of four-component mixture of related molecules.

Another obvious use of HTS for development of ion-exchange chromatography methods is to combine it with mathematical modeling. Mathematical modeling and related tools have been widely used in academia [83,99–103], and numerous papers have been published on model systems [40,54,58,59,97,104–108]. Industry has been reluctant to implement these tools due to conservatism, lack of sufficient accuracy of models, and huge consumption of time and protein material; however, industrial application of models for prediction of separation, profiles, etc., has been published lately [96,109].

Figure 6.23 presents an ion-exchange application of mathematical modeling combined with HTS technique. A 96-well HTS format was employed for measurement of static capacities and adsorption isotherms of a pure target component in a four-component industrial mixture of related polypeptide molecules. Isotherm data were fitted to the SMA formalism [83] and combined with retention data [108] and a mass balance expression, and the complete separation system could be described by characteristic model parameters. Model parameters include resin-specific data such as porosities, and it is essential to have access to this type of data from resin suppliers to pursue the simulation approach going forward. Model parameters were used for simulation and yielded a method with a gradient sequence of optimal conditions with respect to productivity, purity, and yield, presented as dotted lines in Figure 6.23. A verification run in pilot plant scale at conditions found by simulation is presented in Figure 6.23 in full lines. Excellent agreement was obtained between simulation and the experimental verification run as seen from the figure. The current separation task was considered to be difficult and the simulation approach has proven very efficient for development, optimization, and troubleshooting of ion-exchange processes for protein purification.

6.5 CONCLUSION

The biopharmaceutical industry is under constant pressure of delivering more projects through their R&D pipeline to maintain the number of new products placed in the market anually and carry off competition from biogenerics. There is thus an increased demand for material for clinical trials and handling of an increasing number of projects. Basically all purification processes contain at least one ion-exchange step, and there is thus a demand for ever better ion-exchange resins to improve process economy and speed of process development. Focus on resin development will be on performance, however, with equal attention to attributes like consistency in product quality, low batch-to-batch variation, reliable and fast supply, cost, price, and resin lifetime. Manufacturers winning the race will be characterized by supplying products that distinguish themselves from others and by the level of service and support granted. For antibody production, this theme has been obvious for sometimes with generic purification platforms in contrast to the occasional case-by-case approach for new proteins with no similarities to the previous ones for a given company. This platform approach is now also adopted for development of new proteins, at least as much as possible.

APPENDIX

Table A.1 presents paper providing comparison of more than four ion-exchangers. Table A.2 lists suppliers are resins presented in this paper.

TABLE A.1
Papers Presenting Comparison of More than Four Anion- or Cation-Exchange Chromatography Resins, Excluding the Work of this Group

Resin	Functionality	Studies/Parameters	References
Toyopearl SP 550c	Strong and weak CIE	Pore size distribution	[1,4]
Toyopearl SP 650m		Phase ratios	
		Retention/binding strength	
Toyopearl CM 650m			
Fractogel EMD SO_3^- 650m			
Fractogel EMD COO^- 650m			
SP Spherodex M			
CM Spherodex M			
SP Sepharose FF			
CM Sepharose FF			
Q HyperD F	Strong and weak AIE	Dynamic binding capacity	[2]
Toyopearl SuperQ 650		Separation	
Q Sepharose FF		Ionic capacity	
Fractogel EMD Q			
Q Poros II			

(continued)

TABLE A.1 (continued)
Papers Presenting Comparison of More than Four Anion- or Cation-Exchange Chromatography Resins, Excluding the Work of this Group

Resin	Functionality	Studies/Parameters	References
DEAE Spherodex			
QMA Accell			
DEAE Bio-Gel A			
DEAE Cellufine			
DEAE Cellulose			
DEAE Fractogel 650			
DEAE Separon			
HemaBio 1000			
DEAE Sephacel			
DEAE Sepharose CL 6B			
DEAE Sepharose FF			
DEAE Trisacryl			
PAE Matrex			
Whatman DE51	Strong and weak AIE and CIE	Performance Ionic capacity	[3,34]
Whatman DE52		Protein capacity	
Whatman DE53		Packing density	
Whatman QA52		Resolution	
Whatman CM52		Operating pressure	
Whatman SE52			
Whatman SE53			
Express-Ion D			
Express-Ion Q			
Express-Ion C			
Express-Ion S			
DEAE Sephacel			
DEAE Sepharose CL 6B			
DEAE Sepharose FF			
Q Sepharose FF			
Q Sepharose HP			
CM Sepharose FF			
S Sepharose FF			
S Sepharose HP			
DEAE Sephadex A-25			
DEAE Sephadex A-50			
QAE Sephadex A-25			
QAE Sephadex A-50			
CM Sephadex C-25			
CM Sephadex C-50			
SP Sephadex C-25			
SP Sephadex C-50			
Matrex DEAE 200 Cellufine			

TABLE A.1 (continued)
Papers Presenting Comparison of More than Four Anion- or Cation-Exchange Chromatography Resins, Excluding the Work of this Group

Resin	Functionality	Studies/Parameters	References
Matrex DEAE 800 Cellufine			
Matrex CM 200 Cellufine			
Matrex CM 500 Cellufine			
DEAE Thruput			
CM Thruput			
Q Thruput			
DEAE Toyopearl 650 s			
DEAE Toyopearl 650 m			
DEAE Toyopearl 650 c			
CM Toyopearl 650 s			
CM Toyopearl 650 m			
CM Toyopearl 650 c			
SP Toyopearl 550 c			
SP Toyopearl 650 s			
SP Toyopearl 650 m			
SP Toyopearl 650 c			
MacroPrep Q			
MacroPrep High Q			
MacroPrep CM			
MacroPrep S			
MacroPrep High S			
Fractogel EMD TMAE 650			
Fractogel EMD DEAE 650			
Fractogel EMD DMAE 650			
Fractogel EMD SO_3^- 650			
Poros 50 HQ			
Poros 50 HS			
DEAE Trisacryl M			
DEAE Trisacryl Plus M			
CM Trisacryl M			
SP Trisacryl M			
SP Trisacryl Plus M			
DEAE Spherodex M			
CM Spherodex M			
SP Spherodex M			
DEA Spherosil M			
QMA Spherosil M			
DEAE HyperD F			
Q HyperD M			
S HyperD M			
Cellufine	Strong AIE and CIE	NaOH stability	[5]

(continued)

TABLE A.1 (continued)
Papers Presenting Comparison of More than Four Anion- or Cation-Exchange Chromatography Resins, Excluding the Work of this Group

Resin	Functionality	Studies/Parameters	References
Express-Ion SE		Dynanic capacity	
Fractogel EMD SO_3^-		Recovery	
Q HyperD		Flow/pressure properties	
MacroPrep S		Nonspecific binding	
SP Sepharose FF		Separation/resolution	
Sepharose XL			
SP Toyopearl 650 c			
SP Thruput			
Poros HS			
S HyperD			
SP Toyopearl 550 c			
SP Sepharose FF	Strong CIE	Particle size	[6]
SP-PVA-CG1000sd		Pressure drop	
SP-PVA-Source 15 S		Compression	
SP-PVA-Source 30 S		Plate height	
Source 15 S		Pore diffusivity	
Source 30 S		Ionic capacity	
SP-PVA-PLRP4000s 20		Static capacity	
SP-PVA-PLRP4000s 60		Protein recovery	
Poros 20 SP		Resolution	
SP Poros 20 OH		Dynamic capacity	
SP Toyopearl 650 c	Strong CIE	Particle size	[13]
SP Toyopearl 550 c		Porosity	
SP Sepharose FF		Adsorption isotherms/static capacities	
Fractogel EMD SO_3^- M			
Baker Carboxy-Sulfone		Uptake curves	
SP Spherodex M		Pore diffusivities	
Toyopearl SP 550 c	Strong CIE	Retention/binding strength	[25]
Toyopearl SP 650 m			
Fractogel EMD SO_3^- 650 m			
SP Spherodex M			
SP Sepharose FF			
Cellufine Sulfate			
DEAE Sepharose FF	Strong and weak AIE	Retention/binding strength	[35]
Q Sepharose FF		Plate height	
DEAE Toyopearl 650 m			
SuperQ Toyopearl 650 m			
Fractogel EMD DEAE650 m			

TABLE A.1 (continued)
Papers Presenting Comparison of More than Four Anion- or Cation-Exchange Chromatography Resins, Excluding the Work of this Group

Resin	Functionality	Studies/Parameters	References
Fractogel EMD DMAE650 m			
Fractogel EMD TMAE650 m			
DEAE Sepharose FF	Strong and weak AIE	Ionic capacity	[52]
Q Sepharose FF		Retention/binding strength	
Amino Sepharose 6FF		Breakthrough capacity	
Q Sepharose XL			
ANX Sepharose 4FF low			
ANX Sepharose 4FF high			
GMA TEA	Strong and weak AIE,	Separation	[56]
GMA DEA	Weak AIE	Ionic capacity	
GMA EDA			
GMA HEDA			
GMA DETA			
GMA TETA			
Express-Ion D		Titration curves	[57]
DEAE Cellulofine AM		Pressure/flow properties	
DEAE Toyopearl 650 c		Yield	
DEAE Sepharose FF			
(GD/X) + (C300A)			
SuperQ Toyopearl 650 c	Strong AIE	Pore size distribution	[68]
QAE Toyopearl 550 c		Phase ratio	
Source 30Q		Diffusivity	
Q Sepharose FF		Static capacity	
Q Sepharose XL		Dynamic capacity	
Express-Ion Q			
Q-Cellthru BB Plus			
Poros 50 HQ			
SP Sepharose FF	Strong and weak CIE	Static capacity	[93]
CM Sepharose FF		Binding strength	
SP Sepharose XL		Resolution	
CM Toyopearl 650 m			
SP Toyopearl 650 m			
Fractogel EMD SO_3^- Hicap			
Fractogel EMD SE Hicap			
MacroPrep HS			
MacroPrep CM			
Unosphere S			
Fractoprep SP			

TABLE A.2
Resins and Suppliers of Main Paper: Anion-
and Cation-Exchangers

Supplier	Resin
Anion-exchangers	
Applied Biosystems	Poros QE/M
	Poros 50 HQ
	Poros 50 D
Bio-Rad	MacroPrep DEAE Support
	MacroPrep 25Q
	UNO Q-1
	Unosphere Q
GE Healthcare	Q Sepharose XL
	Q Sepharose HP
	Q Sepharose FF
	Capto Q
	Capto DEAE
	Capto adhere[a]
	Source 30Q
	DEAE Sepharose FF
	Mono Q
	ANX Sepharose FF (low sub)
	ANX Sepharose FF (high sub)
Merck	Fractogel EMD DEAE (M)
	Fractogel EMD TMAE 650 s
	Fractogel EMD DMAE (M)
	Fratoprep DEAE
Pall	Q HyperD 20
	DEAE Ceramic HyperD 20
	Q Zirconia
Sterogene	Q-Cellthru Bigbeads Plus
Tessek	Separon HemaBio 1000Q
Tosoh Bioscience	Toyopearl QAE 550 c
	Toyopearl SuperQ 650 c
	TSKGel Q-5PW-HR
	Toyopearl DEAE 650 (M)
	Toyopearl GigaCap Q-650 M
Whatman	Express-Ion Q
Cation-exchangers	
Applied Biosystems	Poros 50 HS
	Poros 20 SP
Bio-Rad	MacroPrep 25 S
	MacroPrep High S
	MacroPrep CM
	Unosphere S

TABLE A.2 (continued)
Resins and Suppliers of Main Paper: Anion- and Cation-Exchangers

Supplier	Resin
	SP Sepharose FF
	SP Sepharose XL
	SP Scpharose BB
	Capto S
	MacroCap SP
	Source 30 S
	CM Sepharose FF
	Capto MMC[a]
	Heparin Sepharose FF
Merck	Fractogel SE HICAP (M)
	Fractogel EMD COO⁻ (M)
	Fractogel EMD SO_3^- (M)
	Fractoprep TMAE
	Fratoprep SO_3
Millipore	CM Cellufine C-500
	Cellufine Sulfate
Pall	CM Ceramic HyperD F
	S Ceramic HyperD F
	Heparin Ceramic HyperD M
	S Ceramic HyperD 20
	CM Hyper Z
	CM Spherodex M
	SP Spherodex M
Tosoh Bioscience	SP Toyopearl 550 c
	TSKGel SP-5PW-HR20
	CM Toyopearl 650 m
	SP Toyopearl 650 m
	SP Toyopearl 650 c
	SP Toyopearl GigaCap 650 m
	SP Toyopearl 550 c
	Heparin Toyopearl 650 m
	Toyopearl GigaCap S-650 M
	Toyopearl Megacap II SP-550EC

[a] Multimodal resin.

ABBREVIATIONS

BSA	bovine serum albumin
C	protein concentration
c_{salt}	salt concentration

CIP	clean-in-place
CV	column volumes
FMC	full manufacturing cost
HBP	heparin binding protein
HCP	host cell proteins
HTS	high-throughput screening
ISEC	inverse size-exclusion chromatography
I_{total}	total ionic strength
k'	retention factor
P	pressure
pI	isoelectric point
Q	binding capacity
SEM	scanning electron microscopy
SMA	steric mass action
SMB	simulated moving bed
t_R	retention time

ACKNOWLEDGMENTS

Donation of resins from GE Healthcare (Uppsala, Sweden), Applied Biosystems (Cambridge, MA), Merck (Darmstadt, Germany), Bio-Rad Laboratories (Hercules, CA), Pall (Cergy-Saint-Christophe, France), Tosoh Bioscience (Philadelphia, PA), Millipore (Billerica, MA), Sterogene (Carlsbad, CA), and Whatman (Maidstone, UK) is highly appreciated.

Experimental work and preparation of figures and tables by Inge Holm Jensen and Randi Holm Jensen; supply of pure proteins from Peter Rahbek Østergaard, Birgitte Silau, Anne Mette Nøhr, and Ole Elvang Jensen; use of simulation program from Lars Sejergaard and Ernst Hansen; and the general support by Inger Mollerup are gratefully acknowledged.

REFERENCES

1. P. DePhillips and A.M. Lenhoff, *J. Chromatogr. A*, 883, 2000, 39.
2. E. Boschetti, *J. Chromatogr. A*, 658, 1994, 207.
3. P. Levison, C. Mumford, M. Streater, A. Brandt-Nielsen, N.D. Pathirana, and S.E. Badger, *J. Chromatogr. A*, 760, 1997, 151.
4. P. DePhillips and A.M. Lenhoff, *J. Chromatogr. A*, 933, 2001, 57.
5. R. Noel and G. Proctor, *Poster Presentation at Recovery of Biological Products X*, Cancun, Mexico, June 2001 (poster D-2).
6. D.C. Nash and H.A. Chase, *J. Chromatogr. A*, 807, 1998, 185.
7. J. Horvath, E. Boschetti, L. Guerrier, and N. Cooke, *J. Chromatogr. A*, 679, 1994, 11.
8. S. Yamamoto and T. Ishihara, *J. Chromatogr. A*, 852, 1999, 31.
9. S. Yamamoto and E. Miyagawa, *J. Chromatogr. A*, 852, 1999, 25.
10. J. Renard, C. Vidal-Madjar, and B. Sebille, *J. Liq. Chromatogr.*, 15, 1992, 71.
11. L. Dunn, M. Abouelezz, L. Cummings, M. Navvab, C. Ordnez, C.J. Siebert, and K.W. Talmadge, *J. Chromatogr.*, 548, 1991, 165.
12. M. McCoy, K. Kalghatgi, F.E. Regnier, and N. Afeyan, *J. Chromatogr. A*, 743, 1996, 221.
13. C. Chang and A.M. Lenhoff, *J. Chromatogr. A*, 827, 1998, 281.

14. Y.-B. Yang, K. Harrison, and J. Kindsvater, *J. Chromatogr. A*, 723, 1996, 1.
15. M. Weitzhandler, D. Farnan, J. Horvath, J.S. Rohrer, R.W. Slingsby, N. Avdalovic, and C. Pohl, *J. Chromatogr. A*, 828, 1998, 365.
16. L.E. Weaver and G. Carta, *Biotechnol. Prog.*, 12, 1996, 342.
17. Y. Hu and P.W. Carr, *Anal. Chem.*, 70, 1998, 1934.
18. M.A. Hashim, K.-H. Chu, and P.-S. Tsan, *J. Chem. Tech. Biotechnol.*, 62, 1995, 253.
19. R. Hahn, P.M. Schulz, C. Schaupp, and A. Jungbauer, *J. Chromatogr. A*, 795, 1998, 277.
20. C.M. Roth, K.K. Unger, and A.M. Lenhoff, *J. Chromatogr. A*, 726, 1996, 45.
21. F. Fang, M.-I. Aguilar, and M.T.W. Hearn, *J. Chromatogr. A*, 729, 1996, 67.
22. C.B. Mazza and S.M. Cramer, *J. Liq. Chromatogr. Rel. Technol.*, 22, 1999, 1733.
23. V. Natarajan and S.M. Cramer, *J. Chromatogr. A*, 876, 2001, 63.
24. V. Natarajan and S.M. Cramer, *Sep. Sci. Technol.*, 35, 2000, 1719.
25. P. DePhillips and A.M. Lenhoff, *J. Chromatogr. A*, 1036, 2004, 51.
26. C.B. Mazza, N. Sukumar, C.M. Breneman, and S.M. Cramer, *Anal., Chem.*, 73, 2001, 5457.
27. J. Hubbuch, T. Linden, E. Knieps, A. Ljunglöf, J. Thömmes, and M.-R. Kula, *J. Chromatogr. A*, 1021, 2003, 93.
28. J. Hubbuch, T. Linden, E. Knieps, J. Thömmes, and M.-R. Kula, *J. Chromatogr. A*, 1021, 2003, 105.
29. N. Lendero, J. Vidič, P. Brne, A. Podgornik, and A. Štrancar, *J. Chromatogr. A*, 1065, 2005, 29.
30. M. Jozwik, K. Kaczmarski, and R. Freitag, *J. Chromatogr. A*, 1073, 2005, 111.
31. A. Staby, M.-B. Sand, R.G. Hansen, J.H. Jacobsen, L.A. Andersen, M. Gerstenberg, U.K. Bruus, and I.H. Jensen, *J. Chromatogr. A*, 1034, 2004, 85.
32. A. Staby, M.-B. Sand, R.G. Hansen, J.H. Jacobsen, L.A. Andersen, M. Gerstenberg, U.K. Bruus, and I.H. Jensen, *J. Chromatogr. A*, 1069, 2005, 65.
33. A. Staby, J.H. Jacobsen, R.G. Hansen, U.K. Bruus, and I.H. Jensen, *J. Chromatogr. A*, 118, 2006, 168.
34. J. Dennis, P. Levison, and C. Mumford, *BioPharm.*, 11, 1998, 44.
35. J.Z. Bai, Characterization of protein retention and transport in anion exchange chromatography, Master thesis, University of Delaware, Newark, DE, Spring 1999.
36. N.B. Afeyan, N.F. Gordon, I. Mazsaroff, L. Varady, S.P. Fulton, Y.B. Yang, and F.E. Regnier, *J. Chromatogr.*, 519, 1990, 1.
37. P.R. Levison, R.M.H. Jones, D.W. Toome, S.E. Badger, M. Streater, and N.D. Pathirana, *J. Chromatogr. A*, 734, 1996, 137.
38. M.T.W. Hearn, A.N. Hodder, F.W. Fang, and M.I. Aguilar, *J. Chromatogr.*, 548, 1991, 117.
39. A. Johnston, Q.M. Mao, and M.T.W. Hearn, *J. Chromatogr.*, 548, 1991, 127.
40. K. Miyabe and G. Guiochon, *J. Chromatogr. A*, 866, 2000, 147.
41. J. Ericsson, E. Berggren, C. Lindqvist, K.-A. Hansson, K. Qvarnström, L. Lundh, and G. Moen, *React. Funct. Polym.*, 30, 1996, 327.
42. A.E. Ivanov and V.P. Zubov, *J. Chromatogr. A*, 673, 1994, 159.
43. C. McNeff, Q. Zhao, and P.W. Carr, *J. Chromatogr. A*, 684, 1994, 201.
44. M.A. Fernandez, W.S. Laughinghouse, and G. Carta, *J. Chromatogr. A*, 746, 1996, 185.
45. M.A. Fernandez and G. Carta, *J. Chromatogr. A*, 746, 1996, 169.
46. I. Lagerlund, E. Larsson, J. Gustavsson, J. Färenmark, and A. Heijbel, *J. Chromatogr. A*, 796, 1998, 129.
47. D. Bentrop and H. Engelhardt, *J. Chromatogr.*, 556, 1991, 363.
48. A.M. Tsai, D. Englert, and E.E. Graham, *J. Chromatogr.*, 504, 1990, 89.
49. M.B. Jensen, Ionbytningskromatografi af Valleproteiner—Isokratisk og lineær gradient eluering, Preliminary master thesis, Technical University of Denmark, Lyngby, June 1999.
50. U.-J. Kim and S. Kuga, *J. Chromatogr. A*, 955, 2002, 191.
51. B.-L. Johansson, M. Belew, S. Eriksson, G. Glad, O. Lind, J.-L. Maloisel, and N. Norrman, *J. Chromatogr. A*, 1016, 2003, 21.

52. M. Andersson, J. Gustavsson, and B.-L. Johansson, *Int. J. Bio-Chromatogr.*, 6, 2001, 285.
53. N. Tugcu, M. Song, C.M. Breneman, N. Sukumar, K.P. Bennett, and S.M. Cramer, *Anal. Chem.*, 75, 2003, 3563.
54. A.K. Hunter and G. Carta, *J. Chromatogr. A*, 971, 2002, 105.
55. P. Arvidsson, F.M. Plieva, I.N. Savina, V.I. Lozinsky, S. Fexby, L. Bülow, I.Y. Galaev, and B. Mattiasson, *J. Chromatogr. A*, 977, 2002, 27.
56. S.-H. Choi, Y.-M. Hwang, and K.-P. Lee, *J. Chromatogr. A*, 987, 2003, 323.
57. K. Sato, Y.-H. Guo, J. Feng, S. Sugiyama, M. Ichinomiya, Y. Tsukamasa, Y. Minegishi, A. Sakata, K. Komiya, Y. Yamasaki, Y. Nakamura, K. Ohtsuki, and M. Kawabata, *J. Chromatogr. A*, 811, 1998, 69.
58. A.K. Hunter and G. Carta, *J. Chromatogr. A*, 897, 2000, 65.
59. A.K. Hunter and G. Carta, *J. Chromatogr. A*, 897, 2000, 81.
60. Y. Yu and Y. Sun, *J. Chromatogr. A*, 855, 1999, 129.
61. J. Thiemann, J. Jankowski, J. Rykl, S. Kurzawski, T. Pohl, B. Wittmann-Liebold, and H. Schlüter, *J. Chromatogr. A*, 1043, 2004, 73.
62. T. Andersen, M. Pepaj, R. Trones, E. Lundanes, and T. Greibrokk, *J. Chromatogr. A*, 1025, 2004, 217.
63. Y. Kato, K. Nakamura, T. Kitamura, T. Tsuda, M. Hasegawa, and H. Sasaki, *J. Chromatogr. A*, 1031, 2004, 101.
64. H. Shen and D.D. Frey, *J. Chromatogr. A*, 1034, 2004, 55.
65. Y. Yao and A.M. Lenhoff, *J. Chromatogr. A*, 1037, 2004, 273.
66. G.-Y. Sun, Q.-H. Shi, and Y. Sun, *J. Chromatogr. A*, 1061, 2004, 159.
67. L. Norling, S. Lute, R. Emery, W. Khuu, M. Voisard, Y. Xu, Q. Chen, G. Blank, and K. Brorson, *J. Chromatogr. A*, 1069, 2005, 79.
68. Y. Yao and A.M. Lenhoff, *J. Chromatogr. A*, 1126, 2006, 107.
69. A. Staby, I.H. Jensen, and I. Mollerup, *J. Chromatogr. A*, 897, 2000, 99.
70. A. Staby and I.H. Jensen, *J. Chromatogr. A*, 908, 2001, 149.
71. A. Staby, R.H. Jensen, M. Bensch, J. Hubbuch, D.L. Dünweber, J. Krarup, J. Nielsen, M. Lund, S. Kidal, T.B. Hansen, and I.H. Jensen, *J. Chromatogr. A*, 1164, 2007, 82.
72. M. Lohrmann, M. Schulte, and J. Strube, *J. Chromatogr. A*, 1092, 2005, 89.
73. M. Bensch, P.S. Wierling, E. von Lieres, and J. Hubbuch, *Chem. Eng. Technol.*, 28, 2005, 1274.
74. K. Rege, M. Pepsin, B. Falcon, L. Steele, and M. Heng, *Biotechnol. Bioeng.*, 93, 2005, 618.
75. J.F. Kramarczyk, High-throughput screening of chromatography resins and excipients for optimizing selectivity, MSc thesis, Tufts University, Medford/Somerville, MA, 2003.
76. L. Thim, S. Bjørn, M. Christensen, E.M. Nicolaisen, T. Lund-Hansen, A.H. Pedersen, and U. Hedner, *Biochemistry*, 27, 1988, 7785.
77. H. Fritz and G. Wunderer, *Arzneim. Forsch. Drug Res.*, 33(I,4), 1983, 479.
78. H. Flodgaard, E. Østergaard, S. Bayne, A. Svendsen, J. Thomsen, M. Engels, and A. Wollmer, *Eur. J. Biochem.*, 197, 1991, 535.
79. I. Mollerup, S.W. Jensen, P. Larsen, O. Schou, and L. Snel, Insulin purification, in: M.C. Flickinger and S.W. Drew (Eds.), *The Encyclopedia of Bioprocess Technology: Fermentation, Biocatalysis and Bioseparation*, Wiley, New York, 1999.
80. E. Boel, T. Christensen, E. Gormsen, B. Huge-Jensen, and B.S. Olesen, in: L. Alberghina, R.D. Schmid, and R. Verger (Eds.), *Lipases: Structure, Mechanism and Genetic Engineering (GBF Monographs, Vol. 16)*, Wiley-VCH, Weinheim, 1990, p. 207.
81. W. Kopaciewicz, M.A. Rounds, J. Fausnaugh, and F.E. Regnier, *J. Chromatogr.*, 266 1983, 3.
82. E. Müller, *J. Chromatogr. A*, 1006, 2003, 229.
83. C.A. Brooks and C.M. Cramer, *AIChE J.*, 39, 1992, 1969.
84. A. Staby, N. Johansen, H. Wahlstrøm, and I. Mollerup, *J. Chromatogr. A*, 827, 1998, 311.

85. J.J. van Deemter, F.J. Zuiderweg, and A. Klinkenberg, *Chem. Eng. Sci.*, 5, 1956, 271.

86. O. Schou, L. Gotfred, and P. Larsen, *J. Chromatogr.*, 254, 1983, 289.

87. L. Hagel, M. Östberg, and T. Andersson, *J. Chromatogr. A*, 743, 1996, 33.

88. Y. Yao, K. Czymmek, M.R. Shure, and A.M. Lenhoff, Analysis of three-dimensional structure of porous chromatographic adsorbents from electron tomography, Poster presented at *PREP'2004*, Baltimore, MD, May 22–26, 2004.

89. A. Jungbauer, *J. Chromatogr. A*, 1065, 2005, 3.

90. E. Müller, J.-T. Chung, Z. Zhang, and A. Sprauer, *J. Chromatogr. A*, 1097, 2005, 116.

91. C. Jung, Y.-P. Lee, Y.R. Jeong, J.Y. Kim, Y.H. Kim, and H.S. Kim, *J. Chromatogr. B*, 814, 2005, 53.

92. X. Li, A.M. Hupp, and V.L. McGuffin, The thermodynamic and kinetic basis of liquid chromatography, in: E. Grushka and N. Grinberg (Eds.), *Advances in Chromatography Vol. 45*, CRC Press, Boca Raton, FL, 2006.

93. A.A. Shukla and X.S. Han, Screening of chromatographic stationary phases, in: A.A. Shukla, M.R. Etzel, and S. Gadam (Eds.), *Process Scale Bioseparations for the Biopharmaceutical Industry*, CRC Press, Boca Raton, FL, 2007.

94. B.-L. Johansson, M. Belew, S. Eriksson, G. Glad, O. Lind, J.-L. Maloisel, and N. Norrman, *J. Chromatogr. A*, 1016, 2003, 35.

95. W. Schwartz, D. Judd, M. Wysocki, L. Guerrier, E. Birck-Wilson, and E. Boschetti, *J. Chromatogr. A*, 908, 2001, 251.

96. J.M. Mollerup, T.B. Hansen, S. Kidal, and L. Sejergaard, A. Staby. *Fluid Phase Equilib.*, 261, 2007, 133.

97. C. Martin, J. Coyne, and G. Carta, *J. Chromatogr. A*, 1069, 2005, 43.

98. H. Charlton, B. Galarza, B. Beacon, K. Leriche, and R. Jones, *Suppl. BioPharm. Int., Adv. Sep. Purif.*, (June 2006) 20.

99. I. Langmuir, *J. Am. Chem. Soc.*, 40, 1918, 1361.

100. P.C. Wankat, *Rate-Controlled Separations*, Chapman & Hall, London, 1994.

101. S. Yamamoto, K. Nakanishi, and R. Matsuno, *Ion-Exchange Chromatography of Proteins, Chromatographic Science Series*, Vol. 43, Marcel Dekker, NY, 1988.

102. D. Farnan, D.D. Frey, and C. Horvath, *Biotechnol. Prog.*, 13, 1997, 429.

103. E. Grushka, *Anal. Chem.*, 44, 1972, 1733.

104. E. Hansen and J. Mollerup, *J. Chromatogr. A*, 827, 1998, 259.

105. D. Karlsson, N. Jakobsson, K.-J. Brink, A. Axelsson, and B. Nilsson, *J. Chromatogr. A*, 1033, 2004, 71.

106. N. Jakobsson, D. Karlsson, J.P. Axelsson, G. Zacchi, and B. Nilsson, *J. Chromatogr. A*, 1063, 2005, 99.

107. S.R. Dziennik, E.B. Belcher, G.A. Barker, M.J. DeBergalis, S.E. Fernandez, and A.M. Lenhoff, *PNAS*, 100, 2003, 420.

108. L. Pedersen, J. Mollerup, E. Hansen, and A. Jungbauer, *J. Chromatogr. B*, 790, 2003, 161.

109. J. Mollerup, T.B. Hansen, S. Kidal, and A. Staby. *J. Chromatogr. A*, 1177, 2008, 200.

110. T. Herrman, M. Schröder, and J. Hubbuch, *Biotechnol. Prog.*, 22, 2006, 914.

7 Advances in Pulsed Electrochemical Detection for Carbohydrates*

William R. LaCourse

CONTENTS

7.1 INTRODUCTION

Human technology excels at its ability to manipulate electrons in the form of electronic circuitry, measuring signals, and processing data. In electroanalytical techniques, chemical analysis is incorporated into the electronics, typically in the form of an electrochemical cell, in which the electrode acts as a transducer to convert

---

* This chapter is being submitted as an invited chapter.

chemical signals in to electronic signals. Similarly, electroanalytical techniques can be classified according to the three fundamental electrical parameters of voltage or potential (E), resistance (R), and current (i). These terms are related via Ohm's law, which is $E = i \cdot R$. The conductance of a solution (G) is the inverse of resistance ($G = 1/R$). Potentiometric-based detection systems measure potential in volts (V) under conditions where i essentially equals zero; conductimetric detectors measure solution conductance in siemens (S); and amperometric detectors measure current in amperes (A) as a function of applied potential. Every imaginable approach to electroanalysis can be traced back to the manipulation of the fundamental parameters of E, i, and R.

Electrochemical detectors following a chromatographic or electrophoretic separation offer many advantages such as high sensitivity, high selectivity, and wide linear range. They are easily miniaturized due to the response being dependent on electrode area and not pathlength as in optical absorbance methods. The detectors are often simple, rugged, and relatively inexpensive. Conductivity detectors have the "potential" (pun intended) to measure all ions in a bulk solution, which also leads to reliance on *a priori* separation and the isolation of the analyte signal via background suppression. Potentiometric techniques often suffer from slow response times, and amperometric detection is sometimes difficult to use due to its heterogeneous detection process. In other words, analytes must diffuse to an electrode surface, which can lead to "poisoning" of the electrode surface. Many amperometric techniques require daily polishing of the electrode surface. Pulsed electrochemical detection (PED) is designed to mitigate these types of problems. Unfortunately, electrochemical detectors are also sensitive to flow rate, mobile phase, and eluent constituents including dissolved oxygen, which is difficult to eliminate or control.

Electrochemical detection (EC) following liquid chromatography (LC) has proven to be a powerful analytical technique for the determination of compounds that are able to be reduced or oxidized. The high sensitivity and selectivity of EC, especially when combined with a separation technique, are ideally suited for complex samples, as evinced by its application to the determination of neurotransmitters in complex biological samples (i.e., brain extracts). Neurotransmitters are typically aromatic compounds (e.g., phenols, aminophenols, catecholamines, and other metabolic amines), which are detected easily by anodic reactions at a constant (direct current [dc]) applied potential at inert electrodes [1,2]. The most common electrode materials are Au, Pt, and C. Electronic resonance in aromatic molecules stabilizes free-radical intermediate products of anodic oxidations, and as a consequence, the activation barrier for the electrochemical reaction is lowered significantly. In contrast, absence of π-resonance for aliphatic compounds results in very low oxidation rates even though the reactions may be favored thermodynamically [3]. Since π-resonance does not exist in polar aliphatic compounds such as carbohydrates, stabilization of reaction intermediates is actively achieved via adsorption at clean noble metal electrodes. Faradaic processes that benefit from electrode surface interactions are described as electrocatalytic processes.

7.2 ELECTROCATALYSIS AT NOBLE METAL ELECTRODES

Figure 7.1 shows the current–potential (i–E) plot for a Au rotating disk electrode (RDE) in 0.1 M NaOH with (·····) and without (-----) dissolved O_2. The observed waves and peaks are attributed as follows:

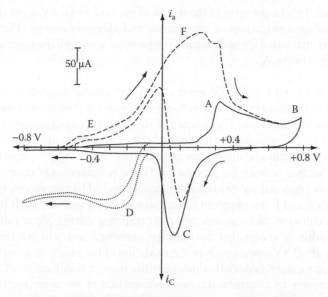

FIGURE 7.1   Cyclic voltammogram of glucose at Au electrode in 0.1 M NaOH. Conditions: 900 rpm rotation speed, 200 mV/s scan rate. Solutions (·····) aerated 0.1 M NaOH, (———) deaerated 0.1 M NaOH, and (-----) 0.2 mM glucose.

- *Signal A.* An anodic peak due to the formation of surface oxide on the positive scan.
- *Signal B.* The onset of O_2 evolution and the upper potential limit for this electrode-supporting electrolyte system.
- *Signal C.* The cathodic removal of the surface oxide formed previously on the positive scan.
- *Signal D.* If dissolved O_2 is present, a cathodic wave (wave *D*) is observed during the positive and negative scans.

Except for hydrogen adsorption and reduction waves, all other features of Pt electrodes are similar to that of Au electrodes in 0.1 M NaOH. However, the wave for dissolved O_2 reduction is well resolved from wave A for a Au electrode; whereas, there is significant overlap of O_2 reduction and oxide formation for a Pt electrode. This difference affords the analyst a potential window free from dissolved oxygen interference, which accounts for the general preference of Au over Pt electrodes for the majority of PED applications.

The current–potential (*i–E*) response for 200 μM glucose (—), a reducing sugar, in the absence of dissolved O_2, is also shown in Figure 7.1. The oxidation of glucose is irreversible as noted only by anodic waves, which are as follows:

- *Signal E.* This wave beginning at −0.6 V on the positive scan corresponds to oxidation of the aldehyde group to the carboxylate anion in this alkaline media, which accounts for the activity of glucose in many solution tests (e.g., Fehling's solution).

- *Signal F.* This larger wave in the region of ca. -0.2 to $+0.4$ V is obtained for the combined oxidations of the alcohol and aldehyde groups. The anodic signal is attenuated abruptly during the positive scan with the onset of oxide formation (wave A).

The signal for ca. $+0.4$ to $+0.6$ V results from the anodic desorption of adsorbed glucose and intermediate products simultaneously with the formation of surface oxide (wave A) on the Au electrode. The absence of signal on the negative scan in the region of ca. $+0.8$ to $+0.2$ V indicates the absence of activity by the oxide-covered electrode surface. Following cathodic dissolution of the oxide on the negative scan to produce wave C, the surface activity for glucose oxidation is immediately returned, and an anodic peak is observed for oxidation of alcohol and aldehyde groups on glucose. Anodic waves E and F are observed to increase in signal intensity with increases in glucose concentration. Similar voltammetric response curves are obtained for all other carbohydrates, except that nonreducing carbohydrates will not show a wave commencing at -0.6 V corresponds to the oxidation of the aldehyde group.

Since electrocatalytic detection mechanisms involve stabilization of intermediates via adsorption to a surface, the electrode surface of the noble metal electrode accumulates adsorbed carbonaceous materials, which eventually foul the electrode surface and results in the loss of its activity [3]. Under constant applied potentials (dc amperometry), noble metal electrodes show high, but transient, catalytic activity, but the activity is quickly diminished due to fouling of the electrode surface. Even for reversible redox couples that are considered to be well behaved, dc amperometry is often accompanied by the practice of disassembling the electrochemical cell and mechanically polishing the working electrode. In this manner, fouling from nonspecific adsorption processes and/or mechanistic consequences is physically removed from the electrode surface. An alternate approach is to combine electrochemical detection with online cleaning. Hence, in order to maintain uniform and reproducible electrode activity at noble metal electrodes for polar aliphatic compounds, PED was developed [4]. A complete description of PED and its application has been published [5].

7.3 FUNDAMENTALS OF PED

From the voltammetric data, it is apparent that surface oxide is reversibly formed and removed by the application of alternating positive and negative potentials, respectively. Oxide-free, or "clean," noble metal surfaces have an affinity to adsorb organic compounds. Upon changing to a more positive potential, electrocatalytic oxidation of adsorbed compounds, is promoted via anodic oxygen-transfer from H_2O by transient, intermediate products in the surface-oxide formation mechanism (i.e., AuOH and PtOH). Any fouling, which results as a consequence of the catalytic detection process or adsorbed compounds, is "cleaned" from the electrode surface by the application of positive and negative potential pulses subsequent to the detection process. The positive potential step results in the formation of stable surface oxide (i.e., AuO and PtO), which forms at the expense of any other surface adsorbed species. The negative potential step is applied to dissolve the surface oxide formed in the positive potential step and induces cathodic "cleaning" of the electrode surface.

When properly applied, the alternating positive and negative potential steps restore the native reactivity of the oxide-free metal surface and give PED its name.

The basis for all PED techniques is electrocatalytic oxidation of analytes at noble metal electrodes followed by a sequence of pulse potential cleaning steps. Three modes of anodic electrocatalytic detection can occur at noble metal electrodes.

- *Mode I: Direct detection at oxide-free surfaces.* At potentials less than ca. +200 mV (Figure 7.1), oxidation of the compound can occur with little or no concurrent formation of surface oxide. The surface-stabilized oxidation results in a product which may leave the diffusion layer readsorb for further oxidation or foul the electrode surface. The baseline signal originates primarily from double-layer charging, which decays quickly to a virtual zero value. All alcohol-containing compounds (e.g., carbohydrates) are detected by Mode I at Au electrodes in alkaline solutions and Pt electrodes in acidic solutions [5].
- *Mode II: Direct oxide-catalyzed detection.* In contrast to Mode I detections, Mode II detections require the concurrent formation of surface oxide. Hence, Mode II detections occur at potentials > ca. +150 mV (Figure 7.1). Oxidation of preadsorbed analyte is the primary contributor to the analytical signal; however, simultaneous catalytic oxidation of analyte in the diffusion layer is not excluded. The oxidation products may leave the diffusion layer or foul the electrode surface. Readsorption of analyte and detection products is attenuated by the surface oxide. The background signal, resulting from anodic formation of surface oxide, is large and has a deleterious effect on quantitation. Advanced waveforms (e.g., integrated pulsed amperometric detection [IPAD]) are required for Mode II detections [6]. Aliphatic amines and amino acids are detected by Mode II at Au and Pt electrodes in alkaline solutions. Numerous sulfur compounds are also detected by Mode II at Au and Pt electrodes in both alkaline and acidic solutions.
- *Mode III: Indirect detections at oxide surfaces.* Essential to Mode I and Mode II detections is the preadsorption of the analyte at oxide-free surfaces at negative potentials prior to electrocatalytic oxidation of the analyte itself. Species which adsorb strongly to the electrode surface and are electroinactive interfere with the oxide formation process. Preadsorbed species reduce the effective surface area of the electrode surface, and the analyte signal originates from suppression of oxide formation. Since the baseline signal results from anodic currents from surface oxide formation at a "clean electrode" surface, a negative peak results. Detection as a result of the suppression of surface oxide formation is known as Mode III detection. Since Mode III does not involve the electrocatalytic oxidation of the analyte, Mode III is only "indirectly" a PED technique. Sulfur-containing compounds and inorganic compounds have been detected by Mode III.

Electrocatalytic-based detection of various members within a class of compounds is controlled primarily by the dependence of the catalytic surface state on the electrode potential rather than by the redox potentials ($E°$) of the reactants. All compounds within a class will give very similar I–E plots, and as a consequence, voltammetric

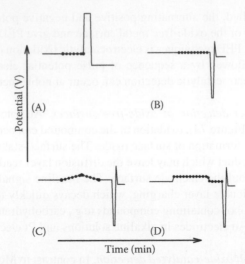

FIGURE 7.2 Generic potential versus time waveforms used in PED. (A) Three-potential pulse waveform, PAD; (B) quadruple pulse waveform, PAD; (C) cyclic scan IPAD or PVD waveform; and (D) square wave IPAD waveform.

resolution of complex mixtures is futile. Therefore, general selectivity is achieved via chromatographic separation prior to electrocatalytic detection. This conclusion does not preclude limited selectivity from control of detection parameters.

7.4 WAVEFORM DESIGN AND RECENT ADVANCES

The original multipotential waveform, known as pulsed amperometric detection (PAD) applied to Au and Pt electrodes, makes use of the three-step waveform illustrated in Figure 7.2A [6–8]. In this waveform, detection occurs at potential E_{det}, during the period t_{det}, with sampling of the Faradaic signal over the time period t_{int} after a delay of t_{del} to overcome capacitance currents. The output signal is either the average of the current (amperes or coulombs per second) or the integrated charge (coulombs) over t_{int}. A sampling period of 16.7 ms or a multiple thereof reduced 60 Hz noise. Thereafter, oxidative cleaning of the electrode occurs at potential E_{oxd} during the period t_{oxd}, followed by reductive reactivation at E_{red} during the period t_{red}. Typically, the total waveform cycle ($t_{total} = t_{det} + t_{oxd} + t_{red}$) is ca. 1 s with a frequency of ca. 1 Hz.

Although the simple three-step waveform gives the highest sensitivity, a four-step waveform known, or "quadruple-potential" waveform (Figure 7.2B), gives better long-term reproducibility [9]. In the quadruple pulse waveform, a large negative potential step E_{red} is applied for a brief period t_{red} after the detection step in order to reduce any partially solvated Au species back to metallic Au. In addition, this step invokes cathodic cleaning of the electrode surface. The negative potential step is following a brief positive potential step (E_{oxd}, t_{oxd}) to activate the electrode surface, which is followed by a potential pulse to effect adsorption and preconcentration of the analyte on the electrode surface (E_{ads}, t_{ads}).

Application of the PED waveforms in Figure 7.2A and B for amines and sulfur compounds requires a choice of E_{det}, which is concomitant with the formation of surface oxides to catalyze the oxidative mechanisms for these compounds. As a result, the large background signal interferes with the measurement of small analytical signals, is sensitive to variations in mobile-phase conditions, and leads to post-peak dipping. An alternate waveform for the detection of amines and sulfur compounds is shown in Figure 7.2C and D [10–12]. In these waveforms, the potential is scanned in a rapid cycle fashion between the values E_{dst} through E_{dmx} to E_{dnd} with a concurrent and continuous electronic integration of the amperometric signal. The values of E_{dst}, E_{dmx}, and E_{dnd} are chosen to correspond to potential regions before oxide formation occurs, maximal mass-transport-dependent signal from the analyte oxidation, and postcathodic dissolution of the surface oxide, respectively. Hence, in theory, the anodic formation of the oxide required to catalytically stimulate the detection process does not contribute to the total integration of the amperometric signal. In other words, analyte and oxide formation signal minus oxide dissolution signal gives the response only for the analyte, which appears to be independent of the oxide. Because the anodic reactions of aliphatic compounds are highly irreversible, there is no cathodic contribution to the total integration from reduction of the detection products during the scan from E_{dmx} to E_{dnd}.

If current is recorded versus the applied potential during the potential scan of E_{det}, three-dimensional (3D) data can be collected, which generate full voltammetric scans for every chromatographic time point. Greater electrochemical selectivity and analyte information can be obtained via the application of "on-the-fly" voltammetry. The generation of current versus potential plots at any time point facilitates the deconvolution of unresolved chromatographic peaks. Voltammetric detection in chromatography has been reviewed [13,14]. This PED technique is known as pulsed voltammetric detection (PVD) [15].

Figure 7.3 shows the surface plots of current versus potential versus time for the separation of lysine, galactosamine, serine, and sucrose using high-performance liquid chromatography (HPLC)-PVD [5]. This plots is background-corrected "on-the-fly." Note that the amine-containing compounds (i.e., a–c) have signal at high potentials; whereas, the compounds with hydroxyl groups of all these compounds are detectable in the oxide-free region of the electrode. At any potential a chromatogram can be extracted to afford you greater selectivity, and at any time point a voltammogram can be extracted to identify or characterize the analyte or peak. The results shown here illustrate the feasibility and doubtless importance of PVD to enhance selectivity and compound characterization, afford a limited degree of functional group identification with peak purity, and allow for quantitation of the compound of interest.

In 2005, Dionex (Sunnyvale, CA) introduced an electrochemical detector capable of collecting information-rich 3D data that allow for signal optimization and compound fingerprinting. The data are viewed in wireframe or isoamperometric displays to study reaction characteristics of analytes. Integration periods for pulsed amperometry are easily optimized postrun without having to reinject samples. When the detector is set to apply a voltage ramp instead of a stepped waveform, the resulting rapid-scan voltammogram gives a specific fingerprint for each compound, similar to

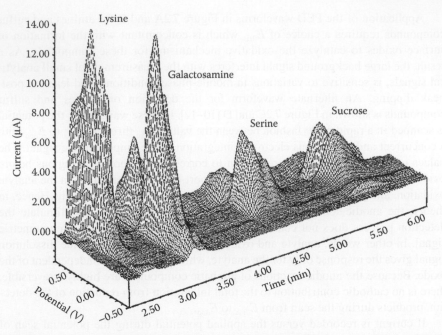

FIGURE 7.3 HPLC-PVD of various mixtures depicted in a surface plot for simple bioactive compounds.

the way a photodiode array provides a unique spectrum for each compound. This fingerprinting capability promises to find applications in compound identification, peak purity assessment, and waveform optimization.

7.5 MICROELECTRODE CONSIDERATIONS

Electrochemical detection, including PED, is ideally suited to the microseparation systems, because detection is based on a reaction at an electrode surface (i.e., $i = n \cdot F \cdot A \cdot D \cdot C^b / \delta$). In contrast, the response in optical detection methods is based upon Beer's law (i.e., $A = a \cdot b \cdot C^b$), which is dependent on the pathlength of the detector cell. In order to maintain the efficiency of microchromatographic and capillary-based separation systems, it is crucial that detection cell volumes also be miniaturized. Since the pathlength of a cell is directly proportional to cell volume, miniaturization is often at odds with optical detection techniques. The loss of response upon miniaturization is further exacerbated by compounds with poor optical detection properties. Present technology allows us to make electrodes very small, and consequently, detector cell volumes can be made similarly small with no decrease in sensitivity. The combination of electrochemical detection systems with microchromatographic and capillary-based separation techniques, which require detection cells of limited volume, offers increased mass sensitivity, higher chromatographic efficiencies, less solvent consumption, and in particular, the ability to analyze samples of limited quantity. PED affords these same advantages to virtually all polar aliphatic compounds, including a limited degree of enhanced selectivity, and most importantly, a self-cleaning working electrode that does not require daily polishing.

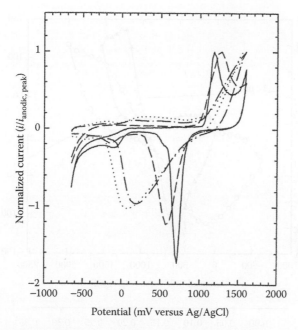

FIGURE 7.4   Voltammetric response of Au electrodes as a function of electrode size. Conditions: supporting electrolyte, 95% 100 mM phosphate buffer (pH 3)/95% CH$_3$CN; rotation rate for RDEs, 900 rpm; scan rate: 10,000 mV/s. Curves: (······) 5 mm RDE; (·····) 3 mm RDE; (-----) 1 mm RDE; and (-----) 50 µm fiber.

Microelectrodes with dimensions of 0.2–50 µm diameters are used extensively for microseparation techniques. In addition to dimensional compatibility [16], microelectrodes have properties of enhanced mass transfer to the electrode [17–19]; low iR-drop [20–22]; and low electrochemical cell time constant [23,24]. The latter two properties allow for fast response of the electrode to a potential pulse or to a potential ramp: scan rates up to 100,000 V/s have been feasible with 7 µm gold disks [23].

Figure 7.4 shows the results from cyclic voltammetry scans (at 10 V/s) with 5, 3, and 1 mm RDEs and a 50 µm × 1 mm Au fiber electrode. In order to present all results for widely differing electrode areas on the same scale, current was normalized by dividing it by the anodic peak current. It is evident that the start of anodic activity due to AuO formation is nearly at the same potential for all electrodes, but the shape of the peak is changed greatly with electrode size, showing widening and smearing effects, to the extent that a peak is not developed for the 5 and 3 mm electrodes. These effects are due to the high iR drop and large cell time constant existing for the large electrodes. Even more exaggerated effects are noted for the AuO reduction peaks. Similar comparisons were made by LaCourse and coworkers (unpublished data) when they exploited the advantages of microelectrodes to perform "in-cell" PVD scans in a microseparation system (see Figure 7.5).

Fast detection electrode response at microelectrodes means that initial Faradaic processes (e.g., capacitive charging) are predominant at a smaller percentage of the detection step time interval. More time is thus allowed for the desired Faradaic processes (analyte oxidation) to predominate. In addition, faster response allows for more

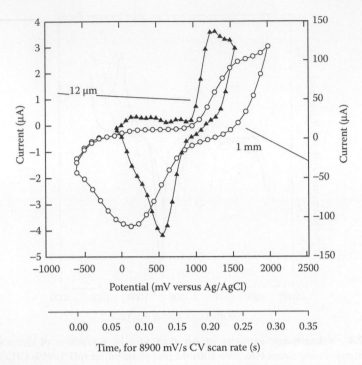

FIGURE 7.5 "In-cell" voltammograms generated using PVD following a microchromatographic separation.

flexibility in waveform design. Unfortunately, the ultimate rate at which PED can be performed is limited by the rate at which oxide can be formed and removed from the electrode surface. Roberts and Johnson [25,26] have speculated on the consequence of increasing the frequency of PED waveforms above the optimal value of 1 Hz prescribed for by HPLC-PED. They studied the kinetics of the oxide formation and dissolution processes at Au electrodes in 0.1 M NaOH and concluded that only 20 ms is required for generation of a monolayer of AuOH at E_{oxd} as the intermediate product in the anodic formation of surface oxide (AuO). Furthermore, only ca. 20 ms is required for the subsequent cathodic dissolution of the monolayer of AuOH at E_{red} of a standard three-potential pulse waveform. Hence, minimal values for t_{oxd} and t_{red} allow a significant increase in waveform frequency without sacrificing the desirable integration time (i.e., signal collection) of 200 ms. Nevertheless, some sacrifice of S/N value is observed when compared to use of larger values of t_{oxd} and t_{red} in normal bore applications of PED.

7.6 ADVANCED APPLICATIONS OF PED FOR CARBOHYDRATES

Oxide-free or Mode I detections are implemented with a three- or four-step potential–time waveform at a frequency of ca. 2–0.5 Hz, which is appropriate for HPLC or high-performance ion-exchange chromatography (HPIEC) applications to maintain chromatographic peak integrity. Optimization of waveform parameters in PAD is best accomplished using pulsed voltammetry (PV). Figure 7.6 shows the "back-

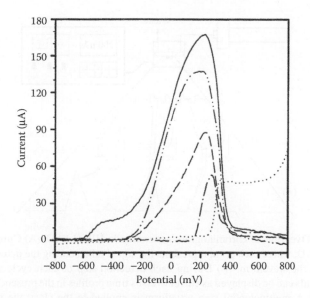

FIGURE 7.6 Pulsed voltammetric response of (———) 0.2 mM glucose, (·······) 0.2 mM fructose, (-----) 0.2 mM sucrose, and (-·-·-) 0.2 mM glycine at a Au electrode in 0.1 M NaOH, background subtracted. Background shown as reference (·····). Conditions: 900 rpm rotation speed.

ground-corrected" i–E_{det} response (positive scan direction) at a Au RDE for equimolar concentration of several carbohydrates in 0.1 M NaOH. As expected, the wave for the aldehyde group (glucose and maltose) begins at ca. −600 mV and quickly plateaus. The peak signal in the region of −200 to +400 mV corresponds to the oxidation of the aldehyde group (glucose and maltose) and the alcohol groups (all carbohydrates). The response for all carbohydrates is inhibited by the formation of surface oxide (not shown) at ca. +400 mV; hence, carbohydrates represent oxide-free detections. A maximum response is obtained at $E_{det} = +180$ mV for all carbohydrates, and the application of this value is universal and results in the highly sensitive detection of virtually all carbohydrates. The advantage of PV is clearly evident when one compares the PV responses in Figure 7.6 with the cyclic voltammetric plot of glucose in Figure 7.1. A detailed description of PV has been published [27]. It is important to note that the waveforms are dependent on the pH, electrode material, and experimental conditions.

Figure 7.7 summarizes the entire process of chromatographic peak formation in HPLC-PAD for a three-step potential waveform. The development of (Figure 7.7A) positive peaks for carbohydrates is illustrated by the (Figure 7.7B) chronoamperometric (i–t) response curves generated following the (Figure 7.7C) potential step from E_{red} to E_{det} in the PAD waveform. The residual current decays quickly, and the baseline signal in HPLC-PAD is minimal for $t_{del} > $ ca. 100 ms. The transient i–t response for the presence of the carbohydrate is dependent on its concentration in the electrochemical cell, and the peak shown is representative of the corresponding anodic signal expected in HPLC-PAD for the value of t_{del} indicated. Compound selectivity is achieved primarily via chromatographic separation.

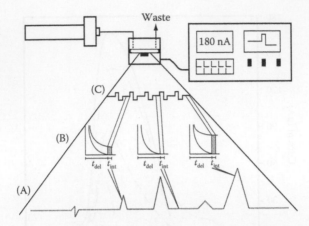

FIGURE 7.7 Overview of chromatogram generation in HPLC-PAD. (A) Chromatogram as generated in PAD, where each point represents the signal taken during the detection step of a single cycle of the waveform. (B) During each detection step of each cycle of the applied waveform, signals can be displayed as current versus time profiles in the presence and absence of analyte. (C) A multipotential step waveform is applied to the electrode that combines amperometric detection with pulsed potential cleaning.

7.6.1 "Fingerprinting" of Bioproducts

High-performance anion-exchange chromatography (HPAEC) is used for the separation of carbohydrates in alkaline media. The weakly acidic sugars are present as anions under these conditions and can be eluted according to the pK_as either isocratically or using an acetate gradient [28]. Virtually all carbohydrates are applicable to this technique. HPAEC followed by PAD has been applied to the direct detection of sugar alcohols, monosaccharides, oligosaccharides, aminoglycosides, amino alcohols, amino acids, and numerous thiocompounds. Reviews of PED/PAD at its applications have been published [5,29,30].

The direct (no derivatization), simple, and sensitive determination of carbohydrates by PED combined with the high selectivity of HPAEC enables this technique to be used for the chemical profiling or fingerprinting of closely related bioproducts such as tobaccos, peptones, or bacterial cell walls.

7.6.1.1 Tobacco Classification

While the tobacco for cigars and cigarettes may be of the same natural origin, the type of tobacco and processing of the tobacco may lead to reproducible changes in the relative amounts of natural constituents (e.g., carbohydrates). For instance, air-cured tobacco is predominantly used in cigars whereas flue-cured tobacco is the predominant tobacco type used (as part of a blend) in cigarettes. The air-drying process of cigars allows enzyme degradation of the plant carbohydrates resulting in tobacco containing a total carbohydrate content of around 3% or less. In addition, the cigar tobacco is put through a fermentation step which further destroys the carbohydrates naturally present in the tobacco leaves. Cigarettes are filled predominantly with flue-cured tobacco at rather high temperatures to dry the tobacco. The flue-curing

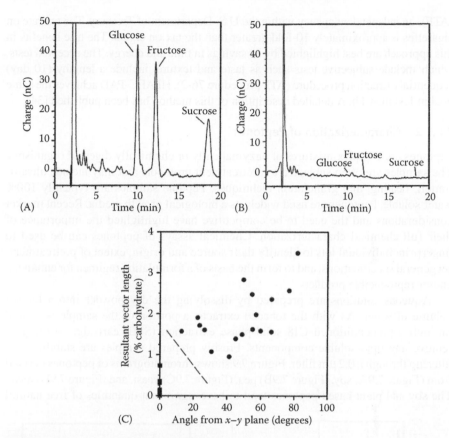

FIGURE 7.8 Chromatograms of extracts from (A) cigarette and (B) cigar tobacco. Vector plot (C) of all (•) cigarettes and (■) cigar tobaccos. An arbitrary (-----) is shown to delineate the two regions of tobacco classification.

process inactivates the enzymes, resulting in tobacco containing ca. 12.5% [31] to 3% [32] of free carbohydrates. This does not include the sugars manufacturers may add to enhance flavor [33]. Therefore, differentiation of tobacco products is possible based on a comparison of carbohydrate content.

Tobacco from cigarettes and cigars were removed from their wrappers and ground. Five hundred milligram of tobacco was shaken for 5 min in 100 mL of water. Two milliliter of the extract was passed through a preconditioned C18 Extract-Clean Catridge (Alltech, Deerfield, IL). The samples were used fresh and stabilized through a 0.2 μm filter. Figure 7.8 shows the elution profile of typical (Figure 7.8A) cigarette and (Figure 7.8B) cigar extracts. Note that the relative amounts of free carbohydrates in the cigarette and cigar extracts are dramatically different. This difference is attributable to the processing of the tobacco as described above. Figure 7.8C shows a vector plot of all (•) cigarettes and (■) cigars tobaccos. Note that cigarette tobacco is easily differentiated from cigar tobacco based on the free carbohydrate content of the tobacco.

Classification of tobacco is crucial to the assignment of the appropriate level of taxation, which is the responsibility of the Bureau of Alcohol, Tobacco, and Firearms

(ATF), an independent agency within the U.S. Department of Treasury. The tax rate on cigarettes is approximately 10-fold greater than the tax on cigars. The true benefits in this approach are best highlighted by its savings in time and energy. The accepted tests, which include subjective tests such as taste and texture, include a lengthy (10 day) sequential extraction procedure (ATF procedure 76-2). HPAEC-PAD achieves the same goal in less than 1 h. A detailed description of this method has been published [34].

7.6.1.2 Characterization of Peptones

Peptones are complex mixtures of enzymatically or chemically digested organisms. The samples are often subjected to pretreatment protocols that may include ultrafiltration, heating, and granulation techniques. Peptone samples are typically 100% water soluble. Peptones are used widely as a biological growth media. Recent market considerations and the need to be competitive have highlighted the importance of their full chemical characterization. Chemical assays of peptones can be used to fingerprint individual lots to identify their source and origin, extent of pretreatment, set general specifications, and to form the basis of a formulation regimen for enhanced and/or reproducible products.

Aqueous solutions are prepared by dissolving the dry powder into a known volume of water. As with the tobacco extracts, a portion of the sample is passed through a preconditioned C18 solid-phase extraction (SPE) cartridge in order to remove any lipid-soluble components. Freshly prepared samples are stabilized by filtering through a 0.2 μm filter. Figure 7.9 shows chromatograms of peptones derived from (Figure 7.9A) soy, (Figure 7.9B) pea, (Figure 7.9C) yeast, and (Figure 7.9D) meat. The soy and plant-based peptones tend to have significant quantities of free natural

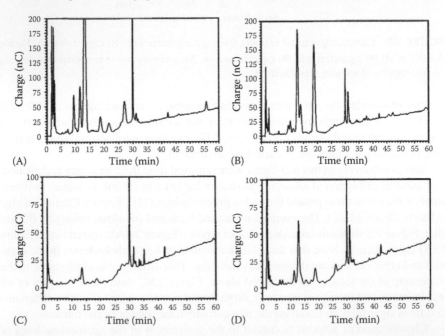

FIGURE 7.9 Chromatograms of (A) soy, (B) pea, (C) yeast, and (D) meat peptones using HPAEC-PAD.

sugars, whereas pea and meat peptones show very little quantities of free natural sugar. It is important to note the difference in scaling between the chromatograms. Also, many of the minor peaks may be attributable to amine- or sulfur-containing compounds. Of consequence is the reproducibility of the pattern. Using internal standard (IS), system suitability parameters for the separation and detection are typically less than 1% relative standard deviation (RSD). This work can form the foundation of a database for peptones.

7.6.1.3   Carbohydrate Analysis of Bacterial Polysaccharides

Capsular polysaccharides and lipopolysaccharides are key virulence factors in bacterial infections in humans [35]. Moreover, the difference between virulent and nonvirulent strains within one species is often associated with the structure and composition of the bacterial polysaccharide. Although carbohydrate composition is easily determined by HPAEC-PAD, the more challenging analytical problem is the determination of the absolute configuration of the carbohydrates. Although majority of the naturally occurring sugars is D-, examples of L-sugars in glycoproteins and glycolipids exist. Bacterial polysaccharides offer both a bigger challenge and an even greater reward; in that, both enantiomers of a particular monosaccharide occur with regular frequency, and occasionally within the same polysaccharide [36–38].

Recently in-line laser polarimetry (OR) has been combined with HPAEC-PAD to determine both the composition and the enantiomeric configuration of component sugars [39]. Figure 7.10A shows both the response of 500 ppm of L- and D-fucose by PAD and OR following retention using HPAEC. Note that PAD is nearly overwhelmed by the high concentration of fucose. PAD is linear in the low- to sub-ppm range. Also, PAD is completely insensitive to the absolute configuration of the carbohydrate. OR clearly shows the configuration of the fucose with a negative peak for the L configuration and a positive response for the D configuration. Sensitivity and peak direction are directly related to the magnitude and sign, respectively, of the specific rotations of each monosaccharide.

The monosaccharide mixtures are prepared by acid hydrolysis at 100°C for 2–8 h of ~1 mg of capsular polysaccharide. After hydrolysis, the samples are cooled and evaporated to dryness with nitrogen gas. The residue is dissolved in water and filtered. Figure 7.10B shows the chromatogram for the monosaccharide mix derived from *Streptococcus pneumoniae* type 12F. The monosaccharides were identified in the order of increasing retention as L-fucose, D-galactosamine, D-galactose, and D-glucose. Although much more sensitive, PAD is blind to the absolute configurations of the sugars. This approach has now been applied to the analysis of the capsular polysaccharides of several Gram-positive and Gram-negative pathogenic bacteria with success [39]. If the technical limitations of the poor sensitivity of the OR detector can be overcome, a new paradigm for carbotyping bacteria will evolve.

7.6.2   *In Vitro* Microdialysis for Carbohydrate Systems

The main achievements of microdialysis (MD) are the facilitation of continuous sampling, online sample cleanup, and the monitoring of small molecules of interest from complex matrices by employing a semipermeable membrane with a specific

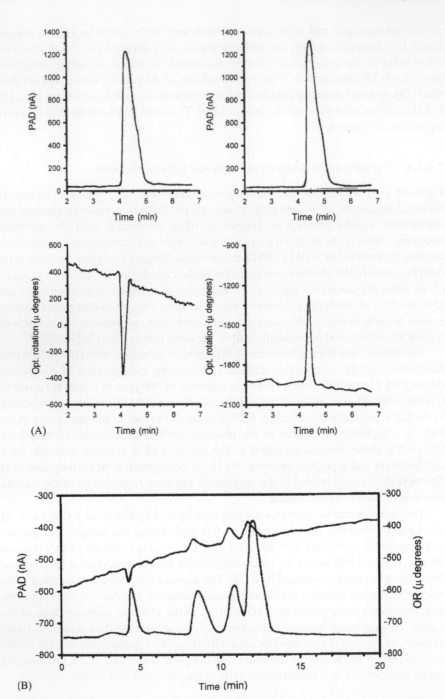

FIGURE 7.10 Chromatograms of 500 ppm (A) ʟ-fucose and ᴅ-fucose. Upper and lower plots are PAD and OR, respectively. HPAEC-OR-PAD of (B) *Streptococcus pneumoniae* type 12 F. Upper and lower traces are OR and PAD, respectively.

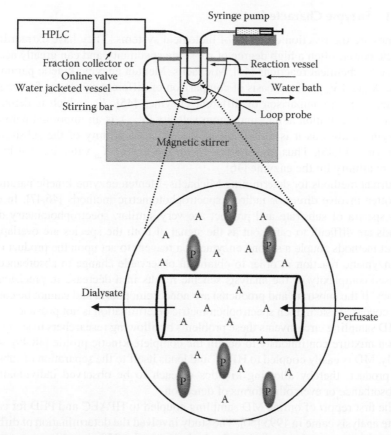

FIGURE 7.11 Schematic of a typical *in vitro* MD sampling system [41]. The inset shows the MD process where A is the analyte being sampled and P is a species (e.g., protein, particle, whole cell) that is restricted by the cutoff of the membrane and excluded from the dialysate.

molecular weight cutoff (MWCO) [40]. This technique, commonly used *in vivo*, has opened up research in the area of pharmacokinetics and biological responses to various stimuli [41,42]. As shown in the inset of Figure 7.11, small analytes (A) are able to pass through the membrane in preference to the large molecules (P), particles, and other entities. The small molecule fraction is collected by a flowing perfusion fluid, which exits as the dialyzate. The dialyzate solution is then available for analysis either by off-line (via fractionation) or online methods using various detection approaches, see Figure 7.11.

As analytes must be recovered across the membrane, the percent recovery can vary greatly depending on the analytes, matrix, temperature, mass transport, perfusion flow rate, membrane type, area, geometry, and thickness. Additionally, complex mixtures of analytes will often have very different recoveries, and the individual recovery value for each analyte may be difficult to obtain. Therefore, many of the *in vitro* MD studies that have been carried out are qualitative, as opposed to providing quantitative results [43].

7.6.2.1 Enzyme Characterization

Enzymes are the reaction catalysts of biological systems. They have extraordinary catalytic power, often a high degree of substrate specificity and can greatly accelerate specific chemical reactions [44]. Michaelis–Menten enzyme kinetic parameters such as K_m and V_{max} offer a basis of comparison of different substrates for the same enzyme, or even comparison of different enzymes [45]. K_m, which is defined as substrate concentration at half maximum velocity (V_{max}), is an important parameter in enzyme studies as it is a measure of the apparent affinity of the substrate for the enzyme [44,45]. Thus, the substrate with the lowest K_m value has the highest apparent affinity for the enzyme [46].

Current methods for determining Michaelis–Menten enzyme kinetic parameters most often involve direct or indirect spectrophotometric methods [46,47]. In cases where spectra of substrate and product are very similar, spectrophotometry direct methods are difficult to carry out as the signal of both the species are overlapping. Indirect methods couple a second enzyme or a reagent to act upon the product of the first enzymatic reaction in order to obtain an observable change in absorbance; the increased complexity of the analysis scheme results in a decrease in precision and accuracy. If the substrate and product(s) are nonchromophoric and cannot be coupled to a secondary system, then spectrophotometric determination is not possible.

MD sampling circumvents these problems by allowing researchers to sample the reaction mixture continuously to obtain the complete kinetic profile [48,49]. Additionally, MD is easily coupled to HPLC and lends itself to the separation of substrate from product, thereby allowing kinetics of each to be observed individually by UV absorbance or even other forms of detection.

The first report of online MD sampling coupled to HPAEC and PED for carbohydrate analysis came in 1995 [50]. The study involved the determination of different oligosaccharides produced during the hydrolysis of 0.25% ivory nut mannan by endomannanase from *Aspergillus niger*. This was a preliminary study and provided no quantitative results, but showed that a small-scale bioprocess (hydrolysis) could be monitored continuously for a period of 32 h.

In 1998, LaCourse and Zook used MD sampling with HPAEC-PED for quantitative monitoring of lactose hydrolysis in skim and whole milk into glucose and galactose via commercially sold Lactaid drops containing β-glucosidase/lactase [51]. MD was carried out using flat membranes, and deoxyglucose was used as an IS for quantitation. Limits of detection were found to be 1–3 ng and recovery data in milk matrix showed an average RSD of 5.0%. Standard first-order exponential decay curves were used to determine the observed rate constants, which were found to correlate well with the manufacturer's values. Additionally, Zook and LaCourse used MD sampling for characterizing the glucose oxidase reaction [40]. The substrate, glucose, and products, gluconolactone and H_2O_2, were monitored by HPAEC-PED. Both cellulose ester (CE) and regenerated cellulose membranes were used, and recoveries using CE were found to be 15–23% higher.

Also in 1998, Zook et al. qualitatively monitored the glucopolymers that were released from the digestion of amylopectin by the enzyme isoamylase [52]. This was also carried out using flat CE membranes with the block design. It was observed that limitations enforced by proper selection of the MWCO could increase sample selectivity for a subgroup of glucopolymers. In 2001, Nilsson et al. carried out similar work on performed enzymatic hydrolysis on potato and corn starches to determine information about molecular structure [53,54]. MD sampling using 30 kDa MWCO polysulfone hollow fiber membranes (concentric probe) was coupled to HPAEC-PED detection for online analysis.

These experiments were the first to show that short-chain fractions of debranched starch for amylopectin can be observed by MD-HPAEC-PED. The chain length distribution patterns of waxy, normal, and two types of high amylose maize were determined. The relative peak areas were plotted after correction with the known extraction efficiency (EF) values, but no IS was used for accurate quantitation. It can be seen that this method has potential applications in fingerprinting of different plant species [54,55].

LaCourse and Modi [56] extended their earlier work for the accurate determination of K_m values using PAD-active, chromophoric substrates, whose enzymatic parameters could be confirmed using standard spectrophotometric assays and literature values. MD was used to monitor the enzymatic hydrolysis of carbohydrate substrates by almond β-glucosidase [57,58] to obtain Michaelis–Menten enzyme constants. The enzyme catalyzes the hydrolysis of a broad array of substrates including the model nitrophenyl glycosides.

The reaction between β-glucosidase and these substrates generates 4-nitrophenol (4NP). The model system for studying enzyme kinetics involves coupling MD to HPLC for separation and simple UV absorbance detection. An IS, 1-(4-nitrophenyl) glycerol, was employed for increased accuracy in quantitation, and the method was used for the direct determination of kinetic constants for these compounds. Thus, for each reaction, the IS and enzyme alone were added to the system prior to the addition of substrate. When the 1-(4-nitrophenyl)glycerol concentration was constant, as observed by constant peak height, substrate was added to the reaction chamber, and the consecutive injections (8 min apart) of the dialysate were continually analyzed. As the reaction progressed, the [S] decreased and [P] increased. Figure 7.12A shows the decrease of 4NP β-D-glucoside and the increase of 4NP in the presence of 12.5 μg/mL β-glucosidase. However, most reactions were carried out using less enzyme (3.3 μg/mL) in order to observe slower reactions and obtain good initial velocity data. Enzyme stability was verified spectrophotometrically using 4NP β-D-glucoside for 8 h (>95% activity), which is longer than the combined time for equilibration and MD experiments. Note the appearance of [P] mirrors the disappearance of [S]; however, at higher substrate concentrations the response for the substrate was out of the linear range of the system and so product concentrations were used to determine enzyme kinetic parameters.

Hanes plots were constructed using substrate concentrations [S] and initial velocity for product formation V_0 in order to determine K_m for the experiments. The slope of the graph yields $1/V_{max}$, and the x-intercept yields $-K_m$. Figure 7.12B shows a Hanes plot of the hydrolysis of 4NP β-D-glucopyranoside from three separate sets of MD experiments run over a period of four weeks. For the three substrates under initial velocity (zero-order) conditions, the K_m and V_{max} values obtained in our laboratory by MD were found to correlate well with literature values and are summarized in Table 7.1. The analytical utility of *in vitro* MD was further demonstrated by its ability to monitor carbohydrate reactions in complex matrices. An application was shown for monitoring the glycoside salicin and its hydrolysis product saligenin in a

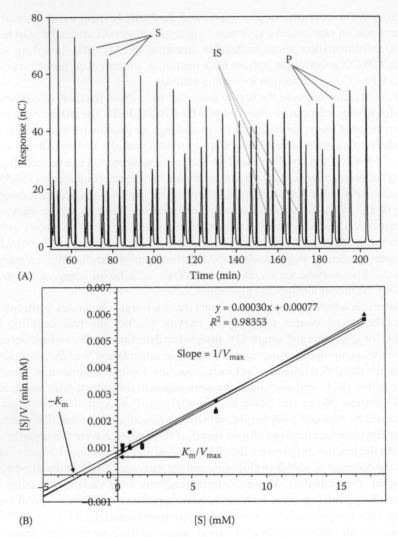

FIGURE 7.12 (A) Chromatogram of enzymatic solution, consecutive injections, duration 100 min. Initial concentrations 12.5 μg/mL of enzyme in 30 mL of 50 ppm IS. 100 ppm 4NP-β-D-glucoside added at $t = 0$. Growth of product, 4NP, and decrease in substrate as reaction progresses. (B) Hanes plot for the determination of K_m for the substrate 4NP-β-D-glucoside by MD-HPLC-UV.

commercially available willow bark product that is used for making tea [56]. Salicin is an analgesic that can be found in many dietary supplements and nutraceutical products, which are sold commercially [59]. The K_m value of salicin was reported for the first time by this new method of *in vitro* MD.

7.6.2.2 Monitoring Carbohydrate-Based Bioprocesses

In 1997, Palmisano et al. monitored *Escherichia coli* fermentations for the determination of glucose using an MD fiber for sampling and a biosensor for detection.

TABLE 7.1

Comparison of Experimental and Literature K_m Values

Substrates	Literature [58] K_m, mM	Experimentally Determined K_m, mM (MD)	K_m, mM (spec)
4NP-β-D-glucopyranoside	2.5	2.6 ± 0.5	2.7 ± 0.4
4NP-β-D-galactopyranoside	15.7	15.2 ± 0.5	15.0 ± 0.3
4NP-β-D-xylopyranoside	3.1	3.3 ± 0.5	2.7 ± 0.6
Salicin	—	20.8 ± 1	22.0 ± 0.3

Since then, improvements have been made, and a quantitative determination of multiple sugars has become possible [60]. In 1998, Palmqvist et al. published a study in which MD was used to follow the hydrolysis of spruce and hardwood spent sulfite liquor by *Saccharomyces cerevisiae* and bakers' yeast [61]. Online sampling of ethanol, glucose, glycerol, and acetic acid was carried out using an MD probe with an MWCO of 5 kDa. The dialysate was delivered to an injection valve for HPLC separation, and compounds were detected using a refractive index detector. Carbon dioxide levels were also monitored with an acoustic gas analyzer. The maximum production of ethanol from the lignocellulosic material is desired since it can be used as a liquid fuel.

In 2002, Rumbold et al. published a similar paper that describes the online monitoring of lignocellulosic hydrolysates by MD coupled to HPAEC followed by PAD and ESI-MS [62,63]. An MD probe equipped with a polysulfone membrane (30 kDa MWCO, 5 mm effective dialysis length) was used to sample enzymatic hydrolysates of dissolving pulp and of sugarcane bagasse. This particular application has potential in processes like prebleaching of paper, production of carbohydrates, etc. Four different enzymes were added to the dissolving pulp, and HPAEC-PAD analysis showed the growth of three products (DP1, DP2, and DP3) over time. After desalting the hydrolysate samples using a cation-exchange membrane, further analysis was carried out by ESI-MS. DP2 was seen to be a mixture of two saccharides that were unresolved by HPAEC-PAD. The identities of DP1 and DP3 were confirmed, by spiking with authentic standards, to be glucose and cellotriose, respectively. One of the DP2s was identified as cellobiose. However, the identity of the coeluting DP2 was not established by the standards available, but it was determined to have a mass to charge (*m/z*) ratio of 203. The same method applied to the enzymatic hydrolysis of sugarcane bagasse showed two pentoses, a hexose, four disaccharides, and two trisaccharides. A similar paper by Okatch et al. published in 2003 describes the use of micro-HPAEC coupled to ESI-MS to characterize the carbohydrates present in legume seeds after enzymatic hydrolysis with endomannanase [64]. The seeds contain galactomannans, which are important as gelling agents for food and cosmetics. Degrees of polymerization were determined between two different bean samples, but no additional structural or quantitative data were given.

In many of these cases quantitative data were not pursued, and the actual concentrations of analytes present in the bioprocess were undetermined. However, these studies show that the combination of chromatography and electrochemical detection with MD sampling is a powerful analysis package for the study of unknown carbohydrates in complex matrices. Additional work is required to make this an accepted analytical tool for quantitative determination of these types of analytes in dynamic enzymatic bioprocesses.

Quantitative determination of *in vitro* MD in dynamic enzymatic processes has been reported by Modi [65]. She used *in vitro* MD to further understand and verify the activity of the amylases in the laundry detergent. A larger branched carbohydrate polymer, amylopectin, was chosen for further qualitative studies. This maize starch is too large to be able to pass through the pores of the MD membrane. The background signal was not found to interfere with the detection of early eluting carbohydrates such as glucose.

When the active Tide detergent (1.5 mg/mL) is added to the amylopectin (2000 ppm), the result is the release of maltooligosaccharides that were monitored over time, see Figure 7.13. It was seen that the recovery of breakdown products of starch could be achieved by this new method. The chromatograms show that a variety of carbohydrates are obtained from the enzymatic hydrolysis of amylopectin, and that this method is amenable to carbohydrate monitoring in this industrial process. However, the time frame for analysis required to see the compounds of interest (1–9 h) was unreasonable as laundry detergent processes currently operate in 15 min or less. Improvements in the chromatography and the use of an internal standard (IS) improved the method to the point the comparisons could be made between different detergent enzymes using a standard test cloth (EMPA Material Science and Technology, Switzerland) in a typical wash cycle. Figure 7.14 shows the comparison of three detergent enzymes, and it is clearly evident that *Bacillus licheniformis* is the most efficient at hydrolyzing starch.

In a related application, Modi [65] applied *in vitro* microdyalsis to monitor the hydrolysis of starch in biofuel production. Corn, wheat, barley, rye, sorghum, and other starch sources are used in the production of ethanol as an alternate energy source to fossil fuels [66]. In these industrial bioprocesses, enzymatic treatment is required to break down the starch into fermentable sugars. This presents a constant source of glucose to the yeast that are used for ethanol production. Currently, most ethanol usage is in the form of an oxygenate or octane booster with blends of around 10% with gasoline. As the availability of natural resources such as oil and natural gas becomes more limited, ethanol fuel will become an even more important avenue for energy production. In the United States, fuel ethanol is made from corn, and 7% of US corn production is used for this purpose.

One glucoamylase enzyme that is useful in the saccharification of corn mash for ethanol production is sold by Novozymes under the name Spirizyme fuel. It is reported to have the highest activity and thermostability of any amylase available to date [67]. This allows ethanol manufacturers to use less of the precious enzymes in their processes. Spirizyme fuel hydrolyzes the 1,4- and 1,6-α linkages in liquefied starch substrates. During hydrolysis, the enzyme acts to remove glucose units from the nonreducing end of the substrate.

The availability and concentration of carbohydrates in enzymatic fermentations can greatly affect product yield. Current methods of sugar analysis required removal

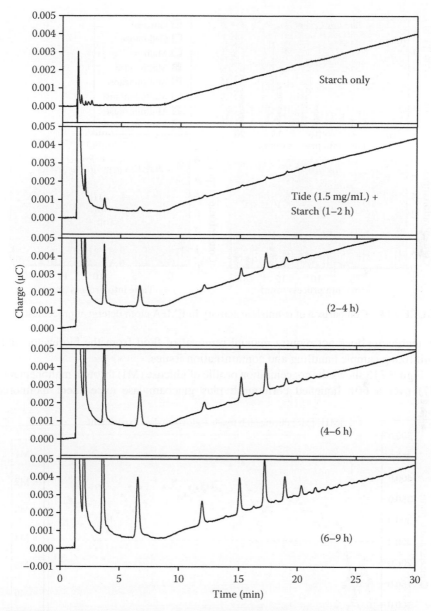

FIGURE 7.13 MD-HPAEC-PED monitoring of amylopectin (2000 ppm) digestion by Tide laundry detergent enzymes using a 3 cm loop probe and perfusion flow rate of 5 μL/min.

of aliquots of fluid from the bioreactor, addition of acid to quench the enzymatic reaction, and then centrifuging, filtering, and diluting each sample prior to the analysis. MD, on the other hand, provides enzyme-free (quenched) sample removed continuously and automatically by a flowing perfusion fluid. The obtained sample typically requires little or no sample preparation prior to analysis. Importantly,

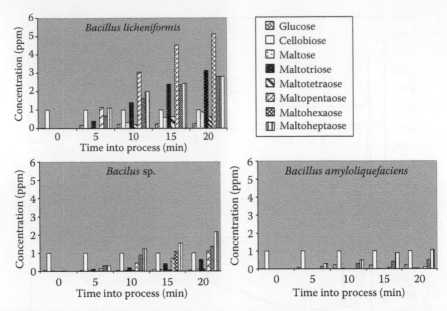

FIGURE 7.14 Comparison of α-amylase activity in EMPA cloth detergent processes.

this technique does not require manual removal of fluid from the bioreactor, thus minimizing sample handling and contamination issues.

Figure 7.15 shows the concentration profile of glucose (M1) through maltoheptaose (M7) over a 60h liquefied corn mash plus glucoamylase experiment monitored

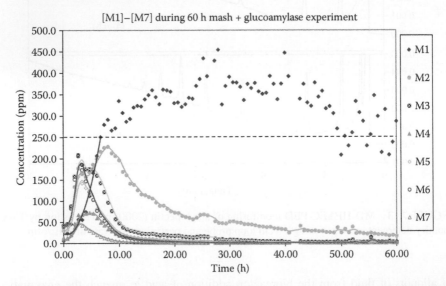

FIGURE 7.15 (See color insert following page 216.) Concentration of M1–M7 over a 60h liquefied corn mash plus glucoamylase experiment monitored by online MD-HPAEC-PED. Data are fit using a moving average of period = 2h. The limit of linearity for glucose is 250 ppm, points outside linear range not included in trend line.

by online MD-HPAEC-PED [65]. Data were fit using a moving average of period = 2 h. The limit of linearity for glucose i was 250 ppm, which points outside linear range not included in trend line. This work clearly shows that the *in vitro* MD as a sampling device coupled to HPAEC-PED for quantitating carbohydrates levels in a complex bioprocess can be performed over an extended length of time.

7.6.3 TOXICOLOGICAL APPLICATIONS: GLUCURONIDES

Alcohol is the most commonly abused substance presented in forensic cases. It is either found in postmortem samples due to alcohol consumption prior to death or from postmortem decomposition [68]. It is very important to differentiate between these two circumstances, by monitoring biomarkers of alcohol consumption. Furthermore, there are several clinical applications to monitor alcohol abuse such as drunk-driving cases and prevention of fetal alcohol syndrome [69].

Ethyl glucuronide (EtG) is a nonvolatile, water-soluble metabolite of ethanol. This highly specific and sensitive biomarker of alcohol consumption has been reviewed elsewhere [70], and its widespread adoption is facilitated with the development of simpler and less expensive methods of analysis. Recently, PED following reversed-phase chromatography has been reported for the simple, sensitive, and direct detection of EtG in postmortem urine samples [69,70]. The LC-PED method isocratically separated EtG and methyl glucuronide (MetG), which serves as an IS, on a C18-bonded phase using a mobile phase consisting of 1% acetic acid/acetonitrile (ACN) (98/2; v/v). Postcolumn addition of NaOH allows for the detection of all glucuronides using PED at a gold working electrode. EtG was found to have a limit of detection of 0.03 μg/mL (7 pmol; 50 μL injection volume) and repeatability at the limit of quantitation of 1.7% RSD. Essential to the method was the development of an SPE procedure using an aminopropyl phase, which was used to remove interferents in urine samples prior to their analysis. Compound recovery following SPE was approximately 50 ± 2%. The forensic utility of this method was further validated by the analysis of 29 postmortem urine specimens, whose results agreed strongly with certified determinations.

More recently, an improved method for EtG was introduced by the LaCourse and Kaushik [71] group in human urine. PV studies revealed that acetonitrile suppressed the signal of species (e.g., glucuronides) that are weakly adsorbed to the gold electrode. Hence, *t*-butanol proved to be a better organic modifier as it does not suppress the signal of the glucuronide at the electrode surface. With improved detection and higher sensitivity for analytes and interferents, the need for a wash step of high solvent strength needed to be incorporated to eliminate the carry over of the matrix components between chromatographic runs. Propyl glucuronide (PG) proved to be a better IS than MetG, which was used in the original method. Figure 7.16 shows the separation and detection of EtG and PG as (Figure 7.16A) standards in water and (Figure 7.16B) extracted from a urine sample. Furthermore, they were able to achieve improved detection limits. SPE recoveries for EtG from Surine have been improved by modifications to the pretreatment of the sample and a decrease in the load volume onto the cartridges. The SPE method used was highly reproducible and required a mere 0.2 mL of sample. This improved method showed that HPLC-PED is a sensitive, selective, and direct method for the determination of EtG, a biomarker of alcohol consumption. HPLC-PED is available to the analyst as an alternative to LC–MS at a fraction of the cost.

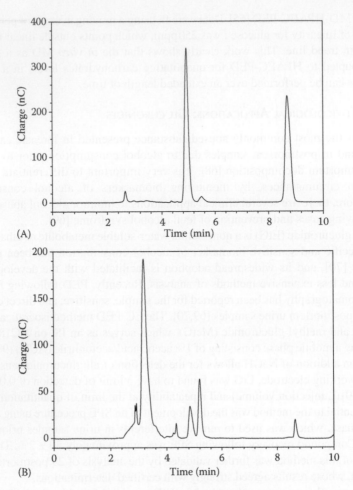

FIGURE 7.16   (A) Chromatogram of EtG and PG at a 10 μg/mL concentration. Conditions: Column, Acclaim Polar-Advantage C16 120 Å Dionex column (250 mm × 4.6 mm, 5 μm), C18 Vydac (7.5 mm × 4.6 mm, 5 μm) with guard; mobile phase: 1% acetic acid/0.5% t-butanol (99.5:0.5, v/v); flow rate, 1.0 mL/min; postcolumn reagent, 600 mM NaOH at 0.5 mL/min. (B) Chromatogram of EtG and PG extracted from urine. Conditions: Analytes were extracted from urine using SPE with aminopropyl cartridges 3 cc, 500 mg sorbent bed, Waters Corporation, Milford, MA. The analytes were diluted into the linear range of the assay for quantitation.

This system has great potential for use in the toxicological and forensic realm, allowing one to sensitively and selectively monitor glucuronides of various compounds in urine matrix. It has the potential to enable toxicologists to screen for the presence of a variety of substances even after the parent compound is eliminated from the body. Figure 7.17 shows a suite of seven different drug glucuronides in a single chromatographic run [72]. The compounds were separated using a gradient and detected with the addition of postcolumn base. The gradient began with 1% ACN for the first 10 min and then gradually increased to 10% ACN from 10 to 30 min; held constant at 10% for

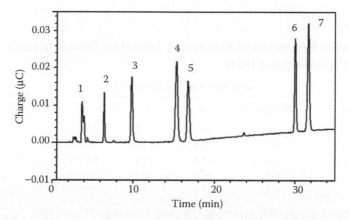

FIGURE 7.17 This is a representative PED chromatogram of a reversed-phase separation of seven glucuronides. The peaks have been identified as (1) MetG (2) EtG (3) M3G (4) acetaminophen glucuronide (5) M3G (6) codeine-6-glucuronide (7) phenyl glucuronide. Conditions: column, Denali C18 100 Å Vydac column (4.6 mm × 250 mm) with guard; mobile phase: 1% acetic acid/ACN gradient; flow rate, 1.0 mL/min; postcolumn reagent, 600 mM NaOH at 0.5 mL/min.

5 min followed by equilibration to 1% ACN from 35 to 40 min. This enabled a baseline separation of all seven analytes. The detector showed good analytical sensitivity toward these analytes, as shown by their limit of detection (LOD) values. The LOD values for MetG, EtG, morphine-3-glucuronide (M3G), morphine-6-glucuronide (M6G), phenyl glucuronide, acetaminophen-glucuronide, and codeine-6-glucuronide were 10, 10, 30, 40, 50, 100, and 100 ng/mL, respectively. There is a great potential for this detection method to be used as a screening tool for glucuronides in complex biological matrices.

7.7 MICROELECTRODE APPLICATIONS IN PED

As discussed earlier in this chapter, the application of an effective PED waveform requires a finite amount of time. The shortest waveform must accommodate both the detection step and pulsed potential cleaning steps, which are dependent upon the rates of oxide formation and dissolution, without affecting peak integrity. Microelectrode applications challenge PED. Microchromatographic and capillary column separations for carbohydrate have been limited or nonexistent due to lack of commercial column technology to perform the separation. The feasibility of PED following microseparations has been proven with the determination of thiocompounds, which can be readily separated using reversed-phase systems.

LaCourse and Owens [73,74] were the first to apply PED following microbore (i.e., 1 mm i.d. [inner diameter] column) and capillary (i.e., 180 μm i.d. column) LC to the determination of thiocompounds (e.g., methionine, cystamine, homocysteine, and coenzyme A and derivatives) under typical reversed-phase conditions. Interestingly, PED enables the direct determination of thio redox couples (i.e., –SH/–S–S–)

TABLE 7.2
Quantitation Parameters of Biologically Important Thiocompounds at a Au Electrode Using IPAD

| Compounds | LOD[a] (pmol) | Linear Range $nC = a \cdot (pmol) + b$ | | | Repeatability %RSD (pmol, n) |
		a	b	R^2	
Cysteine	0.2	56.6	2.24	0.9997	4.6 (2.5, 6)
Homocysteine	0.5	33.8	−13.5	0.9984	3.6 (5, 6)
Methionine	2.0	15.6	−19.4	0.9999	2.5 (20, 6)
GSH	0.5	30.4	−0.14	0.9989	2.1 (5, 6)
GSSG	0.5	24.8	7.99	0.9956	2.2 (5, 6)

Source: From LaCourse, W.R. and Owens, G.S., *Anal. Chim. Acta*, 307, 301, 1995.

[a] Calculated at S/N = 3 from injections within an S/N = 5.

and numerous other sulfur moieties at a single Au electrode without derivatization. Table 7.2 lists the analytical figures of merit for the bioactive compounds as in Figure 7.18. Mass limits of detection were 0.2–0.5 pmol injected except for methionine, which was 2 pmol.

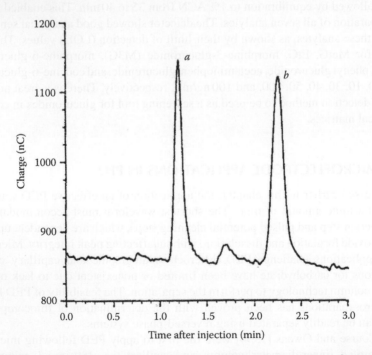

FIGURE 7.18　CLC-PAD separation of the (a) reduced and (b) oxidized forms of DTE at 6 pmol (100 μM) each. (Reprinted from LaCourse, W.R. and Owens, G.S., *Anal. Chim. Acta*, 307, 301, 1995. With permission.)

The only work on PED following capillary liquid chromatography (CLC) was performed by LaCourse and Owens [75]. Dithioerythritol (DTE) in its reduced and oxidized forms was separated on a 180 μm i.d. column from a 60 nL injection and detected using IPAD, see Figure 7.18. The LODs for DTE reduced and oxidized were determined to be 0.3 and 0.1 pmol, respectively. The decreased dispersion in the capillary chromatographic system is reflected in the lower LODs for these compounds as compared to microbore chromatography. The high selectivity of PED for thiocompounds under mildly acidic conditions reduces sample preparation and produces simpler chromatograms of complex mixtures.

7.7.1 Electrophoretic Separations

The lack of electrochemical selectivity in PED begs the need for highly efficient separations, and capillary electrophoresis (CE) is a powerful technique for the separation of compounds of interest in complex sample matrices. As a consequence of the separation mechanism in a capillary tube (i.e., less than 100 μm i.d.), CE produces very narrow electrophoretic bands and high separation efficiencies.

It is well established that the detection of carbohydrates by PAD requires highly alkaline conditions, which may pose a challenge to electrophoretic separations due to its high conductivity. Numerous reviews [76–79] that focus on the detection of carbohydrates in CE, all of which have a section devoted to PED, have been published.

The first application of CE-PAD was for the glucose in blood [80], see Figure 7.19. The level of glucose in blood was determined to be 4.25 ± 0.13 mM,

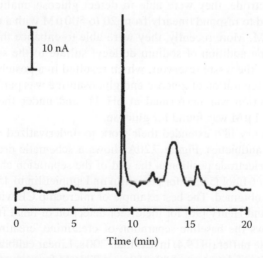

10 nA

FIGURE 7.19 Electropherogram of human blood. Peak at ca. 9 min corresponds to 85 μM glucose. (Reprinted from O'Shea, T.J., Lunte, S.M., and LaCourse, W.R., *Anal. Chem.*, 65, 948, 1993. With permission.)

which agrees well with that reported in the literature [81]. The PAD response for glucose was determined to be linear over the range of 10–1000 μM, and the mass detection limit was determined to be ca. 20 fmol. Lu and Cassidy [82] produced similar results for several wood sugars with separation efficiencies in the range of 100,000–200,000 theoretical plates. Ruttinger and Drager [83] were able to separate and detect a series of polyhydroxyalkaloids known as calystegines in plant extracts. The high resolving power of CE allowed for the separation of calystegines with the same number of hydroxyl groups.

CE-PAD is ideally suited to the separation and detection of charged carbohydrates. LaCourse and Owens [84] have extended the application of IPAD to the direct detection of many polar aliphatic compounds over a wide range of pH conditions following CE. They found that the detection of unsubstituted carbohydrates requires highly alkaline conditions, whereas amine-containing compounds (e.g., glycopeptides, peptides, and amino acids) and thiocompounds can best be detected at mildly alkaline (i.e., pH 9.0) and mildly acidic (i.e., pH 5.5) conditions, respectively. The analytical figures of merit under optimal conditions for glucose, glucosamine, and cysteine are shown in Table 7.3. Mass limits of detection are typically 10 fmol or less. Lunte and coworkers [85,86] focused efforts on the characterization of glycopeptides derived from recombinant proteins. They determined that CE-PAD was a useful alternative to UV detection in the CE analysis of tryptic digests.

7.7.2 Electrophoretic Microchip Separations

The first report of PAD on an electrophoretic chip was presented by Fanguy and Henry [87]. Using a hybrid poly-(dimethylsiloxane) or glass device coupled with a Pt working electrode, they were able to detect glucose, maltose, and xylose. Glucose was found to respond linearly from 20 to 500 μM with a measured detection limit of 20 μM. More recently, they were able to enhance the determination of glucose with the addition of sodium dodecyl sulfate to the separation buffer and a higher pH at the waste reservoir, which resulted in a postchannel pH modification [88]. The separation of glucose and glucosamine was performed at pH 7.1 whereas the detection was performed at pH 11, and under these conditions a detection limit of 1 μM was found for glucose.

Garcia and Henry [89] extended their work to underivatized amino acids and sulfur-containing antibiotics. Figure 7.20A shows a schematic drawing of the CE chip showing the electrode position at the end of the separation channel. Detection limits ranged from 6 fmol (5 μM) for penicillin and ampicillin to 455 fmol (350 μM) for histidine were obtained. The best example of microchip CE with PED was produced by Garcia and Henry [90] for the direct detection of renal function markers. Figure 7.20B shows the baseline separation of creatinine, creatine, and uric acid using 30 mM borate buffer (pH 9.4) in less than 200 s. Linear calibration curves were obtained with limits of detection of 80, 250, and 270 μM for each of the compounds, respectively. The analysis of a real urine sample was presented with validation of creatinine concentration using a clinical assay kit.

TABLE 7.3
Quantitative Parameters of Polar Aliphatic Compounds at a Au Microelectrode in Various Operating Buffers

			Linear Range PAD: $\mu A = a\ (\mu M) + b$ IPAD: $nC = a\ (\mu M) + b$				
Buffer	Compound	Waveform	a	b	R^2	Deviation from Linearity (μM)	LOD[a] (fmol, pg)
NaOH, 16mM pH 12	Glucose	PAD	7.97×10^{-5}	4.28×10^{-4}	0.9998	500	10, 2
		IPAD	2.13×10^{-2}	5.14×10^{-1}	0.9970	500	40, 8
	Cysteine	PAD	—	—	—	—	—
		IPAD	3.41×10^{-1}	1.79×10^{-1}	0.9956	1000	110, 10
Borate, 20mM pH 9.3	Glucosamine	PAD	1.04×10^{-4}	2.92×10^{-4}	0.9988	500	4, 1
		IPAD	3.46×10^{-2}	7.14×10^{-1}	0.9958	1000	6, 1
	Cysteine	PAD	—	—	—	—	160, 20
		IPAD	—	—	—	—	160, 20
Acetate, 10mM pH 5.5	Glucosamine	PAD	1.44×10^{-4}	5.24×10^{-3}	0.9936	500	90, 20
		IPAD	6.44×10^{-2}	1.74×10^{-1}	0.9978	500	30, 7
	Cysteine	PAD	3.90×10^{-4}	4.31×10^{-3}	0.9988	500	20, 2
		IPAD	2.02×10^{-1}	-2.01	0.9851	500	10, 1

Source: From LaCourse, W.R. and Owens, G.S., *Electrophoresis*, 17, 310, 1996.

[a] Calculated at S/N = 3 from injections within an S/N = 5.

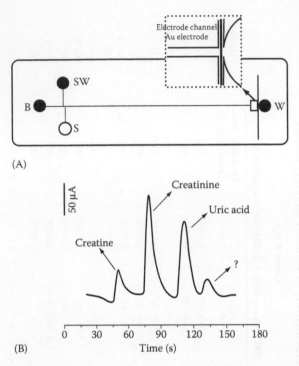

(A)

(B)   Time (s)

FIGURE 7.20 (A) Schematic drawing of CE chip showing the electrode position at the end of the separation channel. (B) Baseline separation of creatinine, creatine, and uric acid using 30 mM borate buffer (pH 9.4) in less than 200 s. (Reprinted from Garcia, C.D. and Henry, C.S., *Anal. Chem.*, 75, 4778, 2003, With permission.)

7.8 CONCLUSIONS

Over the past three decades, PED has matured as an electroanalytical technique. PED excels for the direct, sensitive, and reproducible detection of carbohydrates. Its maturity is reflected in the use of PED in advanced applications such as the "fingerprinting" of bioproducts, enzyme characterization, and bioprocessing. These front-end applications strongly support the rugged and reliable nature of PED following a chromatographic separation, which includes virtually all aqueous-based separations (e.g., ion-exchange, ion-pairing, ion-exclusion, and reversed-phase chromatography). PED is now being applied to forensic and toxicological problems of interest, which include the determination of drug glucuronides in physiological fluids.

PED offers many advantages over alternate detection schemes for LC. Because electrochemical detection relies on reaction at the electrode surface, detector cells can be miniaturized without sacrificing sensitivity. This advantage makes them especially suited for microbore and capillary techniques, *vide supra*. Pulsed potential cleaning eliminates the need for daily polishing of the electrode which renders PED more convenient experimentally than dc amperometry. The sensitivity and selectivity (e.g., sulfur-based compounds under typical reversed-phase conditions) of PED for specific functional groups on the analyte simplifies the analysis of complex (e.g., biological) matrices.

Significant future developments in PED will occur for the detection of amine and sulfur compounds with an emphasis on advanced waveforms (e.g., IPAD, PVD, and 3D amperometry). Applications directed toward peptides, proteins, and macromolecules are anticipated. The impressive accomplishments in HPLC/IHPIEC-PED, thus far, have only accentuated the need for novel chromatographic separations for polar aliphatic compounds, a deeper understanding of PED and its limits, and application of this technology to real-world bioanalytical problems of critical significance.

REFERENCES

1. R.N. Adams, *Electrochemistry at Solid Electrodes*. Marcel Dekker, New York, 1969.
2. P.T. Kissinger, *Laboratory Techniques in Electroanalytical Chemistry*, P.T. Kissinger and W.R. Heineman (Eds.). Marcel Dekker, New York, 1984.
3. S. Gilman, *Electroanalytical Chemistry*, Vol. 2A, J. Bard (Ed.). Marcel Dekker, New York, 1967.
4. S. Hughes, P.L. Meschi, and D.C. Johnson, *Anal. Chim. Acta, 132*, 1981, 11.
5. W.R. LaCourse, *Pulsed Electrochemical Detection in High-Performance Liquid Chromatography*. Wiley Interscience, New York, 1997.
6. D.C. Johnson and W.R. LaCourse, *Anal. Chem., 62*, 1990, 589A–597A.
7. R.W. Andrews and R.M. King, *Anal. Chem., 62*, 1990, 2130.
8. W.R. LaCourse and D.C. Johnson, *Carbohydr. Res., 215*, 1991, 159.
9. R.D. Rocklin, A.P. Clarke, and M. Weitzhandler, *Anal. Chem., 70*, 1998, 1496.
10. G.G. Neuburger and D.C. Johnson, *Anal. Chem., 60*, 1988, 2288.
11. L.E. Welch, W.R. LaCourse, D.A. Mead Jr., D.C. Johnson, and T. Hu, *Anal. Chem., 61*, 1989, 555.
12. A.P. Clarke, P. Jandik, R.D. Rocklin, Y. Liu, and N. Avdalovic, *Anal. Chem., 71*, 1999, 2774–2781.
13. P. Jandik, P.R. Haddad, and P.E. Sturrock, *CRC Crit. Rev. Anal. Chem., 20*, 1988, 1–74.
14. P.R. Haddad and P. Jandik, *Ion Chromatography*, J.G. Tarter (Ed.). Marcel Dekker, New York, 1987.
15. R.T. Kennedy and J.W. Jorgenson, *Anal. Chem., 61*, 1989, 436.
16. R.J. Forster, *Chem. Soc. Rev., 23(4)*, 1994, 289–297.
17. M. Rievaj, S. Mesaros, and D. Bustin, *Coll. Czech. Chem. Commun., 58*, 1993, 2918–2923.
18. M. Rievaj and D. Bustin, *Analyst, 117*, 1992, 1471.
19. J. Wang, E. Sucman, and B. Tian, *Anal. Chim. Acta, 286*, 1994, 189–195.
20. J.W. Bixler, A.M. Bond, P.A. Lay, W. Thormann, P. Van den Bosch, M. Fleischmann, and S.B. Pons, *Anal. Chim. Acta, 187*, 1986, 67–77.
21. S.B. Khoo, H. Gunasingham, K.P. Ang, and B.T. Tay, *J. Electroanal. Chem., 216*, 1987, 115–126.
22. D.L. Luscombe, A.M. Bond, D.E. Davey, and J.W. Bixler, *Anal. Chem., 62*, 1990, 1709–1712.
23. W.F. Strohpen, D.K. Smith, and D.H. Evans, *Anal. Chem., 62*, 1990, 1709–1712.
24. J.O. Howell and R.M. Wightman, *Anal. Chem., 56*, 1984, 524–529.
25. R.E. Roberts and D.C. Johnson, *Electroanalysis, 4*, 1992, 74.
26. R.E. Roberts and D.C. Johnson, *Electroanalysis, 4*, 1992, 269.
27. W.R. LaCourse and D.C. Johnson, *J. Electroanal. Chem., 65*, 1993, 50–55.
28. T.J. Paskach, H.P. Lieker, P.J. Reilly, and K. Thielecke, *Carbohydr. Res., 215*, 1991, 1.
29. D.C. Johnson and W.R. LaCourse, *Electroanalysis, 4*, 1992, 367–380.
30. W.R. LaCourse, *Analysis, 21*, 1993, 181–195.
31. H. Jacin, J.M. Slanski, and R.J. Moshy, *Tob. Sci., 12*, 1968, 136–138.
32. L.A. Elson, T.E. Betts, and R.D. Passey, *Int. J. Cancer, 9*, 1972, 666–675.

33. R.E. Going, S.C. Hsu, R.L. Pollack, and L.D. Haugh, *J. Am. Dent. Assoc.*, *100*, 1980, 27–33.
34. C.M. Zook, P.M. Patel, W.R. LaCourse, and S. Ralapati, *J. Agric. Food Chem.*, *44*, 1996, 1773–1779.
35. J.P. Kamerling, Pneumococcal polysaccharides: A chemical view, in: *Streptococcus pneumoniae*, A. Tomasz (Ed.). Mary Ann Liebert, Larchmont, NY, 2000, pp. 81–114.
36. D. Horton, G. Rodemeyer, and R. Rodemeyer, *Carbohydr. Res.*, *56*, 1977, 129–138.
37. B.A. Dmitriev, Y.A. Knirel, N.A. Kochetkov, N.K. Stanislavsky, and G.M. Mashilova, *Eur. J. Biochem.*, *106*, 1980, 643–651.
38. H. Baumann, A.O. Tzianabos, J.R. Brisson, D.L. Kasper, and H.J. Jennings, *Biochemistry*, *31*, 1992, 4081–4089.
39. C.J.M. Stroop, C.A. Bush, R.L. Marple, and W.R. LaCourse, *Anal. Biochem.*, *303*, 2002, 176–185.
40. C.M. Zook and W.R. LaCourse, *Anal. Chem.*, *70*, 1998, 801.
41. D.K. Hansen, M.I. Davies, S.M. Lunte, and C.E. Lunte, *J. Pharm. Sci.*, *88*, 1999, 14.
42. P.F. Morrison, P.M. Bungay, J.K. Hsaio, I.N. Mefford, K.H. Dykstra, and R.L. Dedrick, *Microdialysis in the Neurosciences*, T.E. Robinson and J.B. Justice Jr. (Eds.). Elsevier, New York, 1991, p. 47.
43. K.L. Snyder, A.Y. Nathan, and J.A. Stenken, *Analyst*, *126*, 2001, 1261.
44. A.G. Marangoni, *Enzyme Kinetics: A Modern Approach*. Wiley-Interscience, Hoboken, NJ, 2003.
45. V. Leskovac, *Comprehensive Enzyme Kinetics*. Kluwer Academic/Plenum Publishers, New York, 2003.
46. K. Tipton, *Enzyme Assays: A Practical Approach*, R. Eisenthal and M.J. Danson (Eds.). IRL Press, Oxford, UK, 1995, p. 1.
47. C.F. Mandenius, L. Bulow, and B. Danielsson, *Acta Chem. Scand. B*, *37*, 1983, 739.
48. J.A. Stenken, D.L. Puckett, S.M. Lunte, and C.E. Lunte, *J. Pharm. Biomed. Anal.*, *8*, 1990, 85.
49. N. Torto and L. Gordon, *Trends Anal. Chem.*, *18*, 1999, 252.
50. N. Torto, T. Buttler, L. Gorton, G. Marko-Varga, H. Stalbrand, and F. Tjerneld, *Anal. Chim. Acta*, *313*, 1995, 15–24.
51. C.M. Zook and W.R. LaCourse, *Curr. Sep.*, *17*, 1998, 41.
52. C.M. Zook, PhD dissertation, University of Maryland, Baltimore County, MD, 1998. Unpublished thesis.
53. N. Torto, L. Gorton, G. Marko-Varga, J. Emneus, C. Akerberg, G. Zacchi, and T. Laurell, *Biotechnol. Bioeng.*, *56*, 1997, 546–554.
54. G.S. Nilsson, S. Richardson, A. Huber, N. Torto, T. Laurell, and L. Gorton, *Carbohydr. Polym.*, *46*, 2001, 59–68.
55. H. Okatch and N. Torto, *African J. Biotechnol.* (online), *2*, 2003, 636–644.
56. S.W. Modi and W.R. LaCourse, *J. Chromatogr. A*, *1118*, 2006, 125–133.
57. J. Schwartz, J. Sloan, and Y.C. Lee, *Arch. Biochem. Biophys.*, *3*, 1970, 122.
58. M.P. Dale, H.E. Ensley, K. Kern, K.A.R. Sastry, and L.D. Byers, *Biochemistry*, *24*, 1985, 3530.
59. L.S. Li, W.D. Huang, Q. He, and S. Ye, *Se Pu-Chinese Journal of Chromatography*, *19*, 2001, 446–448.
60. F. Palmisano, D. Centonze, M. Quinto, and P.G. Zambonin, *Biosens. Bioelectron.*, *11*, 1996, 419–425.
61. E. Palmqvist, M. Galbe, and B. Hahn-Hagerdal, *Appl. Microbiol. Biotechnol.*, *50*, 1998, 545–551.
62. K. Rumbold, H. Okatch, N. Torto, M. Siika-Aho, G. Gubitz, K.-H. Robra, and B. Prior, *Biotechnol. Bioeng.*, *78*, 2002, 821–827.
63. Y.S. Wu, T.H. Tsai, T.F. Wu, and F.C. Cheng, *J. Chromatogr. A*, *913*, 2001, 341–347.
64. H. Okatch, N. Torto, and J. Armateifio, *J. Chromatogr. A*, *992*, 2003, 67–74.

65. S. Modi, *In vitro* microdialysis sampling for monitoring enzyme reactions, Dissertation thesis, University of Maryland, Baltimore County, MD, (2006), *Diss. Abstr. Int., B*, 68(3), 2007, 1608.
66. A.S. Novozymes, *Biotimes*, S. Strand (Ed.). Novozymes A/S, Bagsvaerd, Denmark, 2005, Vol. xx, pp. 1–11.
67. A.S. Novozymes, *Enzymes That Make Glucose from Liquefied Grains*. Bagsvaerd, Denmark, 2005, pp. 1–8.
68. F.M. Wurst, C. Kempter, S. Seidl, and A. Alt, *Alcohol, 34*, 1999, 71.
69. R. Kaushik, W.R. LaCourse, and B. Levine, *Anal. Chim. Acta, 556*, 2006, 255–266.
70. R. Kaushik, W.R. LaCourse, and B. Levine, *Anal. Chim. Acta, 556*, 2006, 267–274.
71. R. Kaushik and W.R. LaCourse, *Anal. Chim. Acta, 576*, 2006, 239–245.
72. R. Kaushik, Dissertation thesis, University of Maryland, Baltimore, MD, 2006. Unpublished thesis.
73. W.R. LaCourse and G.S. Owens, *Anal. Chim. Acta, 307*, 1995, 301–319.
74. G.S. Owens and W.R. LaCourse, *Curr. Sep., 14*, 1996, 82–88.
75. W.R. LaCourse and G.S. Owens, *Anal. Chim. Acta, 307*, 1995, 301–319.
76. S.M. Lunte and T.J. O'Shea, *Electrophoresis, 15*, 1994, 79–86.
77. A. Paulus and A. Klockow, *J. Chromatogr. A, 720*, 1996, 353–376.
78. R.P. Baldwin, *Electrophoresis, 21*, 2000, 4017–4028.
79. W.R. LaCourse, Pulsed electrochemical detection at noble metal electrodes, in: *Carbohydrate Analysis by Modern Chromatography and Electrophoresis*, Ziad El Rassi (Ed.). Elsevier, New York, 2002, pp. 905–945.
80. T.J. O'Shea, S.M. Lunte, and W.R. LaCourse, *Anal. Chem., 65*, 1993, 948–951.
81. R. Berkow (Ed.), *The Merck Manual*, 15th edn. Merck Sharp and Dohme Research Laboratories, Rahway, NJ, 1987, p. 2413.
82. W. Lu and R.M. Cassidy, *Anal. Chem., 65*, 1993, 2878–2881.
83. H.-H. Ruttinger and B. Drager, *J. Chromatogr. A, 925*, 2001, 291–296.
84. W.R. LaCourse and G.S. Owens, *Electrophoresis, 17*, 1996, 310–318.
85. P.L. Weber, T. Kornfelt, N.K. Klausen, and S.M. Lunte, *Anal. Biochem., 225*, 1995, 135–142.
86. P.L. Weber and S.M. Lunte, *Electrophoresis, 17*, 1996, 302.
87. J.C. Fanguy and C.S. Henry, *Analyst, 127*, 2002, 1021–1023.
88. C.D. Garcia and C.S. Henry, *Anal. Chim. Acta, 508*, 2004, 1–9.
89. C.D. Garcia and C.S. Henry, *Anal. Chem., 75*, 2003, 4778–4783.
90. C.D. Garcia and C.S. Henry, *Analyst, 129*, 2004, 579–584.

8 Derivatization Reactions in Liquid Chromatography for Drug Assaying in Biological Fluids

Andrei Medvedovici, Alexandru Farca, and Victor David

CONTENTS

8.1 INTRODUCTION

Pharmaceuticals (active substances and/or their active metabolites) are traced in biological fluids in order to study pharmacokinetics (PK), bioavailability (BA) aspects, and to assess the bioequivalence (BE) between different formulations. Over the last 15 years, analytical chemistry has played an increasingly important role in almost all steps of drug discovery and development. Biological fluids of human or animal origin (blood, plasma, urine, mucus, perspiration, saliva, synovial, etc.) have been intensively studied in order to estimate PK, BA, and BE of different drug formulations. The literature is already overwhelmed, and it is beyond the purpose of this chapter to mention all contributions to this topic. For this reason, this topic has been rarely reviewed, a first attempt dating back to 1983 [1]. As a matter of fact, only a few major works have been cited here to complete the topic related to drug derivatization and its importance [2–4]. It is worth emphasizing here that the quality of data involved in a drug development strategy is highly related to the quality of the analytical processes used to assay target compounds in biological matrices. In pharmaceutics, analytical processes applied for drug assay are commonly based upon a high-performance liquid chromatographic (HPLC) technique. Proper sample preparation procedures should also be applied to achieve cleanup, isolation from the interfering matrices of the target compounds, and their concentration.

The derivatization procedure improves at least one of the principal analytical parameters, namely the detection sensitivity and separation selectivity. The role of derivatization may also respond to some specific goals: (a) increasing stability of the target compounds; (b) increasing recoveries by means of a full accordance between structural properties and proposed isolation methods; and (c) enhancing the selectivity of the sample preparation procedure (even by indirectly acting on the matrix components rather than on the analyte itself).

The choice of the derivatization reaction depends on several process parameters such as functional groups(s) contained within the target compounds, the concentration of these analytes, the complexity of the sample matrix, the detection mode, and the number of produced derivatization artifacts.

Duration is also an important factor to be considered when applying a derivatization-based method to a large-scale analytical assay of drugs in biological samples. Usually derivatization is applied prior to isolation and concentration of the target compounds, which increases the procedure duration and the contribution to systematic/random errors on the analytical results. The use of an internal standard may reduce the systematic errors, but this solution also increases the complexity of the analytical problem.

Chemical modifications are most generally used in trace chemical analysis, regardless of the type of the analytical process being used [5–7]. The importance of derivatization and the basic principles to be considered regarding its analytical application are widely discussed in different books, reviews, and overviews, for example, Refs. [8,9]. For drug assaying in biological samples, some specific features should be taken into consideration (i.e., the possibility of most drugs to be reversibly bound to plasma proteins such as plasma albumin, lipoproteins, and glycoproteins [10]). Therefore, information on the properties of target compounds is necessary before developing a strategy for the derivatization procedure applied to the determination of drugs or their metabolites in biological fluids [11–14].

8.2   DEFINITIONS AND CLASSIFICATION

Derivatization should be defined as a chemical modification brought to the target compound(s) by means of (bio)chemical reagents and/or physical factors (i.e., irradiation, electric fields, and temperature). Classification criteria related to derivatization processes should be considered as answers to the questions "Why?" "When?" and "How?"

The "Why" criterion is related to the main purposes of derivatization: (a) to chemically stabilize the target compounds; (b) to remove or reduce interferences from matrices (acts on method specificity); (c) to generate separable compounds (acts on chromatographic selectivity); and (d) to make analytes detectable (acts on sensitivity).

Some pharmaceutical compounds are well known for their instability in biological media due to fast degradative processes (oxidation, polymerization). Derivatization of the active site immediately after sample collection represents a practical solution [15].

Structural modification of target compounds makes them isolable from the initial matrix. Adsorption on precipitated proteins may be solved by derivatization. Enhancement of the hydrophobic character by blocking polar sites allows extraction of derivatized target compounds in nonmiscible media with increased yields. In rare cases, the purpose of derivatization deals with the removal of endogenous compounds in the initial sample. It has been proved that acetic anhydride, besides its protein precipitation action, may react with interfering compounds from the sample, and consequently induce an improved specificity [16].

Derivatized analytes may generate specific and subtle interactions with the stationary phase used in the separation process, allowing enhanced chromatographic selectivity and resolution. Resolving racemates on achiral stationary phases by changing enantiomers into diastereoisomers prior to separation should be considered a classic example.

More often, derivatization is used to make the target compound detectable. Introduction in the host structure of a chromophore, fluorophore, electrophore, or luminofore allows UV–Vis spectrometric detection (Single Wavelength Detection, SWD; Multiple Wavelength Detection, MWD; Diode Array Detection, DAD), fluorescence detection (FLD), electrochemical detection (ELCD), or chemiluminescence detection (CHLD). Recently, based on the fact that ionization yields depend upon structural properties, derivatization has been extended to mass spectrometric detection (MSD), especially when atmospheric pressure electrospray ionization (AP-ESI) is used. By making a perfect agreement between structural properties of the target compounds and the specific detection system in use, amazing detection limits have been reached (femto to attomoles levels). A recent review [17] discusses strategies for characterization of drugs and metabolites by HPLC/MS/MS in conjunction with chemical derivatization.

The "When" criterion places derivatization in time with respect to the chromatographic separation. Consequently, derivatization may be "precolumn" if the chemical modifications arise prior to injection of the sample onto the chromatographic column or "postcolumn," when the structural changes are obtained between the chromatographic column and the detection system. Usually, postcolumn derivatization is realized when two or more analytes from sample are transformed in the same derivative, or when the derivatization is intended to take place only for a given compound eluting from the chromatographic column.

The third criterion, "How", deals with different practical aspects related to the chemical modification process. From one side, it is possible to discuss derivatization according to the nature of the sites supporting the transformation (belonging to the structure of the target compound considered as substrate) or the organic function from the derivatization agent generating the modification (considered as reagent). Various organic functions (amine, carboxyl, carbonyl, hydroxyl, and thiol) were evaluated in terms of reactivity, stability, detection wavelengths, handling, versatility, and selectivity [18].

On the other side, the "how" question may refer to the conditions in which derivatization is achieved. Derivatization is realized in homogenous or heterogeneous media. Homogenous media can be mainly aqueous (target compound, derivatization reagent and resulting derivative are water soluble; reaction may be performed before, during, or after the removal of a significant part of the matrix, i.e., by protein precipitation) or mainly organic (water-miscible organic solvents containing the reagent are added in major proportion to the biological fluid; the simple addition may generate protein precipitation). Heterogeneous conditions refer to the following alternatives: (a) derivatization occurs at the interface of two nonmiscible media, the substrate is contained in the aqueous phase; the reagent can be added to the aqueous phase or is brought in the organic solvent; the derivative can be kept within the aqueous phase or can be transferred to the organic one; or (b) derivatization occurs at the interface of a solid support, when the reaction is carried out with the two components adsorbed on a solid support; target analyte may be adsorbed initially on the solid support, but adsorption of the reagent on the solid surface is also feasible; such possibilities are usually using known sample preparation techniques such as solid-phase extraction (SPE) or solid-phase microextraction (SPME).

Heterogeneous solid–liquid derivatization may be realized with immobilized reagents placed online, precolumn in the HPLC system. Derivatization procedure is applied only when needed, using a full automation, then switched offline when conversion becomes unwanted. Such experimental setup may also be applied in the offline, precolumn mode, using small, disposable glass or plastic reaction/reagent cartridges. These cartridges are used several times and then discarded or regenerated. Online, solid-phase reagents can be regenerated overnight in an automated way, washed free of excess, unattached reagents, tested for chromatographic sample blanks, and then reused to perform multiple derivatization processes. Both achiral and chiral tags can be placed onto the solid support, using adsorption or, more usually, covalent attachments to perform chemical purity and identification or additional enantiomeric (chiral) determinations via immobilized chiral reagents. Such indirect enantiomeric applications based on the use of inexpensive chiral reagents may never show kinetic resolution, usually display equal detector responses, and only require conventional, normal, or reversed-phase columns for separation of the precolumn, off- or online formed diastereomers from a given racemic sample [19]. As an example, enantiomeric analysis of amphetamine-related designed drugs in body fluids was achieved by using solid-phase derivatization with (–)-1-(9-fluorenyl) ethyl chloroformate (FLEC) and HPLC-DAD [20].

In practice, the elimination of the reagent excess should be attentively considered in order to avoid column overloading effects or interferences during the chromatographic separation. Thereby, derivatization processes carried in heterogeneous conditions are more convenient, allowing an easier separation of the reagent excess from the produced derivatives.

Last but not least, "'How' derivatization is made?" should be discussed as direct or indirect processes. A direct derivatization process means that the target compound acts as a substrate whereas the derivatization agent acts as a reagent. An indirect derivatization process is based on the introduction of a substrate and of a reagent in the reaction media while the target compound acts as a secondary reagent. A classic example of an indirect derivatization is based on the reaction between organic primary amines (substrate) and o-phthaldialdehyde (OPA) (as primary reagent), while compounds containing thiol groups (target compound) are act as secondary reagents.

8.3 KINETIC ASPECTS OF THE DERIVATIZATION

Derivatization reactions are preferred to be fast, although many times they can be slow, especially when the structure of the analyte is drastically modified. Derivatization reactions are also preferred to be quantitative, although in many cases only a reproducible yield should be enough. Undoubtedly, the derivatization kinetics can influence the analytical results [21]; if it is not properly controlled the reproducibility and the linearity of the method can be directly affected. Kinetic studies can be performed separately in the absence of the sample matrix with the aid of spectrometric measurements in order to have a real-time overview on the derivatization yield. However, in case of derivatization applied to biological samples, many competitive reactions may occur, when either analyte or matrix components would possibly react. In this case, the kinetics must be studied on spiked biological samples, and the chromatographic method became the single possibility to study the derivatization product formation. Accordingly, time spent between sample preparation and sample injection should be carefully controlled. If derivatization reaction continues within the chromatographic column with kinetics comparable to partition [22], major disturbances of peak shape may arise. Sometimes, elution conditions may influence on the derivatization route.

If we consider a general derivatization reaction between the target compound A (considered as substrate) and a derivatization reagent R, then several reaction products P_i can be obtained. Among them the main derivatization product being considered as P_1. For instance, derivatization of cycloserine with p-benzoquinone leads to at least three derivatives [23]. In such cases, the choice of the main derivative relates on the detection sensitivity, chromatographic resolution, or on linearity aspects. Sometimes, more than one reagent is used to produce derivatization. In such instances, R should be replaced by R_i. Derivatization reagent may undergo degradation (i.e., hydrolysis) in the reaction medium, producing artifacts D_j. The main derivatization product exhibits its own stability in the reaction medium, resulting in possible degradation by products S_i. These processes are depicted below and act competitively.

$$A + R \xrightarrow{k_1} P_1 + \cdots + P_i \quad \text{(main derivatization reaction)} \tag{8.1}$$

$$R \xrightarrow{k_2} D_1 + \cdots + D_j \quad \text{(formation of artifacts)} \tag{8.2}$$

$$P_1 \xrightarrow{k_3} S_1 + \cdots + S_m \quad \text{(degradation process of the} \tag{8.3}$$
$$\text{main derivatization product)}$$

The reaction rate r depends on all the species involved in the process:

$$r = k_1 \cdot [A] \cdot [R] \tag{8.4}$$

The contribution of reagent R to the reaction rate is seldom taken into consideration owing to the fact that, usually, the reagent R is introduced in large excess to the reaction medium, and thus, the variation of R is less observable. In such a case, the reaction is of first order, and the reaction rate will depend mainly on the concentration of the analyte: $r = k_1 \cdot [A]$, where k is expressed in s^{-1}. Such reaction order is encountered when the derivatization involves the reaction between a specific moiety from the analyte and a functional group belonging to the reagent.

The derivatization reaction rate is defined as the variation of the concentration of the substrate or of the main derivatization product, according to the relations:

$$r_1 = -\frac{d[A]}{dt} = \frac{d[P_1]}{dt} = k_1 \cdot [A]_0 \cdot e^{-k_1 \cdot t} \tag{8.5}$$

$$r_2 = \frac{d[R]}{dt} = k_2 \cdot [R]_0 \cdot e^{-k_2 \cdot t} \tag{8.6}$$

$$r_3 = -\frac{d[P_1]}{dt} = k_3 \cdot [P_1]_0 \cdot e^{-k_3 \cdot t} \tag{8.7}$$

$$[P_1]_0 = k_1 \cdot [A]_0 \cdot e^{-k_1 \cdot t} \tag{8.8}$$

$$r_3 = k_1 \cdot [A]_0 \cdot e^{-(k_1 + k_3) \cdot t} \tag{8.9}$$

According to processes emphasized above, some scenarios are possible. Some of the possibilities are depicted in Figure 8.1. Case (a) considers effective only the derivatization reaction and artifact formation. Case (b) considers degradation of the reagent at major extent, whereas case (c) illustrates the subsequent degradation of the major derivatization product.

Chip-based online nanospray mass spectrometry method enables the study of kinetics of different isocyanate derivatization reactions (propyl, benzyl, and toluene-2,4-diisocyanate) with 4-nitro-7-piperazino-1,3-benzodiazole [24]: rate constants k have been estimated as follows 1.5×10^4, 5.2×10^4, and 2.4×10^4 as $mol^{-1} \times min^{-1}$, respectively.

Some complex derivatization reactions have a reaction rate of the form:

$$-\frac{d[A]}{dt} = k \cdot [A]^n \tag{8.10}$$

where n signifies a reaction order higher than 1. The rate constant k is given by Arrhenius' reaction rate equation:

$$k = A \cdot e^{-(E_a/RT)} \tag{8.11}$$

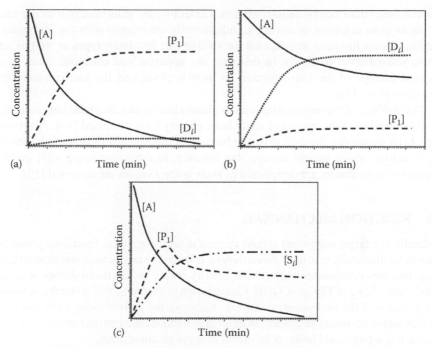

FIGURE 8.1 Results of the derivatization process according to the relative kinetics of the main reaction, degradation of the reagent, and of the main product.

where

E_a is called the activation energy
R is the gas constant
A is a parameter related to the collision number the so-called frequency factor

Therefore, the concentration of the derivatization reagent should be generally 100–1000 times more than the concentration of analyte. The influence of temperature can also be important to the derivatization rate constant, as expressed by Equation 8.11. Thus, a study of the reaction between amphetamine and 2,4-dinitrofluorobenzene (Sanger's reagent), in basic medium, showed that it could last minimum 40 min at 65°C [25].

Stability of the derivatization product is not synonymous with the chemical kinetics. Temperature, pH, or matrix components may affect the time stability of the derivatization product. The stability of the derivatization product in the mobile phase shall be carefully controlled. For instance, a derivatization protocol that exploits the rapid reaction between arenediazonium ions and a suitable coupling agent followed by HPLC analysis of the reaction mixture was used to determine the product distribution and the rate constants for product formation and the association constant of 4-nitrobenzenediazonium ion with β-cyclodextrin, by fitting the experimental data to a simplified Lineaweaver–Burk equation [26]. This protocol is applicable under a variety of experimental conditions providing that the coupling reaction rate is much faster than that of dediazoniation. Generally, the determination of the rate constants from reactions occurring simultaneously on-column with the chromatographic process has been investigated for a variety of first-order types as well as for simple second-order reactions. In this case the apparent rate constants result as a weighed average of the rate constants in mobile phase and the rate constants in stationary phase [22].

The stability of the target compound in plasma matrix and the conditions required for the reliability of the analytical study have already been discussed [10]. However, it has been shown that the derivative can be more stable when compared to the initial target analyte during sample storage; for instance, analytes containing –SH group, susceptible to oxidation, are derivatized as soon as the samples are collected [15].

8.4 REACTION MECHANISMS

Generally, the target compound should contain at least one active functional group in order to be chemically modified. Nevertheless, it should be taken into consideration that many matrices components may exhibit activity owing to functional groups such as $-OH$, $-SH$, $-NH_2$, $-CHO$ or $-COOH$. One solution to overcome this difficulty is to use a high excess of the derivatization reagent. Although the derivatization mechanism is not thoroughly necessary to develop an analytical method based on derivatization, some data on this aspect could however be useful in some circumstances.

Most of the derivatization reactions are nucleophilic substitutions, via unimolecular or bimolecular mechanism. The nature of the molecular mechanism is very important when the derivatization reaction involves asymmetric carbon atoms from the analyte structure or from the derivatization reagent. When derivatization with achiral reagents is used it is likely that the chromatographic separation should be realized on chiral stationary phases. On the other hand, the derivatization with chiral reagents (in pure enantiomeric form) produces diastereoisomers, which can be separated on achiral stationary phases.

The nucleophilic agent may be the analyte or the derivatization reagent. In most cases an sp^3 carbon atom is the center of the nucleophilic attack, but there are many other cases when sp^2 carbon atoms participate to a nucleophilic substitution. For example, the derivatization of lisinopril with 7-chloro-4-nitrobenzo-2-oxa-1,3-diazole at pH 9 was applied for its determination in plasma matrices with the generation of a fluorescent product, using excitation at 470 nm and observing emission at 540 nm [27]. In this case, the nitrogen atom from the analyte molecule carries out a nucleophilic attack to the sp^2 C linked to the chlorine atom (Reaction 8.1).

A similar nucleophilic agent, but with a 7-fluor-substituted atom, has been proposed as a method of choice for quantitation of ABT-089 [2-methyl-3-(2-(S)-pyrrolidinylmethoxy) piridine], which is a new structural type of cholinergic channel modulator [28].

Typical addition reactions to a double bond are used mainly for derivatization of analytes containing reactive double bonds. However, a reactive double bond may be present in the structure of the reagent. For instance, the analytes containing the thiol group may participate to an addition reaction to 1,1-bis-(phenylsulfonyl)ethylene [29] (Reaction 8.2).

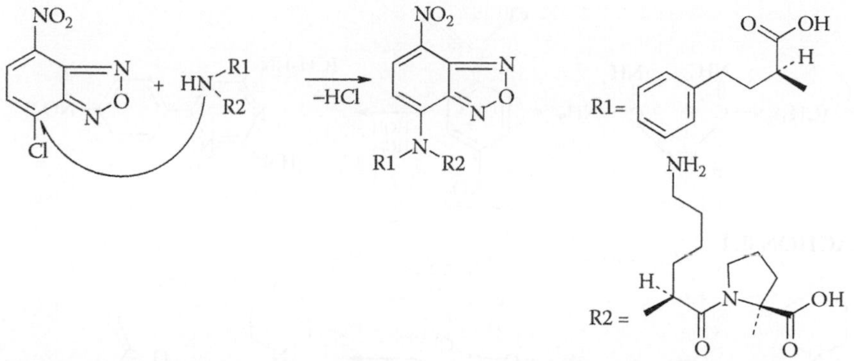

REACTION 8.1

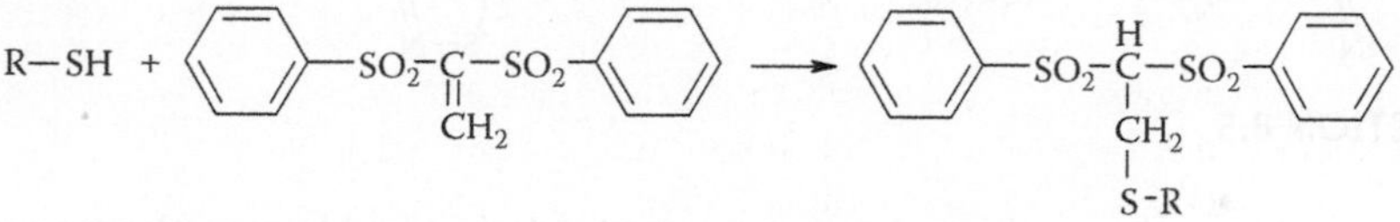

REACTION 8.2

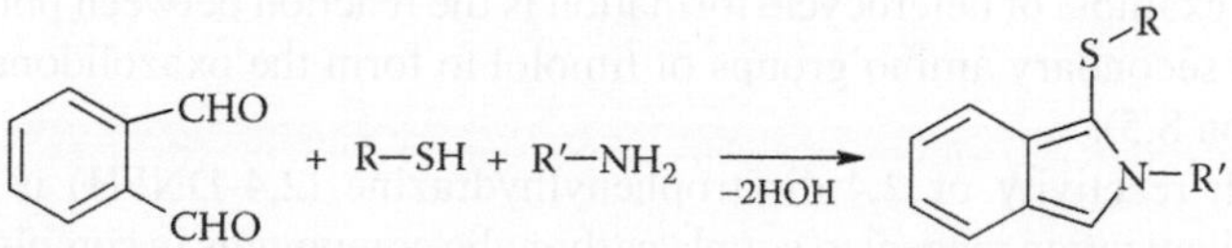

REACTION 8.3

Most complex mechanisms implicate multiple steps in developing the final derivatization product. Such multiple step derivatization takes place, for instance, between primary amines and OPA, in the presence of a thiol compound (i.e., mercaptoethanol is one of the reagents more often used in practice), used to enhance on the fluorescent yield, thus improving on sensitivity (Reaction 8.3).

However, the mechanism is still debatable, as it can be concluded from several experimental data. The irregular behavior of histidine and other compounds containing the moiety $-CH_2-CH(NH_2)-$ and OPA reagent was observed. Under the influence of temperature, this reaction provides more than one derivative product and could be explained by the intramolecular rearrangement with the formation of tautomers [30].

Formation of heterocycles exhibiting increased aromatic character as products of a derivatization reaction is seldom mentioned. Thus, the condensation between compounds containing biguanidine moiety and p-nitrobenzoyl chloride leads to an aromatic ring, which strongly absorb in the UV range. In the case of metformin (an antidiabetic drug) the derivatization reaction is carried out in the presence of NaOH, which allows the formation of acylium ion ($-C^+ = O$), and also transforms the reagent excess into a sodium salt, soluble in the aqueous medium (Reaction 8.4). Consequently, the derivatization product becomes extractable in an organic solvent (dichloromethane), whereas benzoate salt remains in the aqueous phase [31].

REACTION 8.4

REACTION 8.5

For instance, the resulting chromatogram of a derivatized human plasma spiked with metformin overlaid to a blank run is given in Figure 8.2.

Another example of heterocycle formation is the reaction between phosgene with hydroxy and secondary amino groups of timolol to form the oxazolidone derivative [32] (Reaction 8.5).

The high reactivity of 2,4-dinitrophenylhydrazine (2,4-DNPH) is commonly used as a derivatization reagent to couple carbonylic compounds in complex matrices. The reaction is assumed to start either as a nucleophilic attack to the carbon atom

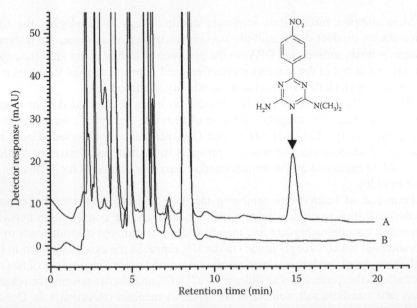

FIGURE 8.2   Two overlaid chromatograms of blank plasma (A) and spiked with metformin (B), derivatized with p-nitrobenzoyl chloride (C18 column, 250×4.6 mm i.d., using a mobile-phase $H_2O/CH_3OH = 65{:}35$; $\lambda = 280$ nm).

REACTION 8.6

or an electrophilic attack of the proton to the oxygen from the carbonyl group and is immediately followed by an elimination reaction, according to the next schema [33] (Reaction 8.6).

It has been proved that the efficiency of this reaction depends on charge distribution on the carbonyl group, which in turn depends on the nature of the two radicals bound to the carbon atom. Thus, molecular orbital calculations for 2,4-DNPH have established the following polarization: $H^{+0.18}-N^{-0.25}-H^{+0.16}$, whereas for simple carbonyl compounds these values have been found to be of similar charge magnitude.

Sometimes, the reaction mechanism may explain some experimental parameters, which must be carefully controlled. Thus, the derivatization of amino group by substitution of a hydrogen atom must be achieved in alkaline conditions; acidic media generate protonation to ammonium ions, reducing their availability to participate to nucleophilic attack of the reagent molecules. For derivatization assisted by liquid–liquid extraction, it can influence the extraction yield, mainly when the product derivative still contains dissociable functional groups.

8.5 DERIVATIZATION FOR IMPROVING THE SELECTIVITY OF THE CHROMATOGRAPHIC SEPARATION

Overall, chemical modifications induced by the derivatization of the analyte significantly affect the selectivity of the HPLC separations. However, components from matrix containing identical functional groups will also be subject to chemical transformations during the derivatization procedure. Briefly, the hydrophobic character of the target compound may be increased or decreased by the derivatization process, depending on the nature of the chemical modification being produced. The character of the derivatization product plays a major role during the chromatographic process, especially when the separation mechanism involves hydrophobic interactions. The octanol or water partition coefficient expressed as a 10-base logarithm (denoted log $K_{o,w}$) is the most important descriptor [34] for the hydrophobic interaction between the analyte and the stationary phase. The value of log $K_{o,w}$ is estimated rather accurately by means of the fragment methodology [35]. According to this methodology, molecular hydrophobicity can be computed by means of the following relationship:

$$\log K_{o,w} = \sum_{i=1}^{n} n_i \log K_{o,w}(i) + \sum_{j=1}^{M} \Phi_j + \zeta \qquad (8.12)$$

where

n_i is the number of fragments of the same type i having log $K_{o,w}(i)$

Φ_j is the factor correction for different groups within the chemical structure of the analyte

ζ is the equation constant

For example, by introducing one group like $-NO_2$ to an aromatic ring, the log $K_{o,w}$ value of the derivatization product decreases with about 0.18 compared to the initial structure. Derivatization with 9-(fluorenyl) methyloxycarbonyl (FMOC) as a fluorophore also increases log $K_{o,w}$ with about 3.45. However, the retention in HPLC is governed not only by hydrophobicity, but also by many other molecular descriptors such as acidity or basicity constants, dipole moment, polarity parameters, or solubility in mobile phase, which can be modified more or less by derivatization [36]. Ionizable functional groups (hydroxyl, thiol, carboxyl, or amino) enhance the molecular polarity and make retention behavior pH dependent. Modification of these groups in esters or amides makes the resulting derivatives less polar. Contrarily, the introduction of nitro or carbonyl moieties makes the final derivatives more hydrophilic, and inherently reduces retention.

Sometimes, it is desirable to control both retention and detection by means of derivatization. Accordingly, the derivatization with FMOC induces an increase in the hydrophobic character and simultaneously makes the product detectable by means of fluorescence. Some other reactive and sensitive fluorescence labeling reagents for the thiol group such as 4-(N-acetylaminosulfonyl)-7-fluoro-1,2,3,-benzoxadiazole and 4-(N-trichloroacetylaminosulfonyl)-7-fluoro-1,2,3-benzoxadiazole considerably reduce the hydrophobic character of the resulting product compared to the parent one. Resulting derivatives are highly water soluble and fluorescent. Detection limits of 25 fmol for homocysteine and 45 fmol for glutathione were obtained [37].

A classic approach to generate enantioselectivity through derivatization refers to the introduction in the target racemate of a group containing an asymmetric carbon atom in pure form. Consequently, after derivatization, enantiomers are transformed to diastereoisomers, being thus separated on achiral stationary phases. During the last few years, the synthesis of chiral stationary phases based π-donors [38–40] or π-acceptors (Pirkle stationary phases), chemically modified polysaccharides [41] and cyclodextrins [42] (and their coating procedures on silicagel), and immobilized macrocyclic antibiotics [43–46] became more and more reproducible, their commercial availability on the market increases and their costs became more affordable. Consequently, use of the chiral derivatization procedures will be probably less frequent in the near future. Some applications of the derivatization procedure designed for chiral discrimination are enlisted in Table 8.1.

8.6 DERIVATIZATION FOR IMPROVING UV–VIS DETECTION

Introduction of a chromophore by the substitution of an active hydrogen atom or by the addition to the host molecule represents the common scenario for making products UV detectable or for enhancing sensitivity. Benzoyl chloride, m-toluol chloride, and p-nitrobenzoyl chloride are the simplest derivatization reagents, introducing an

TABLE 8.1
Derivatization Reactions Inducing Chiral Resolution on Achiral Stationary Phases

Derivatization Reagents	Substrates	Derivatized Groups	Detection	Refs.
(+)-1-(9-Fluorenyl)ethyl chloroformate	Atenolol	Amine	F ex 227 em 310	[47]
	Propranolol	Amine	F ex 260 em 340	[48]
	Reboxetine	Amine	F ex 260 em 315	[49]
	Amphetamine Methamphetamine	Amine	F ex 265 em 330	[50]
	Mefloquine	Amine	UV 263; F ex 263 em 475	[51]
	((R,S)-1-Methyl-8-[(morpholin-2-yl) methoxy]-1,2,3,4 -tetrahydroquinoline	Amine	F ex 260 em 310	[52]
(S)-(−)-α-Methylbenzylamine	Flurbiprofen	Carboxylic acid	UV 245	[53]
(R)-(+)-α-Methylbenzyl isocyanate	Propranolol	Amine	F ex 220 em 300 F ex 228 em 340 F ex 295 em 345 F ex 232 em 340	[54–57]
	(5,6-Dimethoxy-2-[3'-(p-hydroxy-phenyl)-3'-hydroxy-2'-aminotetraline	Amine	Electrochemical screen electrode 0.45V, sample electrode 0.70V	[58]
	Methocarbamol	Alcohol	UV 280	[59]
	Eliprodil	Alcohol	F ex 275 em 336	[60]
	Tocainide	Amine	F ex 220 em 345	[61]
	Metoprolol	Amine	F ex 220 no emission filter	[62]
	Oxprenolol	Amine	F ex 226 em 333 F ex 226 em 340	[63,64]

(continued)

TABLE 8.1 (continued)
Derivatization Reactions Inducing Chiral Resolution on Achiral Stationary Phases

Derivatization Reagents	Substrates	Derivatized Groups	Detection	Refs.
(S)-(+)-1-(1-Naphthyl)ethyl isocyanate	(3-Methylaminomethyl-3,4,5,6-tetrahydro-6-oxo-1H-azepino[5,4,3-cd]indole)	Amine	F ex 320 em 440	[65]
	Acebutolol	Amine	F ex 220 em 389	[66]
	Tertatolol	Amine	F ex 220 em 320	[67]
	Dorzolamide	Amine	UV 252	[68]
	Propranolol	Amine	F ex 305 em 355	[69]
S-FLOPA	Beclobrate	Carboxylic acid	F ex 305 em 355	[70]
	α-Phenylcyclopentylacetic acid	Carboxylic acid	F ex 305 em 355	[71]
(+)-4-(6-Methoxy-2-naphthyl)-2-butyl chloroformate	Metoprolol	Amine	UV 230; F ex 270 em 350	[72]
Diacetyl-L-tartaric anhydride	Propranolol	Alcohol	F ex 290 em 335	[73]
	Halofantrine	Alcohol	UV 254	[74]
Hexachlorobicyclo[2.2.1]hept-5-ene-2-carboxylic acid	Warfarin	Alcohol	F ex 313 em 370, postcolumn reaction, 200 mM NaOH pumped at 0.5 mL/min and flowed through a 1 m reaction coil	[75]
(−)-2-[4-(1-Aminoethyl)phenyl]-6-methoxybenzoxazole	Ibuprofen	Carboxylic acid	F ex 320 em 380	[76]
(R)-(−)-4-(3-Isothiocyanatopyrrolidin-1-yl)-7-(N,N-dimethylaminosulfonyl)-2,1,3-benzoxadiazole	Propranolol	Amine	F ex 460 em 550	[77]

REACTION 8.7

aromatic ring in the structure of the substrate, making UV detection effective. For UV labeling of a target compound containing a carbonyl group, 3,5-dinitrophenyl-hydrazine (3,5-DNPH) or p-nitrobenzylhydroxylamine are probably the most commonly used reagents [78]. Some derivatization reactions produce more drastic changes in the structure of the analyte: a typical example refers to derivatization of an α-amino acid moiety with phenylisothiocyanate (PHI) for UV detection, generating a substituted 3-phenyl-2-thiohydantoin ring (Reaction 8.7).

Generally, PHI is a derivatization reagent designed for compounds containing active hydrogen atoms in their molecule: primary and secondary aromatic amines easily react with PHI to form very stable N-aryl-N'-phenylureas [6]. Such a principle is applied for the determination of gabapentin [79] or pamidronate [80] in human plasma. A similar process can be achieved with 1-naphthyl isothiocyanate (hydrophobic character of the resulting product is enhanced simultaneously). Applications for determination of glucosamine in rat plasma samples [81], tobramycin in human plasma [82], or pamidronate in human urine [83] have been reported.

Other derivatization reagents commonly used for UV labeling of target compounds existing in biological matrices are 4-dimethylaminoazobenzene-4'-sulfonyl chloride (known as Dabsyl chloride) which reacts with primary and secondary amines, thiols, imidazoles, aromatic or aliphatic hydroxyl groups; 4-(dimethylamino) benzaldehyde, also known as Ehrlich's reagent, can react by condensation with primary aromatic amines; 1-fluoro-2,4-dinitrobenzene, also known as Sanger's reagent, is used in precolumn derivatization of amino glycosides, such as amikacin, tobramycin, gentamycin, paromomycin, sisomycin, neamine, neomycin B, and neomycin C, in alkaline media [84].

Condensation reactions with phenylhydrazine are largely used for compounds containing carbonyl groups. Derivatization with 2,4-DNPH can enhance the UV detectability but can also influence the retention process by decreasing hydrophobicity of the target compound. For instance, megestrol acetate, a highly hydrophobic synthetic derivative of naturally occurring steroid hormone (progesterone), and medroxyprogesterone acetate produces poor sensitivity with UV detection due to lack of chromophoric sites and excessive retention in reversed-phase LC separation mechanism. After derivatization with 2,4-DNPH, derivatization products are UV-detected in the low ppm concentration range, and the chromatographic retention becomes reasonable [7].

Derivatization with 2,4-DNPH has been proved an effective approach for determination of propafenone and its metabolite 5-hydroxypropafenone in human plasma samples by means of HPLC/DAD [85]. When target compounds (drug and metabolite) are not subjected to derivatization, they exhibit poor retention due to a reduced hydrophobic character, and consequently, a limited selectivity against the plasma

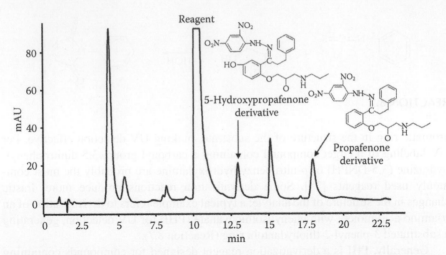

FIGURE 8.3 HPLC-DAD chromatogram of a plasma sample containing propafenone and 5-hydroxypropafenone, derivatized with 2,4-DNPH (column, C18; mobile phase, acetonitrile/aqueous 0.1% H_3PO_4 solution: 75/25 (v/v); $t°C = 25°C$; $\lambda = 375\,nm$) [85].

pattern is achieved. Applying derivatization with 2,4-DNPH in the presence of H_3PO_4 and acetonitrile (as reaction medium), the hydrophobic character of the resulting derivatives is improved and separation against endogenous pattern resolved. The presence of acetonitrile as solvent and H_3PO_4 as catalyst in the condensation process simultaneously acts as a protein precipitation technique, enhancing the overall process selectivity. Results are illustrated in Figure 8.3.

Some other applications reported in literature based on UV labeling are summarized in Table 8.2.

8.7 DERIVATIZATION FOR FLD

Introduction of a fluorophore in the structure of biologically active compounds existing in naturally occurring matrices is probably one of the most common approaches because of the enhanced sensitivity and inherent selectivity of the FLD. Whenever the concentrations of active compounds or their metabolites are at low ppb level and in the absence of a mass spectrometric detector, the first attempt is directed to identify those structural sites on which a fluorescent label can be introduced. Chemical derivatization reactions were recently overviewed with respect on their role in chromatographic and electrophoretic methods for pharmaceutical and biomedical analysis, with a special attention on fluorogenic derivatization [121].

One of the most commonly used reagents for fluorescent derivatization in HPLC is undoubtedly FMOC. Recent applications reported its use in the determination of olpadronate ([3-dimethylamino-1-hydroxypropylidene] bisphosphonate) in plasma and urine [122], azithromycin in plasma, blood, and isolated neutrophils [123], and reboxetine (a new norepinephrine reuptake inhibitor) in human plasma [124].

TABLE 8.2
Derivatization Reactions for UV Labeling

Drug Names	Functional Groups	Reagents	Matrix	Refs.
Penicillin	Amine	Benzoic anhydride and 1,2,4-triazole mercuric chloride	Raw milk	[86]
Amphetamine, methamphetamine	Amine	1,2-Naphthoquinone 4-sulphonate	Urine	[87]
Esmolol	Alcohol	2,3,4,6-Tetra-O-acetyl-β-D-glucopyranosyl isothiocyanate	Plasma	[88]
Oxaliplatin	Metal ion–Pt	N,N-Diethyldithiocarbamate ion	Blood	[89]
Ceftiofur Desfuroyl-ceftiofur	Carboxylic acid	Reduction followed by derivatization with iodoacetamide	Horse plasma, synovial	[90]
Clavulanic acid	Carboxylic acid	Imidazole	Plasma	[91]
Salinomycin, lasalocid	Carbonyl	2,4-DNPH	Poultry feeds	[92]
Perhexiline	Secondary amine	trans-4-Nitrocinnamoyl chloride	Plasma	[93]
Quinic and lactic acids	Carboxylic acid	N,N'-Diisopropyl-O-(p-nitrobenzyl)isourea	Food	[94]
Gabapentin	Amine	2,4,6-Trinitrobenzenesulfonic acid	Blood	[95]
Temafloxacin	Secondary amine	N-1-(2-Naphthylsulfonyl) -2-pyrrolidinecarbonyl chloride	Blood	[96]
Ecgonine methyl ester	Alcohol	4-Fluorobenzoyl chloride	Plasma	[97]
Dihydroqinghaosu	Alcohol	9-Fluoreneacetic acid	Blood	[98]
Polyoxyethyleneglycerol triricinoleate 35	Carboxylic acid	1-Aminonaphthalene	Blood	[99]

(continued)

TABLE 8.2 (continued)
Derivatization Reactions for UV Labeling

Drug Names	Functional Groups	Reagents	Matrix	Refs.
Tiaprofenic acid	Carboxylic acid	L-Leucinamide	Plasma	[100]
Allantoin	Carbonyl	2,4-DNPH	Blood	[101]
4-Hydroxycyclophosphamide	Carbonyl	2,4-DNPH	Blood	[102]
Pregnanolone	Ketone	2,4-DNPH	Blood	[103]
Spectinomycin	Ketone	2,4-DNPH	Plasma	[104]
Ethosuximide	Amide	4-(Bromomethyl)-7-methoxycoumarin	Blood	[105]
Ifosforamide	Halogen	Sodium diethyldithiocarbamate	Plasma	[106]
Captopril	Thiol	2,4′-Dibromoacetophenone	Plasma	[107–109]
Captopril	Thiol	1-Benzyl-2-chloropyridinium bromide	Plasma	[110,111]
Glutathione	Thiol	Iodoacetic acid	Blood	[112]
Tiopromin	Thiol	p-Bromophenacyl bromide	Plasma	[113]
Busulfan	Sulfonate	Sodium diethyldithiocarbamate	Plasma, blood	[114,115]
Artemisinin	Lactone	Degradation	Blood	[116]
Isoniazid	Hydrazine	Cinnamaldehyde	Blood	[117]
Hydralazine and dihydralazine	Hydrazine	2-Hydroxy-1-naphthaldehyde	Plasma	[118]
Isoniazid, monoacetylhydrazine	Hydrazine	Salicylaldehyde	Plasma, blood	[119,120]

Dansyl chloride is another classic derivatization regent, successfully used for assaying neuroactive steroids, alphaxalone, and pregnanolone, in plasma [125]. OPA was used for the determination of sphingosine 1-phosphate in plasma [126], histamine, and flurazepam [127] and for the determination of 1- and 2-adamantanamine in human plasma [128], whereas its homologous naphthalene-2,3-dicarboxaldehyde was successfully used for labeling intracellular reduced glutathione [129].

The histamine content of tears of healthy sex- and aged-matched subjects and patients affected by allergic or nonallergic inflammatory ocular diseases was determined through a precolumn derivatization reaction with OPA and HPLC-FLD [130]. It has been turned out that the tear histamine content is low (around 2 ppb) for all healthy patients, irrespective of age and sex, but can increase 10 times in those affected by allergic or *Haemophilus influenzae*-associated conjunctivitis.

Fluoxetine, a potent selective 5-hydroxytryptamine (serotonin) reuptake inhibitor prescribed for the therapy of mental depression, has been determined in rat plasma samples using a derivatization method with a fluorescent reagent bearing a benzofurazan moiety, according to the following reaction (Reaction 8.8).

The enantiomeric ratios are not modified by derivatization in the derivative products, which have been separated on an amylose-based chiral column using a column-switching HPLC method [131].

Fluorescamine reacts with the primary amine groups generating fluorophores that can be excited at 390 nm to generate emission around 470 nm. Thus, vigabatrin (4-amino-5-hexenoic acid) and gabapentin, 2-[1-(aminomethyl) cyclohexyl] acetic acid, can be determined at low concentrations (Reaction 8.9) [132].

Several other experimental methods designed for fluorescence labeling are summarized in Table 8.3.

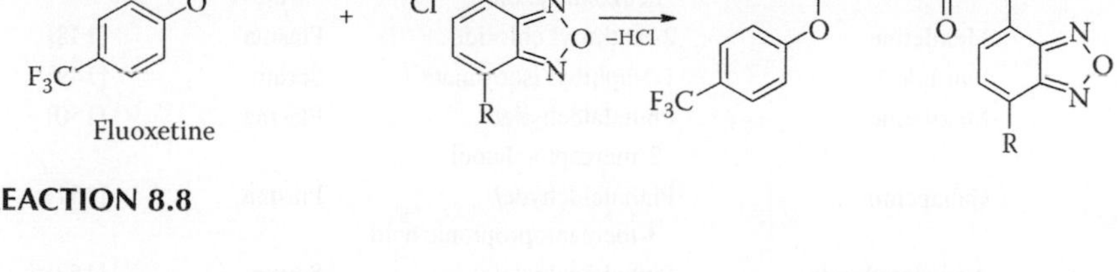

Fluoxetine

REACTION 8.8

Gabapentin   Fluorescamine

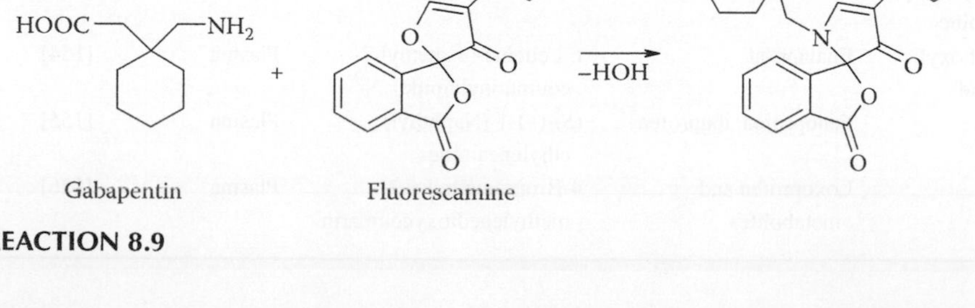

REACTION 8.9

TABLE 8.3
Derivatization Reactions that Introduce a Fluorogenic Group

Functional Groups	Drug Names	Reagents	Matrix	Refs.
Amine	Sodium alendronate	2,3-Naphthalene-dicarboxaldehyde	Plasma	[133,134]
	Catecholamines	1,2-Diphenylethylenediamine	Urine	[135]
	Vigabatrin	4-Chloro-7-nitrobenzofuran	Plasma, urine	[136]
	Duloxetine	Dansyl chloride	Blood	[137]
	Remikirene	9-Fluorenylmethyl chloroformate	Plasma, blood	[138]
	Gentamicine	9-Fluorenylmethyl chloroformate	Plasma	[139]
	Amikacin	1-Methoxycarbonylindolizine-3,5-dicarbaldehyde	Blood	[140]
	Tenofovir	Chloroacetaldehyde	Plasma	[141]
Secondary amine	Lisinopril, insulin	7-Chloro-4-nitrobenzo-2-oxa-1,3-diazole	Plasma	[27,142]
	Fenfluramine, fluoxetine, norfluoxetine	Dansyl chloride	Plasma	[143]
	Methotrexate	Cerium(IV) trihydroxy peroxide	Blood	[144]
	Ranitidine	Phthalaldehyde/ 2-mercaptoethanol/ postcolumn	Biological fluids	[145]
	Pilocarpine	4-(Bromomethyl)-7-methoxycoumarin	Blood	[146]
	Ethambutol	4-Fluoro-7-nitro-2,1,3-benzoxadiazole	Human plasma, urine	[147]
	Mexiletine	2-Anthroyl chloride	Plasma	[148]
	Ibutilide	1-Naphthyl isocyanate	Serum	[149]
	Mexiletine	Phthalaldehyde/ 2-mercaptoethanol	Plasma	[150]
	Gabapentin	Phthalaldehyde/ 3-mercaptopropionic acid	Plasma	[151]
	Anticonvulsants	Phthalaldehyde/ 2-mercaptoethanol	Serum	[152]
Amine + secondary amine	Cidofovir	2-Bromoacetophenone	Plasma	[153]
Carboxylic acid	Enalaprilat	L-Leucine-(4-methyl-7-coumarinylamide)	Plasma	[154]
	Fenoprofen, ibuprofen	(S)-(−)-1-(Naphthyl) ethyleneamine	Plasma	[155]
	Loxoprofen and metabolites	4-Bromomethyl-6,7-methylenedioxycoumarin	Plasma	[156]

TABLE 8.3 (continued)
Derivatization Reactions that Introduce a Fluorogenic Group

Functional Groups	Drug Names	Reagents	Matrix	Refs.
	Ibuprofen	5-Bromoacetyl acenaphthene	Blood	[157]
		(R)-(+)-1-(1-Naphthyl) ethylamine	Serum, urine	[158]
	Fatty acids	4-(2-Carbazolylpyrrolidin -1-yl)-7-(N,N-dimethylaminosulfonyl)-2,1,3-benzoxadiazole	Blood	[159]
		5,6-Dimethoxy-2-(4-hydrazinocarbonylphenyl) benzothiazole	Blood	[160]
		2-(5-Hydrazinocarbonyl-2-furyl)-5,6 dimethoxy benzothiazole	Human serum	[161]
		4-Aminomethyl-6,7-dimethoxycoumarin	Blood	[162]
		N-(4-Bromomethyl-7-hydroxy-2-oxo-2H-6-chro menyl)bromoacetamide	—	[163]
	Simvastatin	1-(Bromoacetyl)pyrene	Plasma	[164]
Aldehyde	Pyridoxal	Semicarbazide	Human plasma, postcolumn	[165]
Ketone	δ-Aminolevulinic acid	Acetylacetone	Plasma, urine	[166]
	Tacrolimus	Dansyl hydrazine	Whole blood	[167]
Thiol	Captopril, cysteine, glutathione	2-Chloro-4,5-bis(p-N,N-dimethylamino sulfonylphenyl)oxazole	Urine	[168]
	Penicillamine	2-(4-N-Maleimidephenyl) -6-methylbenzothiazole	Blood	[169]
	Glutathione	Monobromobimane	Human plasma	[170]
	Sodium mercaptoundeca hydrodecaborate	Monobromobimane	Rat urine and plasma	[171]
	Mesna	Monobromobimane	Blood	[172]
β-Lactame	Cefaclor	4-(2'-Cyanoisoindolyl) phenylisothiocyanate	Blood	[173]
Hydroxyl	Propranolol, hydroxy -propranolol	2,3,4,6-Tetra-O-acetyl-β-glucopyranosyl isothiocyanate	Human plasma	[174]
	Estradiol	1-Pyrenesulfonyl chloride	Serum	[175]
	Moxidectin	Aromatization	Plasma	[176]
	Digoxin and metabolites	1-Naphthoyl chloride	Human serum	[177]
	Propranolol	Phosgene	Plasma	[178]

8.8   DERIVATIZATION FOR CHLD

The state of the art and the main advantages of chemiluminescence-related derivatization for biologically active compounds separated by means of HPLC or capillary zone electrophoresis (CZE) have been recently reviewed [179]. Some catecholamines (norepinephrine, epinephrine, and dopamine) have been extensively studied as substrates for chemiluminescence derivatization after HPLC separation. A sample preparation method based on SPE for isolation of the target compounds from plasma matrices combined with an HPLC separation and CHLD was developed as an experimental approach. After a reversed-phase mechanism-based separation, on-column fluorogenic derivatization with ethylenediamine was carried out, followed by a postcolumn peroxyoxalate chemiluminescent reaction using bis[4-nitro-2-(3,6,9-trioxadecyloxycarbonyl)phenyl]oxalate and hydrogen peroxide [180]. Later, the same procedure has been fully automated [181] and applied to the same analytes in human and rat plasma samples, allowing detection limits in the low fmole interval. Alternative chemiluminescence mechanism was studied for the same target compounds (norepinephrine, epinephrine, and dopamine) and isoproterenol, based on the peroxyoxalate reaction [182]. Substrates derivatized with 1,2-diarylethylenediamines, were separated by reversed-phase HPLC and were detected after a postcolumn derivatization reaction, using bis[4-nitro-2-(3,6,9-trioxadecyloxycarbonyl)] phenyl] oxalate and hydrogen peroxide. Detection limits of tens of amoles when using injection volume of 100 µL are claimed.

Development of illegal markets for chemical cocktails designed for growth promotion in food-producing animals evidenced substances such as synthetic corticosteroids inducing serious risks for human health. Determination of these residues in food required development of rapid and sensitive analytical methods for their detection. HPLC [183] or flow injection [184] techniques with chemiluminescence-based detection for monitoring synthetic corticosteroids (prednisolone, betamethasone, dexamethasone, and flumethasone) in bovine urine have been developed using a classical solution, the luminol system. Sulfonamide residues in milk were derivatized with fluorescamine, separated by HPLC and then postcolumn reacted with bis[4-nitro-2-(3,6,9-trioxadecyloxycarbonyl)]phenyl] oxalate, using imidazole as catalyst [185].

4,5-Diaminophthalhydrazide, a luminol-type chemiluminescence derivatization reagent, was found to be a highly sensitive chemiluminescence derivatization reagent for α-dicarbonyl compounds in HPLC [186]. This reagent reacts with α-keto acids or α-dicarbonyl compounds in dilute hydrochloric acid media, in the presence of β-mercaptoethanol, yielding highly chemiluminescent quinoxaline derivatives, producing light emission through the interaction with hydrogen peroxide and potassium hexacyanoferrate (Reaction 8.10).

Analytical methods based on these derivatization reactions have been developed for the determination of 3α,5β-tetrahydroaldosterone in human urine and dexamethone in human plasma after oral administration, using beclomethasone as internal standard [187–189]. Detection limit of 1 fmol can be reached.

A similar reaction takes place between aromatic and aliphatic aldehydes with 4,5-diaminophthalhydrazide. The derivatives can be separated by RP-LC and become highly chemiluminescent by postcolumn oxidation with hydrogen peroxide

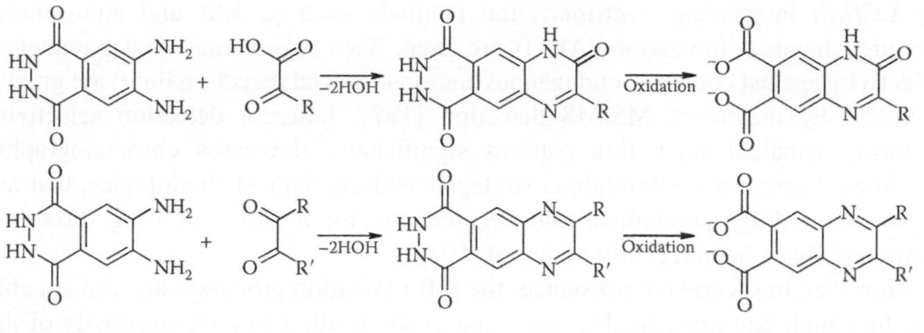

REACTION 8.10

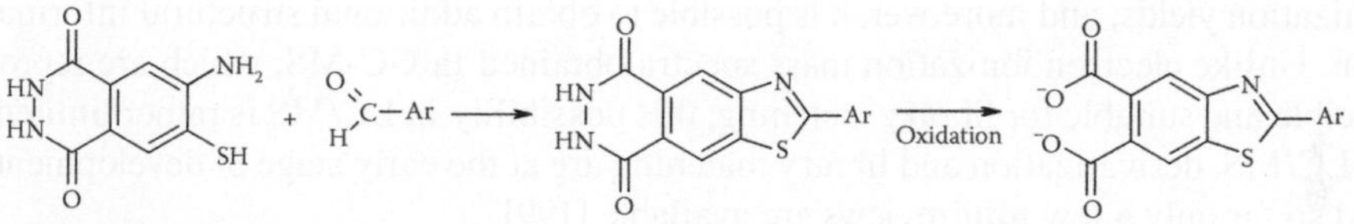

REACTION 8.11

and alkaline hexacyanoferate [190]. 5-Amino-4-thio-phthalhydrazide reacts only with aromatic hydrocarbons (Reaction 8.11) [191].

Maprotiline in plasma was determined by HPLC with CHLD after a single-step extraction with a mixture of *n*-hexane and isoamyl alcohol, followed by conversion in to chemiluminescent derivatives by reaction with 6-isothiocyanatobenzo[g] phthalazine-1,4(2H,3H)-dione, in the presence of triethylamine [192]. Propentofylline in rat brain microdialysate was determined by means of peroxyoxalate CHLD, using 4-(N,N-dimethylaminosulfonyl)-7-hydrazino-2,1,3-benzoxadiazole [193].

4-(N,N-dimethylaminosulfonyl)-7-fluoro-2,1,3-benzoxadiazole was used as derivatization reagent for the secondary amine moiety in ebiratide found in plasma samples [194], immediately after isolation on octadecyl silicagel SPE cartridges. After separation on a C18 column, the derivatized compound in the effluent was mixed with 100 mM hydrogen peroxide containing 0.5 mM bis[4-nitro-2-(3,6,9-trioxadecyloxycarbonyl)phenyl] oxalate pumped at 1.2 mL/min, and the mixture was put through a coil at 30°C to the detector. CHLD was achieved, having a detection limit around 250 fmol.

Electrogenerated chemiluminescence derivatization reagents were applied for determination of fatty carboxylic acids, ibuprofen, and amines [195,196]. Thus, 2-(2-aminoethyl)1-methylpyrrolidine, N-(3-aminopropyl)pyrrolidine, and 3-isobutyl-9,10-dimethoxy-1,3,4,6,7,11β-hexahydro-2H-pyrido(2,1-α) isoquinolin-2-ylamine were found to be selective and sensitive for the HPLC assay of above-mentioned compounds using electrogenerated CHLD with tris(2,2′-bipyridine) ruthenium(II).

8.9 DERIVATIZATION FOR IMPROVING MASS SPECTROMETRY DETECTION

Liquid chromatography coupled with tandem mass spectrometry (MS/MS) is an analytical technique ideally suited for drug assay and characterization in complex matrices.

For LC/MS interfacing, soft-ionization methods such as ESI and atmospheric pressure chemical ionization (APCI) are used. Two major analytical parameters (selectivity against coeluting endogenous components and detection limit) are greatly improved by means of MS/MS detection [197]. Inherent detection selectivity tolerating nonideal separation patterns significantly decreases chromatographic run times. Current high throughput strategies and efficient methodologies, that are employed in drug metabolism and PK screens, for a series of drug discovery compounds have been recently reviewed [198].

However, in several circumstances the soft-ionization processes are still not able to induce high and reproducible ionization yields leading to poor sensitivity of the MS detection. In such cases, derivatization of the target compounds may improve ionization yields, and moreover, it is possible to obtain additional structural information. Unlike electron ionization mass spectra obtained in GC-MS, which are reproducible and suitable for library matching, this possibility in LC/MS is rather limited. In LC/MS, derivatization and library matching are at the early stage of development, and so far only a few minireviews are available [199].

APCI is a gas-phase ionization method, generating satisfactory results in terms of sensitivity and robustness for highly apolar compounds. Any changes improving volatility of the target compounds are beneficial for such ionization mechanism.

ESI strongly depends on the solution chemistry of the detected compounds. The introduction of permanently charged moieties or readily ionized functional groups may greatly improve the ionization yields, and consequently the sensitivity of the MS detection. True enhancement may arise by introducing moieties with high proton or electron affinity. Modifications of the mobile phase, by pH adjustment and/or adduct formation, are an alternative possibility for sensitivity improvement. Coordination of transitional metal ions by the target compounds immediately before introduction into the ionization area may be considered as an in situ postcolumn derivatization, leading to coordination ion spray working mode [200,201].

In case of the biological matrices, the sensitivity of detection may be controlled by inhibiting matrix-induced ion-suppression mechanisms through derivatization or mobile-phase additivation. In a recent review [202], MS detection-oriented derivatization has been discussed together with the use of additives for mobile phases in order to enhance the sensitivity of APCI/MS detection, particularly focusing on the applications involving small molecules in biological samples.

Derivatization methods for HPLC/MS applications referring to drugs in biological fluids have been developed in the previous years [203]. This recent review also focuses on description of some functional groups designed for the enhancement of the structural information content, mainly for separation and quantitation of chiral compounds.

Derivatization enhances the ionization efficiencies of steroids, leading to high sensitivity and specific MS detection. The introduction of moieties with proton affinity or electron affinity improves analyte response in positive and negative APCI/MS, respectively [204]. Norethindrone and ethinyl estradiol at low pg/mL were extracted from human plasma matrix with n-butyl chloride and then derivatized with dansyl chloride in order to enhance the sensitivity of HPLC-MS/MS determinations [205].

REACTION 8.12

A new derivatization reagent, 2-hydrazino-1-methylpyridine, was successfully used for the LC-ESI-MS analysis of oxosteroids. It has been proved that the reagent reacts quantitatively with oxosteroids at 60°C within 1 h and the resulting derivatives can provide a 70–1600-fold higher sensitivity compared to nonderivatized oxosteroids [206,207]. Determination of testosterone concentrations in rat plasma has been reported, using HPLC with APCI/MS detection and based on ethyloxime and acetyl-ester derivatization [208]. Dansyl chloride, known as a fluorescence reagent, was used for derivatization of propofol in blood, followed by HPLC-ESI/MS/MS assay [209]. A sensitive HPLC/particle beam/mass spectrometry assay for the determination of all-*trans*-retinoic acid and 13-*cis*-retinoic acid in human plasma based on derivatization with pentafluorobenzyl bromide was also reported [210]. Propylene glycol is a commonly used vehicle for aerosol dosage formulations. Therefore, its determination in plasma and lung tissue is very important for new drug development. The propylene glycol determination in low concentration in biological matrices was possible by HPLC/tandem mass spectrometry after derivatization with benzoyl chloride (Reaction 8.12) [211].

A well-known fluorescence labeling reagent, namely 9-fluorenylmethyl chloroformate, has been successfully used for HPLC with electrospray–MS detection of (*R*)-2-amino-3-(3-hydroxypropylthio) propionic acid, known as fudosteine, in human plasma [212]. Monobromobimane, proved to be a very sensitive fluorescence derivatization reagent for captopril [15], has also been used for mass spectrometry detection of this drug in human plasma samples [213]. Although majority of the derivatization products have a fragmentation pattern involving the new chemical bond, the –C–S– bond formed between captopril and monobromobimane is stable during electrospray interface. The MS and MS-MS spectra of the derivative (Figure 8.4) show a fragmentation process that is different to the process occurring in derivative between internal standard and monobromobimane (Figure 8.5).

Ziprasidone, a novel atypical antipsychotic agent, has been recently approved for the treatment of schizophrenia. The major metabolite of this drug, resulted from reductive cleavage of the benzisothiazole ring, was characterized by LC-MS/MS, hydrogen/deuterium (H/D) exchange, and chemical derivatization with *N*-dansyl-aziridine [214]. H/D exchange, the simplest derivatization, can be used for determination of the number and position of H/D-exchangeable functional groups on the analyte structure.

8.10 DERIVATIZATION FOR ELCD

Not all compounds lead themselves to ELCD. To achieve such detection, a derivatizing agent may be used to form or add an electroactive group to the analyte molecules

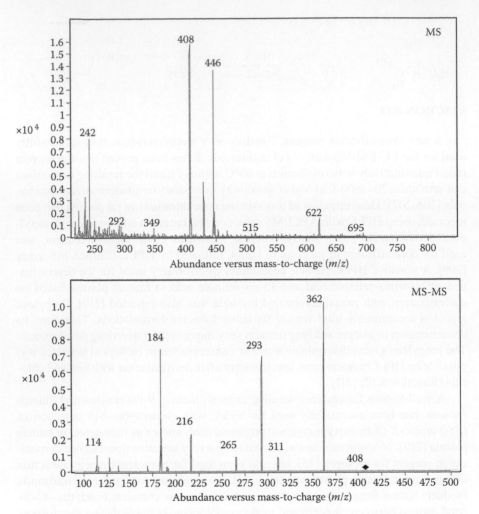

FIGURE 8.4 MS and MS/MS (precursor ion m/z = 408) spectra of derivative between captopril and monobromobimane.

that will oxidize (or reduce) under electrochemical conditions. Sometimes, generation of electroactive sites is achieved by means of physical processes such as photo irradiation [215–217] and solid-phase catalytic reduction, or enzymatic processes [218,219]. An overview of the applications involving ELCD of derivatized biologically active compounds is given in Table 8.4.

8.11 PRECOLUMN VERSUS POSTCOLUMN DERIVATIZATION

Many of the described derivatization methods are achieved in the precolumn mode (before chromatographic separation). Consequently, the derivatization reaction may be performed manually, or automatically (off-line or online), before chromatographic injection. This mode allows flexible working conditions (reaction time, solvent,

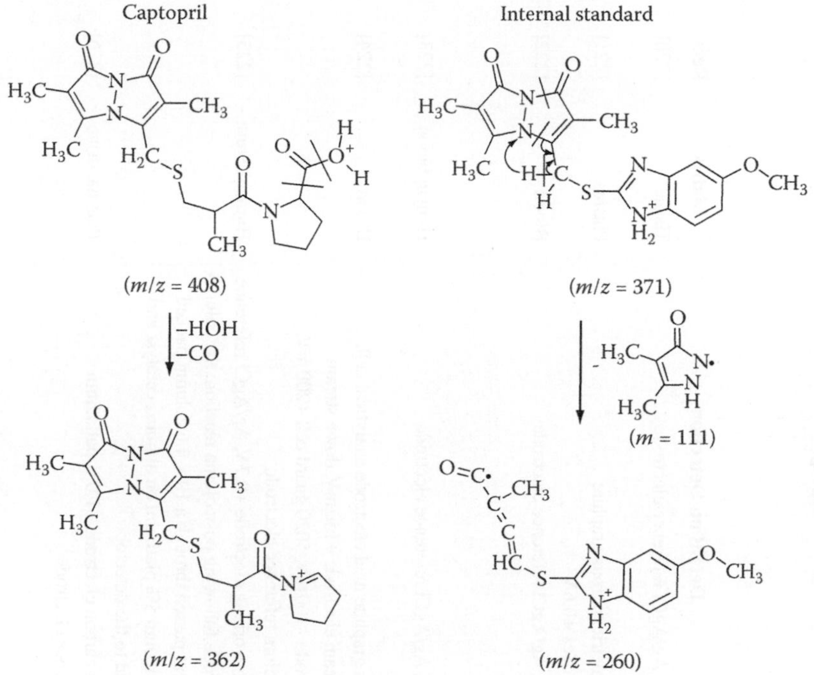

FIGURE 8.5 MS fragmentations of derivatives between captopril and internal standard (2-mercapto-5-methoxybenzimidazol) with monobromobimane.

elimination of the reagent excess, additional analytical operations such as liquid extraction and/or sample concentration).

In the postcolumn mode, the reaction is performed automatically by adding a derivatization reagent after the chromatographic column, and before detection, by means of a secondary high-pressure pump. This approach requires heavier equipment, but automation considerably reduces the number of manipulations. In this case, the derivatization reaction must be fast and compatible with the mobile-phase composition. Postcolumn derivatization may introduce severe chromatographic problems if it is not achieved properly. For instance, reaction coils are needed to achieve postcolumn reactions, by delaying detection in time against the reagent addition in the column flow. Large dead volumes of the reaction coils will determine the loss of chromatographic efficiency and peak spreading. Low internal volumes are recommended for reaction coils, but lower internal diameters of the tubing combined with increased lengths will produce high pressure drop on the postcolumn reactor, and consequently transfer a constraint on the chromatographic system.

Among the above described derivatization reactions, those based on CHLD or ELCD are more often carried out in a postcolumn mode. Also, several derivatization reactions for FLD are possible only in the postcolumn design (derivatization lead to the same reaction product for all compounds subjected to the chemical transformation, or resulting compounds are unstable under elution conditions). Thus, a simple procedure for the determination of imidacloprid and its main metabolites was

TABLE 8.4
Applications of Derivatization Allowing ELCD

Derivatized Functions	Derivatization Reagents	Derivatized Compound(s)	Details on Detection	Matrix	Refs.
Carboxylic acid	4-Aminophenol	Fatty acids, bile acids, prostaglandins	0.75 V, Ag/AgCl reference electrode	Tissue	[220]
	1-(2,5-Dihydroxyphenyl)-2-bromoethanone	Bile salts	Coulometric, porous graphite electrodes +0.6 V	Plasma	[221]
Thiol	N-(4-Anilinophenyl) maleimide	(2R,4R)-2-(2-Hydroxy phenyl)-3-(3-mercapto propionyl)-4-thiazo lidinecarboxylic acid	+1.0 V, Ag/AgCl reference electrode	Blood	[222]
	N-(4-Dimethylamino phenyl) maleimide	Captopril	+0.9 V, Ag/AgCl reference electrode	Human blood	[223]
	N-(Ferrocenyl) maleimide	Glutathione, L-cysteine	Porous graphite dual electrode analytical cell, upstream electrode +150 mV, down stream electrode −100 mV, 5020 guard cell +200 mV, palladium reference electrode	Blood	[224]
Alcohol	2-[2-(Isocyanate)ethyl]-3-methyl-1,4-naphtho quinone	Cholesterol, cholestanol	Carbon working electrode +0.7 V, Ag/AgCl reference electrode following postcolumn reaction; the column effluent passed through a 10 × 4.6 column packed with 10 mm 5% platinum on alumina catalyst and flowed to the detector	Human serum	[225]
	Salicyl chloride	Testosterone	Glassy carbon electrode +1.0 V, palladium reference electrode	Plasma, urine	[226]

Functional group	Reagent	Analyte	Detection conditions	Matrix	Reference
Secondary amine	2,3,4,6-Tetra-O-acetyl-β-D-glucopyranosyl isothiocyanate	Fenoldopam	Guard electrode +0.2 V, working electrode 1 −0.20 V, working electrode 2 +0.20 V	Human plasma	[227]
	2,4-Dinitrofluorobenzene	Renin inhibitors	First electrode 300 mV, second electrode 550 mV, guard cell 800 mV (before injector)	Plasma	[228]
Amine, carboxylic acid	6-Aminoquinolyl-N-hydroxysuccinimidyl carbamate	Aminoacids, peptides	Glassy carbon working electrode +1.1 V, stainless steel counter electrode, Ag/AgCl reference electrode	Blood	[229]
	Phenyl isothiocyanate	Glycine	Glassy carbon electrode +1.10 V, Ag/AgCl reference electrode	Plasma	[230]
	Phenyl isothiocyanate	Aminoacids	Glassy carbon electrode +1.10 V, Ag/AgCl reference electrode	Plasma, urine	[231]
Amine	Salicyl chloride	Organic amines	Glassy carbon working electrode +1.2 V, Ag/AgCl reference electrode, carbon-filled PTFE auxiliary electrode	Plasma, urine	[232]
	Phthalaldehyde/t-butanethiol	Baclofen	Guard cell +1.2 V, glassy carbon working cell, screen electrode +0.2 V, quantifying electrode +0.7 V	Human plasma	[233]
	2,5-Dihydroxyphenylacetic acid, 2,5-bis-tetrahydropyranyl ether p-nitrophenyl ester; homogentisic gamma-lactone tetrahydropyranyl ether	Dipeptide isoleucine leucine methyl ester	Detection potential of +200 mV versus Ag/AgCl ([Cl⁻] = 3 M)	Plasma	[234]
Ketone	4-Nitrophenylhydrazine	17 Ketosteroid sulfates	0.8 V, Ag/AgCl reference electrode (divert column effluent from detector for 5 min after injection.)	Serum	[235]

(continued)

TABLE 8.4 (continued)
Applications of Derivatization Allowing ELCD

Derivatized Functions	Derivatization Reagents	Derivatized Compound(s)	Details on Detection	Matrix	Refs.
Peptidic	Cu^{2+}	Peptide TP9201	Cell detector with dual glassy carbon electrodes, 16 mm PTFE gasket, upstream electrode 410 mV, downstream electrode 85 mV, Ag/AgCl reference electrode, following postcolumn reaction; the column effluent mixed with reagent pumped at 0.1 mL/min and the mixture flowed through a 0.25 mm i.d. coil of knitted PTFE tubing (residence time 1.1 min) at 50° to the detector; reagent was 1.2 M pH 9.9 carbonate buffer containing 0.5 mM copper sulfate and 3 mM disodium tartrate	Blood	[236]
Imine	Reduction	Progabide	Glassy carbon electrode 1 V +850 mV, Ag/AgCl reference electrode	Plasma	[237]
—	Immobilized enzyme reactor	Choline	Analytical cell +300 mV, following an enzyme reactor; the reactor was a 30 × 2.1 mm, 7 mm Aquapore AX-300 anion-exchange cartridge, inject slowly 50 mL 100 U/mL choline oxidase and catalase, wash with mobile phase for several minutes before use, reload after 100 samples	Biological fluids Blood	[238] [239]
—	Solid phase reactor; reduction	Idebenone	Glassy carbon working electrode +0.7 V, Ag/AgCl reference electrode following postcolumn reaction; the column effluent flowed through a 10 × 4.6 mm column packed with 10 μm 5% platinum on alumina catalyst to the detector; purge catalyst column with H_2O at 10 mL/min for 5 min before use		[240]

developed recently by means of HPLC with electrochemical detector and postcolumn photochemical derivatization of the target analytes [241].

Fluorescence may be induced by a structural modification achieved through UV irradiation in a quartz coil positioned after the chromatographic column. Determination of fenbufen and its metabolites in serum by reversed-phase HPLC using SPE and online postcolumn irradiation for fluorescence determination was reported [242]. The column effluent flowed through a knitted PTFE coil irradiated with an UV-lamp to achieve derivatization. Determination of cysteine, homocysteine, and glutathione in concentrations higher than 40 ppb was possible by using Cd^{2+} and hydroxyquinoline-5-sulfonic acid in a noncomplexing buffer at pH 10 [243].

Although, the derivatization of amino-containing compounds with o-phthalaldehyde and thiol are usually carried out in precolumn mode, some reported applications are based on postcolumn derivatization. For instance, a fully automated HPLC analyzer for methylated L-arginine metabolites (N,N-dimethyl-, N-methyl- and N,N'-dimethyl-L-arginine) from rat plasma samples was developed [244].

8.12 ANALYTICAL PARAMETERS

The validation of an analytical procedure based on derivatization should represent a documented evidence of its suitability for the intended purpose. Such a process follows the same rules and concepts generally applied to procedures, which are not based on derivatization for assaying of drugs and their metabolites in biological samples. The number of parameters (p_i) could be somewhat higher in comparison with procedures based on simple protein precipitation or extractions, and consequently the number of validation protocols increases. Each parameter p_i $(i = 1,2,..., n)$ influences the information outcome $(I_{outcome})$ from the analytical process, according to a general relation denoted by

$$I_{outcome} = f(p_1, p_2, ..., p_n) \tag{8.13}$$

Variation of a setup p_i value within its normal interval of variation $(P_2^{set-up} \pm \Delta_{p_i})$ should not determine the variation of the information outcome which is beyond its normal interval of variation $(I_{outcome} \pm 3\sigma_{method})$, where σ_{method} represents the estimated value of the standard deviation of the entire analytical process applied to the drug determination in biological samples.

Analytical characteristics of the derivatization process are the following: derivatization yield, selectivity, sensitivity, kinetics, number of artifacts, and time stability of the derivative product [12,13]. The experimental parameters (derivatization reagent concentration, reaction medium, pH, reaction time, stability, extraction parameters, and influence of the matrix) must be validated, and the results must be properly interpreted in order to be reliable for the intended purpose of the final application [245]. However, linearity domain, intra and interday precision, or limit of quantitation (LOQ) are treated in the same way [246] as treated for the rest of nonderivatization HPLC methods applied to the determination of drugs or their metabolites in biological samples. While the LOQ can be improved by derivatization, some other statistical parameters, such as precision or accuracy, are not necessary improved.

Injection volume is of general interest in HPLC analysis [247], and therefore, still plays a particular role in this case.

Accurate quantitation in chromatography requires the addition of a known amount of an internal standard to an accurately measured aliquot of the sample being analyzed. The internal standard corrects for losses during subsequent sample preparation steps and provides a known amount of material to be measured. A suitable internal standard should be structurally related to the analyte and should be selectively separated against the target compound and the endogenous pattern of the matrix. Current experience in this field leads to the conclusion that an internal standard participating to the derivatization reaction is not absolutely necessary for a derivatization-based assay; it can be added only to control the systematic errors involved in different sample preparation stages of the entire analytical procedure and conveniently not to take part in chemical transformations. The addition of an internal standard that participates at the derivatization reaction may result in a difficult chromatographic separation and loss of accuracy and precision.

Derivatization procedures applied to the BA/BE studies do not avoid some necessary stages of the validation of the entire analytical process, such as: (a) freeze and thaw stability of derivatives obtained by derivatization of target compounds immediately after blood sampling from volunteers; (b) long-term stability of derivatives in plasma samples; (c) short-term stability for derivatized samples; (d) postpreparative stability; and (e) stability of stock solutions (internal standard and reagent).

Automatization is of general concern in analytical chemistry, especially when a large number of samples are investigated, such as in case of BE studies. A robotic sample preparation method combined with HPLC-FLD for the determination of the major component of ivermectin in human plasma has been reported. The authors equipped a Cyberlab C-300 with customized software and hardware in order to perform all the semiautomated liquid–solid phase loading and hexane elution on cartridges. Under the automated precolumn derivatization conditions, both ivermectin and internal standard of the extracted samples were chemically modified to highly fluorescent derivatives [248].

A fully automated and highly sensitive method with a semimicrocolumn LC system was developed for the assay of catecholamines in rat plasma samples [249]. Automated online extraction of these neurotransmitters in diluted plasma samples using a precolumn packed with strong acidic cation exchanger was coupled with their separation on a semimicrocolumn, fluorogenic derivatization with ethylenediamine, and finally postcolumn peroxyoxalate CHLD using bis[2-(3,6,9-trioxadecanyloxycarbonyl)-4-nitrophenyl]oxalate and hydrogen peroxide. In this way, the authors obtained very low limits of detection, below 1 fmol for norepinephrine, epinephrine, and dopamine.

8.13 CONCLUSIONS

In the laboratory practice, derivatization procedures are used to solve real problems related to the analytical process. Derivatization is a tool for making compounds separable or detectable with respect to the required limits or the available equipments. Derivatization induces increased stability in unstable target compounds or eliminates interferences from complex matrices. Derivatization represents a "crisis solution."

Derivatization processes are of an extreme variety. Structural modifications are obtained by means of chemical interactions and/or physical factors. Derivatization is an occasion for the analyst to return to basic background knowledge such as organic synthesis, reaction mechanisms, and kinetics.

Derivatization "cohabitates" well with other analytical procedures such as protein precipitation, liquid–liquid extraction, SPE, supercritical fluid extraction, and membrane processes. Derivatization forces the automation of the sample preparation process, closely dealing with robotics and computer-assisted programming.

Derivatization creates a world: a world of chemical structures simply designed to interact with specific sites and functional groups belonging to substrates. Derivatization may be a game where sometimes substrates are changing roles with the reagents or playing the secondary role of a third partner. Derivatization is never applied for its intrinsic beauty. The process itself is a challenge synonymous to enhancement.

REFERENCES

1. J. A. F. de Silva, *Pure Appl. Chem.*, *55*: 1905, 1983.
2. M. J. Humphrey and D. A. Smith, *Xenobiotica, 22*: 743, 1992.
3. J. H. Lin and A. Y. H. Lu, *Pharmacol. Rev., 49*: 403, 1997.
4. P. R. Chaturvedi, C. J. Decker, and A. Odinecs, *Curr. Opin. Chem. Biol., 5*: 452, 2001.
5. J. M. Rosenfeld, *TrAC Trends Anal. Chem.*, *22*: 785, 2003.
6. G. Lunn and L. C. Hellwig, *Handbook of Derivatization Reactions for HPLC*, John Wiley & Sons, New York, 2005.
7. S. C. Moldoveanu and V. David, Sample preparation in chromatography, *J. Chromatograph. Libr.*, Amsterdam, 65, 2002.
8. G. Lunn and N. R. Schmuff (Eds.), *HPLC Methods for Pharmaceutical Analysis*, Wiley-Interscience, New York, 1997.
9. G. Lunn (Ed.), *HPLC Methods for Pharmaceutical Analysis*, Wiley-Interscience, New York, 2000.
10. W. Grimm, Stability testing of clinical trial materials. In *Drug Stability. Principles and Practices*, 3rd ed., J. T. Carstensen and C. T. Rhodes (Eds.), Marcel Dekker, New York, 2000.
11. V. P. Shah, K. K. Midha, J. W. Findlay, H. M. Hill, J. D. Hulse, I. J. McGilveray, G. McKay, and A. Yacobi, *Pharm. Res., 17*: 1551, 2000.
12. Y. Vander Heyden, A. Nijhuis, J. Smeyers-Verbeke, B. G. M. Vandeginste, and D. L. Massart, *J. Pharm. Biomed. Anal., 24*: 723, 2001.
13. J. Ermer, *J. Pharm. Biomed. Anal., 24*: 755, 2001.
14. B. K. Matuszewski, M. L. Constanzer, and C. M. Chavez-Eng, *Anal. Chem., 75*: 3019, 2003.
15. F. Tache, A. Farca, A. Medvedovici, and V. David, *J. Pharm. Biomed. Anal., 28*: 549, 2002.
16. V. David, C. Barcutean, I. Sora, and A. Medvedovici, *Rev. Roum. Chim., 50*: 269, 2005.
17. D. Q. Liu and C. E. C. A. Hop, *J. Pharm. Biol. Anal., 37*: 1, 2005.
18. T. Toyo'oka, *J. Biochem. Biophys. Methods, 54*: 25, 2002.
19. C. X. Gao and I. S. Krull, *J. Pharm. Biomed. Anal., 7*: 1183, 1989.
20. J. Verdu-Andres, P. Campins-Falco, and R. Herraez-Henandez, *Chromatographia, 60*: 537, 2004.
21. H. A. Mottola and D. Perez-Bendito, *Anal. Chem., 66*: 131R 1994.
22. R. Thede, Rate constants: Determination from on-column chemical reactions, In *Encyclopedia of Chromatography*, J. Cazes (Ed.), Taylor & Francis, Boca Raton, FL, 2006, on-line.
23. V. David, M. Ionescu, and V. Dumitrescu, *J. Chromatogr. B, 761*: 27, 2001.

24. B. M. Liesener, R. E. Oosterbroek, W. Verboom, U. Karst, A. van der Berg, and D. N. Reinhouldt, *Anal. Chem., 77*: 6852, 2005.
25. H. Hegedus, A. Gergely, T. Veress, and P. Horvath, *Analysis, 27*: 458, 1999.
26. C. Bravo-Diaz and E. Gonzalez-Romero, *J. Chromatogr. A, 989*: 221, 2003.
27. A. A. El-Eman, S. H. Hansen, M. A. Moustafa, S. M. El-Ashry, and D. T. El-Sherbiny, *J. Pharm. Biomed. Anal., 34*: 35, 2004.
28. Y. H. Hui, S. Carrol, and K. C. Marsh, *J. Chromatogr. B, 695*: 337, 1997.
29. V. Cavrini, R. Gotti, V. Andrisano, and R. Gatti, *Chromatographia, 42*: 515, 1996.
30. A. Csampai, D. Kutlan, F. Toth, and I. Molnar-Perl, *J. Chromatogr. A, 1031*: 67, 2004.
31. F. Tache, V. David, A. Farca, and A. Medvedovici, *Microchem. J., 68*: 13, 2001.
32. J. D. Gilbert, T. V. Olah, M. J. Morris, A. Bortnick, and J. Brunner, *J. Chromatogr. Sci., 36*: 163, 1998.
33. J. Z. Dong and S. C. Moldoveanu, *J. Chromatogr. A, 1027*: 25, 2004.
34. C. Hansch and S. M. Anderson, *J. Org. Chem., 32*: 2583, 1967.
35. W. M. Meylan and P. H. Howard, *J. Pharm. Sci., 84*: 83, 1995.
36. V. David and A. Medvedovici, *J. Liq. Chromatogr. Related Technol., 30*: 761, 2007.
37. K. Okabe, R. Wada, K. Ohno, S. Uchiyama, T. Santa, and K. Imai, *J. Chromatogr. A, 982*: 111, 2002.
38. R. Thompson, *J. Liq. Chromatogr. Related Technol., 28*: 1215, 2005.
39. Y. Zheng and J. N. Fang, *Chin. J. Anal. Chem., 32*: 685, 2004.
40. P. Borman, B. Boughtflower, and K. Cattanach, *Chirality, 15*: S1, 2003.
41. M. L. de la Puente, C. T. White, and A. Rivera-Sagredo, *J. Chromatogr. A, 983*: 101, 2003.
42. N. Grobuschek, M. G. Schmid, J. Koidl, and G. Gubitz, *J. Sep. Sci., 25*: 1297, 2002.
43. Z. Juvancz and J. Szejtli, *TrAC Trends Anal. Chem., 21*: 379, 2002.
44. C. Karlsson, L. Karlsson, and D. W. Armstrong, *Anal. Chem., 72*: 4394, 2000.
45. Q. Sun and S. V. Olesik, *J. Chromatogr. B, 745*: 159, 2000.
46. L. A. Svensson, J. Donnecke, K. E. Karlsson, A. Karlsson, and J. Vessman, *Chirality, 12*: 606, 2000.
47. M. T. Rosseel, A. M. Vermeulen, and F. M. Belpaire, *J. Chromatogr., 568*: 239, 1991.
48. A. Roux, G. Blanchot, A. Baglin, and B. Flouvat, *J. Chromatogr., 570*: 453, 1991.
49. E. Frigerio, E. Pianezzola, and M. Strolin Benedetti, *J. Chromatogr. A, 660*: 351, 1994.
50. A. Hutchaleelaha, A. Walters, H. H. Chow, and M. Mayersohn, *J. Chromatogr. B, 658*: 103, 1994.
51. Y. Bergqvist, M. Doverskog, and J. Al Kabbani, *J. Chromatogr. B, 652*: 73, 1994.
52. C. Boursier-Neyret, A. Baune, P. Klippert, I. Castagne, and C. Sauveur, *J. Pharm. Biomed. Anal., 11*: 1161, 1993.
53. M. P. Knadler and S. D. Hall, *J. Chromatogr., 494*: 173, 1989.
54. S. Laganiere, E. Kwong, and D. D. Shen, *J. Chromatogr., 488*: 407, 1989.
55. H. G. Schaefer, H. Spahn, L. M. Lopez, and H. Derendorf, *J. Chromatogr., 527*: 351, 1990.
56. H. Spahn-Langguth, B. Podkowik, E. Stahl, E. Martin, and E. Mutschler, *J. Anal. Toxicol., 15*: 327, 1991.
57. C. Pham-Huy, A. Sahui-Gnassi, V. Saada, J. P. Gramond, H. Galons, S. Ellouk-Achard, V. Levresse, D. Fompeydie, and J. R. Claude, *J. Pharm. Biomed. Anal., 12*: 1189, 1994.
58. I. Rondelli, F. Mariotti, D. Acerbi, E. Redenti, G. Amari, and P. Ventura, *J. Chromatogr., 612*: 95, 1993.
59. S. Alessi-Severini, R. T. Coutts, F. Jamali, and F. M. Pasutto, *J. Chromatogr., 582*: 173, 1992.
60. B. Malavasi, M. Ripamonti, A. Rouchouse, and V. Ascalone, *J. Chromatogr. A, 729*: 323, 1996.

61. R. A. Carr, R. T. Foster, D. Freitag, and F. M. Pasutto, *J. Chromatogr., 566*: 155, 1991.
62. M. M. Bhatti and R. T. Foster, *J. Chromatogr., 579*: 361, 1992.
63. M. E. Laethem, M. T. Rosseel, P. Wijnant, and F. M. Belpaire, *J. Chromatogr., 621*: 225, 1993.
64. M. E. Laethem, R. A. Lefebvre, F. M. Belpaire, H. L. Vanhoe, and M. G. Bogaert, *Clin. Pharmacol. Ther., 57*: 419, 1995.
65. W. Naidong, R. H. Pullen, R. F. Arrendale, J. J. Brennan, J. D. Hulse, and J. W. Lee, *J. Pharm. Biomed. Anal., 14*: 325, 1996.
66. M. Piquette-Miller, R. T. Foster, F. M. Pasutto, and F. Jamali, *J. Chromatogr., 526*: 129, 1990.
67. T. Lave, C. Efthymiopoulos, J. C. Koffel, and L. Jung, *J. Chromatogr., 572*: 203, 1991.
68. B. K. Matuszewski and M. L. Constanzer, *Chirality, 4*: 515, 1992.
69. H. Spahn-Langguth, B. Podkowik, E. Stahl, E. Martin, and E. Mutschler, *J. Anal. Toxicol., 15*: 209, 1991.
70. S. Mayer, E. Mutschler, and H. Spahn-Langguth, *Chirality, 3*: 35, 1991.
71. B. Liebmann, S. Mayer, E. Mutschler, and H. Spahn-Langguth, *Arzneimittelforschung, 42*: 1354, 1992.
72. R. Büschges, R. Devant, E. Mutschler, and H. Spahn-Langguth, *J. Pharm. Biomed. Anal., 15*: 201, 1996.
73. W. Lindner, M. Rath, K. Stoschitzky, and G. Uray, *J. Chromatogr., 487*: 375, 1989.
74. D. R. Brocks, M. J. Dennis, and W. H. Schaefer, *J. Pharm. Biomed. Anal., 13*: 911, 1995.
75. S. R. Carter, C. C. Duke, D. J. Cutler, and G. M. Holder, *J. Chromatogr., 574*: 77, 1992.
76. J. Kondo, N. Suzuki, H. Naganuma, T. Imaoka, T. Kawasaki, A. Nakanishi, and Y. Kawahara, Biomed. *Chromatogr., 8*: 170, 1994.
77. T. Toyo'oka, M. Toriumi, and Y. Ishii, *J. Pharm. Biomed. Anal., 15*: 1467, 1997.
78. A. Medvedovici, V. David, F. David, and P. Sandra, *Anal. Lett., 32*: 581, 1999.
79. Z. Zhu and L. Neirinck, *J. Chromatogr. B, 779*: 307, 2002.
80. R. W. Sparidans, J. Den Hartigh, W. M. Ramp-Koopmanschap, R. H. Langebroek, and P. Vermeij, *J. Pharm. Biomed. Anal., 16*: 491, 1997.
81. A. Aghazadeh-Habashi, S. Sattari, F. Pasutto, and F. Jamali, *J. Pharm. Pharm. Sci., 5*: 176, 2002.
82. C. H. Feng, S. J. Lin, H. L. Wu, and S. H. Chen, *J. Chromatogr. B, 780*: 349, 2002.
83. R. W. Sparidans, J. Den Hartigh, P. Beinjnen, and P. Vermeij, *J. Chromatogr. B, 696*: 137, 1997.
84. L. K. Sorensen, B. M. Rasmussen, J. O. Boison, and L. Keng, *J. Chromatogr. B, 694*: 383, 1997.
85. A. Medvedovici, unpublished results.
86. P. Campis-Falco, A. Sevillano-Cabeza, C. Molins-Legua, and M. Kohlmann, *J. Chromatogr. B, 687*: 239, 1996.
87. Y. H. Tang, Y. He, T. W. Yao, and S. Zeng, *J. Biochem. Biophys. Methods, 59*: 159, 2004.
88. P. Ehrsson and I. Wallin, *J. Chromatogr. B, 795*: 291, 2003.
89. S. De Baere, F. Pille, S. Croubels, L. Ceelen, and P. De Backer, *Anal. Chim. Acta, 512*: 75, 2004.
90. K. M. Matar, E. M. Nazi, Y. M. El-Sayed, M. J. Al-Yamani, S. A. Al-Suwayeh, and K. I. Al-Khamis, *J. Liq. Chromatogr. Related Technol., 28*: 97, 2005.
91. A. K. Mathur, *J. Chromatogr. A, 664*: 284, 1994.
92. G. Dusi and V. Gamba, *J. Chromatogr. A, 835*: 243, 1999.
93. N. Grgurinovich, *J. Chromatogr. B, 696*: 75, 1997.
94. J. Lu, M. Cwik, and T. Kanyok, *J. Chromatogr. B, 695*: 329, 1997.
95. C. M. Stevenson, L. L. Radulovic, H. N. Bockbrader, and D. Fleisher, *J. Pharm. Sci., 86*: 953, 1997.

96. M. Matsuoka, K. Banno, and T. Sato, *J. Chromatogr. B, 676*: 117, 1996.
97. L. Virag, B. Mets, and S. Jamdar, *J. Chromatogr. B, 681*: 263, 1996.
98. O. R. Idowu, J. M. Grace, K. U. Leo, T. G. Brewer, and J. O. Peggins, *J. Liq. Chromatogr. Related Technol., 20*: 1553, 1997.
99. A. Sparreboom, O. van Tellingen, M. T. Huizing, W. J. Nooijen, and J. H. Beijnen, *J. Chromatogr. B, 681*: 355, 1996.
100. M. Vakily and F. Jamali, *J. Pharm. Sci., 85*: 638, 1996.
101. J. Lagendijk, J. B. Ubbink, and W. J. H. Vermaak, *J. Chromatogr. Sci., 33*: 186, 1995.
102. M. Johansson and M. Bielenstein, *J. Chromatogr. B, 660*: 111, 1994.
103. D. J. Jones, A. R. Bjorksten, and D. P. Crankshaw, *J. Chromatogr. B, 694*: 467, 1997.
104. S. D. Burton, J. E. Hutchins, T. L. Fredericksen, C. Ricks, and J. K. Tyczkowski, *J. Chromatogr., 571*: 209, 1991.
105. S. H. Chen, H. L. Wu, J. K. Wu, H. S. Kou, and S. M. Wu, *J. Liq. Chromatogr. Related Technol., 20*: 1579, 1997.
106. G. P. Kaijser, J. H. Beijnen, E. Rozendom, A. Bult, and W. J. M. Underberg, *J. Chromatogr. B, 686*: 249, 1996.
107. A. Jankowski, A. Skorek, K. Krzysko, P. K. Zarzyxki, R. J. Ochocka, and H. Lamparczyk, *J. Pharm. Biomed. Anal., 13*: 655, 1995.
108. K. Li and J. H. Zhou, *Biomed. Chromatogr., 10*: 237, 1996.
109. M. Bahmaei, A. Khosravi, C. Zamiri, A. Massoumi, and M. Mahmoudian, *J. Pharm. Biomed. Anal., 15*: 1181, 1997.
110. E. Bald, S. Sypniewski, J. Drzewoski, and M. Stepien, *J. Chromatogr. B, 681*: 283, 1996.
111. S. Sypniewski and E. Bald, *J. Chromatogr. A, 729*: 335, 1996.
112. T. Yoshida, *J. Chromatogr. B, 678*: 157, 1996.
113. T. Huang, B. Yang, Y. Yu, X. Zheng, and G. Duan, *Anal. Chim. Acta, 565*: 178, 2006.
114. K. I. Funakoshi, K. Yamashita, W. F. Chao, M. Yamaguchi, and T. Yashiki, *J. Chromatogr. B, 660*: 200, 1994.
115. J. MacKichan and T. P. Bechtel, *J. Chromatogr., 532*: 424, 1990.
116. H. A. C. Titulaer, J. Zuidema, P. A. Kager, J. C. F. M. Wetsteyn, C. B. Lugt, and F. W. H. M. Merkus, *J. Pharm. Pharmacol., 42*: 810, 1990.
117. N. Sadeg, N. Pertat, H. Dutertre, and M. Dumontet, *J. Chromatogr. B, 675*: 113, 1996.
118. J. Manes, J. Mari, R. Garcia, and G. Font, *J. Pharm. Biomed. Anal., 8*: 795, 1990.
119. A. Walubo, K. Chan, and C. L. Wong, *J. Chromatogr. B, 567*: 261, 1991.
120. A. Walubo, P. Smith, and P. I. Folb, *J. Chromatogr. B, 658*: 391, 1994.
121. S. Görög, *Fresenius' J. Anal. Chem., 362*: 4, 1998.
122. R. W. Sparidans, J. den Hartigh, S. Cremers, J. H. Beijnen, and P. Vermeij, *J. Chromatogr. B, 738*: 331, 2000.
123. E. Wilms, H. Trumpie, W. Veenendaal, and D. Touw, *J. Chromatogr. B, 814*: 37, 2005.
124. M. A. Raggi, R. Mandrioli, G. Casamenti, V. Volterra, and S. Pinzauti, *J. Chromatogr. A, 949*: 23, 2002.
125. S. A. G. Visser, C. J. G. M. Smulders, W. F. T. Gladdines, H. Irth, P. H. van der Graaf, and M. Danhof, *J. Chromatogr. B, 745*: 357, 2000.
126. J. J. Butter, R. P. Koopmans, and M. C. Michel, *J. Chromatogr. B, 824*: 65, 2005.
127. J. James, D. Lowe, and H. T. Karnes, *Pharm. Res., 2*: S21, 1992.
128. Y. Higashi and Y. Fujii, *J. Chromatogr. B, 799*: 349, 2004.
129. L. Diez, E. Martenka, A. Dabrowoska, J. Coulon, and P. Leroy, *J. Chromatogr. B, 827*: 44, 2005.
130. I. Venza, M. Visalli, G. Ceci, and D. Teti, *Ophthalmic Res., 36*: 62, 2004.
131. X. Guo, T. Fukushima, F. Li, and K. Imai, *Analyst, 127*: 480, 2002.
132. A. A. Majed, *J. Liq. Chromatogr. Related Technol., 28*: 3119, 2005.
133. W. F. Kline, B. K. Matuszewski, and W. F. Bayne, *J. Chromatogr., 534*: 139, 1990.
134. W. F. Kline and B. K. Matuszewski, *J. Chromatogr., 583*: 183, 1992.

135. F. A. J. van der Horn, F. Boomasma, A. J. Man in't Veld, and M. A. D. H. Schalekamp, *J. Chromatogr., 563*: 348, 1991.
136. S. Erturk, E. S. Aktas, and S. Atmaca, *J. Chromatogr. B, 760*: 207, 2001.
137. J. T. Johnson, S. W. Oldham, R. J. Lantz, and J. Delong, *J. Liquid Chromatogr. Relat. Technol., 19*: 1631, 1996.
138. P. Coassolo, W. Fischli, J. P. Clozel, and R. C. Chou, *Xenobiotica, 26*: 333, 1996.
139. D. A. Stead and R. M. E. Richards, *J. Chromatogr. B, 675*: 295, 1996.
140. S. Oguri and Y. Miki, *J. Chromatogr. B, 686*: 205, 1996.
141. R. W. Sparidans, K. M. Crommentuyn, J. H. Schellens, and J. H. Beijnen, *J. Chromatogr. B, 791*: 227, 2003.
142. C. Yang, H. Huang, H. Zhang, and M. Liu, *Anal. Lett., 39*: 2463, 2006.
143. R. F. Suckow, M. F. Zhang, and T. B. Cooper, *Clin. Chem., 38*: 1756, 1992.
144. S. Emara, S. Razee, A. Khedr, and T. Masujima, *Biomed. Chromatogr., 11*: 42, 1997.
145. P. Viñas, N. Campillo, C. López-Erroz, and M. Hernández-Córdoba, *J. Chromatogr. B, 693*: 443, 1997.
146. S. Merali and A. B. Clarkson, *J. Chromatogr. B, 675*: 321, 1996.
147. M. Breda, P. Marrari, E. Pianezzola, and S. Benedetti, *J. Chromatogr. A, 729*: 301, 1996.
148. D. K. Kowk, L. Igwemezie, C. R. Kerr, and K. M. McErlane, *J. Chromatogr. B, 661*: 271, 1994.
149. C. L. Hsu and R. R. Walters, *J. Chromatogr. B, 667*: 115, 1995.
150. V. L. Lanchote, P. S. Bonato, S. A. C. Dreossi, P. V. B. Gonçalves, E. J. Cesarino, and C. Bertucci, *J. Chromatogr. B, 685*: 281, 1996.
151. G. Forrest, G. J. Sills, J. P. Leach, and M. J. Brodie, *J. Chromatogr. B, 681*: 421, 1996.
152. U. H. Juergens, T. W. May, and B. Rambeck, *J. Liquid Chromatogr. Relat. Technol., 19*: 1459, 1996.
153. E. J. Eisenberg and K. C. Cundy, *J. Chromatogr. B, 679*: 119, 1996.
154. F. Levai, C. M. Liu, M. M. Tse, and E. T. Lin, *Acta Physiol. Hung., 83*: 39, 1995.
155. Y. Y. Lau, *J. Liquid Chromatogr. Relat. Technol., 19*: 2143, 1996.
156. H. Naganuma and Y. Kawahara, *J. Chromatogr. B, 530*: 387, 1990.
157. L. A. Gifford, F. T. K. Owusu-Daaku, and A. J. Stevens, *J. Chromatogr. A, 715*: 201, 1995.
158. S. C. Tan, S. H. D. Jackson, C. G. Swift, and A. J. Hutt, *Chromatographia, 46*: 23, 1997.
159. T. Toyo'oka, M. Takahashi, A. Suzuki, and Y. Ishii, *Biomed. Chromatogr., 9*: 162, 1995.
160. M. Yamaguchi, S. Hara, and K. Obata, *J. Liquid Chromatogr., 18*: 2991, 1995.
161. M. Saito, T. Ushijima, K. Sasamoto, Y. Ohkura, and K. Ueno, *J. Chromatogr. B, 674*: 167, 1995.
162. K. Sasamoto, T. Ushijima, M. Saito, and Y. Ohkura, *Anal. Sci., 12*: 189, 1996.
163. E. Gikas, M. Derventi, I. Panderi, A. Vavayannis, M. Kazanis, and M. Parissi-Poulou, *J. Liquid Chromatogr. Relat. Technol., 25*: 381, 2002.
164. H. Ochiai, N. Uchiyama, K. Imagaki, S. Hata, and T. Kamei, *J. Chromatogr. B, 694*: 211, 1997.
165. H. Mascher, *J. Pharm. Sci., 82*: 972, 1993.
166. H. Oishi, H. Nomiyama, K. Nomiyama, and K. Tomokuni, *J. Anal. Toxicol., 20*: 106, 1996.
167. A. J. Beysens, G. H. Beuman, J. J. van der Heijden, K. E. J. Hoogtanders, O. M. Steijger, and H. Lingeman, *Chromatographia, 39*: 490, 1994.
168. T. Toyo'oka, H. P. Chokshi, R. S. Givens, R. G. Carlson, S. M. Lunte, and T. Kuwana, *Biomed. Chromatogr., 7*: 208, 1993.
169. A. I. Haj-Yehia and L. Benet, Z. *Pharm. Res., 12*: 155, 1995.
170. C. S. Yang, S. T. Chou, L. Liu, P. J. Tsai, and J. S. Kuo, *J. Chromatogr. B, 674*: 23, 1995.

171. K. Abu-Izza and D. R. Lu, *J. Chromatogr. B, 660*: 347, 1994.
172. B. Stofer-Vogel, T. Cerny, M. Borner, and B. H. Lauterburg, *Cancer Chemother. Pharmacol., 32*: 78, 1993.
173. M. Kai, H. Kinoshita, and M. Morizono, *Talanta, 60*: 325, 2003.
174. S. T. Wu, Y. P. Chang, W. L. Gee, L. Z. Benet, and E. T. Lin, *J. Chromatogr. B, 692*: 133, 1997.
175. K. H. DeSilva, F. B. Vest, and H. T. Karnes, *Biomed. Chromatogr., 10*: 318, 1996.
176. M. Alvinerie, J. F. Sutra, M. Badri, and P. Galtier, *J. Chromatogr. B, 674*: 119, 1995.
177. M. C. Tzou, R. A. Sams, and R. H. Reuning, *J. Pharm. Biomed. Anal., 13*: 1531, 1995.
178. K. Stoschitzky, S. Kahr, J. Donnerer, M. Schumacher, O. Luha, R. Maier, W. Klein, and W. Lindner, *Clin. Pharmacol. Ther., 57*: 543, 1995.
179. T. Fukushima, N. Usui, T. Santa, and K. Imai, *J. Pharm. Biomed. Anal., 30*: 1655, 2003.
180. S. Higashidate and K. Imai, *Analyst, 117*: 1863, 1992.
181. P. Prados, S. Higashidate, and K. Imai, *Biomed. Chromatogr., 8*: 1, 1994.
182. G. H. Ragab, H. Nohta, M. Kai, Y. Ohkura, and K. Zaitsu, *J. Pharm. Biomed. Anal., 13*: 645, 1995.
183. B. I. Vazquez, X. Feas, M. Lolo, C. A. Fente, C. M. Franco, and A. Cepeda, *Luminiscence, 20*: 197, 2005.
184. Y. Iglesias, C. Fente, B. I. Vazquez, C. Franco, A. Cepeda, and S. Mayo, *Anal. Chim. Acta, 468*: 43, 2002.
185. J. J. Soto-Chinchilla, L. Gamiz-Gracia, A. M. Garcia-Campana, K. Imai, and L. E. Garcia-Ayuso, *J. Chromatogr. A, 1095*: 60, 2005.
186. J. Ishida, S. Sonezaki, and M. Yamaguchi, *J. Chromatogr. A, 598*: 203, 1992.
187. T. Nakahara, J. Ishida, M. Yamaguchi, and M. Nakamura, *Anal. Biochem., 190*: 309, 1990.
188. J. Ishida, T. Nakahara, and M. Yamaguchi, *Biomed. Chromatogr., 6*: 135, 1992.
189. J. Ishida, S. Sonezaki, M. Yamaguchi, and T. Yoshitake, *Anal. Sci., 9*: 319, 1993.
190. M. Yamaguchi, M. Isokane, and J. Ishida, *Anal. Sci., 11*: 569, 1995.
191. H. Yoshida, R. Nakao, H. Nohta, and M. Yamaguchi, *J. Chromatogr. A, 898*: 1, 2000.
192. J. Ishida, N. Horike, and M. Yamaguchi, *J. Chromatogr. B, 669*: 390, 1995.
193. Y. Hamachi, M. N. Nakashima, and K. Nakashima, *J. Chromatogr. B, 724*: 189, 1999.
194. Y. Hamachi, K. Nakashima, and S. Akiyama, *J. Liq. Chromatogr. Related Technol., 20*: 2377, 1997.
195. H. Morita and M. Konishi, *Anal. Chem., 74*: 1584, 2002.
196. H. Morita and M. Konishi, *Anal. Chem., 75*: 940, 2003.
197. A. Medvedovici, F. Albu, C. Georgita, and V. David, *Biomed. Chromatogr., 19*: 549, 2005.
198. Y. Hsieh and W. A. Korfmacher, *Curr. Drug Metab., 7*: 479, 2006.
199. J. M. Halket, D. Waterman, A. M. Przyborowska, R. K. P. Patel, P. D. Fraser, and P. M. Bramley, *J. Exp. Bot., 56*: 219, 2005.
200. A. Medvedovici, K. Lazou, A. d'Oosterlinck, Y. Zhao, and P. Sandra, *J. Sep. Sci., 25*: 173, 2002.
201. P. Sandra, A. Medvedovici, Y. Zhao, and F. David, *J. Chromatogr. A, 974*: 231, 2002.
202. S. Gao, Z. P. Zhang, and H. T. Karnes, *J. Chromatogr. B, 825*: 98, 2005.
203. J. M. Halket and V. G. Zaikin, *Eur. J. Mass Spectrom., 11*: 127, 2005.
204. T. Higashi and K. Shimada, *Anal. Bioanal. Chem., 378*: 875, 2004.
205. W. Li, Y. H. Li, A. C. Li, S. Zhou, and W. Naidong, *J. Chromatogr. B, 825*: 223, 2005.
206. T. Higashi, A. Yamauchi, and K. Shimada, *J. Chromatogr. B, 825*: 214, 2005.
207. T. Higashi, A. Yamauchi, K. Shimada, E. Koh, A. Mizokami, and M. Namiki, *Anal. Bioanal. Chem., 382*: 1035, 2005.
208. N. M. Watanabe, H. Ochiai, and K. Yamashita, *J. Chromatogr. B, 824*: 258, 2005.
209. F. Beaudry, S. A. Guenette, A. Winterborn, J. F. Marier, and P. Vachon, *J. Pharm. Biomed. Anal., 39*: 411, 2005.

210. P. A. Lehman and T. J. Franz, *J. Pharm. Sci., 85*: 287, 1996.
211. S. Gao, D. M. Wilson, L. E. Edinboro, G. M. McGuire, S. G. P. Williams, and H. T. Kernes, *J. Liq. Chromatogr. Related Technol., 26*: 3413, 2003.
212. F. Xu, H. Jiao, Y. Tian, B. Zhang, and Y. Chen, *J. Mass Spectrom., 41*: 685, 2006.
213. A. Medvedovici and F. Albu, unpublished results.
214. Z. Miao, A. Kamel, and C. Prakash, *Drug Metab. Dispos., 33*: 879 2005.
215. C. M. Selavka, I. S. Krull, and I. S. Lurie, *J. Chromatogr. Sci., 23*: 499, 1985.
216. N. Simeon, R. Myers, C. Bayle, M. Nertz, J. K. Stewart, and F. Couderc, *J. Chromatogr. A, 913*: 253, 2001.
217. A. H. Scholten, P. L. Welling, U. A. Brinkman, and R. W. Frei, *J. Chromatogr., 199*: 239, 1980.
218. J. L. Meck and C. Eva, *J. Chromatogr., 317*: 343, 1984.
219. N. Kaneda, M. Asano, and T. Nagatsu, *J. Chromatogr., 360*: 211, 1986.
220. S. Ikenoya, O. Hiroshima, M. Ohmae, and K. Kawabe, *Chem. Pharm. Bull., 28*: 2941, 1980.
221. W. J. Bachman and J. T. Stewart, *J. Liquid Chromatogr., 12*: 2947, 1989.
222. K. Shimada, M. Tanaka, T. Nambara, and Y. Imai, *J. Pharm. Sci., 73*: 119, 1984.
223. K. Shimada, M. Tanaka, T. Nambara, Y. Imai, K. Abe, and K. Yoshinaga, *J. Chromatogr. B, 227*: 445, 1982.
224. K. Shimada, T. Oe, and T. Nambara, *J. Chromatogr., 419*: 17, 1987.
225. M. Nakajima, S. Yamato, H. Wakabayashi, and K. Shimada, *Biol. Pharm. Bull., 18*: 1762, 1995.
226. R. Wintersteiger and M. J. Sepulveda, *Anal. Chim. Acta, 273*: 383, 1993.
227. V. K. Boppana, L. Geschwindt, M. J. Cyronak, and G. Rhodes, *J. Chromatogr., 592*: 317, 1992.
228. J. Leube and G. Fischer, *J. Chromatogr. B, 665*: 373, 1995.
229. G. D. Li, I. S. Krull, and S. A. Cohen, *J. Chromatogr. A, 724*: 147, 1996.
230. R. A. Sherwood, E. M. Bayliss, and O. Chappatte, *Clin. Chim. Acta, 203*: 275, 1991.
231. R. A. Sherwood, A. C. Titheradge, and D. A. Richards, *J. Chromatogr., 528*: 293, 1990.
232. R. Wintersteiger, M. H. Barary, F. A. El-Yazbi, S. M. Sabry, and A. A. M. Wahbi, *Anal. Chim. Acta, 306*: 273, 1995.
233. L. Millerioux, M. Brault, V. Gualano, and A. Mignot, *J. Chromatogr. A, 729*: 309, 1996.
234. M. J. Rose, S. M. Lunte, R. G. Carlson, and J. F. Stobaugh, *Anal. Chem., 71*: 2221, 1999.
235. K. Shimada, M. Tanaka, and T. Nambara, *J. Chromatogr., 307*: 23, 1984.
236. S. J. Woltman, J. G. Chen, S. G. Weber, and J. O. Tolley, *J. Pharm. Biomed. Anal., 14*: 155, 1995.
237. W. Yonekawa, H. J. Kupferberg, and T. Lambert, *J. Chromatogr., 276*: 103, 1983.
238. P. Padovani, C. Deves, G. Bianchetti, J. P. Thenot, and P. L. Morselli, *J. Chromatogr., 308*: 229, 1984.
239. T. Fossati, M. Colombo, C. Castiglioni, and G. Abbiati, *J. Chromatogr. B, 656*: 59, 1994.
240. H. Wakabayashi, M. Nakajima, S. Yamato, and K. Shimada, *J. Chromatogr., 573*. 154, 1992.
241. M. Rancan, A. G. Sabatini, G. Achilli, and G. C. Galletti, *Anal. Chim. Acta, 555*: 20, 2006.
242. M. Siluveru and J. T. Stewart, *J. Chromatogr. B, 682*: 89, 1996.
243. S. Pelletier and C. A. Lucy, *J. Chromatogr. A, 972*: 221, 2002.
244. Y. Dobashi, T. Santa, K. Nakagomi, and K. Imai, *Analyst, 127*: 54, 2002.
245. L. Huber, *LC-GC International*, Feb. 96, 1998.
246. M. Otto, *Chemometrics: Statistic and Computer Application in Analytical Chemistry*, Wiley-VCH, Weinheim:1999, p. 84.

247. M. Li, Y. Alnouti, R. Leverence, H. Bi, and A. I. Gusev, *J. Chromatogr. B, 825*: 152, 2005.
248. Y. K. Hsieh, L. Lin, W. Fang, and B. K. Matuszewski, *J. Liq. Chromatogr. Related Technol., 26*: 895, 2003.
249. K. Takezawa, M. Tsunoda, K. Murayama, T. Santa, and K. Imai, *Analyst, 125*: 293, 2000.

9 Countercurrent Chromatography: From the Milligram to the Kilogram

Alain Berthod

CONTENTS

9.1   INTRODUCTION

First proposed in 1966 by Ito [1], countercurrent chromatography (CCC) gradually begins to be known by the separation community. However, it is still associated with numerous negative ideas such as incredibly complicated apparatuses, difficult operation, poor efficiency, and lengthy procedures. This is slowly evolving. It is very simple to present the CCC technique: CCC is just a liquid chromatography (LC) technique working with a support-free liquid stationary phase [2].

The very name of the technique is confusing since there is absolutely no fluid flowing in a countercurrent fashion. The naming came from the Craig machine that worked with two liquid phases and a countercurrent distribution (CCD) method [3,4]. The name CCC was pointed out as inappropriate in the 1972 Gordon conference but Craig himself defended it stating that the CCD and CCC principles were similar [5]. In more than 40 years, the CCC naming and the CCC acronym were used in more than 2500 scientific communications, articles, and books [6]. They will stay and will be used throughout this chapter.

In an earlier chapter in this book series, we presented the bases of CCC focusing on its preparative side and applications [7]. In this chapter, we shall focus on the twenty-first century CCC developments actualizing the recent applications.

9.2   "KEEP IT SIMPLE" OR THE BASES OF THE CCC TECHNIQUE

CCC works with a liquid stationary phase. The main advantage of this configuration is that the solutes can access the volume of the stationary phase and not just its surface (solid stationary phase). Because the stationary phase is liquid the CCC is a preparative technique. It is possible to load much larger amounts of material on a CCC column than on a prep LC column with silica stationary phase of similar size [7].

The advantages of a liquid stationary phase are so important that it is worth the trouble to work with the rotating CCC columns. Edward (Fang Te) Chou (1946–2004), the founder of Pharma-Tech Inc., a company that built a line of excellent hydrodynamic (HD) CCC columns, always recommended to "keep it simple." Yes, the CCC columns are a problem compared to high-performance liquid chromatography (HPLC) columns, but try to work with a silica mobile phase. Besides the first advantage of a liquid stationary phase, column loading and the preparative capability of CCC, the other main advantages are listed hereafter.

9.2.1   Simple Retention Mechanism

In a CCC column of internal volume V_C, there are only a volume V_M of mobile phase and a volume V_S of stationary phase. We have the trivial relationship:

$$V_C = V_M + V_S \qquad (9.1)$$

The only driving force responsible for solute retention is liquid–liquid partitioning between the two phases. The solute distribution ratio, also called partition coefficient, K_D, is defined as the ratio of the solute concentration in the stationary phase over that in the mobile phase. Since there is no silanol interaction, no size-exclusion, no strong adsorption, pH problems, or other phenomena that are encountered when working with a solid silica stationary phase, the CCC retention equation is simply:

$$V_R = V_M + K_D V_S \qquad (9.2)$$

The first consequences of the simple CCC retention equation are reproducibility and ease of scale-up. The solute K_D distribution ratio depends only on the biphasic liquid system used, not on the CCC column volume. If the same biphasic liquid system is used, the same solutes will have the same K_D ratio. It is possible to measure the stationary phase volume, V_S, retained in the CCC column and then to predict the solute retention volumes, V_R, using Equation 9.2. Scaling-up means that a larger column will be used. This has no effect on the solute K_D ratios. It will be still possible to predict solute retention volumes if the volume of liquid stationary phase retained in the large column is known. However, Equation 9.2 shows that the liquid phase volumes in the CCC column matters greatly to the solute retention volume, V_R. This will be illustrated later.

9.2.2 STATIONARY PHASE RETENTION IS PARAMOUNT

Column manufacturers sell gas chromatography (GC) or HPLC columns with a test chromatogram and often with a little vial containing the solute mixture used for the test. Injecting the mixture using the indicated conditions (mobile phase composition and temperature program) should reproduce the test chromatogram. If the test chromatogram is not obtained, it means that something is wrong either in the system or with the column. This procedure is not possible in CCC because the stationary phase is not independent from the mobile phase. The two liquid phases are chemically linked by the phase diagram. Any change in the composition of one phase will modify the composition of the other phase. The volume ratio of the two phases may change as well as their composition. The two liquid phases are also linked by Equation 9.1. A special stationary phase retention parameter, Sf, was introduced in CCC to compare the performance of different CCC columns between them or to study a particular liquid system in a given column:

$$Sf = V_S / V_C \qquad (9.3)$$

The resolution power of a CCC column depends critically on its ability to retain a large percentage of stationary phase (high Sf). The chromatographic resolution factor depends on two parameters only: the distance between two peaks and the peak width at base, W

$$Rs = \frac{V_{R_2} - V_{R_1}}{\dfrac{W_2 + W_1}{2}}$$ (9.4)

Introducing N, the number of theoretical plates:

$$N = 16(V_R / W)^2$$ (9.5)

and using Equations 9.1 through 9.4, it is possible to express the resolution equation as

$$Rs = \frac{\sqrt{N}}{4} \frac{K_{D2} - K_{D1}}{\dfrac{1 - Sf}{Sf} + \dfrac{K_{D2} + K_{D1}}{2}} = Sf \frac{\sqrt{N}}{4} \frac{K_{D2} - K_{D1}}{1 - Sf[1 + (K_{D2} + K_{D1})/2]}$$ (9.6)

The resolution factor and the stationary phase retention ratio are directly related. This is illustrated by Figure 9.1.

Figure 9.1 shows the separation of the same mixture of seven test solutes (Table 9.1) with the same liquid system and the same CCC column. The difference between the chromatograms is the amount of stationary phase retained in the column. A dramatic decrease in resolution is observed between the chromatogram with $Sf = 90\%$ with seven peaks baseline resolved and the chromatogram with $Sf = 10\%$ where only two peaks are distinguishable.

As already stated, the solute distribution ratios (or partition coefficients) depend on the liquid system used and neither on the CCC column used and nor on the amount of liquid stationary phase retained in the CCC column. It was recently proposed to use the solute K_D values instead of the solute retention volumes in retention plots [8,9]. Equation 9.2 shows that the solute K_D is related to the retention volumes by

$$K_D = \frac{V_R - V_M}{V_C - V_M} = \frac{V_R - V_M}{V_S}$$ (9.7)

It is straightforward to use Equation 9.7 and convert the retention volumes in K_D values. Figure 9.2 shows the chromatograms of Figure 9.1 expressed as K_D plots. In such plots, the position of the peak centroid for each compound does not change. However, the peak width broadens dramatically as the V_M volume increases or as the V_S volume decreases:

$$\frac{dV_R}{dK_D} = V_S$$ (9.8)

9.2.3   Partition Coefficients versus Retention Factors

The CCC column used in the Figure 9.1 had an efficiency of about 300 theoretical plates. This is considered as a ridiculously low efficiency by many chromatographers.

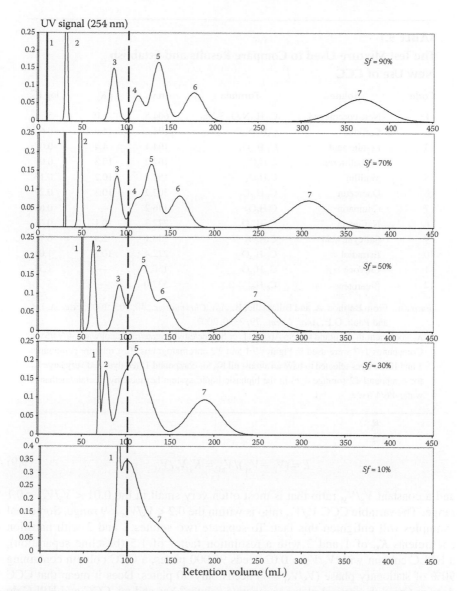

FIGURE 9.1 The importance of the amount of stationary phase retained in a 100 mL CCC column (vertical dotted line). Chromatograms of the same mixture of seven solutes (listed in Table 9.1) obtained with the same CCC column and the same biphasic liquid system: hexane/ethyl acetate/methanol/water 4:6:4:6 v/v. Aqueous lower mobile phase, 2 mL/min, rotor rotation 800 rpm. Sf ratios obtained by filling the column with the required amount of upper organic phase.

This efficiency is sufficient to obtain baseline peak separation if the column holds enough stationary phase. HPLC and CCC are not competitors. Both techniques do not work in the same range of solute polarity. HPLC uses solute retention factors noted k:

TABLE 9.1

The Test Mixture Used to Compare Results and Establish New Use of CCC

Code	Solute	Formula	m.w.	pK_a	log K_D^a
1	New coccine red	$C_{20}H_{11}N_2O_{10}S_3Na_3$	604.5	1.0	−5
2	Caffeine	$C_8H_{10}N_4O_2$	194.2	14.0	−0.6
3	Ferulic acid	$C_{10}H_{10}O_4$	194.1	4.4	−0.07
4	Umbelliferone	$C_9H_6O_3$	162.2	10.3	0.06
5	Vanillin	$C_8H_8O_3$	152.2	10.2	0.15
6	Quercetin	$C_{15}H_{10}O_7$	302.2	10.3	0.27
7	Coumarin	$C_9H_6O_2$	146.2	—	0.60
8	Narigenin	$C_{15}H_{12}O_5$	272.3	10.1	0.65
9	Salicylic acid	$C_7H_6O_3$	138.1	2.9	0.78
10	Estradiol	$C_{18}H_{24}O_2$	272.4	10.5	1.04
11	Carvone	$C_{10}H_{14}O$	150.2	—	1.38
12	β-carotene	$C_{40}H_{56}$	537.0	—	5

Sources: From Berthod, A. and Billardello, B., *Adv. Chromatogr.*, 40, 503, 2000; Friesen, J.B. and Pauli, G.F., *Anal. Chem.*, 79, 2320, 2007.

[a] K_D coefficient in the liquid system Arizona L or hexane/ethyl acetate/methanol/water 4/6/4/6 v/v. Compounds 1–7 were used in Figures 9.1 and 9.2 chromatograms. Test mixture compounds 1 and 12 are dyes selected to have an almost nil K_D for compound 1 (red dye) and very large K_D for compound 12 (orange dye) in the biphasic liquid system hexane/ethyl acetate/methanol/ water 4/6/4/6 v/v.

$$k = (V_R - V_M)/V_M = K_D V_S/V_M \tag{9.9}$$

and a constant V_S/V_M ratio that is most often very small in the $0.01 < V_S/V_M < 0.1$ range. The variable CCC V_S/V_M ratio is within the $0.2 < V_S/V_M < 9$ range. Some real examples will enlighten this fact: To separate two solutes 1 and 2 with partition coefficients K_D of 1 and 2 with a resolution factor of 1.5 (baseline separation), a HPLC column with $V_S/V_M = 0.02$ needs 95,000 plates, a CCC column containing 90% of stationary phase ($V_S/V_M = 9$) needs only 90 plates. Does it mean that CCC does not need theoretical plates to separate solutes? Yes and no. CCC and HPLC do not work in the same polarity domains. For HPLC, the two solutes with K_D of 1 and 2 are actually two solutes with retention factors of 0.02 and 0.04 (Equation 9.9). These two solutes are considered as very poorly retained by the column and elute too close to the hold-up volume to be correctly separated. For CCC, the same two solutes have retention factors of 9 and 18 with absolutely no problem to separate them with 90 plates. Of course, 200, 500, or 2000 plates would even be better (Equation 9.6), but these efficiencies are not needed.

The solute retention factor is the important parameter in HPLC because the V_S/V_M ratio is constant. Since it is also small, the solute distribution ratio, K_D, must be significant. The HPLC column works well with compounds having retention factors

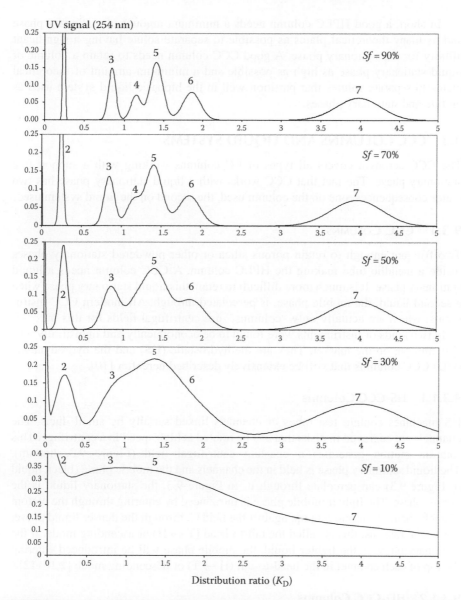

FIGURE 9.2 K_D plots of the Figure 9.1 chromatograms. The retention volumes were converted in K_D values using Equation 9.7. All compounds stay at the same K_D but their representative peaks broaden dramatically as the amount of stationary phase inside the column decreases.

between 0.2 and 10, meaning K_D ratios between 10 and 500. The solute distribution ratio is the important parameter in CCC because the V_S/V_M ratio is variable. The CCC column works well with compounds having partition ratios K_D between 0.1 and 10 meaning retention factors between 0.9 and 90 with an Sf of 90% ($V_S/V_M = 9$). Clearly the HPLC and CCC polarity ranges are complimentary.

In short, a good HPLC column needs a minimum amount of stationary phase and as many theoretical plates as possible to separate solute having a significant affinity for the stationary phase. A good CCC column needs to retain a volume of liquid stationary phase as high as possible and a minimum amount of theoretical plates to separate solutes that partition well in the biphasic liquid system used as mobile and stationary phases.

9.3   CCC COLUMNS AND LIQUID SYSTEMS

The CCC acronym covers all types of LC columns working with a support-free stationary phase. The fact that CCC works with a liquid stationary phase has two major consequences, one on the column used, the second on the liquid system used.

9.3.1   CCC COLUMNS

Two frits are enough to retain porous silica or other powdered stationary phases inside a metallic tube making the HPLC column. A CCC column needs a liquid stationary phase. It is much more difficult to retain this liquid stationary phase when a second liquid, the mobile phase, is percolated through. All modern CCC instruments, which are actually only "columns," use centrifugal fields for this purpose. Only two kinds of instruments were found to be economically and industrially reliable and put on the market. They are the hydrostatic (HS) and the hydrodynamic (HD) CCC columns that will be extensively described hereafter [10].

9.3.1.1   HS-CCC Columns

HS machines contain test tubes or channels linked serially by small ducts. The channels are engraved in disks that can be rotated at high speed in a centrifuge. This rotation motion generates a constant centrifugal field (Figure 9.3, bottom). The liquid stationary phase is held in the channels and the mobile phase (black liquid in Figure 9.3) can percolate through it. In Figure 9.3, the stationary liquid is the denser phase. The lighter mobile phase is introduced by entering through the bottom of each channel and moving up, against the field G, through the denser liquid phase. For trivial reasons, this is called the tail-to-head (T → H) or ascending mode. If the stationary phase is the lighter liquid, the mobile phase will be introduced entering the top of each channel in the head-to-tail (H → T) or descending mode [2,10–12].

9.3.1.2   HD-CCC Columns

HD machines contain a rotor with one or several spools coiled with tubing. A gear arrangement produces a planetary motion of the spools around the central axis. This planetary motion generates a highly variable centrifugal field. When two immiscible liquid phases are introduced in the spool, they are submitted to the variable field which produces a succession of mixing and decantation zones in each successive turn of the tubing (Figure 9.3, top). Because of the thread of the coiled tube, the zones move toward the on side of the coil called "head," the high-pressure side (Archimedean forces [13]). For reasons not yet well understood, it is observed that one liquid phase is retained while the other phase percolates through it. By analogy

Mixing zone

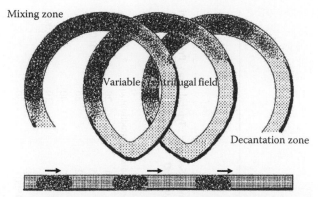

Variable centrifugal field

Decantation zone

The two liquid phases contact in a succession of moving, mixing and settling zones

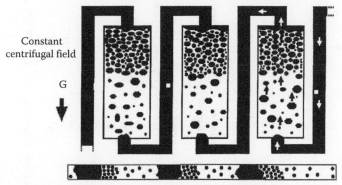

Constant centrifugal field

G

The two liquid phases contact inside the fixed channels only not inside the ducts

FIGURE 9.3   Top: Scheme of HD principle. The two liquid phases contact inside an open tube coiled on a bobbin rotating with a planetary motion. Bottom: Scheme of the hydrostatic principle, HS. The mobile phase is shown in black flowing in the T→H or ascending direction.

with the HS apparatuses, one end of the tubing is called "head," the other end is called "tail" and the motion of the mobile phase is similarly termed H→T (or descending if the mobile phase is the denser phase) or T→H (or ascending if the mobile phase is the lighter phase) [2,10–12].

9.3.1.3 Commercially Available CCC Columns

Table 9.2 lists the commercially available CCC columns along with the direction and Web sites of the producers. The prices were those posted on the Web site www.cherryinstruments.com valid for the US market and years 2007–2008. In most cases, the desired CCC column can be tailored on specifications.

Major developments were made on large columns for preparative kilogram-scale production. Reliable HD large CCC columns were developed by the UK team of Pr. Ian Sutherland at the University of Brunel, West London and the Dynamic Extractions Company was created to commercialize the columns.

TABLE 9.2

Some Commercially Available CCC Columns

Model Name	Volume (L)	Size (L×W×H, cm)	Weight (kg)	Maximum Rotation (rpm)	Maximum Pressure (kg/cm²)	Maximum Flow Rate (mL/min)	Maximum Sample Load (g)	Typical Run Time (min)	Price Level (US$)
HD Columns									
Dynamic Extractions 890, Plymouth Rd, Slough, Berkshire, SL1 4LP, UK, Tel: +44 1753 696979, Fax: +44 1753 696976, www.dynamicextractions.com									
Spectrum	0.02					1	0.1	20	
	0.05	46 × 52 × 48	65	2100	8	5	0.5	20	~100,000
	0.125					10	1.2	30	
Midi	0.04					1	0.2	40	
	0.5	58 × 66 × 56	85	1400	6	100	5	10	~130,000
	1					100	20	20	
Maxi	5	200 × 270 × 150	500	850	6	1500	1500	30	–
Quattro AECS P.O. Box 80, Bridgend, South Wales, CF31 4XZ, UK, Tel +44 1656 782985, Fax: +44 1656 789 282, www.ccc4labprep.com									
Quickprep MK5	0.02					1	0.5	40	
	0.1	42 × 42 × 45	80	860	8	5	4	40	~30,000
	0.85					30	30	80	

Tauto Biotech 326, Aidisheng Road, Zhangjiang High-Tech Park, 201203 Shanghai, China, Tel: +86-21-51320588, Fax: +86-21-51320502, www.tautobiotech.com

TBE 20A	0.02	33 × 60 × 50	60	1500	15	1	0.5	40	30,000
TBE 300A	0.26	53 × 52 × 33	95	900	10	2	5	200	35,000
TBE 1000A	1	62 × 91 × 104	400	600	12	10	20	200	44,000

HS Columns

Armen Instruments 15 Rue Ampère, Z.I. de Kermelen, 56890 Saint Ave, France, Tel +33 297 618 400, Fax: +33 297 618 500, www.armen-instrument.com

Elite CPC100	0.1	48 × 48 × 47	60	2500	140	10	2	20	~50,000
Elite CPC250	0.25	48 × 48 × 47	60	2500	140	25	5	20	~60,000
Elite CPC1000	1	56 × 68 × 56	80	2500	140	100	20	20	~100,000

Kromaton SEAB 9 Rue Alexander Fleming, 49066 Angers, France, Tel +33 241 774 148, Fax: +33 241 739 623, www.kromaton.com

FCPC C50	0.05	33 × 52 × 60	45	2000	60	10	1	10	32,000
FCPC A200	0.2	72 × 68 × 48	65	2000	60	20	5	20	50,000
FCPC A1000	1	72 × 68 × 48	80	1500	50	30	30	60	63,000
FCPC A5000	5	100 × 67 × 117	410	1200	50	150	150	60	250,000

Note: US prices were obtained from the CCC Web site: www.Cherryinstruments.com, Delivery in continental USA, US$ value September 2007.

FIGURE 9.4 The Maxi and the Spectrum HD-CCC columns from the Dynamic Extraction Ltd. British company. Pr. Ian Sutherland poses inside the rotor to give the scale. The Maxi is capable to process 1.5 kg of injected material per run. The Spectrum HD-CCC column can work with different coil volumes between 20 and 125 mL. (Photograph from Dynamic Extraction.)

Figure 9.4 shows the Maxi, an 18 L HD column that was used to process kilogram amounts of raw material. Figure 9.5 shows the industrial prototype of a 12.5 L HS column made by Armen Instruments in France. The smaller Elite CPC 250 model by the same company is also shown.

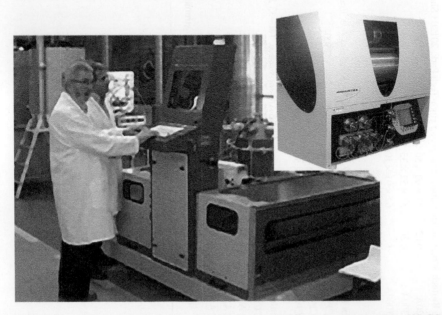

FIGURE 9.5 Industrial 12.5 L Armen Elite Continuum FCPC and analytical Elite HS-CCC (250 mL) columns. The 12.5 L Armen FCPC is the largest HS-CCC column ever built. Both columns are delivered with a fully operational pumping system. (Photography from Armen Instruments.)

All the CCC columns presented in Table 9.2 and Figures 9.4 and 9.5 are actually only containers. The mobile phase and the liquid stationary phase that will be used in the column are both part of the same biphasic liquid system.

9.3.2   LIQUID SYSTEMS

The liquid system is another major difference between HPLC and CCC. In HPLC the operating mode, normal-phase mode or reversed-phase mode is first selected, next the corresponding column is chosen between a reduced number of solid stationary phases and last the mobile phase composition is adapted to the sample. The procedure is completely different in CCC. Selecting a liquid system implies selecting the mobile phase and the liquid stationary phase at the same time. Any change in the composition of the mobile phase will change the composition of the stationary phase: CCC gradient elution is risky and only possible in limited occasions.

The choice of the solvent system fully depends on the sample to be separated. A good way is to look for the best solvent for the sample [7]. Once this solvent is found, a biphasic liquid system can be formed with it using two immiscible solvents. Most often water is part of the system as the most polar, available, cost-effective, and environment-friendly solvent. A wide range of liquid systems has been described in review articles [14,15]. A system was recently found to be the most commonly used to fractionate natural products; it is the heptane/ethyl acetate/methanol/water system [16].

9.3.2.1   Heptane/Ethyl Acetate/Methanol/Water System

The hexane/ethyl acetate/methanol/water system was first proposed by Oka in 1991 [17]. Margraff standardized the use of this system and defined a range of 24 proportions coded by alphabetic letters from A to Z (and no I and no O) [18]. He demonstrated that if the components that should be separated are located in the ethyl acetate less polar upper phase of the two-solvent ethyl acetate–water system A and in the methanolic more polar lower phase of the two-solvent methanol–heptane system Z, then, there is necessarily a four-solvent biphasic system between composition A and Z in which the components will be equally distributed [14,16–18]. The AZ range of biphasic liquid system became very popular known as the "Arizona" liquid system from the AZ postal abbreviation of the US state, Arizona [19].

It was recently demonstrated that changing hexane for heptane, a less toxic alkane, had minor effect on the Arizona system [19]. Table 9.3 lists the AZ compositions in v/v and percentage compositions. The Reichardt polarity of the upper organic and the lower aqueous phase are also indicated. In this normalized scale, obtained by measuring the bathochromic shift of the absorbance maximum of the Reichardt dye (a pyridinium-N-phenoxide betaine dye) 100 is assigned to water and 0 to tetramethylsilane [20]. Table 9.3 shows that the polarity of the compositions changes monotonously, the polarity of the two phases decreasing together as the amounts of water (lower phase) and ethyl acetate (upper phase) both decrease similarly. However, the solute partitioning between the two phases can change dramatically when the liquid composition change from one letter to the next one [16–19]. Since ethyl acetate is slowly hydrolyzed by water in acetic acid and ethanol, the water-rich polar compositions A through H should be prepared daily.

TABLE 9.3
The A to Z Compositions of the "Arizona" Range of Biphasic Liquid Systems

Letters	v/v				Initial % v/v				Up/Low. Phase Ratio	Reichardt Polarity Low/Up
	Heptane	Ethyl Acetate	Methanol	Water	Heptane	Ethyl Acetate	Methanol	Water		
A	0	1	0	1	0.0	50.0	0.0	50.0	0.88	100/50
B	1	19	1	19	2.5	47.5	2.5	47.5	0.92	90/51
C	1	9	1	9	5.0	45.0	5.0	45.0	0.965	88/52
D	1	6	1	6	7.1	42.9	7.1	42.9	0.96	85/53
F	1	5	1	5	8.3	41.7	8.3	41.7	0.95	84/53
G	1	4	1	4	10.0	40.0	10.0	40.0	0.95	83/53
H	1	3	1	3	12.5	37.5	12.5	37.5	0.945	82/53
J	2	5	2	5	14.3	35.7	14.3	35.7	0.91	80/54
K	1	2	1	2	16.7	33.3	16.7	33.3	0.88	79/55
L	2	3	2	3	20.0	30.0	20.0	30.0	0.84	78/55

M	5	5	6	22.7	22.7	27.3	27.3	27.3	0.80	77/54
N	1	1	1	25.0	25.0	25.0	25.0	25.0	0.70	76/53
P	6	6	5	27.3	27.3	22.7	22.7	22.7	0.69	77/54
Q	3	3	2	30.0	30.0	20.0	20.0	20.0	0.68	77/52
R	2	2	1	33.3	33.3	16.7	16.7	16.7	0.68	77/51
S	5	5	2	35.7	35.7	14.3	14.3	14.3	0.70	77/51
T	3	3	1	37.5	37.5	12.5	12.5	12.5	0.735	77/51
U	4	4	1	40.0	40.0	10.0	10.0	10.0	0.76	76/50
V	5	5	1	41.7	41.7	8.3	8.3	8.3	0.78	76/40
W	6	6	1	42.9	42.9	7.1	7.1	7.1	0.775	76/28
X	9	9	1	45.0	45.0	5.0	5.0	5.0	0.77	75/26
Y	19	19	1	47.5	47.5	2.5	2.5	2.5	0.71	74/25
Z	1	1	0	50.0	50.0	0.0	0.0	0.0	0.45	73/23

Note: The upper over lower phase volume ratios and Reichardt polarity index measured with the Reichardt dye (2,6-diphenyl-4-(2,4,6-triphenylpyridinio)phenolate, CAS 10081-39-7) are given for the freshly prepared liquid systems. Heptane can be replaced by hexane, isooctane, and/or petroleum ether with minimum polarity changes [19].

9.3.2.2 Wide Variety of Possible Liquid Systems

The "Arizona" system is extremely useful and the most used for natural products [14,16,19]. The accessible polarity range with biphasic liquid systems is much wider than that of this quaternary system. Table 9.4 lists some biphasic liquid systems used in CCC. The less polar liquid systems are the waterless systems such as heptane/dimethylsulfoxyde, heptane/acetonitrile, or heptane/methanol. The more polar systems are the aqueous two-phase systems (ATPSs) such as the phosphate/polyethylene glycol (PEG) ATPS or the dextran/PEG ATPS biphasic systems.

The chlorinated solvents were very popular in CCC because they have excellent solvent properties and they are denser than water. The methanol/chloroform/water system was for a long time the favorite system of Ito [10,12,21]. Today, the toxicity

TABLE 9.4
Nonchlorinated Biphasic Liquid Systems Used in CCC

No	Less Polar Solvent	Intermediate Solvent	Polar Solvent	Applications
Waterless Biphasic Liquid Systems				
1	Heptane	—	DMSO	Petroleum products
			Furfural	Essential oils
			NMP	
			DMF	
			DMA	
2	Heptane		ACN	Triglycerides, vegetable
			MeOH	oils, animal fat
Systems with Intermediate Polarities				
3	Heptane	MeOH/AcOEt, See Table 9.3	Water	Natural products
4	AcOEt		Water	Natural products
5	MTBE or MIBK		Water	Natural products
6	MTBE or AcOEt	BuOH or ACN	Water	Polyphenols
Polar Biphasic Systems				
7	BuOH		Water	Polar natural products (e.g., saponins)
8	Octanol		Water	$K_{o/w}$ measurements
9	BuOH or PeOH	PrOH or EtOH or MeOH	Water	Amino-acids
10	BuOH	AcOH	Water	Polar compounds
11	Phosphate Dextran	PEG	Water	Proteins

Note: ACN, acetonitrile; AcOEt, ethyl acetate; AcOH, acetic acid; BuOH, butanol; DMA, dimethylacetamide; DMF, dimethylformamide; DMSO, dimethylsulfoxyde; EtOH, ethanol; MeOH, methyl isobutyl ketone; MIBK, methanol; MTBE, methyltertiobutyl ether; PEG, polyethylene glycol; PeOH, pentanol; PrOH, *n*-propanol. Heptane is listed as the less toxic alkane; it can be chemically replaced by hexane or other alkanes.

of the chlorinated solvents and environmental concerns are reasons explaining why their use is avoided. Alternative solvents were found [22]. However, nitriles, esters, ethers, and/or ketones that have similar polarities and solvent capabilities as chlorinated solvents lack the high density. They are all lighter than water. Without the chlorinated solvents, the organic phases in the CCC biphasic liquid systems listed in Table 9.4 are always the upper phases.

It is possible to perform chemical reactions in the liquid stationary phase. Acid–base reactions are used to perform "pH-zone refining" CCC. In this CCC way, a sharp change of pH is produced using an acidic mobile phase and a basic stationary phase (or *vice versa*) and performing displacement chromatography of ionizable solutes [23]. Complexation reactions were used to separate inorganic ions [24,25]. A wide range of chelating agents can be dissolved in e.g., a heptane liquid phase and contacted with a buffered aqueous phase containing metal cations. The apolar chelates are partly soluble in the heptane phase which allows for separation.

9.4   OPERATING A CCC COLUMN

Once the liquid system is selected, the stationary phase, i.e., the denser or the lighter liquid phase, must be selected. Reversed-phase mode will work with the organic liquid stationary phase and the polar aqueous mobile phase, and *vice versa* for the normal-phase mode [22].

9.4.1   Column Preparation

The CCC column is first filled with the stationary phase. The maximum flow rate can be used and it is not necessary to put the rotor in rotation. Once the column is full of stationary phase, it is observable at the column exit where drops of stationary phase are visible. Then, the centrifugal field must be established putting the rotor in rotation at the desired rpm value. A graduated cylinder is placed at the column exit to collect all effluents. The mobile phase is introduced in the CCC column in the right direction: descending or head-to-tail if the mobile phase is the denser phase, ascending, or tail-to-head if the mobile phase is the lighter phase.

As the mobile phase enters the column, it establishes equilibrium with the stationary phase in the first turns of the coil (HD columns) or the first channels (HS columns) and stationary phase is displaced and seen leaving the column. As the equilibrium between phases progress in the column, more stationary phase is collected in the graduated cylinder at the column exit. At a point, a drop of mobile phase is collected at the column exit meaning that the whole CCC column is equilibrated, ready for sample injection. The volume of collected stationary phase corresponds to V_M the volume of mobile phase inside the column. The volume of stationary phase remaining in the column, V_S, is $V_C - V_M$ and the *Sf* factor can be calculated (Equation 9.3).

9.4.2   Stationary Phase Retention

As illustrated by Figures 9.1 and 9.2, it is of paramount importance to retain a maximum volume of stationary phase inside any CCC column. However, the HS and HD columns do not behave similarly.

9.4.2.1 HD Columns

It was demonstrated that the HD column retained an amount of stationary phase that decreased with the square root of the mobile phase flow rate [13,26]:

$$Sf = A - B\sqrt{F} \tag{9.10}$$

in which A and B are constants, A being close to 100% [25]. Wood et al. demonstrated that the B gradient could be expressed as [13,27]

$$B = \frac{800}{\pi d_C^2} \sqrt{\frac{\mu_M}{\omega^2 R(\rho_L - \rho_U)}} \tag{9.11}$$

where
   d_C is the internal diameter of the coiled tube
   μ_M is the mobile phase viscosity
   ω is the rotor rotation speed
   R is the rotor radius (distance between the spool axis and the rotor axis)
   ρ_L and ρ_U are respectively the lower phase and upper phase density

Equation 9.10 shows that a high mobile phase flow rate produce low stationary phase retention in the HD-CCC column. Equation 9.11 shows that an HD column coiled with large bore tubing, rotating rapidly in a large rotor machine and with a liquid system with a significant density difference between phases will hold the liquid stationary phase tightly. High flow rates will then be possible.

9.4.2.2 HS Columns

Figure 9.3 (bottom) shows that the ducts interconnecting channels contain only mobile phase. Then there is a maximum amount of retained stationary phase: the column volume, V_C, minus the interconnecting duct volume. This later volume makes 10–15% of the HS column volume; so the maximum Sf value for a HS column is between 85% and 90%. Equation 9.10 does not apply for HS columns [28]. A slight linear decrease of Sf is observed when the flow rate is increased up to a particular high value: the flooding flow rate [28,29]. Above this flow rate, all the liquid stationary phase is pushed out of the HS column. The flooding flow rate depends on the rotor rotation speed and on the biphasic liquid system used.

9.4.3 COLUMN DRIVING PRESSURE AND SCALING-UP

Another major difference between the two types of CCC columns is the driving pressure. The pressure drop, ΔP, in operating a CCC column at equilibrium is the sum of a HS term and an HD term [2,28]:

$$\Delta P = C \Delta \rho \omega^2 Sf + D \mu_M F \tag{9.12}$$

C and D are constants depending on the column geometry.

   The hydrostatic term is the dominant term with HS columns. It is produced by the HS pressure which is produced by the two immiscible phases present in each channel (Figure 9.3, bottom). At the beginning of an experiment, the CCC column is full of stationary phase, then the hydrostatic term of Equation 9.12 is nil since there is a single phase in each channel. As the mobile phase moves in the HS column, equilibrating more and more channels, the driving pressure increases. This can be used to observe the column equilibration. The final pressure can reach values as high as 50–60 kg/cm^2 (5–6 MPa or 700–850 psi). If the driving pressure is seen going too high, the rotor rotation speed must be reduced before the mobile phase flow rate.

It was demonstrated that HD-CCC columns could be considered as working with a constant pressure drop [13]. The Hagen–Poiseuille equation in open tubes with laminar flow is

$$\Delta P = 8\mu_M LF/(\pi r^4) \tag{9.13}$$

where r is the internal tubing radius. It shows that the driving pressure needed to push a liquid with a viscosity, μ_M, in a tube of constant internal radius, r, increases linearly with the flow rate. However, in the case of the internal tubing of an HD-CCC column, the radius r is not constant. The amount of stationary phase retained in the column depends on Sf as shown by Figure 9.6. In a given tube, the apparent bore, radius r, available for the mobile phase increases if there is less stationary phase retained in the tube [13]. The Hagen–Poiseuille relationship (Equation 9.13) shows that the column pressure drop may remain constant when the flow rate increases. The driving pressure depends on the liquid phase density differences and the rotor rotation speed.

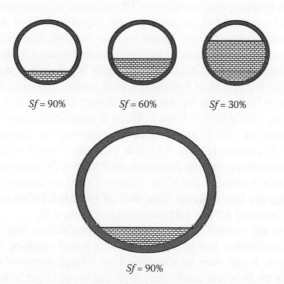

FIGURE 9.6   The concept of "Constant Pressure Drop Pump" with the HD-CCC columns. The mobile phase is shown in bricked area. The stationary phase is shown in open area, the apparent mobile phase bore increases as the stationary phase volume (and Sf factor) decreases. In the large bore tube, the mobile phase sees a similar bore with a high Sf factor (90%) than in the smaller tube with a 60% Sf factor.

TABLE 9.5
Liquid Stationary Phase Retention and Flow Rate with Three HD Columns

Column[a]	Spectrum 20 mL			Midi 125 mL			Maxi 5000 mL		
Tubing Bore (mm)	0.8			1.6			10		
(rpm)	B	F(85%) (mL/min)	t(V_C) (min)	B	F(85%) (mL/min)	t(V_C) (min)	B	F(85%) (mL/min)	t(V_C) (min)
600	—	—		17.2	0.76	164	1.7	81	62
800	—	—		12.9	1.35	93	1.3	144	35
1000	30.6	0.24	83	10.3	2.11	59	1	225	22
1500	20.4	0.54	37	6.9	4.75	26	0.7	—	—
2000	15.3	0.96	21	5.2	8.45	15	0.5	—	—

Sources: Data from Wood, P.L., Hawes, D., Janaway, L., and Sutherland, I.A., *J. Liquid Chromatogr.,* 26, 1373, 2003; Du, Q., Wu, C., Qian, G., and Ito, Y., *J. Chromatogr. A,* 835, 231, 1999.

Note: Liquid system, heptane/ethyl acetate/methanol/water 16/6/10/10 v/v; mobile phase, denser aqueous phase in the descending head-to-tail direction; stationary phase, upper organic phase. B is the gradient of the $Sf = f(F)$ relationship (Equation 9.10). $F(85\%)$ is the flow rate in mL/min producing a stationary phase retention factor of 85% at the indicated rotor rotation speed. $t(V_C)$ is the time needed to pass one column volume of mobile phase at the indicated flow rate.

[a] See Table 9.2 for full technical details of Dynamic Extraction columns.

This constant pressure drop property of HD columns is interesting when considering scaling-up for preparative chromatography. Equation 9.11 of the B gradient of the loss of stationary phase versus square root of the flow rate indicates that B is inversely proportional to the square of the coiled tube diameter. If the tube diameter is bigger, the column volume is bigger. However, it is possible to work with a high flow rate since the B gradient of Equation 9.10 is small (Equation 9.11). Table 9.5 lists the Equation 9.10 B values for the three different Dynamic Extraction HD columns for which the tubing bore is known. These B values were used to calculate the flow rate producing an 85% Sf factor, a value producing an acceptable resolution power for the CCC column (Figure 9.1). It is possible to work with a flow rate 10 times higher when comparing the Spectrum and Midi CCC columns just because the tubing bore was doubled in the Midi column. The Maxi column was made with 10 mm ID tubing, six times bigger than that of the Midi column. The flow rate producing an Sf factor of 85% is 100 times larger (Table 9.5).

These results are extremely important when production and throughput are considered. The separation can be optimized on a small machine to find the best liquid combination. It can then be transposed on a large machine with large bore coiled tube. Since the flow rate can be very high due to the larger tube bore, the separation can be done in a reasonable amount of time. It can even be done faster on the large column than on the smaller one. Table 9.5 lists the time needed to pass one column volume of mobile phase, $t(V_C)$, with a flow rate maintaining 85% of liquid stationary phase in the HD column. For example, this time is shorter than 1 h with the 20 mL Spectrum column that can rotate at high speed (2000 rpm) (20 min with a flow rate of 1 mL/min). The large Maxi column cannot rotate that fast, but at 800 rpm the time for one column volume is very similar (22 min with a flow rate of 225 mL/min, Table 9.5). The mass load in the 5 L column is in the kilogram range, four orders of magnitude higher than the mass load of the smaller 0.02 L column (Table 9.2) [30].

9.4.4 USING THE LIQUID NATURE OF THE STATIONARY PHASE

The liquid nature of the stationary phase allows for original uses of a CCC column that are just impossible in classical LC, i.e., with a solid stationary phase. Giddings studied the chromatographic process and demonstrated that the solute bandwidth inside the column, expressed using σ, the standard deviation, depends only on the band position, x, and on the column height equivalent to a theoretical plate, H, not on the solute affinity for the stationary phase [31]:

$$\sigma = \sqrt{xH} \qquad (9.14)$$

This result has little interest in classical LC with solid stationary phase since the solutes must be collected outside the column. In CCC, it has great interests since it is possible to extrude the column content as all phases are liquids.

9.4.4.1 The Dual-Mode Method: Regular Use of the Liquid Nature of the Stationary Phase

The dual-mode method was described very early [32]. It simply exchanges the liquid phase role: the liquid stationary phase becomes the mobile phase and *vice versa*. There are solutes still in the CCC column after the volume V_{CM} of mobile phase has been used in the classical mode. They are eluted after the phase inversion with the simple retention equation [33]:

$$V_R = V_{CM}(1 + 1/K_D) \qquad (9.15)$$

The solute retention volume is made of a volume V_{CM} of one phase of the liquid system used and a volume $V_{DM} = V_{CM}/K_D$ of the other phase of the biphasic liquid system. The dual-mode method was used to measure partition coefficient with the solute K_D, coefficient of the eluted after-phase inversion is simply the ratio of the two volumes used [33,34]:

$$K_D = V_{DM}/V_{CM} \qquad (9.16)$$

In this method, both the nature and the flowing direction of the mobile liquid phase are changed. The descending or T→H rule for the denser mobile phase and the ascending or H→T rule for the lighter mobile phase are respected. The solutes eluted in the classical mode have increasing K_D coefficient. The solutes eluted after-phase reversal elute in decreasing K_D order.

9.4.4.2 Elution–Extrusion Method

The elution–extrusion CCC method (EECCC) was recently proposed to enhance the polarity window in a single CCC run [35]. The EECCC method starts with a regular elution of the solutes for a V_{CM} volume of mobile phase. Next, it changes the nature of the mobile phase, but not the flowing direction. Pushing the liquid that was initially the stationary phase in the "wrong" direction induces the extrusion of all solutes present in the CCC columns maintaining their peak width [36]. EECCC chromatograms show peaks with increasing peak widths during the classical mode, V_{CM}, elution and peaks becoming sharper and sharper during the extrusion part [9,35]. The EECCC method was used to scan rapidly the content of natural product extracts [37]. The extract is injected and one column volume of mobile phase is passed in the elution mode followed by a one column volume of the other liquid phase that extrudes the whole column content. The elution step displaces the solutes with K_D up to 1. The extrusion step displaces all other solutes with $1 < K_D < \infty$. With two column volumes of phases, it is certain that all components of the extract are flushed out of the column.

9.4.4.3 Back-Extrusion Method

The back-extrusion CCC method (BECCC) is the last logical step in using the capabilities of a liquid stationary phase [38]. The BECCC method also starts with a regular elution with a V_{CM} volume of mobile phase. Next, it changes the flowing direction of the mobile phase, but not its nature. Pushing the mobile phase in the "wrong" way produces a sudden break in the liquid phase balance inside the CCC column. The initial stationary phase accumulates at the column exit carrying all contained solutes. The break of the phase equilibrium is so fast that the solutes distributed between the two phases will see their portion contained in one phase separated from that contained in the other phase. The BECCC should not be done with a small V_{CM} volume or "echo" peaks will show in the mobile phase following the extruded stationary phase [38].

Table 9.6 compares the eight possible ways to run a CCC column with a set of solutes having widely differing polarity. The four modes: classical mode, dual mode, EECCC, and BECCC can be used in the normal-phase mode or in the reversed-phase mode. These various ways to use a CCC column compensate for the lack of gradient elution commonly used in HPLC. The major difference is that the three methods that use the liquid nature of the stationary phase ensure for a complete elution of the whole sample. No compounds can stay irreversibly adsorbed inside a CCC column.

9.4.5 Detection

Microdroplets of the liquid stationary phase are often present in the mobile phase effluent. These droplets are responsible for spikes in UV detectors. They also completely blind refractive index detectors that are not usable in CCC. As commonly done in the early CCC works, it is always possible to collect column effluents in a fraction collector and analyze each tube. The chromatogram is constructed plotting the results versus time. Continuous UV detection is possible working with short light path length cells used in preparative LC. Since CCC is a preparative technique, the concentration in the effluent is often high so that a short light path length with reduced sensitivity is not a problem. The solute concentration can be actually so high that it is necessary to avoid detecting at the solute maximum absorption wavelength.

TABLE 9.6
Chromatographic Figures of Merit for the Separation of Five Compounds Using All Possibilities of a Liquid Stationary Phase in CCC

Lower Mobile Phase[a]

Solute	Reversed-Phase Elution				Dual-Mode			EECCC			BECCC		
	K_D	V_R (mL)	W_b (mL)	R_S	V_R (mL)	W_b (mL)	R_S	V_R (mL)	W_b (mL)	R_S	V_R (mL)	W_b (mL)	R_S
1—catechol	0.07	37.2	13	—	37.2	13	—	37.2	13	—	37.3	12	—
2—benzoic acid	0.24	56	20	1.1	56	20	1.1	56	19.7	1.1	56	20	1.2
3—benzaldehyde	0.98	138	48	2.4	138	48	2.4	138	48.3	2.4	139	48	2.4
4—anisole	3.9	461	162	3.1	281	40	3.1	306	35.4	3.1	285	32	2.1
5—cumene	19	2129	750	3.7	236	15	3.7	352	16	1.7	237	14	3.2

Upper Mobile Phase[b]

Solute	Normal-Phase Elution				Dual-Mode			EECCC			BECCC		
	K'_D	V_R (mL)	W_b (mL)	R_S	V_R (mL)	W_b (mL)	R_S	V_R (mL)	W_b (mL)	R_S	V_R (mL)	W_b (mL)	R_S
1—catechol	14	1720	747	—	240	24	—	348	22.0	—	241	20	—
2—benzoic acid	4.2	517	224	2.5	277	33	1.3	310	40.8	1.3	276	32	1.4
3—benzaldehyde	1.02	142	61.8	2.6	142	61.8	2.3	142	61.8	2.3	142	60	2.9
4—anisole	0.26	51.5	22.3	2.2	51.5	22.3	2.2	51.5	22.3	2.2	51.5	23	2.2
5—cumene	0.05	27.3	11.8	1.4	27.3	11.8	1.4	27.3	11.8	1.4	28	10	1.4

Notes: Liquid system, Arizona R or hexane/ethyl acetate/methanol/water 2/1/2/1% v/v; mobile phase flow rate, 3 mL/min; machine volume, V_C = 140 mL; rotor rotation, 650 rpm; V_R, retention volume; W_b, peak width at base; plate number $\bar{N}$ = 16 $(V_R/W_b)^2$ or about 130 plates (reversed-phase mode) or 85 plates (normal-phase mode); R_S, resolution factor. The classical mode was performed for V_{CM} = 224 mL in every configuration in normal- and reversed-phase mode. Experimental data from Ref. [38].

a Aqueous mobile phase in the descending or head-to-tail flowing direction, V_M = 29.5 mL, V_S = 110.5 mL, Sf = 79%; EECCC, extrusion with the organic liquid phase flown in the head-to-tail direction BECCC, extrusion by switching the aqueous mobile phase from the head-to-tail to the tail-to-head direction.

b Organic mobile phase in the ascending or tail-to-head flowing direction, V_M = 21 mL, V_S = 119 mL, Sf = 85%; the indicated K'_D correspond to $1/K_D$ with the lower phase mobile. EECCC, extrusion with the aqueous liquid phase flown in the tail-to-head direction; BECCC, extrusion by switching the organic mobile phase from the tail-to-head to the head-to-head direction.

The UV detector is not saturated if the selected wavelength is set apart from the absorption maximum.

The evaporative light scattering detector (ELSD) nebulizes the mobile phase into fine droplets and dries them out. If there is a solid compound contained in the mobile phase, it is detected as microparticles by light scattering. When the mobile phase contains microdroplets of stationary phase, all solvents are evaporated giving a neat signal. This detector works well with both diluted and concentrated solutions. The ELSD has two drawbacks: (a) it is a destructive detector; it is not possible to collect the detected compounds; (b) it is not perfectly proportional to the mass of solute present in the mobile phase; several small crystals diffuse more light than a single large one.

A new detector, the charged aerosol detector patented by ESA Bioscience (Chelmsford, MA) as the Corona CAD detector uses a corona needle discharge to ionize the solute contained in the mobile phase after evaporation. It is claimed that this new detector is as sensitive as a UV detector without needing chromophore [39]. It is also extremely sensitive and reproducible with a maximum dynamic range. Since it evaporates the mobile phase before detection, it should not be sensitive to the microdroplets of stationary phase. The drawbacks are also its destructive character and its price.

Modern powerful mass spectrometers coupled to a CCC column can be used. The working flow rate is too high for direct MS introduction so part of the effluent should be derived into the MS detector. This is easily done and allow for rapid compounds identification following CCC separation [40].

9.5 APPLICATIONS

9.5.1 Natural Products Isolation and Purification

The isolation and purification of natural products of botanical or animal origin is, by far, the main field of application of CCC. Figure 9.7 shows the results of an electronic search with the words: "countercurrent chromatography" in the SciFinder and Scopus search engines with the 277 articles on the January 2000–September 2007 time period, and 146 articles (53%) described the separation and/or purification of natural products by different CCC columns and liquid systems. The second topic is method development that includes new CCC columns, theory of the method and new and original uses of CCC columns. Organic chemicals purified by CCC included dies and purification of synthetic compounds. Inorganic cation separations cover the separation of radioactive uranides, processing of nuclear wastes, and environmental heavy metal pollutants. The other topics include enantiomer separation, industrial development, original biphasic liquid systems, and review articles. The complete list of the 277 articles was prepared in the form of an Excel spreadsheet that can be requested from the author [41].

9.5.2 Case Study of Honokiol Purification

Honokiol is the major active constituent found in the barks of *Houpu*, a traditional Chinese medicine vegetable whose official name is *Magnolia officinalis Rehd. et Wils* [42].

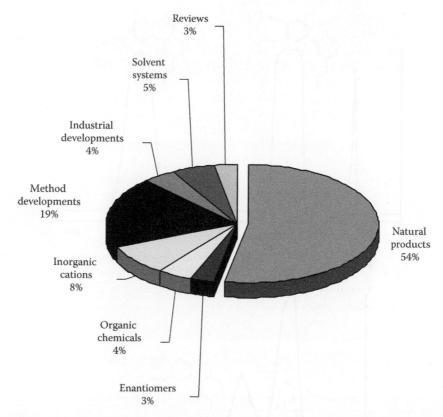

FIGURE 9.7   Topics treated in 277 articles dealing with CCC published between 2000 and 2007.

Honokiol induces apoptosis and/or inhibits the growth of leukemia cells HL-60 and Molt 4B, colon cancer cells RKO, lung cancer cells CH27 [43]. It has anti-inflammatory, analgesic, and anti-angionesis activities. Its anti-oxidative activity was found to be 1000-fold greater than that of vitamin E [42]. This biphenol is found in significant amount in the plant barks, but always associated with its isomer magnolol. The ethanol extract of *M. officinalis* bark contains more than 90% of magnolol and honokiol that are difficult to separate by preparative LC.

CCC was found to be very successful in fractioning this extract with the liquid system Arizona S (hexane/ethyl acetate/methanol/water, 5/2/5/2 v/v, Table 9.3). In this system the partition coefficients, K_D (org/aq), of honokiol and magnolol were, respectively, 0.37 and 1.1 [42,43]. Using a 240 mL HD-CCC column rotating at 800 rpm, an aqueous lower phase flow rate of 2 mL/min and injecting 10 mL of a 15 g/L extract solution (150 mg injected), a Chinese team was able to obtain 80 mg of 99.2% pure honokiol and 45 mg of 98.2% pure magnolol in 150 min (Figure 9.8, top) [42]. The honokiol throughput is 32 mg/h or 0.19 g/day. The honokiol solvent consumption is 4 mL/mg. The resolution factor between the two isomers is 3.3 for a column showing only 320 plates (magnolol peak) or even 160 plates (honokiol peak, Figure 9.8, top).

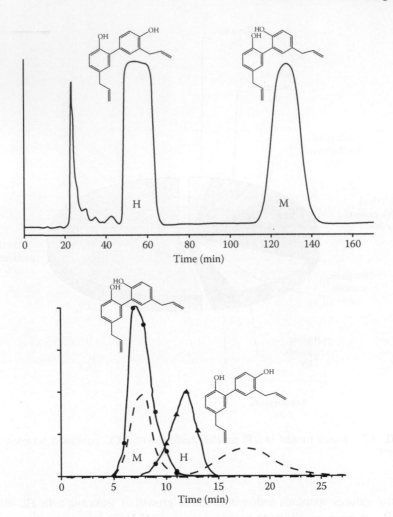

FIGURE 9.8 Preparative separation of honokiol (H) and magnolol (M) by two different HD-CCC columns and the same Arizona S liquid system (hexane/ethyl acetate/methanol/water 5/2/5/2 v/v). Top: column 230 mL, rotation speed 800 rpm, mobile phase lower phase at 2 mL/min, $Sf = 80\%$, injection 10 mL of 15 g/L solution in lower phase, separation duration 150 min, with 300 mL mobile phase. (Adapted from Wang, X., Wang, Y., Geng, Y., Li, F., and Zheng, C., *J. Chromatogr. A*, 1036, 171, 2006.) Bottom: column 4600 mL, rotation speed 600 rpm, mobile phase upper phase at 600 mL/min. Full line: $Sf = 40\%$, injection 172 mL of 250 g/L solution in upper phase, separation duration 15 min with 9000 mL mobile phase. Dotted line: $Sf = 80\%$, injection 172 mL of 75 g/L solution in upper phase, separation duration 25 min with 15,000 mL mobile phase. (Adapted from Chen, L.J., Zhang, Q., Yang, G., Fan, L., Tang, J., Garrard, I., Ignatova, S., Fisher, D., and Sutherland, I.A., *J. Chromatogr. A*, 1142, 115, 2007.)

The honokiol value is so high that gram amounts were desired. The British team owning the large 5 L HD-CCC Maxi column decided to scale-up the results published by the Chinese team [43]. They used the same Arizona S liquid system to fractionate larger amounts of *Houpu* ethanol extract but selected the normal-phase way (upper organic mobile phase and lower aqueous stationary phase). In the normal-phase way, honokiol elutes last but are located in the organic phase that is easier to evaporate and recycle than the aqueous phase. Figure 9.8, bottom, shows the separation of 43 g of extract injected as 172 mL of a concentrated 250 g/L solution in the upper mobile phase. In less than 20 min (yield 84%) [43] 19.4 g of 99% pure magnolol and 17 g of 98% pure honokiol were recovered. The throughput for honokiol is theoretically 51 g/h or 306 g/day, 1.6 thousand times higher than the one with the Figure 9.8 top CCC column. The solvent consumption is 0.6 mL/mg or seven times lower.

Interestingly, the injection of the viscous concentrated 250 g/L extract solution produced a leak of aqueous stationary phase. The *Sf* factor dropped to 40%. This low *Sf* factor explains the resolution factor of only 0.95. The theoretical plate number is estimated as about 40 plates. These two values may appear low for LC chromatographers. However, since the goal is reached, honokiol is purified in the required amount and purity, no effort was done to increase *Sf* (and consequently the resolution factor). Such improvement in resolution deteriorates the throughput because the retention volume of honokiol increases. The dotted line in Figure 9.8, bottom, shows the separation of 13 g of extract injected as 172 mL of 75 g/L solution. This experiment produced 4.4 g of 100% pure magnolol and 5.4 g of 100% pure honokiol (yield 76%) in less than 30 min (throughput 11 g/h). This less concentrated solution displaced less stationary phase during the injection time. The *Sf* factor was 80% producing a resolution of 1.4 between the two compounds, more than two times higher than the 0.6 value obtained with *Sf* = 40% (Equation 9.6). The theoretical plate count was again about 40 plates. It could be possible to reduce the loss of stationary phase during the 250 g/L injection by dissolving the *Houku* ethanolic extract in a mixture of Arizona S upper and lower phase and not just the upper phase.

9.6 CONCLUSION

Significant progresses were achieved in CCC at the dawn of the twenty-first century. It was demonstrated that a CCC column needs to retain a large amount of liquid stationary phase to have some resolution capability. The retention of stationary phase is much more important than the plate number for a CCC column. The definitive preparative character of the technique was used by research teams developing large and reliable CCC columns to reach the kilogram per day level and to work in industrial environment. New designs for the cells of the HS columns were developed taking into account the Coriolis effect and using it. New large volumes and reliable HS columns appeared on the market. Major advances were made with HD-CCC columns. It was demonstrated that this type of columns works like a constant pressure pump. The main consequence is that larger columns made with larger bore tubing retain well the liquid stationary phase allowing for very high flow rates (L/min) and for very high production rate and throughput (kg/day). The association of very specific biphasic liquid systems with the CCC technique has tremendous capabilities. Aqueous two-phase solvent systems are very useful in protein and biological purification. Their use as liquid systems in CCC columns may allow manufacturing of high-value pharmaceutical new products [44].

SYMBOLS

A	constant in Equation 9.10, usual value 1% or 100%
B	gradient of the loss of liquid stationary phase with the square root of the flow rate (Equation 9.10)
C	geometrical constant in Equation 9.12
D	geometrical constant in Equation 9.12
d_C	internal tubing diameter
F	mobile phase flow rate
H	height equivalent to a theoretical plate
H→T	the head-to-tail or descending flowing mobile phase direction for the denser liquid phase
k	retention factor
K_D	solute distribution ratio or partition coefficient
N	theoretical plate number
r	tubing radius = $d_C/2$
R	column rotor radius, distance between the central axis and the coil axis in HD-CCC column
Rs	resolution factor
Sf	liquid stationary phase retention percentage in a CCC column
T→H	the tail-to-head or ascending flowing mobile phase direction for the upper liquid phase
V_C	column volume
V_M	mobile phase volume in an equilibrated CCC column
V_R	solute retention volume
V_S	liquid stationary phase volume
W	peak width at base
X	solute position inside the CCC column

GREEK LETTERS

Σ	standard deviation
μ_M	mobile phase viscosity
ΔP	column pressure drop
P	liquid phase density
ω	rotor rotation speed

ACRONYMS

BECCC	back extrusion countercurrent chromatography
CAD	charged aerosol detector
CCC	countercurrent chromatography
CCD	countercurrent distribution
DM	dual-mode elution
EECCC	elution–extrusion countercurrent chromatography
ELSD	evaporative light scattering detector

HD   hydrodynamic
HS   hydrostatic
HPLC   high-performance liquid chromatography
UV   ultraviolet

ACKNOWLEDGMENTS

Continuous support for this work by the French National Center for Scientific Research, CNRS UMR5180, Universite de Lyon, Laboratoire des Sciences Analytiques, Pierre Lanteri, is gratefully acknowledged.

REFERENCES

1. Y. Ito, M. Weinstein, I. Aoki, R. Herada, E. Kimura, and K. Nunigaki, *Nature*, 212: 985, 1966.
2. A. Berthod, *Countercurrent Chromatography, the Support-Free Liquid Stationary Phase*, D. Barcelo (Ed.), *Comprehensive Analytical Chemistry*, Vol. XXXVIII, Elsevier, Amsterdam, 2002.
3. L.C. Craig, *Anal. Chem.*, 22: 1346, 1950.
4. L.C. Craig, *Meth. Med. Res.*, 5: 3, 1952.
5. Y. Ito, Foreword in *Countercurrent Chromatography, the Support-Free Liquid Stationary Phase* (A. Berthod), *Comprehensive Analytical Chemistry*, D. Barcelo (Ed.), Vol. XXXVIII, Elsevier, Amsterdam, 2002, pp. xix–xx.
6. Scifinder, ACS, and Scopus, Elsevier, consulted with the search words "countercurrent chromatography", Oct. 2007.
7. A. Berthod and B. Billardello, *Adv. Chromatogr.*, 40: 503–538, 2000.
8. J.B. Friesen and G.F. Pauli, *Anal. Chem.*, 79: 2320, 2007.
9. A. Berthod, J.B. Friesen, T. Inui, and G.F. Pauli, *Anal. Chem.*, 79: 3371, 2007.
10. Y. Ito, Countercurrent chromatography, overview, in *Encyclopedia of Chromatography*, J. Cazes (Ed.), Marcel Dekker, New York, 2001, pp. 241–252.
11. Y. Ito and W.D. Conway, *High-Speed Countercurrent Chromatography, Chemical Analysis*, Vol. 132, Wiley, New York, 1996.
12. Y. Ito, *Sep. Purif. Rev.*, 34: 131, 2005.
13. P.L. Wood, D. Hawes, L. Janaway, and I.A. Sutherland, *J. Liq. Chromatogr.*, 26: 1373, 2003.
14. J.H. Renault, J.M. Nuzillard, O. Intes, A. Maciuk, in *Countercurrent Chromatography, the Support-free Liquid Stationary Phase*, A. Berthod (Ed.), *Comprehensive Analytical Chemistry*, Vol. XXXVIII, Chap. 3, Elsevier, Amsterdam, 2002, pp. 49–83.
15. A.P. Foucault and L. Chevolot, *J. Chromatogr. A*, 808: 3, 1998.
16. I.J. Garrard, *J. Liquid Chromatogr. Relat. Technol.*, 28: 1923, 2005.
17. F. Oka, H. Oka, and Y. Ito, *J. Chromatogr.*, 538: 99, 1991.
18. R. Margraff, in *Centrifugal Partition Chromatography*, A. Foucault (Ed.), *Chromatographic Science Series*, Vol. 68, Marcel Dekker, New York, 1994, pp. 331–350.
19. A. Berthod, M. Hassoun, and M.J. Ruiz-Angel, *Anal. Bioanal. Chem.*, 383: 168, 2005.
20. C. Reichardt, *Chem. Rev.*, 94: 2319, 1994.
21. W.D. Conway, Y. Ito, and A.M. Sarlo, *J. Liq. Chromatogr.*, 11: 107, 1988.
22. A. Berthod, *J. Chromatogr.*, 550: 677–693, 1991.
23. Y. Ito, K. Shinomiya, H.M. Fales, A. Weisz, and A.L. Scher, in *Modern Countercurrent Chromatography*, W.D. Conway and R.J. Petroski (Eds.), *ACS Symposium Series*, Vol. 593, Chap. 14, ACS, Washington, WA, 1995, pp. 156–183.

24. T.A. Maryutina, P.S. Fedotov, and B.Ya. Spivakov, in *Countercurrent Chromatography*, J.M. Menet and D. Thiebaut (Eds.), *Chromatographic Science Series*, Vol. 82, Marcel Dekker, New York, 1999, pp. 171–221.
25. Y.R. Jin, L.X. Zhang, L.Z. Zhang, and S.J. Han, in *Countercurrent Chromatography, the Support-free Liquid Stationary Phase*, A. Berthod (Ed.), *Comprehensive Analytical Chemistry*, Vol. XXXVIII, Chap. 9, Elsevier, Amsterdam, 2002, pp. 261–300.
26. Q. Du, C. Wu, G. Qian, and Y. Ito, *J. Chromatogr. A*, 835: 231, 1999.
27. I.A. Sutherland, Q. Du, and P. Wood, *J. Liq. Chromatogr. Related T.*, 24:1669, 2001.
28. A.P. Foucault (Ed.), *Centrifugal Partition Chromatography*, *Chromatographic Science Series*, Vol. 68, Marcel Dekker, New York, 1994.
29. A. Berthod and D.W. Armstrong, *J. Liq. Chromatogr.*, 11:567, 1988.
30. I.A. Sutherland, *J. Chromatogr. A*, 1151: 6–13, 2007.
31. J.C. Giddings, *Unified Separation Science*, John Wiley & Sons, New York, 1991, pp. 97–101.
32. I. Slacanin, A. Marston, and K. Hostettmann, *J. Chromatogr.*, 482: 234, 1989.
33. R.A. Menges, G.L. Bertrand, and D.W. Armstrong, *J. Liq. Chromatogr.*, 13: 3061, 1990.
34. S.J. Gluck, and E.J. Martin, *J. Liquid Chromatogr.*, 13: 3559, 1990.
35. A. Berthod, M.J. Ruiz-Angel, and S. Carda-Broch, *Anal. Chem.*, 75: 5886, 2003.
36. A. Berthod, *J. Chromatogr. A*, 1126: 347, 2006.
37. A. Berthod, M. Hassoun, and G. Harris, *J. Liq. Chromatogr. Related T.*, 28: 1851, 2005.
38. Y. Lu, Y. Pan, and A. Berthod, *J. Chromatogr. A*, 10: 1189, 2008.
39. http://www.coronacad.com/CAD_Overview.htm# consulted on October 2007.
40. L.J. Chen, H. Song, D.E. Games, and I.A. Sutherland, *J. Liq. Chromatogr. Related Technol.*, 28: 1993, 2005.
41. Excel file can be requested by e-mail at berthod@univ-lyon1.fr.
42. X. Wang, Y. Wang, Y. Geng, F. Li, and C. Zheng, *J. Chromatogr. A*, 1036: 171, 2006.
43. L.J. Chen, Q. Zhang, G. Yang, L. Fan, J. Tang, I. Garrard, S. Ignatova, D. Fisher, and I.A. Sutherland, *J. Chromatogr. A*, 1142: 115, 2007.
44. I.A. Sutherland, *Curr. Opin. Drug Disc. Dev.*, 10: 540, 2007.

10 Hyphenated Techniques in Thin-Layer Chromatography

Simion Gocan

CONTENTS

10.1   INTRODUCTION

The introduction of hyphenated techniques in thin-layer chromatography (TLC) was a substantial step forward for the detection/identification and quantification of compounds from very complex mixtures. Identification and quantification of unknown compounds from a mixture is one of the fundamental problems in analytical chemistry. It is well known that TLC provides insufficient information to obtain a reliable identification of the compound using only retention values. This problem has been solved due to the remarkable progress made in the last two decades by coupling TLC with other detection methods such as ultraviolet/visible (UV/Vis) spectrometry, fluorescence spectrometry (FS), infrared spectrometry (IRS), Raman spectrometry (RS), photoacoustic spectrometry (PAS), and mass spectrometry (MS). Thus, the amount and quality of information from the analysis of compounds in mixtures has increased. These spectral methods can be coupled with TLC either "offline" or "online." The term online can be used instead of the term *in situ*. Where an authentic reference substance is available, confirmation of the structure of a chromatographically separated unknown compound is possible. As a rule, a single physical method is sufficient for this purpose, e.g., infrared (IRS) or MS. If no reference substance is available, then an attempt can be made to identify the structure with the aid of a spectroscopic method, e.g., nuclear magnetic resonance (NMR), in addition to using IR or MS. High-performance thin-layer chromatography (HPTLC) when compared with high-performance liquid chromatography (HPLC) is more flexible with regard to detection strategies associated with the thin-layer technique, including detecting microchemical reactions occurring on the surface of the thin layer, biomonitoring, and the possibility of using surface-enhanced spectroscopy for the identification of compounds.

Here, we cite only important reviews that cover the literature with respect to all combinations of TLC or HPTLC with spectral methods. The offline coupling between TLC and UV/Vis and the range of IR and MS are reviewed in Szekely [1]. Other reviews of the same topic are reported in Siouffi et al. [2], while Touchstone [3] discussed in detail about the developments in TLC detection, visible spectrometry, densitometry, TLC-IR, and TLC-MS. An overview of the use of TLC in drug analysis in biological fluids and the capabilities of HPTLC with scanning densitometry and TLC-MS are briefly discussed in Wilson [4]. A review of advancements in coupling TLC with a variety of spectroscopic techniques was given in Somsen et al. [5]. Studies of combined TLC-MS with different ionization techniques are discussed here with regard to analysis of industrial chemicals, pesticides, dyes, drugs, and natural products. Coupling of TLC with all spectrometric techniques—Fourier transform infrared spectrometry (FTIR), near-IR, PAS, and RS—is also presented and evaluated. Fluorescence line-narrowing spectrometry is also described. Gocan and Cimpan [6] reported the possibility of coupling TLC and HPTLC with spectrometric methods to improve the selectivity and sensitivity of compounds' detection and identification. Coupling TLC or HPTLC with spectroscopic methods, such as UV/Vis, IR, FTIR, RS, PAS, and MS or MS-MS is particularly described. TLC-IR or TLC-MS can be practiced either offline or online. Through these coupling methods, planar chromatography has acquired a new dimension. These methods were discussed detail with their applications. In a relatively recent review by Wilson [7], the practicalities of TLC-MS, both manual and instrumental methods, and the use of direct ionization mode fast atom bombardment (FAB) or liquid secondary ion (LSI) methods are described. Recent developments, however, include the use of matrix-assisted laser desorption ionization (MALDI), surface-assisted laser desorption ionization (SALDI), and the development of a TLC-electrospray (ES) interface. These techniques with their applications are described.

The objective of this chapter is to present advances in qualitative and quantitative TLC analysis by using offline or online spectroscopic methods—UV, FTIR, Raman and surface-enhanced resonance Raman scattering (SERRS), MS, or MS-MS, and in the end, the barriers to the widespread use of hyphenated techniques in TLC are discussed.

10.2 SPOT IDENTIFICATION BY R_F VALUES AND COLOR REACTION

10.2.1 Color Reaction

Compound identifications can be obtained by using a specific reagent. Generally, the reagent is specific for a class of compounds or for a functional group. Jork et al. [8,9] published a monograph on the reagents used in TLC for component identification.

A relatively recent overview was published by Cimpan [10] with regard to the identification of compounds by pre- and postchromatographic derivatization, and reagents used for nondestructive postchromatographic derivatization were listed. Furthermore, a selection of more representative reagents used for the postchromatographic derivatization in TLC of different classes of compounds together with their respective comments was also presented. More information about the specific conditions and layers involved can be found in the references cited.

Usually, for the identification of a compound in the same classes of compounds, one must use at least three different reagent systems. Kocjan [11] applied 11 visualizing agents for the detection of five quinones after separation by TLC. Some of those reagents were specific only for two or three quinones, while the other reagents were able to detect all of them, which were visualized through different colored spots on the chromatographic plate. Detection limits for these quinones were from a range of 0.1–5 µg per spot.

Bladek [12–15] developed an interesting method for visualizing compounds from a thin-layer chromatographic plate by using liquid crystals. The method comprises transferring the organic compounds from the chromatographic plate to a liquid crystal layer. This method was successfully applied for pesticide detection.

10.2.2 RETENTION DATA, R_F OR hR_F

Retention data, R_F or hR_F, and relative retention, $R_{rel} = R_{Fi}/R_{Fst}$, values are widely
used for the identification of unknown compounds by comparing these values with a
value obtained for a known standard under the same chromatographic conditions.
However, practically this is not very simple, because the mixtures of compounds are
complex and are often with similar chemical structures.

For increasing the identification probability, Nyiredy et al. [16] considered three
developments necessary for different eluent systems. Only in this case, can we
consider that the equality between the retention hR_F value for the standard compounds and the unknown sample compound is true. However, the strength of the
eluent systems must be different. Practically, the total selective factors and the total
solvent strength must be different:

$$(S_F)_1 \neq (S_F)_2 \neq (S_F)_3 \quad \text{and} \quad (S_T)_1 \neq (S_T)_2 \neq (S_T)_3 \tag{10.1}$$

The total selectivity factor, S_F, is defined as the sum of the individual values of the
components, S_f, multiplied with their volume fraction, φ

$$S_F = \sum \varphi_i (S_f)_i \tag{10.2}$$

where $(S_f)_i$ is the individual solvent strength of component i ($i = 1, 2, ..., n$), and the
individual selective factor for one solvent is the ratio of proton acceptor to proton
donor values, given in Snyder [17]: $S_f = X_e/X_d$.

By analogy with the S_F value, the total solvent strength can be calculated as

$$S_T = \sum \varphi_i S_i \tag{10.3}$$

where
   φ_i is the volume fraction of component i in the mixed eluent
   S_i is the individual solvent strength of component i

The identification probability, I_P, can be defined as the area of the triangle formed
by three points with the following coordinates: $(S_T)_1, (S_F)_1; (S_T)_2, (S_F)_2; (S_T)_3, (S_F)_3$.
This dimensionless value of the triangle can be calculated according to the rules of
analytical geometry.

On the bases of voluminous experimental data, the following criteria were
derived:

1. For routine laboratory work, the I_P value should be between 0.1 and 0.5.
2. For the identification of possibly new, naturally occurring compounds, the
   value of I_P should be greater than 0.5.

The system was investigated using various classes of compounds (13) and the mobile
phases were obtained from the literature; however, all separations were repeated in
the laboratory [16]. For different substance classes, the I_P values varied between 0.01
and 1.21, and were generally between 0.1 and 0.4, with the exception of neutral and
basic substances from the Merck Tox Screening System [18], for which the I_P values
were greater than 1.0.

10.2.3 R_{rel} AND COLOR CODES

It is well known that TLC or HPTLC offer extra advantage of applying pre- and postchromatographic derivatization on the thin-layer plate, and this may increase the identification power.

De Zeeuw et al. [19] proposed five chromatographic systems in combination with four color reactions. The observed color for an unknown compound was compared with a chart containing 21 different colors that had been coded numerically. These codes were subdivided into six groups, which differ by eight digits. The first three groups contained only one code each: 00, when no spot is seen; 08, for black spots; 16, for white spots; 31–34, for various shades of green spots; 42–49, for gray to blue to violet to red spots; and 57–62, for orange to yellow to brown spots. The color code and the corresponding R_F values were expressed as R_{rel} [20]. The color codes and R_{rel} values are the inputs for the database. The experimental R_{rel} values and color codes can be arranged as a basic matrix with 9 columns and 99 rows. The 99 rows represent the number of drugs used for the experiment. To compare the suitability of the techniques and the system, the mean list length (MLL) concept was developed. The MLL represents the mean number of candidates that would qualify for the identification of an unknown compound when a given technique/system is used, individually or in combination. A compound is precisely identified when the MLL value is 1.00. Thus, the MLL represents an objective criterion to establish the identification capacity of a single system, combination of systems, or even combination of different analytical techniques, such as gas chromatography (GC), TLC, and UV spectrometry [19,21]. The MLL calculation for the matrix of R_{rel} values, color codes, and the 99 drugs was performed using the Tox program, in Turbo Pascal under MS-DOS [19]. Even when the best combination of the three independent solvent systems along with the color codes was used, the MLL value of 1.22 was slightly higher than the optimal value of 1.00, and thus, the 99 compounds in the drug set cannot be unequivocally identified [19].

Finally, the data should be outlined as listed by the computer, in ranking order, on the basis of a probability and similarity index. The index shows concordance between observed and listed data in the database [22]. However, the rank order is of supplementary assistance in the final indication process [19].

10.3 TLC-UV/VIS AND FLUORESCENCE SPECTROMETRY

In the last two decades, modern instruments for the qualitative and quantitative evaluation of thin-layer plates were developed to increase performance and optimize computer data handling.

Diffuse reflectance spectrometry has been used for the quantification of spots in the thin-layer plate, and Ebel and Wuther brought out a review [23] in the early years of the development of TLC techniques. The instruments for scanning thin-layer chromatograms by diffuse reflectance rely on photospectrometer principles.

The photodiode array detector was maneuvered by modifying a Schoeffel TLC scanner SD 3000. A 250 W halogen lamp was used. An optical system focused the monochromatic light to the TLC plate. While the slit width was defined by the monochromator, the slit height was changed by an optical way instead of one aperture, to avoid decreases in the intensity of light. The scattered light from the plate was focused onto the photodiode array. The scanner was modified by stepper motors to scan under computer control [24].

Beroza et al. [25] constructed an instrument for scanning thin-layer chromatograms using diffuse reflectance. The principal component of the instrument consists of a fiber-optic scanning head. It consists of a randomly mixed bundle of glass fibers. Half of them conduct light to a small defined area on the thin-layer plates that are scanned at a constant rate, and the other half conduct the reflected light to a photosensitive cell that has its response recorded. A dual beam operating a second fiber-optic, which scans the blank area adjacent to the spots, is used to correct for background differences on the plate. The performance of this instrument was demonstrated in the determination of 13 chlorinated insecticides, in amounts ranging from 0.1 to 32 µg, chromatographed either singly or in mixtures on silver nitrite-impregnated alumina plates, and six thiophosphate pesticides.

The coupling of TLC with photodiode arrays in the early 1980s was described by Ebel and Wuther [23]. Multiwavelength scanning is used in cases where it is not possible to measure all compounds at the same wavelength or in cases of incomplete separations [26].

Another earlier paper [27] dealt with the development of a powerful TLC scanner that combines the advantages of diode-array detection (DAD) with the simplicity of optical signal transmission by fiber-optic bundles. The apparatus comprised a double-arm fiber-optic bundle "Y," which is used both to direct the UV light (deuterium lamp) on the TLC plate and collect the scattered light, and to guide it to a diode-array detector supported by a personal computer (PC) equipped with commercially available software. The spectra can either be stored or printed. The spectra may be recorded from 200 to 600 nm with a wavelength precision of 0.1 nm and a resolution of 2.4 nm. Furthermore, the description of a fiber-optic-based fluorescence instrument for *in situ* quantitative scanning in TLC has also been reported [28–31].

Spangenberg and Klein [32] offer a brief review of the historical development of densitometry scanning in TLC during 1962–1989. A diode-array scanner has the capability of simultaneous quantification of TLC or HPTLC plates at different wavelengths. With this scanner it is possible to perform fluorescent measurements without filter or special lamps. Online spectrometry of HPTLC plates was carried out with a specially designed diode-array spectrophotometer (J&M Analytische, Aalen, Germany) that scans in the range of 197–621 nm with an average optical resolution of more than 2.0 nm. A homemade reflection device was attached comprising a bundle of optical fibers arranged in the form of "Y" for transporting light of different wavelengths from a deuterium lamp to the HPTLC plate and back to the diode-array detector. To obtain dense light intensities, the light-emitting and light-detecting fibers were arranged parallel to each other, because only in this arrangement, the Lambert cosine law predicts an optimal response. The plate can be moved along *X–Y* coordinates. For instrumental control and data acquirement, own laboratory-developed software was used.

A new scanner capable of measuring TLC or HPTLC plates simultaneously at different wavelengths without damaging the plate surface was developed [32,33]. Fiber optics and special fiber interface were used in combination with a diode-array detector, which enables the use of chemometric methods in HPTLC evaluation. The reflected light intensity, I_0, from one clean plate was compared with the reflected intensity light, I_{rem}, from a sample spot that contained benzo[a]pyrene. The absolute value of the difference spectra, $D(\lambda)$, between sample spectra, $I_{rem}(\lambda)$, and reference spectra, $I_0(\lambda)$, from a clean plate can be obtained using the software. The intensity of the reflected light from the sample spot was found to be obviously less than that reflected from the clean plate, valid between 200 and 400 nm. However, between 400 and 500 nm, the situation was observed to be the contrary, because the benzo[a]pyrene spot surprisingly emitted more light than the clean thin-layer plate. To obtain a lamp-independent spectrum, it is recommended to use corrected remission spectrum as remission values, $R(\lambda) = I_{rem}(\lambda)/I_0(\lambda)$. Furthermore, the transformation of remission data into a mass-dependent signal is possible using the mathematical transformation, $-\ln R(\lambda) = A(I_{rem}, \lambda)$.

As demonstrated earlier using suitable wavelengths, it is possible to quantify the 16 incompletely separated polycyclic aromatic hydrocarbons (PAHs) by scanning them at a suitable wavelength in one track.

A homemade diode-array scanner was used for the quantification, and for the first time, it enabled simultaneous measurement at different wavelengths. With this new scanner, we can perform fluorescence evaluation without additional equipments. The new scanner performances were evaluated by the separation and quantification of caffeine and quinine in beverages [34].

In addition, the performance of the scanner was evaluated through the study of linearization models for absorption and fluorescence quantification [34], and methods for reduction of errors in quantitative analysis at smaller values of reflectance, where peak height was used for calibration [35].

Furthermore, Ebel et al. [36,37] discussed background correction methods in reflectance and absorption spectra, as well as the normalization spectra. A validation procedure for the TLC scanner was presented by them, including the determination of signal-to-noise (S/N) ratio and instrument sensitivity.

UV/Vis spectra are usually obtained in the scattered reflection mode from thin-layer plates, and are specific for each substance. The compound identification is, in fact, a simple comparison of the *in situ* spectra obtained from the thin-layer plate in the same condition for unknown and standard substances. When standard substances are not available, the comparison can be performed using a library spectrum. Both R_F values and spectra for the unknown and standard compounds must be similar for proper identification.

Currently, quantitative determinations using TLC and HPTLC are performed in exclusivity using modern densitometry. Numerous papers on this topic are being published; hence, those who are interested can refer the journals specific to this subject as well as books on thin-layer chromatography.

10.3.1 Modern Scanners

Modern scanners are most advanced for the qualitative and quantitative evaluation of thin-layer chromatograms of dimensions up to 200 mm × 200 mm. The diode-array scanner with its corresponding software can rapidly scan a complete spectrum of all the compounds on the thin-layer plate with simultaneous acquisition in the range from 190 to 1000 nm (pixel resolution of 0.8 nm). Deuterium, tungsten-halogen, and high-pressure mercury lamp can be used as the UV/Vis light source. The optical and mechanical resolutions on the plate are up to 200 and 20 μm, respectively. Furthermore, the scanning speed of 1–100 mm/s is programmable and a spectrum recording up to 100 nm/s is achieved.

The following are the advantages of using the modern scanner:

1. Simultaneous measurement of all the compounds on the thin-layer plate at different wavelengths in real time is possible.
2. Spectral information can be used for confirmation of the purity of the peaks of interest.
3. Overlap peaks can be practically resolved into individual components by dual wavelength measurements.
4. Scanner is freely programmable with respect to the geometrical parameters of the coordinate (X–Y).

Furthermore, the modern scanner provides various facilities for data presentation, such as

1. Component spectrum intensity (AU) = f (wavelengths in nanometer) *in situ*
2. Component identification and confirmation by comparing the unknown spectra from the TLC plate with all spectra from the library, and the hit list exhibits five matching spectra with the best match in the first position (library match is shown in percent)
3. Color contour plot, chromatogram (densitogram), and component spectrum together with a remission color scale
4. Colored 3D representation of the densitogram

10.3.2 VIDEO DOCUMENTATION AND DENSITOMETRY

Video documentation of densitometric evaluation of thin-layer chromatograms was introduced in 1970s. In the subsequent years, the principle and instrumentation of this technique were developed [38,39]. The instrument consisted of four principal parts: optical unit, the TV camera, the electronic unit, and the calculator [39].

Furthermore, another study [40] presented the theoretical and practical aspects of using a TV-type detector in densitometry, describing the work principles, spectral sensitivity, and detection limit. For amino acids stained with ninhydrin, the detection limit lies in the range of 2–5 nmol with respect to the layer quality and spot shape.

In an earlier study, a homemade, inexpensive high-speed densitometer using an Apple II computer, along with a black-and-white video camera and an image digitizing board, was described. By adding a very fast coprocessor to the computer, measurement of a typical thin-layer spot can be obtained in 30–40 s [41]. For multiple readings of a single 500 ng/spot of charged lipids, the coefficients of variation (CV) were about 0.5%.

For the purpose of documentation and video densitometry, the Camag system consists of the following subensemble:

1. Camag Reprostar 3/Transilluminator-versatile lighting module and camera stand
2. Color three charge-coupled device (CCD) high-resolution camera with zoom objective
3. Computer running Microsoft Windows with frame grabber and monitor
4. Powerful VideoStore software
5. Color printer

The HPTLC-densitometry method yields the most selective spectra and sensitive densitograms, and hence accurate and precise results are obtained. Currently, the following types of densitometers are commercially available:

1. Slit-scanning densitometers (Camag, Muttenz, Switzerland; Desaga, Heidelberg, Germany, and Shimadzu, Corporation, Japan)
2. Diode-array densitometers (J&M, Aalen, Germany)
3. Video densitometers (Camag, Muttenz, Switzerland, and Desaga, Heidelberg, Germany)

Nowadays, a modern video densitometer is an indispensable tool in plant extraction analysis. With this instrument, we can obtain colored photographs of the chromatogram, which can be saved in memory storage devices. This colored photograph is considered as the fingerprint of the tincture (alcoholic extract from a plant) of the respective plant. In addition, we can also use video densitometry for the chromatogram.

10.3.3 Reflectance Measurement

10.3.3.1 Theoretical Considerations

Any problems in measuring the transmittance or absorbance of clear solutions in the UV/Vis range is validated using the Boguert–Lambert–Beer law. However, in TLC, this becomes much more complicated. *In situ* densitometry is a simple method of quantification by measuring the optical density of the separated spots directly on the plate. Quantitative evaluation of thin-layer chromatograms by optical methods is based on a differential measurement of light emerging from the sample-free and sample-containing zones (spot) of the plate [42].

The theoretical approach of reflectance measurement in quantitative analysis in TLC and HPTLC was carried out by Ebel and Post [43,44], and much later by Spangenberg [45]. Initially, they discussed the basic principles of reflectance measurement on the basis of the well-known Kubelka–Munk equation in terms of homogeneity of light absorption. In the approach described, the adsorbent layer was divided into multisublayers. The incident light was reflected in different multiple directions from each particle in the layer. If the analyte substance was present in the layer, then absorption also occurred. From the incident light beam will be result the diffuse light beam reflectance, which is often called remittance.

To simplify this scattering effect, only the perpendicular incoming light was considered. In this case, a set of two differential equations can be written

$$-\frac{\delta I}{\delta x} = -2(a+s)I + 2sJ \tag{10.4}$$

$$\frac{\delta J}{\delta x} = -2(a+s)J + 2sI \tag{10.5}$$

where

I is the intensity of the incident light

J is the reflectance intensity

a is the coefficient of adsorption

s is the coefficient of reflectance (scattering coefficient)

If a sample is illuminated by a perpendicular and parallel light flux of intensity, I_0, and if there is no scattering in the sample ($s = 0$) and no fluorescence, then the level of the sample as transmitted light, I_T, is reduced in intensity, because a lot of usual processes occur during plate illumination. In this case, a fraction of scattered light, J, is emitted as reflectance light upward from the plate surface, and the diffuse reflectance light from the plate will be $R_0 = J/I_0$.

From the two differential equations derived, when the light passes a distance d through the thin-layer plate, Boguert–Lambert–Beer law is achieved as

$$\ln\frac{I_0}{I_T} = 2ad \tag{10.6}$$

The positive values are obtained for $I_T < I_0$, which indicates that light is absorbed through the layer.

According to the differential equations, the diffuse reflectance of an infinite thick layer, R_∞, can be expressed as

$$R_\infty = \frac{J}{I_0} \tag{10.7}$$

The linear relationship between the adsorption coefficient and the light intensity reduction is given as the well-known Kubelka–Munk equation:

$$R_{KM} = \frac{(1-R_\infty)^2}{2R_\infty} = \frac{a}{s} \tag{10.8}$$

where

R_∞ is the total reflectance of an infinite layer

a is the absorption coefficient

s is the scattering coefficient

The so-called linear equation (Equation 10.8) is only valid for infinite thickness and for an ideal scattering medium without any regular reflection.

Spangenberg et al. [34] considered that no light is lost from the HPTLC plate other than the loss by sample absorption. As a result, the plate absorption intensity becomes zero, but in fact, the light intensity is replaced by the reflected light intensity of the plain plate. Hence, the whole loss of light on the plate surface can be assumed to be zero. In these circumstances, the corrected remission term R must be used in the Kubelka–Munk equation:

$$KM\left(I_{rem}, \lambda\right) = \frac{[(1 - R_{corr}(\lambda)]^2}{2R_{corr}(\lambda)} = ma(1 - a_m) \tag{10.9}$$

where

I_{rem} is the only scattered light emitted from the plate surface as remitted light
$R_{corr}(\lambda)$ is the ratio between $I_{rem}(\lambda)$ and reflected light intensity of the plain plate, I_P
m is the mass of the absorbing species
a_m is the mass absorption coefficient

The mass absorption coefficient, a_m, is a wavelength-dependent function. However, in this case, the reference light intensity at all wavelengths must be measured to know where to set the reference and at which wavelengths the sample should be evaluated. The use of reference light from a plate position free from sample is necessary for background correction to improve the baseline stability to perform calibration curves without intercepts.

It is obvious that the Boguert–Lambert–Beer law does not apply in TLC, because the medium is intensely light scattering owing to the complex processes involved in transmittance and reflectance of scattered radiation. However, reliable data indicate that linear calibration can be applied only for the lowest concentration. Also, the special case of the Kubelka–Munk theory poses limitations for a number of reasons, such as partial transparency of the layer and no diffuse illumination of the samples. As a result, significant nonlinearity has been found, especially in the low-concentration range. From the extension of the simultaneous transmittance and reflectance techniques, the question whether Boguert–Lambert–Beer and Kubelka–Munk classical theories would complement each other arises.

Treiber [46] estimated the calibration curves in accordance with the Boguert–Lambert–Beer law, including the Kubelka–Munk function for intensely scattering and infinitely thick media, written in terms of intensity of the light, and after a few mathematical operations to obtain:

$$K_x C = K_R \left[\frac{I_0}{I_x} + \frac{I_x}{I_0} - 2 \right] + K_T \ln \frac{I_0}{I_x} \tag{10.10}$$

where

C is the concentration of the substance chromatographed in weight per surface unit
K_x is a constant depending on the substance chromatographed
K_R and K_T are the constants depending on the properties of the adsorbent layer
I_x is the intensity of the light leaving the sample
I_0 is a constant, maximal light intensity on the adsorbent layer free from any substance chromatographed, thus $0 \le I_x \le I_0$ is the possible range

The first term from Treiber's equation (Equation 10.10) represents the Kubelka–Munk function and the second term represents the Lambert–Beer law. These theoretical considerations were verified experimentally. A series of standard amounts (0–10μg) of sulfadiazine were prepared on the same thin-layer chromatographic plate. For the three functions mentioned earlier, the following correlation coefficients were found under 2D scanning: Lambert–Beer, r = .99585; Kubelka–Munk, r = .99778; Treiber, r = .99991. In the first two cases, the parameters were highly dependent on the different regions of the concentration range considered.

However, classical Kubelka–Munk theory with uniform sample distribution is valid only for isotropic scattering, and the chromatographic thin layer is not an ideal isotropic scattering medium, as it is asymmetric. In this case, the classical Kubelka–Munk theory must be extended to a situation where scattering is asymmetric.

Spangenberg [45] demonstrated that for asymmetric scattering, the classical Kubelka–Munk equation should be extended by a backscattering factor k, which can be expressed as

$$\frac{(k-R)}{R} + R(1-k) = \frac{ma_m}{(1-a_m)} \tag{10.11}$$

where

$$R = \frac{J}{J_0}$$

in which J is the diffused light reflected from the plate surface and is called the reflectance, while J_0 is the intensity of the light reflected plain from a plate, and can be easily measured at a region of the plate without any compounds.

This permitted to obtain a linear calibration function from a nonlinear calibration function through appropriate selection of a backscattering factor k. When absorption alone is taken into account, the backscattering factor k will be considered as 1. If the backscatter factor $k = 0$ is taken into account, then the absorption is not considered, and Equation 10.11 gives the relation for fluorescence [33]. If the backscattering factor is taken as $k = \frac{1}{2}$, then it results in the Kubelka–Munk equation (Equation 10.9).

Treiber [46] used a combination of the Kubelka–Munk formula and Lambert–Beer expression and obtained a formula equivalent to Equation 10.11, with a backscattering factor of $k = 0.6$ without any theoretical explanation. However, it was observed that curved Kubelka–Munk plots can indeed be transformed into perfect linearity.

These theoretical considerations were experimentally verified by HPTLC separation of caffeine and densitometry determination, *in situ* [45]. HPTLC was carrier out on silica-gel K60 plates (10 cm × 10 cm) with and without a fluorescent dye, obtained from Merck (Germany). For calibration purpose, 10.711 mg of caffeine was dissolved in 10 mL methanol, and the solution was diluted in methanol. Samples 1–10 μL were applied onto the plate as 7 mm bands using a Desaga AS 30 device (Germany). The plates were developed in a Camag horizontal chamber to a distance of 45 mm from the origin, with isopropanol–cyclohexane–25% NH_3 (7:2:1, v/v) as the mobile phase. A Tidas TLC 2010 system (J&M Aalen, Germany) was used for direct reflection spectrometry of HPTLC plates. Caffeine densitograms were obtained from the spectra in the wavelength range of 271.5–279.8 nm, because the absorption spectrum is maximum in this region. The peak areas were calculated using laboratory-written integration software.

The calibration plots for caffeine in the range from 1 ng to 10 µg per spot for nonspherical silica-gel without a fluorescent indicator were demonstrated [45], and the raw data were evaluated using the Kubelka–Munk equation [45]:

$$KM(J_0 \lambda) = \frac{(1-R)^2}{2R} = \frac{ma_m}{1-a_m} \qquad (10.12)$$

For caffeine in the range of 1–107.11 and up to 214.22 ng, we calculated the correlation coefficient, R^2, on the numerical values taken from Tables 2 and 3 in [45], and the values are presented in Table 10.1 after statistical processing. The R^2 values show much more objective deviation from the linearity of the calibration plot than that described in simple terms.

The raw data were evaluated using the Kubelka–Munk equation (Equation 10.12), but in the range up to 107.11 ng of caffeine calibration plot no. 1 from Table 10.1, a significant deviation from linearity ($R^2 = 0.8297$) was observed and performing quantitative determination in this range using a linear calibration plot was not indicated. The same situation was exhibited for the calibration plot no. 2 for $k = 55$ below 107.11 ng of caffeine, where the curved graph was fully compensated if Equation 10.11 was used with a backscattering factor, $k = 0.68$ and $k = 0.9$. However, the best linearity ($R^2 = 0.9967$) was obtained for calibration plot no. 7, when backscattering factor, $k = 0.9$, was used in Equation 10.11. The calibration plot nos. 1, 2, and 5 supported high changes in the slope function for a range of concentrations of caffeine. In these cases, a polynomial function is more than adequate for the purpose of evaluation. The calibration plot nos. 3, 4, 6, and 7 obtained using the Kubelka–Munk transformed Equation 10.11, employing backscattering factors, k, furnished a straight line.

TABLE 10.1

Dependence on the Backscattering Values Chosen from the Peak Areas Obtained for Different Amounts of Caffeine of Range 1.077–107.11 ng and 1.077–214.22 ng

Type of Particle	Nonspherical Particle				Spherical Particle with Fluorescent Dye		
KM(k)	0.50	0.55	0.68	0.90	0.50	0.75	0.90
Calibration plot number	1	2	3	4	5	6	7
Range (ng)	**Correlation Coefficient, R^2** **Slope from $y = a_1 x + a_0$**						
1.077–107.11	0.8297	0.9678	0.9876	0.9914	0.9643	0.9977	0.9942
	0.0011	0.0033	0.0092	0.0189	0.0008	0.0072	0.0118
1.077–214.22	0.9362	0.9908	0.9966	0.9958	0.9821	0.9970	0.9983
	0.0017	0.0036	0.0091	0.0178	0.0010	0.0079	0.0121

It can be noted that the slopes calculated in different ranges of concentrations are practically constant for a certain k, as shown in Table 10.1. Thus, it can be concluded that the correlation coefficient is a very strong function of a range of analyte concentrations and the backscattering factor, k, as observed in Table 10.1.

A new formula was presented for transforming fluorescence measurement in accordance with the Kubelka–Munk theory [47]. In a majority of densitometry instruments, the light reflected from a clean plate, J_0, is combined with the reflected light from the TLC spot, J, which can be expressed as

$$A(\lambda) = -\log \frac{J}{J_0} = -\log R \tag{10.13}$$

where

$A(\lambda)$ is called absorbance and is positive if more light is absorbed by a substance spot than a clean TLC plate surface [42]

R is the corrected remission term

The fluorescence measurement may be obtained by direct measurement of the intensity of the emitted fluorescent light [47]. When a diode-array scanning is used, the fluorescence information can be extracted from the raw data using the equation:

$$F(\lambda) = \frac{(J_F - J_0)}{1000} \tag{10.14}$$

where

J_F is the emitted fluorescent light intensity, which is corrected only for the intensity of reflectance J_0, measured from the clean plate surface

$F(\lambda)$ is positive if the intensity of fluorescence emitted by the surface spot is greater than that from a clean HPTLC plate [42]

The fluorescence signals, absorption signals, and data from a selected reference are combined into one equation [47]:

$$\frac{J_F^2}{2J_0 J} = \frac{m a_m q^2}{1 - a_m} \tag{10.15}$$

where q is the quantum yield.

The squared fluorescent light intensity, J_F^2, divided by the intensity of the light remitted, J_0, from the clean plate measurement at the wavelength of the fluorescence and divided by the light intensity, J, reflected from the sample spot at the absorption wavelength, is directly proportional to the amount of sample, m, in the layer, sample adsorption coefficient, a_m, and the squared quantum yield, q. Equation 10.15 shows a huge advantage in fluorescence measurement in comparison with adsorption spectrometry—the measurement signal increases with increasing illumination power.

Only diode-array techniques can measure all the required data simultaneously to linearize the fluorescence data accurately. The new theory of HPLC quantification of the analgesic flupiritine was tested by several spectra presented, densitograms, and calibration plots, and their results confirmed the accuracy of earlier equations [47].

A sample of urine was mixed with sulfuric acid and extracted with diethyl ether. The aqua phase was alkalinized and extracted with ethyl acetate, while the organic phase was evaporated to dryness and residue was dissolved in methanol–ethyl acetate. HPTLC was performed on a silica-gel K60 F_{254} as the stationary phase. The plate was developed in a saturated chamber using ethyl acetate–methanol–25% NH_3 (85:10:5, v/v) as the mobile phase. After development, the plate was protected by dipping it in paraffin–hexane (1:3) for 2 s. Dipping increases the fluorescence by tenfold and preserves the stability of the fluorescence intensity for hours. The whole track of 450 spectra in the range of 180–610 nm was scanned over a distance of 45 mm [47]. The absorbance and fluorescence spectra, and Kubelka–Munk and fluorescence spectra for flupiritine of one urine sample have been presented. The reference spectrum was measured from a position on the plate that is free from the sample. Equation 10.13 sets the peak and baseline to positive values, but cannot separate the sample and background signals, which makes peak integration difficult. However, the effect of baseline subtraction by the corrected Kubelka–Munk evaluation is equipped for easy integration.

Generally, fluorescence analysis at high concentrations is not suitable, because often calibration curves are not linear and even run to saturation. The calibration curve for flupiritine in the range of 300–5000 ng was evaluated using Equation 10.14. As is shown, there exists linearity up to 3000 ng, and in this range limited quantification is possible with a good precision and reproducibility.

A theoretical basis for a linear relationship between the absorption signal and the amount of sample can be observed in the Kubelka–Munk equation. By applying the new theory [47] of fluorescence evaluation according to Equation 10.15, the transformed measurement data were found to depend only on the amount of flupiritine applied on the layer. This was verified when all the data were transformed in fluorescence range, J_F (410–450 nm), for sample reflectance, J (320–328 nm), and for reference data, J_0 (320–328 nm) using Equation 10.15. Furthermore, the calibration plot was found to be perfectly linear over the whole range (300–5000 ng) of flupiritine.

This new Equation 10.11 comprises the equation for absorption or fluorescence in a diode-array TLC chromatographic determination. These theoretical considerations were experimentally demonstrated [45] by calibration plots for caffeine separation on silica-gel K60 nonspherical and spherical particles, with and without a fluorescence dye. A curvature with a typical s-form was obtained for the Kubelka–Munk equation used for estimation, with $k = 0.5$. However, evaluation using $k = 0.9$ gave a straight line. In this way, the backscattering factor $k = 1/2$ was demonstrated to be responsible for the shape of the calibration curve. Furthermore, such nonlinear Kubelka–Munk calibration plots often occur in practice. This situation can be avoided by a good selection of k values in Equation 10.11.

10.3.3.2 Practical *In Situ* TLC-UV/Vis

An overview of the calibration and evaluation in quantitative analysis in TLC and HPTLC is presented in Ebel [44]. Initially, the problems regarding linear regression and some procedures of linearization in TLC determinations were discussed. Generally, for a low amount of compound in a spot, a linear calibration between peak heights h or peak area A and concentrations of the compound can be obtained. In agreement with the Kubelka–Munk equation, it may be concluded that there is a linear relationship between the squares of peak, h, or A, and the amount of substances per spot. After analyzing the nature of errors in quantitative HPTLC in detail, the HPLC and HPTLC methods using the same phenomena (absorption in UV) can be directly compared in terms of precision and accuracy. However, in HPTLC, one can eliminate the source of error in spotting the compound, by spotting the same sample twice. In routine analysis by UV/Vis spectrophotometry, HPLC and HPTLC have comparable precision and accuracy, and the value of the percentage relative to the standard deviation is ca. 0.6%–1.5%, depending on whether the experiment is performed correctly. Also, the specific range in which the response is linear depends on the amount of the sample. For instance, estradiol benzoate is in the range of 95.0–630.0 ng [48] and aceclofenac is in the range of 1–10.0 µg [49] on silica-gel plates.

The *in situ* spectra are significantly influenced by the layer structure. The spectra of aromatic compounds are similar on the silica-gel layers and ion-exchange resins with 20% silica-gel, but are significantly different on alumina [50]. Advances in obtaining the structural information for sample identification on separated compounds by TLC-UV spectrometry are presented in a detailed review in Poole and Poole [51].

Furthermore, another earlier experimental work confirmed that the determination range in which response was linear for progesterone was 25–154 ng/zone at 254 nm, and for over a wide concentration range of 25–500 ng/zone; the calibration plot was described as a 2D linear regression (parabolic regression) [52]. Chromatography was performed on silica-gel 60F$_{254}$ HPTLC plates from Merck (Darmstadt, Germany). Samples were applied using an automatic TLC Sampler 3 Camag (Muttenz, Switzerland). Plates were developed in a Camag AMD2 automatic multiple development chamber with toluene-2-propanol (9:1, v/v) as the mobile phase, without chamber saturation and with 10 min drying time. HPTLC densitometry was performed with a Camag TLC Scanner 3, and operated in reflectance mode.

Olah et al. [53] performed a selective extraction of a caffeic acid derivative from *Orthosiphon stamineus* Benth. (Laminaceae) leaves. Rosmarinic acid was identified by TLC on silica-gel plates with three different mobile phases and three different visualization systems. The confirmation of rosmarinic acid was made by comparing the R_F values, color, and *in situ* UV spectra of the standard and the sample, after TLC separation in the same condition.

The TLC methods were optimized for the separation of the principal polar aromatic flavor compounds to determine the botanical source and quality of cinnamons commercial importance [54]. Eugenol and cumarin can be easily differentiated by their *in situ* UV spectra of cinnamaldehyde and 4,2-methoxycinnamaldehyde, which can be used to determine the identity and purity of the separated zone. In the acid-containing fraction after the separation of the diol layer, none of the 12 candidate compounds matched with the chromatographic properties (R_F) or the UV absorption properties of the unknown compound. Thus, the identification of the unknown compound must be continued using other modern methods of investigation.

The separation methods of ascorbigen and 1'-methylascorbigen, and other possible molecules with different substituents on the indole and ascorbic acid skeleton by normal and reversed-phase TLC were described. The identification of ascorbigen and 1'-methylascorbigen was performed by comparing the UV spectrum *in situ* and in solution [55].

TLC is a highly versatile and low-cost technique and is still very popular in analytical toxicology laboratories. Its usefulness in combination with UV detection for systematic toxicological analysis purposes has been demonstrated. Ojanpera et al. [56,57] extracted the basic and amphoteric drugs from urine or enzyme-digested liver sample using ion-pair extraction. Separation of basic drugs was carried out on silica-gel TLC plates using toluene–acetone–94% ethanol–25% ammonia (45:45:7:3, v/v) as the mobile phase. Densitometry was performed at 220 nm *in situ* spectra, recording in the range of 200–400 nm. Identification was performed by correlation of corrected R_F values and spectra [57]. A first TLC system (hydroxyzine, lidocaine, codeine, and morphine) was carried out on RP-18 silica-gel with methanol–water–37% hydrochloric acid (50:50:1, v/v). A second TLC system (codeine, promazine, clomipramine, cocaine) was separated on a normal silica-gel plate with toluene–acetone–94% ethanol–25% ammonia (45:45:7:3, v/v) as the mobile phase. Densitometry by absorbance in UV at 220 nm was performed. Novel software was used for the processing of qualitative data from the two parallel TLC analyses simultaneously, based on the comparison of the libraries of corrected R_F values and *in situ* UV spectra [57].

Spangenberg et al. [58] elaborated on a new criterion for identification of forensic drugs. About 33 compounds with benzodiazepine properties were analyzed by HPTLC on silica-gel plates after prewashing with methanol followed by dichloromethane–methanol (19:1, v/v). The developments were carried out in three optimized mobile phases:

1. Dichloromethane–methanol (19:1, v/v)
2. Ethyl acetate–cyclohexane–25% ammonia (50:40:0.1, v/v)
3. Cyclohexane–acetone–methyl *t*-butyl ether (3:2:1, v/v)

Detection was performed with a diode-array densitometer that identifies all the compounds with high accuracy of <20 ng. An algorithm for spectral recognition combined with R_F values obtained in three separations with different mobile phases into one fit factor has been presented. This set of data is unique for each compound investigated and enables unequivocal identification.

A simple, rapid, and reliable HPTLC-UV online method has been developed for the identification and quantification of 11-nor-Δ^9-tetrahydrocannabinol-9-carboxylic acid in urine. The best results were obtained on HPTLC silica-gel 60 WRF$_{254S}$ layers with dichloromethane–*n*-hexane–methanol (7:2:1, v/v) as the mobile phase. Measurements were performed at 210 nm and the spectra were recorded over the range of 200–339 nm. The identification was carried out by using a spectral library [59].

Butz and Stan [60] analyzed 265 pesticides by TLC with automated multiple development (AMD) using a universal gradient based on dichloromethane. The pesticides were obtained in concentration from 1000 mL drinking water. The concentration was determined using solid-phase extraction (SPE). The compound identification was performed considering the migration distance, UV spectra, and detection limits.

A method based on TLC was developed for the simultaneous analysis of residues of 2,4,6-trinitrotoluene and its metabolites [61]. Chromatography was performed on silica-gel GF_{254} with benzene–diethyl ether–methanol (60:3:1, v/v) as the mobile phase. Identification of separation zones of the compound was made by comparison of R_F values, colors obtained after spraying the chromatogram with 5% ethylenediamine in ethanol, and *in situ* spectra with those of the standards.

An efficient HPTLC-UV/FTIR coupling procedure for separation and rapid identification of flurazepam hydrochloride and its related substance in bulk powder and capsules was reported [62]. The chromatographic conditions were optimized: silica-gel 60 with 50% magnesium tungstate and a mixture of methanol–ethylacetate–toluene–NH_3 (10:10:80:1, v/v) were used as stationary and mobile phases. The measurement of compounds after the development of the plate was carried out online by HPTLC-UV/FTIR using the peaks in the UV, Gram–Schmidt, or window chromatograms. Furthermore, unambiguous identification can be obtained by postrun extraction of the DRIFT spectra and compared with the reference spectra in the library.

10.3.3.3 Practicalities of TLC-Fluorodensitometry

Fluorodensitometry is an attractive method for qualitative identification and quantitative determination. Fluorescence is the property of some atoms and molecules to absorb light at a particular wavelength and to subsequently emit light of longer wavelength after a brief interval, termed the fluorescence lifetime. So far, the most frequently applied excitation wavelengths are the Hg emission lines at 254 and 366 nm, and the silica-gel layer shows considerable absorbance at these wavelengths. Furthermore, silica-gel absorbance over 366 nm can be practically negligent [63]. Therefore, silica-gel is a suitable chromatographic layer for separation of compounds that exhibit fluorescence.

The fluorescence response for substances on a thin layer can be far less than expected from the measurement made in a solution. During the adsorption of the excitation energy onto the sorbent layer, a part of this energy is lost as heat to the surroundings, reducing the fluorescent signal. Another cause for the quenching of the fluorescence signal is the interaction with oxygen or reaction of the solute with oxygen to produce new products with a diminished yield of fluorescence. In many cases, the emission signal can be enhanced by application of a viscous reagent to the layer before scanning the TLC plate. Some of the well-known viscous reagents include paraffin and silicon oil of different viscosities, Triton X-100 (isooctylphenoxypolyethoxyethanol), glycerol, Foblin (poly[perfluoroalkyl ether]), poly(ethylene glycol) 4000 (PEG), and the mixtures PEG and silicon oil. In the fluorescence enhancement mechanism, a fraction of solute is transferred into the reagent's liquid phase where fluorescence quenching is less severe than in the adsorbed solute phase. In some favorable cases, the use of a fluorescence-enhancing reagent can increase the response by as much as 10–200-fold [64–67].

Fluorodensitometric determination of fluorescence emitted by flavones and flavonols *in situ* on silica-gel 60 (without fluorescence indicator) plates can be improved by dipping the developed plates in appropriate solution of reagents [68]. About 14 flavone and 26 flavonol derivatives were developed with toluene–ethyl formiate–formic acid (5:4:1, v/v) for aglicones, and ethyl acetate–formic acid–water (100:10:5, v/v) for glycosides. After dipping the developed and dried plate into a solution of 1% (m/v) diphenylboric acid 2-aminoethyl ester (Naturstoffreagens A, NA) in diethyl ether–methanol (2:1, v/v) to form flavonoid–NA chelates, the plate was dipped into a second fluorescent intensifier solution that consists of a mixture of 60% (v/v) silicon oil in diethyl ether, 10% (m/v) PEG in diethyl ether–methanol (2:1), or 60% (v/v) viscous paraffin in diethyl ether. Silicon oil or paraffin is most appropriate for lipophilic flavonoids, while the more hydrophilic PEG is most suitable for less lipophilic aglycones as well as glycosides. In the case of luteolin-5-glucoside with fluorescence intensifier PEG, the factor of increase in intensity is 62.1. The reference standard has to be measured on the same plate as the experimental sample, because *in situ* fluorodensitometry measurements are difficult to reproduce. Consequently, only the fluorescence intensity of substances with the same R_F values can be compared. At the end of this chapter, a brief discussion about the correlation between fluorescence and molecular structure is presented.

Butler et al. [69–72] developed an RP-HPTLC for qualitative identification and determination of PAH using fluorescence scanning densitometry. The separation of PAHs was performed on RP-18 silica using methanol–water (10:3, v/v) as the mobile phase. In this study [69], the following three aspects were mainly explored:

1. Quantification by fluorescence scanning at 254, 266, 313, and 365 nm
2. Determination of sample identity
3. Test of peak homogeneity using normalized emission–response ratios

A standard mixture of PAHs (benzo[a]pyrene, fluoranthene, benzo[g,h,i] perylene) was separated by RP-HPTLC on silica RP-18 with methanol–dichloromethane–water (20:3:3, v/v) as the mobile phase. This chromatographic system was used for the optimization of a scanning densitometer for fluorescence detection at 366 nm in HPTLC [70]. The procedure for the accurate measurement of emission response ratio for PAH's identification using fluorescence scanning densitometry is reported in Ref. [71]. The two-point calibration method was applied for the determination of PAHs after separation by RP-HPLC on silica RP-18 [72].

Furthermore, a laser-based indirect fluorometric detection method for TLC was also described [73–75]. This technique can be easily applied in TLC for detection and quantification measurements, because of its 2D-scanning capability. The detection mechanism is similar to that of liquid chromatography, and a constant background signal is maintained when the analytes are absent. When the analyte is eluted, displacement of the eluent causes a change in the background signal. In this way, the analytes as anions [75], cations, and nonelectrolytes [74] can be detected indirectly in TLC. Since this method is based on the indirect fluorescence mode, universal detection is possible without derivatization.

In indirect fluorometric detection, the limit of detection should be approximated by the following equation:

$$C_{\lim} = \frac{C_F}{DR \cdot TR} \qquad (10.16)$$

where
$C_{\lim}$ is the molar concentration of analytes at the detection limit
C_F is the molar concentration of the fluorophore
DR is the dynamic reserve (which is defined as the ratio of background S/N level)
TR is the transfer ratio (which is defined as the number of fluorescing molecules transferred by an analyte molecule)

Equation 10.16 shows that a larger DR and a smaller C_F are needed to decrease the detection limit. This indicates a more stable background signal and the need for the realization of a lower concentration of the fluorophore. Furthermore, a large transfer ratio, TR, can improve the detection limit; however, it results in chromatographic conditions [74,75].

Chromatographic separations were carried out using a K_6 silica-gel plate (Whatman) that was pretreated with 2×10^{-6} M Nile Blue A perchlorate in methanol for 20 min. The test sample contained orange G and crocein orange G. A mixture format of 2-butanol–acetone–water (75:15:10, v/v) was used as the mobile phase. Nile Blue A was chosen as the fluorophore, because it can be excited by a He–Ne laser [74]. A schematic diagram of a laser-based indirect fluorometric detector for TLC [74], based on the indirect fluorescence mode and universal detection without derivatization, was presented. It takes only 35 s for acquiring a data array of 256×64 points. In this experiment, a detection limit of 6 pg and a linearity over two orders of magnitude was obtained.

In the 1990s, an interesting method, namely TLC with supersonic jet fluorometric detection, was described [76]. A mixture sample (PAHs) was deposited in a straight line on a sheet with Kieselgel 60 Merck, in normal phase on a polyester film and was developed with hexane as the mobile phase. This sheet was mounted on a sliding sheet roller attached to a supersonic jet nozzle. The first dye laser beam (275 nm) was introduced from a small through-hole to desorb the compound on the TLC sheet. The vaporized compound was introduced into a carrier gas of argon, and the gas was then expanded into a vacuum to form a supersonic jet. The compound molecules were detected by fluorescence induced by the second dye laser beam (366–375 nm). Through this technique, it is possible to directly measure the chromatogram and the excitation spectrum of the separated compounds on the TLC sheet. The detection limits are about 10 ng in spectrometric and chromatographic measurements; however, a value of 4.4 µg is necessary for recording a spectrum.

In addition, another method to obtain a fluorescence-quenching zone was reported [77]. A quantitative method using silica-gel HPTLC plates (Whatman) with a fluorescent indicator were developed using acetonitrile–water (7:3, v/v) as the mobile phase. The standard samples consisted of 1.00–4.00 µg of creatine. After development, the mobile phase was evaporated using a hair dryer for 10 min. Subsequently, it was heated at 160°C on a Camag plate heater for 5 min for thermochemical activation of fluorescence-quenching zones. The quantification was made by scanning *in situ* at 254 nm. Accuracy was found to be within 0.2% of the certified values, and precision was 3–4% of the relative standard deviation.

The advantage of coupling online liquid chromatography and TLC for identification of PAHs by fluorescence excitation and emission was reported in 1980s [78,79]. At that time, it represented a progress, but now, another possibility to perform separation of PAHs, e.g., by 2D HPTLC, has been reported.

10.3.3.4 Practicalities of TLC-Videodensitometry

In the past, there were some studies [80–83] comparing the performances of the videodensitometer with those of the scanning densitometer on another part with the CCD camera.

However, a new TLC method with densitometry and videoscanning was elaborated for the quantitative determination of some drugs [80,81]. The active substance (fenofibrate or genfibrosil) was extracted from the tablets with methanol. Chromatographic separation was performed on HPTLC diol F_{254} plates using hexane–tetrahydrofuran (8:2, v/v) as the mobile phase. Densitometric assay and videoscanning quantitation were performed at 227 and 254 nm, respectively. Calibration plots were constructed in the range of 5–30 μg/spot for both the drugs. The calibration data were tested on three regression models, and the optimum model was selected. The quadratic model was selected for videoscanning, and nonlinear $y = ax^m + b$ was selected for the densitometric method, and in all cases, $R^2 > 0.997$. The recovery function was sufficiently linear in all the cases, with an intercept approximately at zero and the slope being very close to 1. Stastically, no significant difference was detected between the precision and accuracy for densitometric or videoscanning methods.

The performances of a CCD camera (video documentation system) were compared with a slit-scanning densitometer to validate the TLC method for caffeine [82]. The separation of caffeine by HPTLC was carried out on silica-gel with dichloromethane–methanol (9:1, v/v) as the mobile phase. The slit-scanning densitometer was operated at 270 nm. From a practical point of view, no difference was observed between the determined values of the validation parameters by both methods.

The CCD camera and image-analysis techniques were used for the evaluation of the TLC plate [83]. A computer program was developed to enable the use of inexpensive image-generation systems, such as CCD cameras, webcams, or flat-bed scanners for the quantitative evaluation in UV or white light of TLC separation. The performance of a CCD for optical detection of sample bands in thin-layer chromatograms was also investigated [84]. The CCD camera was mounted on a modular system configured for mass spectrometric analysis of the same TLC plates, and the sensitivity of detection under several modes of illumination for model compounds was studied.

The slit-scanning densitometers and diode-array densitometers versus video densitometers have much more facilities with respect to qualitative analysis, e.g., performing compound spectra analysis *in situ*; verification of purity of the peak; obtaining simultaneous densitograms at multiwavelength. All these facilities are significant in the identification of unknown compounds.

10.4 TLC-IR OR FTIR

10.4.1 OFFLINE TLC-IR OR FTIR

An interesting idea was put forward in the field of TLC-IR about 40 years ago, which opened the possibility of coupling TLC separation with IRS. Initially, the methods involved scarping the sample together with support from the TLC plate and the component transfer from the spot to a transparent substance such as KBr or AgCl for IR. However, these methods are laborious and consist of three steps:

1. The spot of interest has to be precisely located using a nondestructive method.
2. Extraction of the substances from the spot with an appropriate solvent should be carried out, followed by filtration or centrifugation, evaporation, and dissolving the residue in a smaller amount of solvent.

3. Impregnation of a small quantity of KBr or AgCl with concentrated solution, evaporated solvent, and subsequent preparation of a pellet suitable for IR analysis should be carried out.

Indeed, the identification of compounds after separation by IR or FTIR spectrometry significantly increases the possible applications of TLC. By the hyphenation of TLC-FTIR, identification of each compound from unknown mixtures can be performed, for which standards are unavailable [1,85,86].

The direct transfer technique is described in Rice [87]. In this method, the sample is separated on TLC plates in the usual manner. After locating the spot on the plate, it is lightly outlined into a teardrop shape. The TLC thin-layer support around the outlined spot was removed, and the glass adjacent to the point of the teardrop was cleaned. A quantity of 15–20 mg of KBr was added to the cleaned area on the plate, with a microspatula. The KBr was formed into a line 0.2 cm wide and 0.6 cm long, such that one end of the line is in contact with the teardrop. Then, an appropriate solvent is added drop-by-drop using a microsyringe. The rate of solvent must be controlled so that it does not flow onto the glass from the edges of the spot. As the sample moves across the spot, it would become concentrated at the teardrop, and then would be transferred on the pile of KBr powder. When the solvent front and sample reach the pile of KBr powder, the contact point of KBr and TLC support must be broken, to prevent the flow-back of the solvent onto the TLC support. After evaporation of the solvent, 5–7 mg of KBr is pressed into a micropellet and IR spectra are obtained in the usual manner. A typical spectrum of an acetone adduct of BLE-25 was obtained.

In addition, a similar method of approximation has been described in Ref. [88], in which the location of the separation spot was detected by UV radiation at 254 or 350 nm. The spots were outlined by scratching the support. The TLC support was removed around the spot, and the glass around the support was carefully cleaned. In this case, a wall of KBr powder was built around the tip of the spot on an open space of 1–2 mm left between the spot and the KBr wall. An appropriate solvent in which the compound is readily soluble was added to the spot. After the solvent front passed through the spot, it spread rapidly over the glass surface and was absorbed by the KBr powder. A part of the KBr powder was transferred to a micro-KBr disk, and the IR spectra were obtained in the usual manner.

However, the direct elution technique is sensitive to the following:

1. Type of compound eluted—the compound that forms strong hydrogen bonds shows a poor recovery.
2. Different support materials used show different recoveries.
3. Contamination of the spectrum of the eluted compound with that of the support material. All the three supports (silica-gel, aluminum oxide, and cellulose) show their strongest absorption around 1100 cm^{-1}, but silica-gel is much stronger. The compounds that are very polar show a low percentage of recovery.

TLC separation of benzo(a)pyrene and dimethyl benzantracene were carried out on silica-gel with benzene–hexane (1:9, v/v) as the mobile phase. Detection was performed by UV, and a transfer method is reported in Issaq [89]. An Eluochrom (Camag) device was used for the *in situ* elution of a compound from the spot to a small quantity of KBr, followed by solvent evaporation, and compressing the powder as a micropellet. However, about 5 μg of substance is necessary for an IR spectrum.

Other methods consist of scarping the spot and the layer, and placing a small amount of KBr powder tamped down in a special device on the top of a hypodermic needle. The KBr serves as a filter to prevent the silica-gel from passing through the needle. Using a syringe (1 mL), an adequate solvent can elute the compound, drop-by-drop, from the layer to the 10 mg of KBr. Each drop is allowed to evaporate and 20 drops are usually sufficient. The KBr that contains the compound will be homogenized and pressed as a 1.5 mm diameter pellet. Excellent IR spectra can be obtained from 5 to 30 μg of compound [90].

Goodman [91] developed another transfer method using a TLC plate and another plate with a KBr layer. The spot was identified on the thin-layer chromatographic plate and the plate was rotated 90° as in bidimensional development, and eluted again. The KBr plate was placed one above the other in the direction in which the spot was moving. The detection limit was 50 μg/spot.

Garner and Packer [92] described an absorption method using a KBr pyramid. In their method, the spot together with the support was removed from the chromatographic plate and introduced into a glass cylinder. The solvent dissolved the substance from the layer and migrated through the KBr pyramid toward its top, where the substance was concentrated by solvent evaporation. The pyramid top, containing the analyzed compound, was cut, dried, and used to make a micropellet. Good IR spectra were reported for 10 μg of substance and a recovery of 50%–80% was obtained. This technique can be applied using the commercial Wick Stick.

Using this procedure, Alt and Szekely [93] successfully identified the components of an indigo dye by TLC-IR spectroscopy. The TLC separation was performed on a silica-gel plate with a high-boiling eluent. The spot to be analyzed was scarped off the TLC plate and transferred into a vial, and put on a Wick Stick, following the addition of about 0.2 mL of a high-boiling solvent, e.g., 1-bromonaphthalene, and the vial was heated to 170°C in an oven. A blue main spot ($R_F = 0.7$) was identified in the IR spectrum as indigotin and the spot ($R_F = 0.6$) was recognized as indirubin.

The separation and identification possibilities of some metal–diethyldithiocarbamate (DEDTC) complex have been investigated using the TLC-IR sequential system. The IR spectra were detected after preconcentration by the Wick-Stick procedure. It was concluded that an efficient and successful qualitative analysis is possible for the overlapped spots using an IR spectrometer as the TLC detector [94].

Chalmers et al. [95] described a method for substance transfer from the spot to a KBr pellet. Alkali halide pellets were prepared by compressing 0.7 g of dried KCl powder in a 13 mm die at a pressure of ~500 psi. A spot from aluminum-backed silica-gel TLC plates containing F_{254} was cut out and a metal backing was placed downward in an adequate glass sample tube. Then, a KCl pellet was placed centrally on the spot and about 2 mL of chloroform was carefully pipetted down the inside of the tube. After about 1 h, when the chloroform was evaporated, the pellet was removed for examination by diffuse reflectance-FTIR (DR/FTIR). Spectra obtained from standard substances with additives like Irganox 1010, Irganox 1330, Topanol

OC, and stilbenequinone, and extracted from polypropylene that were previously separated by TLC, showed excellent similarities. This technique offers a simple and effective way of identifying TLC fraction for quantities of about 40 μg of substances (and probably for 10–20 μg) in favorable conditions.

Iwaoka et al. [96] performed a complex study regarding the substance of the spot transfer via a silica-gel layer–organic solvent–KBr micropyramid support. In their study, the stability of a TLC plate with silica-gel $60F_{254}$ (Merck) was investigated. Five micrograms of the silica was scarped from the plate, ground, and extracted with different nonpolar and polar solvents. The extract was then dried, mixed with KBr powder, and pressed into a micropellet, with the absorbance at 1095 cm^{-1}.

For the nonpolar series of compounds, the best solvent for extraction is chloroform > methylene chloride > ether > hexane > benzene = ethyl acetate > ethanol > acetone, and for polar series the best compound is $CH_3CN > H_2O + AcOH > H_2O + n\text{-}BuOH + AcOH > > > H_2O + NH_4OH$.

The recovery rates for the transfer substances (Thymol Blue CS-045) were 85%–100%. The study possessed numerous data for achieving the geometrical form of pellet. The first step after chromatographic separation on thin-layer plates is the determination of the exact position of each spot using a nondestructive method. The spots were bordered by two lines traced in the layer to prevent transversal diffusion of the solvent during the second elution. The direction of the second elution was perpendicular to that of the first elution, as in 2D TLC. The KBr pyramids were placed on the side of the plate that will be reached by the elution front after the second elution, with their axes perpendicular to the direction of migration. Excellent FTIR spectra were obtained for several aminopyrines using this technique.

Shafer et al. [97] described a sample transfer technique from the thin-layer plate to an IR transparent material finely ground IR-transmitting glass composed of germanium, antimony, and selenium; Glass no. 1173 (Texas Instruments, Inc.). They used a 2 cm × 8 cm silica-gel 60 on an aluminum backing (Alltech Associates) thin-layer plate for the separation of a dye mixture. The transfer plate was an aluminum strip in which a row of 35 small holes (1.2 mm in diameter and 4.0 mm deep) was packed with an IR-transparent material. A bundle of glass fibers held at the base of each hole were in contact with the powder. After development, the chromatographic plate was rotated by 90° so that the longer side was in the horizontal position. The aluminum strip was attached to the top side in contact with the thin-layer plate by pressing the side of the glass fiber against the stationary phase to ensure good contact with the plate. The thin-layer plate was developed with a suitable eluent (dichloromethane), which enabled the transfer process in about 2 min. The aluminum band was detached from the thin-layer plate and dried at 40°C. Diffuse reflectance measurements were performed on an FTIR spectrometer equipped with a DR accessory. The aluminum strip was moved by a continuous translational process through the focus of the beam. The chromatogram was reconstructed from the integrated absorbance in a window region set from 1650 to 1350 cm^{-1} for the separation of a dye mixture. It was observed that the percentage of IR radiation reflected from the stationary phase of a TLC plate was significantly lesser than that reflected from an IR-transparent powder. This method has the advantages of the lower possibility of sample loss, decomposition, or contamination, and saves time over conventional methods (i.e., removal of the stationary phase containing the spot, followed by extraction and concentration of the extraction). The most important factor is that the spectra of the compound are identical to the published reference spectra, in contrast to the case of either DR or photoacoustic spectra measured *in situ*. Furthermore, the spectra for samples at a concentration of $40\,\mu g/\mu L$ are easily identified.

The color pigments of *Trichoderma harzianum* fermentation broth were separated on an RP-18 W/UV$_{254}$ plate (Macherey-Nagel) with two-step gradient elution of water–acetone. The main fractions were tentatively identified by FTIR spectrometry. The FTIR measurements were unable to identify the exact chemical structure of the main pigment fraction, but the presence of OH, = CH, and C = O was confirmed [98].

The separation of color pigments of chestnut sawdust was performed on various direct TLC on silica, alumina, and diatomaceous earth, and by RP–TLC on the same adsorbent impregnated with paraffin oil and developed by multistep gradient with water-THF [99]. The plates were evaluated by a dual wavelength TLC CS-930 Shimadzu scanner at 340 nm, and by a Nicollet Magna 750 FTIR spectrometer. The impregnated silica-gel layer showed a better scale of separation. Only the presence of the carbonyl group in the main fraction was verified by online FTIR, but offline FTIR spectra suggested that each pigment fraction contained –OH and –COO groups, and that they are probably tannic acid-like compounds.

The soluble color pigments of raisin were separated by RP–TLC, and TLC-FTIR with online and offline coupling was used for the identification of the main fraction. However, online TLC-FTIR cannot be used for the identification, because of the strong adsorbance of the stationary phase. On the other hand, combined offline TLC-FTIR with retention behavior of main pigment fraction indicated that it is a polymer, caramel-like compound formed of fructose and erythrose monomers [100].

10.4.2 ONLINE TLC-IR OR FTIR

Measurement of the IR spectrum of a TLC spot can be performed online (*in situ*) or offline, after the spot has been transferred to an IR-transparent substrate. As offline methods are time-consuming and introduce the possibility of loss, decomposition, or contamination, online measurements have been the preferred approach to TLC-IR, till date. Furthermore, the number of applications for the use of DRIR has grown substantially, since 1975.

A recent review by Cimpoiu [101] discussed the problems regarding HPTLC-FTIR. First, the principles and instrumentation were presented, followed by the theoretical aspects of qualitative and quantitative IR analysis. Instrumentation was presented and explicated with a diagram of a Bruker HPTLC-FTIR unit. The theoretical aspect focused on the Kubelka–Munk equation and interpretation of interaction between radiation and the sample. The possibilities of presentation of the spectra were also shown.

An optical system for diffuse reflectance infrared Fourier transform (DRIFT) measurement was also described [102]. The optical configuration of this diffuse reflectometer is as follows: radiation from the interferometer is focused on the sample by an off-axis paraboloidal mirror, and the diffused reflected radiation is collected by the ellipsoid mirror and focused onto the detector. The nature of the matrix affects the detection limit for microsampling by DRIFT spectrometry, but strong bands of adsorbed compounds can be seen even for strongly absorbing matrices such as silica-gel. Detection limits of submicrogram of compounds chromatographically separated on TLC plates were also reported.

In 1990s, another report [103] covered the development of HPTLC-FTIR online coupling. However, the major problem was on the quantitative determination of the compounds. The initial time was for the assessment of peak area. The peak area can be prepared in the Gram–Schmidt trace or window diagram, or by the evaluation of Kubelka–Munk spectra with the integration of their strongest bands. The Gram–Schmidt traces indicate the changes in absorbance over the whole spectra region and are suitable for rapid determination. The determination of the peak area in the window chromatogram is appropriate for the quantification of individual compounds. This method has a better S/N, but has a poorer precision. More precise results can be obtained using the assessment Kubelka–Munk spectra. The limit of identification and determination is 10 times higher than those obtained by densitometry.

Frey and Kovar [104] studied the possibilities and limitation of assays by online coupling TLC and FTIR spectroscopy. The method enabled the evaluation of chromatograms and spectra measured at the region of highest solute concentration. The problems and sources of error that arise during the recording of chromatograms were also investigated. The methods were compared from the perspective of precision, selectivity, and time consumed. The method was more appropriate for the identification of substances. However, quantification is particularly useful for substances with a loss of absorption in the UV region, especially when the precision required is not too high. The limit of identification and determination was found to be about 10 times higher than that of densitometry; however, it was 10 times lower than that for measurement in the near-IRS.

Direct densitometry on TLC plates by Fourier transform near-infrared (FT-NIR) spectrometry was also described [105]. A mixture of three phospholipids, L-α-phosphatidylcholine (PC), L-α-phosphatidylethanolamine (PE) (both from soybeans), and synthetic L-α-phosphatidic acid, were separated on silica-gel HPTLC plates (Merck). A mixture of chloroform–methanol–acetic acid–water (170:25:25:6, v/v) was used as the mobile phase. The plate with separated zones of phospholipids was fixed on the stage of a microscope linked with an FT-NIR spectrometer and scanned with a narrow beam of light. The NIR chromatogram and spectra of the separated zones were then obtained by the diffuse transmission method. They had strong absorption bands between 4400 and 4200 cm^{-1}. A typical NIR chromatogram was performed by scanning the HPTLC plate and recording the integrated absorbance signal between 4600 and 4000 cm^{-1}. The absorption spectra of each zone were obtained from 100 scans between 4600 and 4000 cm^{-1}. Calibration curves for three phospholipids were obtained from the height of one of the two main peaks (4340.7 and 4273.0 cm^{-1}) and for the derivative spectra from the height of one peak (4340.7 cm^{-1}). The detection limit was estimated to be <1 μg for these three phospholipids.

However, the first approach to such measurements was developed by Percival and Griffiths [106], who prepared a TLC plate directly on an IR-transparent window (on 2.54 mm diameter AgCl), which contained a thin layer (of 100 μm) of silica-gel or alumina. The compounds studied included amino acids and Stahl's test dye solution.

The IR spectra for compounds in quantities of microgram per spot were obtained for the separated compounds in the transmittance mode.

The various effects of different matrices on the DR-IR spectra made the quantitative analysis difficult and hence the sample should be prepared, as far as possible, in a reproducible way [107]. Qualitative analysis can be accomplished, but the changes in the spectra owing to the interaction between the adsorbed substances and the layer must be taken into account.

Otto et al. [108] studied the *in situ* spectra of acetylsalicylic acid (100 µg) and caffeine (10 µg) mixture after separation on a thin-layer plate of cellulose with methanol–water (10:1, v/v) as the mobile phase. For comparison, the corresponding absorbance spectra of the pure compounds measured from KBr disks were provided. For a sample quantity of up to 40 µg, a linear relationship was observed, e.g., 1650 cm^{-1} absorption band in the *in situ* DR spectra. On the silica layer, it was observed that the detection limit was significantly influenced in the regions where the layer showed a strong absorbance. For a proper identification, both the spectra, for sample and for reference substances, have to be recorded on the same plate layer.

Zirconium oxide was tested as a stationary phase for improved TLC-detection by DRIFT spectroscopy (DRIFTS) for the determination of the dyes [109]. The thin-layer plates (microscope slides 75 × 22 mm) of silica, alumina, and zirconium were made locally. A 1 g to 2 mL ratio of zirconia to water was found to give more useful slurry. The silica and alumina plates were prepared by various methods. The number of OH groups on the surface of zirconia was considerably less than that on the surface of silica or alumina. The zirconia showed a strong IR absorbance, which was evident only below 800 cm^{-1}. In the DRIFTS, the presence of the OH bands at 3500 and 1650 cm^{-1}, respectively, was more evident in the spectrum.

The spectrum of zirconia showed significantly higher IR reflectivity than that of either silica or alumina. The strong IR absorption was observed below 2000 cm^{-1} in silica, and below 1600 cm^{-1} in alumina, but, on the contrary, zirconia in the region of 2000–1100 cm^{-1} had a relatively low IR absorption. These results clearly indicate the possibility of online qualitative identification of dyes on zirconium using DRIFTS spectra.

Bouffard et al. [110] developed a new microchannel TLC with IR microspectroscopic detection. This new TLC technique was termed "microchannel TLC." The TLC separation was performed in channels with a dimension of 400 µm × 200 µm × 5 cm, packed with a zirconium TLC stationary phase. The stationary phase was packed in the microchannels by pouring zirconium–methanol slurry. After methanol evaporation, the excess zirconium that overfilled the grooves was eliminated. Typically, 0.1 µL of the sample was applied onto an end of a channel. A chloroform–methanol (90:10, v/v) mixture was used as mobile phase for the development. Some samples used were organic dyes, such as methyl orange, and fluorescein gave R_F values of zero, while rhodamine B and methyl red gave R_F values of 0.1 and 0.4, respectively, and both methylene green and methylene blue gave R_F values of 1.0. From the R_F values, retention appeared to be dependent on the acidic functional groups present in the structure of these compounds. Subsequent IR microspectroscopic detection of organic dyes separated in these microchannels provided excellent DR spectra, and an improvement in the detection limit by a factor of 500 times over previous TLC work using microscopic slides [109]. In this way, four polyaromatic hydrocarbons were separated and identified using hexane as the mobile phase, where two of the hydrocarbons were isomers, 7,8-benzoquinoline ($R_F = 0.32$), and phenanthrene ($R_F = 0.1$), while the other two were similarly structured compounds, phenanthrene ($R_F = 1$) and 1,10-phenanthroline ($R_F = 0$). The DRIFTS spectra were also presented and discussed. This technique offered several attractive advantages than conventional TLC, such as the requirement of smaller amount of stationary phase and sample, and also microchannels with thin layer can be made locally.

An interesting idea was to combine the silica-gel with a weak IR-active, reflection-enhancing material like magnesium tungstate, for obtaining the quality of DRIFT spectra [111]. The optimum proportion of 50% magnesium tungstate resulted in one sorbent layer with separation performance practically identical with that of the silica-gel $60 F_{254 s}$, but the retention properties were different. By using this optimized sorbent thin-layer, the S/N for caffeine, paracetamol, and phenazone were increased by factors of 3.4, 3.1, and 2.3, respectively, when compared with pure silica 60 KG 60, and the detection limits were reduced by a factor of 3.7 for caffeine and 2.3 for both paracetamol and phenazone.

Zuber et al. [112] presented a method for direct analysis of TLC spots by diffuse reflectance fourier transform infrared (DRFTIR) spectrometry in ≤30 min. The spot (2–8 mm diameter) can be placed manually under the IR beam (1 mm diameter). The precise position of the spots on the thin-layer plate can be obtained with a UV lamp or by using a second plate as reference, and eluted under similar conditions when sprayed with a color reagent. After development, all solvents were carefully removed from the plate prior to spectra recording. A reference laser aided in spot alignment, and the plate contributions to background were subtracted from the resultant IR spectra by using the FTIR data system. To obtain good results, surface characteristics of each plate must be identical. Generally, the online method has one major disadvantage, even with the correction of layer absorption; it is very difficult to obtain reliable spectra in the regions where the layer shows strong IR absorption. For example, silica-gel showed strong absorption in the $3700–3100 cm^{-1}$ and $1650–800 cm^{-1}$ (especially $1300–1100 cm^{-1}$) regions, such that information about the sample cannot be obtained. Analtech Avicel (microcrystalline cellulose) thin-layer had a lower IR absorption in the region $1800–1600 cm^{-1}$ when compared with the silica-gel layer, and substances that contain carbonyl groups can also be detected. It is important to compare the IR spectra of a substance only if they were recorded from the same layer; otherwise, the band intensities dramatically change. The thin-layer width and the granule dimension did not significantly influence the FTIR spectra.

Beauchemin and Brown [113] proposed a method for *in situ* quantitative analysis of diazonaphthoquinones by RP-HPTLC using the same thin-layer plate (5 × 10 cm) for the background spectrum and the sample spectrum. The advantages of this technique are that it is rapid (DR analysis takes about 2 min) and more reproducible than the two-plate methods. For separation of these polar substances, RP-HPTLC plates were used (RP-18, F_{254}, Merck), and the mobile phase was acetonitrile–water (70:30, v/v). Developed plates were dried for 5 min at 45°C and the FTIR reflectance accessory was directly inserted. The substance spectrum was recorded in the middle of 2–8 mm diameter spots, on an area of 1–2 mm diameter. Qualitative analysis was accomplished between 2800 and $1300\,cm^{-1}$, which was the region of the greatest stationary phase reflectivity. The substance spectrum was recorded in the middle of 2–8 mm diameter spots on an area of 1–2 mm diameter. The strongest IR absorbance of the diazonaphthoquinone was that of the diazo group (C = N = N), which occurred at $2162\,cm^{-1}$ near the maximum diffuse reflectance of the reversed-phase silica-C18. In the case of *in situ* measurements, the interaction between the adsorbed substance and the layer led to significant changes of the maximum absorbance position for several functional groups. An example is diazo-naphthochinonsulfonic ester, which showed a doublet at 2164 and $2124\,cm^{-1}$ for the diazo group on a RP-C18 layer. The strong diazo vibrational doublet permitted quantification of the analytes; however, weaker bands in the fingerprint region were obscured. Precision of the overall technique (chromatographic and spectrometric) was 3% of the relative standard deviation when commercially available plates were used. For the adsorbance plot of the integrated area of the diazo group versus that of diazonaphthoquinone esters in the range of 1.1–17.7 µg, the correlation coefficient was 0.998. The *in situ* degradation of diazochinonic ester on RP-HPTLC plates can be observed owing to the decrease in the "azo" absorption bands at 2132–$2180\,cm^{-1}$, and the appearance of maximum absorption at $1710\,cm^{-1}$ corresponding to the newly formed carbonyl group. This technique is suitable for the determination of photochemically labile compounds.

In another study [114], the same technique of transfer of TLC sheet by direct insertion into the FTIR spectrometer accessory was described. The method was applied in the determination of caffeine in a mixture containing phenacetin and noscapine, resulting in a detection limit of 2 µg. The transfer required only 2 min, thus preventing sample loss and reducing data collection time.

Bauer and Kovar [115] studied the interaction of acids and bases with the binder in the precoated HPTLC plates, by online HPTLC-FTIR that led to greatly altered spectra. It was demonstrated that the cause was an acid–base reaction between the substance and the binder present in the precoated layer. Water-resistant, acid-stable fluorescent indicator silica-gel 60 WRF_{254S} HPTLC plates were used for HPTLC-FTIR analysis of substances with acidic properties, because there was no possibility of reaction between polyacrylic acid as binder and acids, and better DRIFT spectra were obtained. Polyacrylic acid reacted with bases, resulting in negative peak at $1723\,cm^{-1}$ and an interfering carboxylated band between 1560 and $1570\,cm^{-1}$. To avoid this situation, it was preferred to use plates of silica-gel 60 F_{254} layers for the analysis of basic substances. The standard silica-gel 60 F_{254} contained a polyacrylate salt as binder. This can be protonated by acidic substances to yield a band in the region of 1515–$1535\,cm^{-1}$ and a negative peak between 1570 and $1590\,cm^{-1}$. To avoid this situation, plates with silica-gel 60 WRF_{254S} layers can be used.

A scheme was presented for the identification of 16 substituted arylalkylamines (amphetamines), including stimulant and hallucinogenic drugs. For the identification of the compounds, the R_F values in six mobile phase and UV, visible, and IR spectra were used [116].

After numerous studies, Kovar et al. [117–125] developed online coupling of HPTLC and FTIR spectroscopy in combination with AMD. It was demonstrated by several applications: benzodiazepines and edetic acid (EDTA) [117]; amphetamine and determination of caffeine in the range from 100 to 50 µg [118]; forensic samples were identified on the basis of R_F values and DRIFT spectra [119]; edetic acid (EDTA) in surface water [120];

limits of identification were defined for hexobarbital (55 ng), phenobarbital (55 ng), caffeine (30 ng), salicylic acid (220 ng), and ascorbic acid (240 ng) [121]; lysergic acid diethylamide (LSD), 1-(1,3-benzodioxol–5yl)-N-methylbutan-2-amine (MBDB), and atropine in real sample were identified on the basis of R_F values and DRIFT spectra [122]; adenosine in biological sample [123]; N-ethyl-3,4-methylenedioxyamphetamine and its major metabolites in urine were determined in the range between 0.1 and 8.2 μg/mL [124]; chlordiazepoxide in bulk powder and its tablets were identified by calculating quasi-absorbance spectra and in comparison with reference spectra of the library [125].

Identification of new phthalazine derivative was performed by HPTLC-FTIR spectrometry [126]. The separation was carried out on silica-gel $60 F_{254}$ (Merck) using toluene–chloroform–methanol (70:20:1, v/v) as mobile phase. The plate was developed twice in the same mobile phase. The FTIR spectra of the compounds were recorded in the range of 4000–400 cm^{-1} using a Bruker EQUINOX 55 spectrometer with an Attenuated Total Reflectance. The identification was performed by comparison of standard spectrum with sample spectrum after subtracted spectrum of silica-gel. More peaks were easily recognized in the region of 1300–650 cm^{-1}.

10.5   TLC-PHOTOACOUSTIC SPECTROMETRY

PAS is a technique for obtaining spectral information from strong dispersive, opaque, or weak adsorptive samples. Photoacoustic absorbance spectra are generated by an amplifier with synchron signal detection as a function of the incident radiation wavelength. The IR spectra can be obtained using photoacoustic spectroscopy. Usually, the sample is placed in a constant volume cell filled with an inert gas, and the incident IR monochromatic radiation reaches the sample through the NaCl window. The sample molecules are excited to superior levels of vibrational and rotational energy by absorbing radiant energy. During relaxation back to the fundamental state, the absorbance energy is transformed into heat so that the temperature and the pressure of the gas in the cell is increased. The change in gas pressure determines a corresponding change of a microphone capacity. The microphone output signal is amplified as a function of incident radiation wavelength.

The analysis of the sample on TLC plates by PAS was first discussed in the literature in 1975 by Rosencwaig and Hall [127]. In addition, Castieden et al. [128] demonstrated the possibility of the quantitative analysis of fluorescein samples on TLC plates. After the development of the plate, the analyte spot was scarped from the thin-layer plate and placed in a sealed PAS cell for analysis. A linear calibration was obtained for the determination of fluoresceine in the range of 0.2–2 μg, and a limit of detection of 20 ng of fluoresceine in 4.5 mg of silica-gel was reported.

Grey et al. [129] described a photoacoustic spectrometer assembled from available components with a sample cell utilizing an expensive microphone and integral preamplifier. The performance characteristics and representative spectra for several solids and solutions (rose Bengal, malachite green [MG]) were given. Suggestions for improvement of spectrometer performance by utilizing dual channel operation and Hadamard transform techniques were also included.

Fishman and Bard [130] described an open-ended photoacoustic spectroscopy cell for TLC and other applications. The photoacoustic cell was made of aluminum and the bottom part was carefully polished for a good tightening with the layer. The light entered the cell through a flexible fiber-optic probe. The other end of the 51 cm long probe was fixed at the exit slit of the monochromator. The optical fiber bunch was fixed on the cell by silicon rubber. The microphone was completely closed in the cell by ring-shaped fittings and it was electrically isolated from the cell. The entire volume of the photoacoustic cell was $1.13 \, cm^3$, and the area exposed to light was $0.35 \, cm^2$. A micrometer drive from the specimen stage of a microscope was used to scan the PAS cell over the area of the analyte spot to analyze these samples.

A special method with open photoacoustic cell that can be directly applied on the plate layer was also presented. The magnitude of the PAS signal at a wavelength of 541 nm, corrected for the background signal from the blank silica-gel, can be used in quantitative determination. The nonlinear calibration curve for rose Bengal was obtained. This nonlinearity (e.g., deviation from Beer's law) can probably be attributed to the thermal saturation effect present in PAS. The smallest mass of rose Bengal dye analyzed was $0.4 \, \mu g$ with an S/N of 3.3. The mass of the palmitic acid present was proportional to the area under the curve obtained by plotting PAS signal magnitude versus linear distance along the plate. This curve had the characteristics of a Gaussian peak. In this way, a linear calibration curve ($R^2 = 0.996$) was obtained over the range $10-80 \, \mu g$ of palmitic acid. Several cosmetic ants and $Ru(bipyridine)_3^{2+}$ complexes were analyzed from an ion-exchange membrane using this technique.

Lloyd et al. [131] described two procedures to obtain FTIR-PAS of sample *in situ* on TLC plates. The first procedure involves a microphonic detection system, which is a simple extension to the IR of the TLC microphonic PAS techniques. The second technique, which is novel for FTIR-PAS, uses a piezoelectric transducer as a PAS detector. Using both the methods, PAS of tetraphenylcyclopentadienone were produced on three types of commercial TLC plates, aluminum-backed, precoated with silica-gel F_{254}, and neutral alumina Type E (Merck) and plastic (poly(ethylene tetraphthalate))-backed Eastman sheets with 13181 silica-gel layer. An extension of the photoacoustic detection system to the quantitative analysis of TLC adsorbed samples was also considered.

PAS has a common problem with DRIFTS—in the region where silica-gel shows strong absorption, it is difficult to obtain significant information about the sample. Regions of laser absorption provide IR-spectral information on the adsorbate, and interaction between adsorbed species and the silica-gel substrate [132].

The application of PAS on TLC remained at its developed state for 82 years, because of the competition owing to the development of other spectral methods with better performance.

10.6 TLC-RESONANCE RAMAN SPECTROMETRY

10.6.1 TLC-Surface-Normal Resonance Raman Spectrometry

Raman scattering is a phenomenon that was discovered in 1928 by Sir Chandrasekhara Venkata Raman long before the necessary technology was available for simple implementation. With the development of simple yet powerful spectrometers, Raman spectroscopy was finally ready to take its rightful place as an important tool in the TLC arsenal. Raman combined the advantages of FTIR and NIR spectroscopy without the limitations traditionally associated with these techniques. Raman spectra, as well as IR spectra, can be determined by molecular vibration, except in the excitation procedure of molecules. Raman spectroscopy is a useful technique, especially for bands with very symmetric linkages, which are weak in an IR spectrum (e.g., $-S-S-$, $-C-S-$, $-C = C-$), because the absorbed energy is not sufficient to determine a dipole-moment formation in the molecule. To obtain a Raman spectrum, a change in the polarizability of electrons, at the same time as the vibrational excitation, is sufficient. When a molecule has a high IR absorption, then it will show a weak Raman effect and vice versa. In Raman spectra, the bands are narrower and combination bands are generally weak. The standard spectral range reaches well below $400 \, cm^{-1}$, making the technique ideal for both organic and inorganic species. Nowadays, Raman spectroscopy can be used for both qualitative and quantitative applications, because the spectrum is very specific and chemical identification can be performed using search algorithms against digital databases.

The use of *in situ* identification by IRFT Raman spectroscopy of TLC separated compound is imperiously necessary to know the thin-layer spectrum. On this basis, we can decide to select the thin layer with additional lesser backgrounds. Everall et al. [133] initially evaluated the TLC thin-layer Merck plates, namely, alumina F_{254}, silica-gel, cellulose, and polyamide F_{254}. The FT Raman spectra were recorded with a Perkin-Elmer 1725x spectrometer equipped with a polarized Nd:YAG laser and an InGaAs detector. Obviously, the silica-gel plate is the most suitable thin layer because it has the weakest background.

Another interesting conclusion was that it is not possible to obtain a spectrum from the aluminum plate, because of the intense background fluorescence (which evidently did not result from the incorporation into the thin-layer plate, since the fluorescor-doped polyamide gave an acceptable spectrum). The cellulose and polyamide thin-layer both gave strong Raman spectra, which makes them unsuitable for *in situ* identification of adsorbed compounds. The detection of polypropylene additives, namely Topanol OC (3,5-di-*tert*-butyl-4-hydroxytoluene), Irganox 1010 (3,5-di-*tert*-butyl-4-hydroxyphenylpropionate), and erucamide (*cis*-13-docosenamide), which are used in the plastic industry, can be cited as examples. The erucamide and Irganox 1010 produced reasonable quality spectra obtained from a sample of about $3 \, \mu g/mm^2$ in the most favorable case. Another conclusion was that the background fluorescence either from the TLC thin-layer plates with fluorescence indicator or the adsorbate does not cause any difficulty. Similarly, staining with iodine to identify positions of the spots of compounds did not degrade the FT Raman spectra.

In a brief review, Sollinger and Sawatzki [134,135] presented the coupling of TLC with Fourier transform RS (FTRS), with particular reference to the Bruker IFS 28 FT-NIR instrument equipped with FRA 106 Raman module and a computer-controlled TLC table for precise positioning of plates. This system was used to quantify polyphenols separation. The chromatographic separation of a mixture of catechol, resorcinol, quinol, pyrogallol, and phloroglucinol was performed on silica-gel $60 F_{254}$ and silica-gel $60 F_{254}$ (an acid-stabile fluorescent indicator) plates. Toluene–methanol–acetic acid (45:16:4, v/v) was used as mobile phase. A laser (Nd:YAG) beam (1000 mW) from 1064 nm and FT-Raman spectra was acquired at $4 \, cm^{-1}$ resolution. The resulting spectra enabled the component analyzed to be identified. The detection limit for resorcinol was found to be $5\mu g$. The spectra obtained were comparable with the Raman spectra of pure samples from conventional libraries for compound identification in TLC-FTRS. For obtaining 3D Raman spectra, a computer-controlled TLC-scanning stage over the surface of the plate, as well as a contour plot showing the spots, were used. In this way, it is possible to identify partially overlapping spots. The use of specialized TLC-Raman plates (silica-gel $60F_{254}$) was also discussed, and these were expected to enhance the sensitivity of the method, but showed a high background signal.

Resonance Raman spectroscopy in conjunction with TLC was used for separation, detection, and characterization of trance levers of metalloporphyrins [136]. The separations were accomplished on silica-gel K5 plates (Whatman) as stationary phase. Nickel uroporphyrin (NiUroP) and copper uroporphyrin (CuUroP) were developed with ethylene glycol–isopropanol (6:4, v/v), and nickel protoporphyrin dimethyl ester (NiPPDME) developed with toluene–ethyl acetate (6:4, v/v). Chromatographic separation showed the presence of three to four compounds in all pure metallouroporphyrin samples, and a single spot was found for NiPPDME. A backscattering geometry and an unfocused laser beam (514.5 nm) were used for the sample separation by TLC plate, and a 90° geometry was used for the solution spectra. The quality of resonance Raman spectra obtained from TLC plates was often superior to that obtained from analog solution samples, and some changes in the relative intensity of peak were also observed. NiUroP and CuUroP can be identified by change in the intensities and small shifts in the peak positions.

In a relatively recent work by Cimpoiu et al. [137], separation and identification of eight hydrophilic vitamins using TLC method and Raman spectroscopy were reported. The TLC separation was accomplished on plastic-backed HPTLC sheets precoated with silica-gel $60F_{254}$ (Merck) by programmed multiple development with different mixtures of methanol–benzene. The Raman spectra were obtained on a Bruker (Ettlingen, Germany) EQUINOX 55 spectrometer IFS 66 v interferometer, equipped with an FT Raman FRA 106 module and OPUS software. The sample (powdered vitamins and vitamins on a support) was excited by the 1064 nm lain laser at 380 mW. A good separation of eight hydrophilic vitamins was achieved. Identification of the separated compounds was carried out by comparing the Raman spectra obtained experimentally with the spectra found in the spectral library.

10.6.2 TLC-SURFACE-ENHANCED RESONANCE RAMAN SPECTROMETRY

To improve the microsurface-enhanced Raman spectroscopy (Micro-SERS), Koglin [138] used silver colloids incorporated in HPTLC plates. The silver colloids were prepared according to the published procedure [139] by reduction of $AgNO_3$ with $NaBH_4$. The yellow-brownish silver colloidal solution was stored as such at 5°C for weeks. To the HPTLC plates of silica-gel 60 (5 cm × 5 cm) without fluorescent indicator (Merck), 20 pg of adenine or 50 pg of 7-methyl-guanine was applied as a spot. After drying, HPTLC plates were sprayed to wetness with silver colloidal solution. Micro-Raman SERS-HPTLC spectra were obtained with a fully computerized S.A. MOLE-S 300 spectrometer. The laser excitation line of 514.5 nm was used.

The sensitivity permits the acquisition of Raman spectra from HPTLC spots down to 1 μm in size, and it was possible to identify organic substances (7-methyl-guanine and adenine) in the picogram and fentogram region of mass.

The nucleic purine derivatives were directly analyzed on HPTLC by surface-enhanced resonance Raman scattering (SERRS) spectrometry [140]. The HPTLC plate of silica-gel 60 (E. Merck) was used. The nucleic purine derivative was examined using a mixture of chloroform–methanol–ammonia (60:20:1, v/v) as mobile phase. After drying, HPTLC plate was sprayed to wetness with silver colloidal solution, prepared following the procedure described in Ref. [139], and spots arising on the HPTLC plate were used to locate the separated nucleic purine derivatives. The HPTLC-SERRS spectra were measured in a typical 90° scattering arrangement. Plates were analyzed by a computer-controlled double-beam spectrometer. The diameter of the analyzed spot was 2 mm and the light beam dimensions were 1.5 mm × 0.1 mm. Spectra were obtained with the scanning speed (accumulation) of 2 cm⁻¹/1.5 s for a substance quantity per spot from 10 to 60 ng. The identification of nucleic purine derivatives was based on the R_F values and the specific Raman frequency measured with a precision of ±4 cm⁻¹. In the case of partially overlapping spots with very close R_F values, a second specific frequency was considered. Guanine and adenine provided characteristic information about the methylated centers of nucleic purine derivatives, in the region of 600–700 cm⁻¹.

A similar system [141] was described for the identification of highly fluorescent molecule (e.g., acridine orange, Glu-P-2 of 2-aminofluorene) on HPTLC plates by SERRS. The plates were analyzed at room temperature by a computer-controlled double-beam spectrometer. HPTLC-SERRS spectra were measured in a typical 90° arrangement. The detection range was in nanograms to picograms. The same micro-SERS method [142] has been used for in situ study of the absorption behavior of the cationic surfactant molecule, cetylpyridinium chloride, on HPTLC plates. Also, it was possible for the analysis of chromatographic spot of cationic surfactants in quantities down to subnanogram levels. The advances in HPTLC and SERS coupling were discussed [143] in terms of surface-enhanced Raman scattering spectroscopy, detection limits subnanogram level, and practical applications of HPTLC-SERS spectrometry.

A Raman spectrometer usually has a monochromatic excitation system and a system to analyze the diffuse light positioned perpendicularly to the incident light direction. Soper et al. [144] described an instrument that had a monochromatic source at 514.5 nm. The incident light reached the thin-layer plate under an angle of 45°, led by an optical fiber bunch, which was fixed with an X–Y–Z system. The diffuse light was focused on the entrance slid of a double-grid monochromator and analyzed by a complex optic system. The monochromator was computer controlled. The detection system was a photomultiplier and an amplifier/discriminator photon counter. Pararosaline acetate was separated by HPLC and the effluent was deposited spot-by-spot on a silver (Ag sol)-activated thin-layer plate. SERRS spectroscopy was incorporated as a detector for an LC-coupled TLC system. The silver sol was prepared according to the citrate reduction procedure [145]. Optical fibers carried the excitation light to the TLC plate and the scattered Raman radiation to the spectrometer for analysis. The detection limit was 750 fmol and a linear response from 10⁻⁵ to 10⁻⁷ M was obtained.

Somsen et al. [146] reported a study regarding the potential of TLC-SERR to distinguish between compounds that have nearly similar electronic structures. For this purpose, aminotriphenylmethane dyes were considered as the best selection. In this study, four standard dyes, namely the diamino derivatives MG, brilliant green (BG), triamino derivatives crystal violet (CV), and ethyl violet (EV), were selected. TLC separation was accomplished on aluminum-backed silica-gel 60 layer (Merck), without fluorescence indicators. The silver sol was prepared according to the citrate reduction procedure [145]. Ascending development was carried out using 1-butanol–acetic acid–water (50:5:10, v/v) to separate CV and EV, and 1-butanol–acetic acid–water (50:10:20, v/v) to separate MG and BG. The silver-gel solution was applied to each analyte spot using the TLC applicator. Raman spectra were recorded on a Perkin-Elmer Model 2000 FT Raman spectrometer equipped with an Nd:YAG laser and an InGaAs detector. For silver, the maximum enhancement occurred between 500 and 600 nm, but the resonance enhancement was obtained when the excitation frequency coincided with absorption bands of the analyte.

To establish the relevant SERR spectral characteristics of the aminotriphenyl-methane dyes, silver was added to the standard solutions and analyzed in a cuvet. In this case, good-quality SERR spectra could be recorded for each day. The intense and large band at about $1380\,cm^{-1}$ was observed in all spectra. The dyes can be distinguished, because the dialkylaminogroups induced specific change in the triph-enylmethane skeleton vibration. For instance, the peak at $1070\,cm^{-1}$ in the EV spectrum was absent in the CV spectrum. The dyes MG and BG were distinguished by bands profile in the $1250–1100\,cm^{-1}$ region, and the presence of a peak at $525\,cm^{-1}$ in the spectrum of MG, which were all absent in BG.

To explore the potential of SERR detection of dyes on silica-gel, 5 ng of CV per spot (diameter, 1 mm) was examined directly without chromatographic development and silver sol. In these experimental conditions, Raman scatter was not observed for the CV spot, whereas the spectral background was significantly increased. Since both the Raman scatter and the fluorescence background of the bare TLC silica-gel thin-layer were negligible, large signal was probably a result of the CV fluorescence. By application of silver sol on dye spot, however, intense Raman bands were observed for CV. For the aqueous solution SERR spectra, the TLC-SERR spectra had a considerable nonzero baseline, again indicating the presence of (remaining) sample fluorescence.

After TLC separation of two mixtures that contained either CV and EV or MG and BG, symmetrical spots that were well resolved could be observed, but it also contained minor spot, showing that only BG is pure. After silver sol application, the SERR spectra were recorded. Thus, it was possible to compare these spectra with those obtained in aqueous solution with silver sol SERR spectra to observe the same peak position, but there were notable differences in the relative peak intensities. The separated dyes can be identified on the basis of comparison with their aqueous standard solution SERR spectra. The limits of identification of the TLC-SERR method (ca. 5 ng applied) were sufficient for acquisition of spectra of impurities present in the (certified) dye standard.

Near-IR SERS spectra were obtained for 2,2'-bipyridine, 5,6-benzoquinoline, pyrene, anthracene, and fluorenthene on SERS preactivated TLC plate by FTRS. The preactivated plates were based on a procedure of dipping the plates into solution of silver oxalate and the subsequent pyrolysis of adsorbed silver oxalate forming silver particles on the plates. The effects of the type of TLC plates, the time of immersion in the silver oxalate solution, and the time of pyrolysis were discussed with respect to the SERS activity of the systems formed [147].

The silver was put on TLC silica-gel thin-layer plates by using a physical procedure into evaporation chamber. By coupling the AMDs with SERS, it was possible to detect and measure the substance in the nanogram range [148].

Wang et al. [149–151] succeeded in performing bidimensional separation of amino acids using TLC-Fourier transform surface-enhanced Raman scattering (FT-SERS) spectroscopy, and reported the qualitative identification and quantitative determination of detection limits for tryptophan and histidine of 8 µg [149]. Also, hyoscine and scopolamine were identified by TLC separation coupled with SERS [150], and yohimbine in *Rauvolfia verticulata* (Lour) Bail was also determined [151].

10.7   TLC-MASS SPECTROMETRY

10.7.1   INTRODUCTION

In the past few decades, considerable efforts have been made to combine TLC with the existing MS methods. An MS instrument consists of the following components:

1. *Ion sources*: Electron ionization (EI), chemical ionization (CI), field ionization (FI), electrospray ionization (ESI), LSI, MALDI, SALDI, field desorption (FD), FAB, and atmospheric pressure ionization (API)
2. *Mass analyzer*: sector, time-of-flight (TOF), quadrupole (Q), quadrupole ion trap (QIT), linear quadrupole ion trap (LQIT), Fourier transform ion cyclotron resonance (FTICR), or orbitrap (OT)
3. *Detector*
4. *Data and analysis*: data representation and data analysis

To obtain a mass spectrum from a substance present in a spot on a TLC plate, it is necessary to obtain ions. In the late 1980s and early 1990s, improvement in ionization techniques facilitated the ionization of large biomolecules and accurate mass determination of this species. The ionization techniques must be selected in function of the molecules' properties. In these circumstances, for small volatile molecules, EI, CI, and FI are recommended, but for small or large nonvolatile molecules, ESI, FD, FAB, MALDI, and SALDI are recommended.

The following are related to the term, laser desorption/ionization (LDI):

1. MALDI that uses liquid and particulate matrix is called SALDI.
2. MALDI using a biochemical affinity is called surface-enhanced laser desorption/ionization (SELDI).
3. MALDI in which the matrix is covalently linked to the target surface is called surface-enhanced neat desorption (SEND) (after IUPAC Recommendation 2006).

MALDI is now a method widely used in biological research. It has the advantage of higher throughput than ESI; however, it also has the disadvantage of being limited to low charge states. Biochemists are becoming more and more interested in simple, high-performance instrumentation, and MALDI has become the preferred ionization method for many of them [152,153].

The initial attempts to couple TLC with MS (TLC-MS) were carried out over 38 years ago and were presented in a review by Kaiser [154]. The new development of coupling these two techniques has begun after 1980, and was based on a new generation of MS with reduced dimensions and reasonable prices, and on the improvement of the ionization sources and interfaces. A data system was dedicated to a secondary-ion mass spectrometer (SIMS) instrument, specially built for analysis in planar chromatography [155,156]. The data system was based on a microcomputer, plug-in data acquisition and control cards, and a commercial scientific software system. The system included storage space for large arrays of data and real-time control of instrument parameters, and provided enhancement of spectral quality by increasing the S/N ratio by background subtraction, signal average, and digital filtering techniques. The data acquisition was carried out at higher scan rates of the MS.

In 1983, Henion and Maylin [157] used TLC-tandem MS (TLC-MS-MS) for determination of drugs in biological samples. The applications of FAB-MS-MS in combination with planar chromatography (TLC or HPTLC) are reviewed in Wilson and Morden [158]. The examples include the identification of various synthetic compounds (polymer additives and pharmaceuticals) and natural products (steroids, glycolipids) by TLC-MS-MS. In addition, a tandem of MS that consists of two MS (MS1 and MS2) was described. Between the analyzer MS1 and MS2, there existed a collision gas cell.

The two most-common MS-MS include: (1) the triple quadrupoles in both the analyzers, MS1 and MS2, are quadrupoles, designated as Q1 and Q2, and the collision cell, where fragmentation is induced, is an Rf-only gas cell, designated q; (2) the magnet sector–quadrupole hybrid, in which the magnetic sector (MS1) exists, usually is in front of the quadrupole analyzer (MS2). The collision region is usually an Rf-only gas cell, designated q, situated immediately before MS2. In such an instrument, MS1 comprises a magnetic sector, B, to effect mass focusing, and an electric sector, E, for energy focusing. If the electric sector precedes the magnetic sector, then MS1 is defined as being EB, the so-called forward geometry; however, in reverse geometry, MS1 is designed BE. Thus, a forward geometry hybrid MS–MS instrument would have the designation EBqQ, and reverse geometry would be BEqQ. Both types of instruments can be operated in three principal modes: (1) product-ion scanning; (2) precursor-ion scanning; and (3) neutral mass-loss scanning. The advantages of MS-MS over simple MS techniques were also discussed.

10.7.2 TLC-MS Transfer Offline Methods by Solvent Extraction

Preparative TLC was the first solution to extract an analyte from a separated spot. However, this procedure is time consuming, especially when more compounds have to be extracted, and moreover, the compounds are exposed to contamination during the extraction. The first method consists of scraping the analyte from the TLC plate, extraction, and conventional MS analysis by EI, CI, or other methods.

The first device, called the "Eluchrom" for quantitative extraction of spot, is introduced in Esteban [159]. Using this device, a circle of adsorbent layer can be removed to allow a tight sealing of an extraction head to the supporting material that contains the spot.

A similar technique to Wick-Stick [92] for IR spectrometry has been used for MS, but the prism was made of ammonium-chloride powder [160]. The drug spots were localized, and the areas of interest were scarped off and transferred to a container in which a triangle-shaped prism from ammonium chloride was placed. A suitable solvent was used for desorption of the analyte from silica-gel, and by migration into the prism, the solvent was evaporated into the tip of the prism, producing the concentration of analyte. Subsequently, the tip of the prism was cut off, powdered, and subjected to MS. The mass spectra were recorded under EI source conditions, and had the appearance of CI spectra owing to the hydrogen chloride formed by thermal decomposition of the ammonia-chloride matrix.

Anderson and Busch [161] described a homemade microsolvent extraction device, which is presented in Figure 10.1.

The dimension of the sample extraction and transfer probe should be much smaller than the size of the TLC spot itself. A suitable solvent for TLC-ESI-MS is delivered through a capillary tube to an analyte spot or band on the developed TLC plate. The solvent diffuses to a short distance through the spot, and desorbs the analyte into the solution before being drawn up by capillary action into a sheath of absorbent material contained in a further tube surrounding the solvent delivery capillary. The analyte is then recovered from the sorbent with methanol as the extraction solvent (Figure 10.1A).

In addition, a microcapillary array geometry was described for using with bands rather than spots. A modified extraction device was also constructed that allowed the use of local heating to facilitate the extraction process, whilst at the same time minimized the possibility of spot diffusion. In this case, the probes were modified by the

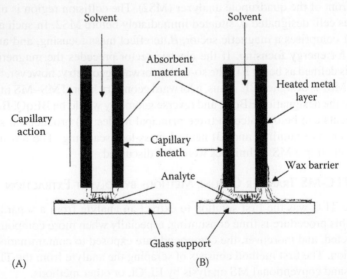

FIGURE 10.1 Schematic diagram of a small capillary probe tip. (A) For integral analyte extraction and solution storage from TLC plate prior to ESI-MS; (B) modified probe tip for channeling extraction. (Adapted from Anderson, R.M. and Busch, K., *J. Planar Chromatogr.*, 11, 336, 1998.)

addition of an outermost metal sheath that carried the wax, and which also protected the fragile extraction capillary inside it. The outer metal sheath was isolated with a small amount of wax, which when placed in contact with the surface of a heated TLC plate, melted and formed a ring, which prevented cross-diffusion of the solvent into adjacent sampling areas. As the TLC plate remained warm, the solvent itself was warm and extracted the analyte molecules from the silica-gel with greater efficiency. At the same time, the solvent moved more naturally into the cooler absorbent material (Figure 10.1B).

An example includes the phosphonium salt for which the coupling of ESI-MS with TLC could be demonstrated. A diluted solution of the ethoxytriphenylphosphonium chloride salt (cation mass 307 Da) was separated on a C_{18} reversed-phase TLC plate with methanol–water (80:20, v/v) as mobile phase. The R_F value was 0.67. The amount of the sample applied on a spot was approximately 0.5 ng. Methanol (1 μL) was used as extraction solvent in a probe of the extraction device described earlier. The amount recovered was diluted to 1.9 mL in methanol to bring the concentration within the usual range (estimated as 2 pg), which was detected with ESI-MS. The solvent very efficiently extracted the material from the thin-layer plate; however, the positive ion ESI-MS background mass spectrum became complex, and it was impossible to identify every ion in this mass spectrum. In this particular background mass spectrum, there existed a low background level in the range m/z 300–310. The positive ion ESI-MS of the cationic ethoxytriphenylphosphonium salt was 307 Da and, as expected, an intense signal was obtained in the positive ion ESI-MS. In this particular case, 2 pg on the plate could be detected by ESI-MS. The limit of detection of 50 pg is considered reasonable for a large range of organic compounds separated by TLC and analyzed by ESI-MS, provided that there is no specific mass interference in the background mass spectrum. However, the fragmentation in ESI-MS is generally low, because ESI is considered to be a soft ionization method.

Product ion MS–MS spectra are recorded when a mass-selected parent ion is given additional energy by collision with a relatively high pressure of neutral target gas, transforming ion kinetic energy into internal energy. The product ion results from collision-induced dissociation (CID); however, when it results from mass-selected parent ion, the interpretation becomes easy. The amount of fragmentation can be varied as the collision energy is varied into a range of 5–100 V, with additional fragmentation induced at the higher collision energies. In this particular case, product ion MS-MS spectrum of the cation at m/z 307 derived from standard ethoxytriphenylphosphonium salt, m/z 307, loss of a methoxy group produces m/z 277, and the doublets at m/z 183, 185 and 261, and 263 correspond to the aromatic portion of the molecule. By comparing the positive ion ESI-MS-MS mass spectra of standard ethoxytriphenylphosphonium salt with TLC separated ethoxytriphenylphosphonium salt, the identification of spectra becomes evident for compound identification.

For preparative or analytical TLC, a solution to extract a substance from a separated spot was prepared. The hydroxyl and nonpolar metabolites of amitriptyline were extracted from plasma and separated by TLC on silica-gel GF_{254} with chloroform–methanol–acetic acid (95:5:11, v/v) as mobile phase. The zones were removed from the plate and extracted with an appropriate solvent for direct-introduction in MS [162,163]. Capsaicin was extracted from red hot pepper (*Capsicum annuum*) fruit and from commercial hot pepper powder, and separated on silica-gel 60 TLC with concentration zone (Merck), with toluene–acetone–chloroform (45:30:25, v/v) as mobile phase. The spot corresponding to capsaicin was removed from the plate and was extracted with chloroform. The extracted compound was analyzed by GC-MS [164].

A method for offline coupling of TLC with electron impact ionization-MS (TLC-EI-MS) for routine determination of 151 pesticides in toxicology and forensic medicine was also elaborated [165]. TLC separation was performed on Silica P_{254} (Merck) plates (20 cm × 10 cm) in combination with six different mobile phases:

1. Methanol–25% aqueous ammonia (100:1.5, v/v)
2. Cyclohexane–toluene–diethylamine (75:15:10, v/v)
3. Chloroform–methanol (90:10, v/v)
4. n-Hexane–acetone (80:20, v/v)
5. Toluene–acetone (95:5, v/v)
6. Chloroform–acetone (50:50, v/v)

The results obtained with these mobile phases were similar to those described earlier [20]. The reactions were performed with the usual reagents. For EI-MS, a Finnigan MAT 212 (Bremen, Germany) instrument equipped with a spectrosystem SS 300 was used.

The spot or band to be investigated by MS was marked, scratched from the plate, extracted with methanol–dichloromethane under sonication and centrifuged. The supernatant was evaporated and the residue was dissolved in methanol, and was transferred to a quartz MS crucible, and the solvent was evaporated again. The dry sample was transferred to the MS. The hR_F data of 151 pesticides were listed together with eight peak mass spectra under EI. This screening technique can be used in practice as an alternative to GC-MS. In the case of intoxication in emergency medicine, currently, there is the possibility of making quick analyses.

A simple and effective method for the analysis of sucrose esters-insecticides from the surface of tobacco plant leaves was developed using offline TLC-MS [166]. HPTLC silica-gel F_{254} analytical plates were developed with n-hexane–ethyl acetate (1:3, v/v), and preparative experiments were performed on TLC silica-gel $60F_{254}$ plates of 0.5 mm thickness (Merck) using the same mobile phase, but in a different ratio (1:4, v/v). Twenty microliters of the surface extract solution in 80% (v/v) methanol–water of one oriental tobacco-type Prilep P-23 sample was applied in a line on a preparative TLC plate. Five bands were scarped, extracted in methanol, centrifuged, and the clear solution was directly injected in MS system Finnigan MAT (San Jose, California) with ion-trap MS after ESI in positive mode. The extract obtained contained sucrose esters, $C_{29}H_{48}O_{15}$, and was effective against *Myzus perscae* (Sulzer) in laboratory and field experiments.

TLC-MS identification of coumarines from *Cnidium monnieri* L. Cousson was also reported [167]. The powdered fruit sample was extracted and purified on a silica-gel column to obtain a mixture of coumarins. The separation of coumarins was performed on silica-gel H plate with toluene–ethyl acetate (9:1, v/v) as mobile phase. Six bands were isolated and scarped off the plates for direct MS analysis of the coumarins: band (a) contained osthol; (b) bergapten; (c) a mixture of imperatorin and xanthotoxin; (d) a mixture of isopimpinellin and alloisoimperatorin; (e) 6-methoxy-8-methylcoumarin; and (f) contained xanthotoxol. The structures were confirmed by their EI mass spectra.

An application of californium-252 plasma desorption MS for drug monitoring after separation by TLC was investigated [168]. Plasma from patients receiving etoposide was mixed with teniposide (internal standard [IS]) and extracted with chloroform at pH 4.5. The organic phase was evaporated by freeze-drying and the residue was dissolved in acetone. TLC separation was performed on silica-gel $60\,F_{254}$ with a mixture of chloroform, ethyl ether, and acetonitrile. The compounds were extracted into acetone and then evaporated. The residue was dissolved in chloroform and analyzed by TOF-MS. The spectra were obtained for both positive- and negative-ion modes. The calibration plot of etoposide was rectilinear in the range from 0.1 to $10\,\mu g/mL$, and the detection limits were $<0.1\,\mu g/mL$.

The separation, identification, and determination of the tocochromanols (mono-, di-, and tri-methylated tocols and tocotrienols) by TLC-MS have been described in a previous study [169]. The monomethyl compounds were separated by TLC on silica-gel G with light petroleum–ethyl ether (41:9, v/v) as mobile phase. A VG ZAB-2F mass spectrometer was used with the probe at 55°C, nitrogen as collision gas, and 70 eV EI. The mass-analyzed ion kinetic-energy spectrum was obtained by CID. The spectra were signal-averaged from three to five voltage scans of ions selected by the magnetic sector. The results obtained for each subgroup were described and discussed. Furthermore, occurrence of δ-tocopherol and γ-tocopherol in cyanobacteria was established.

Acidic glycosphingolipid (GSL) component was extracted from the mycelium form of the thermally dimorphic mycopathogen, *Sporothrix schenckii* [170]. The separation was performed on HPTLC silica-gel 60 plates (Merck) using chloroform–methanol–water (60:40:9, v/v) containing 0.002% (w/v) $CaCl_2$ as mobile phase. The detection was made by Bial's orcinol reagent. ESI-MS was carrier out by using an ESI-MS and tandem ESI-MS-CID-MS in the positive ion mode with Li^+ adduct on a PE-Sciex (Concord, Ontario, Canada) API-III spectrometer using a standard IonSpray source. IonSpray voltage was set at 5 kV and the sample was introduced by direct infusion of glycosylinositol phosphorylceramide samples dissolved in methanol. Two fractions from the mycelium (Ss-M1 and Ss-M2), having the highest R_F values on HPTLC analysis, were isolated and their structures were elucidated by ^{13}C and ^1H NMR spectroscopy, and ESI-MS with lithium adduct of molecular ions.

A combined strategy of preparative HPTLC and nano-ESI quadrupole TOF-MS was established for the structural characterization of immunostained GSLs in silica-gel extract [171]. Preparative HPTLC on silica-gel with chloroform–methanol–water (30:60:8, v/v) as mobile phase was used. The immunostained TLC bands were extracted immediately after the completion of the development, and were subjected to nanoelectrospray low-energy CID MS without further purification. The GLS species investigated were isomeric monosialogangliosides. The resulting mass spectra showed that only analytical quantities of approximately 1 µg of a single GLS within a complex mixture were required for the structure determination by MS and MS-MS. All species studied were detected as singly charged deprotonated molecular ions, and neither buffer-derived salt adduct nor those coextracted was observed.

The determination of caffeine (a model substance) in different soft drinks using a TLC-SPE-APCI-MS (atmospheric pressure chemical ionization) was reported [172]. The repeatability, recovery, limit of detection and quantification, and linearity of classical densitometry, image analysis, and TLC-MS were compared. The TLC was performed on glass-backed HPTLC plates coated with silica-gel $60 F_{254}$ (Merck, Germany). A mixture of dicloromethane–methanol (90:10, v/v) was used as mobile phase. After development, the plate was dried and the spot was marked, and the stationary phase was scraped from the plate and transferred to empty SPE cartridge. Initially, a frit was placed at the bottom of the cartridge, followed by the adsorbent and finally, another frit was placed. To ensure constant elution volume, caffeine was eluted with methanol (1.0 mL) using a Rapid Trace (Zymark, New Jersey) SPE Workstation. Mass spectrometric analysis was accomplished by LCQ-MS^n ion-trap mass spectrometric detector. The APCI interface was used for sample introduction into the MS. The total ion current mass spectrum and full-scan MS-MS spectrum were obtained for an analytical standard of caffeine. The analyte base peak $[M + H]^+$ at m/z 195.0 fragment were obtained in the ion trap. Twenty-five percent collision energy was applied for fragmentation of base ion $[M + H]^+$. After collision fragmentation, the fragment $[M + H - (CH_3-N-CH = N)]^+$ at m/z 138 was monitored as the caffeine peak, to avoid matrix interference. The results from validation showed that the proposed offline TLC-MS combination is sensitive and repeatable. Acceptable MS and MS-MS spectrum was obtained from 20 ng caffeine.

The use of MS-MS for the confirmation of drugs in plasma and urine samples, screened by TLC was one of the first applications of this technique. The goal was to explore whether this combination can provide a rapid assay capable of detection and identification of the presence of administered drug in urine and blood with a high degree of precision, which is of principal interest to the drug-testing laboratories as well as the clinical chemistry laboratories. In an earlier study [157], the possibilities of using TLC-MS-MS in the detection and confirmation of caffeine and nicotine in human urine and butrophanol, betamethasone, and clenbuterol in equine urine were discussed. The preparative TLC was performed on silica-gel $60 G_{254}$ TLC plates (Merck). The separation of the drugs was performed for each of them on another mobile phase. After development, the spot removal was accomplished with a glass micropreparative probe. This was made at home by using a Pasteur pipette and gently tapping into the polypropylene wicks to form porous polypropylene plugs. The large diameter end of a Pasteur pipette was drawn out to make it symmetrical at the center. This tool was used to gently scrape off the TLC spot of interest. A vacuum was created at the opposite end of this micropreparative TLC probe, such that the silica particles from the TLC plate remained in the central, larger diameter region of this tool. When the spot was removed completely, the micropreparative TLC probe was charged with a suitable solvent to elute the organic compound from the silica spot for subsequent analysis by MS-MS.

The MS used in this study was a TAGA6000 triple quadrupole mass spectrometer system equipped with an API source. All samples were introduced into the API source on glass surfaces in a direct insertion probe. The liquid extracts were deposited from a syringe onto a glass surface in which a small heating source was embedded. In the case of silica-gel, this was introduced by placing a few of the particles in a capillary glass tube with a closed end, and heating this tube in the direct insertion probe. Full-scan mass spectra and CID spectra of the parent MH$^+$ ion were obtained (except for betametazone, where the CID spectrum of [MH$^+$ − 60]$^+$ was obtained). The API mass spectrum of standard caffeine showed the protonated molecular ion at m/z 195, which was the base peak with only minor fragment ions at lower mass. By using MS-MS, the unique ion, characteristic of the molecular weight of the compound, which may be dissociated by collision with neutral gas molecules can be examined, thus providing a CID mass spectrum of the [M + H]$^+$ ion. The daughter ion spectrum included abundant fragment ions at m/z: 138, 110, 83, 69, 56, and 42 that are characteristics of the structure of caffeine.

The full-scan API spectrum of solvent extract sample of human urine from a person who was known to drink coffee and smoke cigarettes (no TLC cleanup) showed several abundant ions in the molecular weight region of caffeine (m/z 195), in addition to ions at m/z: 163, 177, 237, 255, 353, 371, and 389. Clearly, this is a "missed" API mass spectrum, indicating protonated molecular ions of several organic components that exist in raw urine organic solvent extract. Hence, the CID mass spectrum of [M + H]$^+$ ion or m/z 195 was carried out with raw urine solvent organic extract, and one spectrum that could be practically identified with the standard caffeine was obtained. In the samples cleaned up by preparative TLC, the components observed at higher masses in full-scan API mass spectrum of raw urine solvent organic extract were not observed when a spot "eluted TLC scrape" of the same R_F as standard caffeine was analyzed. However, in this case, an abundant ion at m/z at 195 was observed at lower mass ions. This "mixed" mass spectrum was not sufficient for identifying caffeine in urine sample. The CID mass spectrum of ion at m/z at 195 was identical with standard caffeine, and thus, caffeine in this eluted TLC scrape spot was identified. It is interesting to note that the observed [M + H]$^+$ ion m/z 163 suggested the presence of nicotine.

Another experiment was performed by direct desorption of organic substance isolated from a TLC scrape spot. However, this technique did not require elution of a drug from the silica-gel surface, and can be used only for organic substances present in high level or is not thermally labile. The full-scan API mass spectrum of a TLC scraped spot was observed at an R_F identical to that of standard caffeine. The silica recovered from the surface of the TLC plate was placed directly on the probe tip, inserted into the API source, and heated to desorb any volatile substances from the surface of the silica-gel. The spectrum showed several abundant ions, including m/z 195. The CID mass spectrum was similar to the standard caffeine spectrum.

Full-scan API mass spectrum of the analgesic, butorphanol, standard sample showed only the protonated molecular ion at m/z 328. This simple mass spectrum verified the molecular weight of the drug, but did not provide unequivocal identification. However, the CID mass spectrum of the m/z 328 ion of standard butorphanol revealed numerous fragments of ions that include m/z 328, 185, 157, 131, 69, and 41.

In contrast to the caffeine experiments described earlier, neither a full-scan API nor a CID mass spectrum of a raw extract of equine urine containing administered butorphanol could identify this drug in a complex matrix in nanogram per milliliter level. However, an eluted TLC scrape spot of butorphanol was analyzed by TLC-MS-MS, and the full-scan API mass spectrum showed an abundant molecular ion m/z 328.

The CID mass spectrum of this molecular ion m/z 328 showed identical spectrum with the standard butorphanol CID mass spectrum. On placing the TLC scrape of silica spot containing butorphanol from the equine urine organic extract into the API source, no significant ion m/z 328 similar to the CID data was observed.

The administration of the betamethasone (MW, 392) (corticosteroid) to horses is followed by rapid metabolism and excretion of the parent drug $[M + H]^{+1}$ m/z 393. The possibility of identification of the betamethasone by TLC–MS–MS was investigated. The full-scan API mass spectrum of standard betamethasone showed (m/z 393) and m/z 363, m/z 351 and the principal ion at m/z 333. The structurally significant $[M + H - 60]^+$ ion (m/z 333) resulted from standard betamethasone $[M + H]^{+1}$ ion m/z 393 by fragmentation (loss of $C_2H_4O_2$) of a base peak, which is a convenient parent ion for the CID experiment. This CID daughter ion mass spectrum provides sufficient specific information for the identification of the betamethasone in mixture m/z: 333, 171, 147, and 121 ions. Thus, equine urine obtained after betamethasone administration was analyzed by TLC. The eluted spot was removed from a TLC plate at an R_F that was the same as that of standard betamethasone. The full-scan API mass spectrum showed several ions in addition to m/z 333. However, when a CID experiment was carried out on the m/z 333 ion, the daughter ion spectrum was practically identical with that obtained from the standard betamethasone.

The other drug investigated was clenbuterol (respirator stimulant), administered to horses in a range from 0.1 to 0.2 mg. The full-scan API mass spectrum of standard clenbuterol showed the base peak $[M + H]^+$ at m/z 277, and this drug could be readily documented. The CID daughter ion spectrum for m/z 277 showed fragment ions at m/z: 277, 259, 203, 168, 132, and 57. The full-scan API mass spectrum was obtained from an eluted TLC scrape spot of an equine urine extract obtained after intravenous administration of 0.2 mg of clenbuterol. The developed spot was removed from the TLC plate at the same R_F values as standard clenbuterol. The full-scan API spectrum showed a relatively small m/z 277 ion with other ions of m/z: 331, 359, and 391. However, if CID was carried out on the m/z 277 ion, the daughter ion mass spectrum was obtained, which agreed well with the CID mass spectrum obtained from standard clenbuterol, allowing unequivocal identification of clenbuterol from the eluted TLC spot from the plate.

Thus, the combination of TLC screening followed by MS-MS allowed the rapid identification of organic substances in raw organic extract of urine when the concentrations of these substances were relatively high.

10.7.3 TLC-MS TRANSFER OFFLINE METHODS BY PROBE TIP

10.7.3.1 TLC-FAB-MS

The offline methods are especially suited for situations where the requirement is essentially one of the target compound analysis, i.e., when spectral data are not required for all the compounds separated on any individual track, and only information about the nature and identity of specific bands are needed. Thus, the simplicity of the offline methods is recognized as being advantageous.

For concentration of the analytes from the spot into a smaller area, Chang et al. [173] developed a simple technique that can be applied in the laboratories without

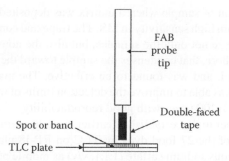

FIGURE 10.2 Stylized diagram for transferring TLC spot to the FAB probe tip. (Adapted from Chang, T., Lay, J., and Francel, R., *Anal. Chem.*, 56, 109, 1984.)

any cost (Figure 10.2). After completion of the TLC separation, this technique involved the following procedures:

1. Spots were visualized on the plate, usually by UV/fluorescence.
2. Trapezoidal shape was scribed around the spot or band of interest.
3. FAB probe tip was covered with a strip of double-faced masking tape, and the tip of the probe was pressed against the TLC spot or band of analyte to transfer the coated adsorbent (Figure 10.2).
4. Strong solvent, such as methanol or dichloromethane and FAB matrix liquid (glycerine or thioglycerine), was added to the TLC adsorbent, which was adhered to the probe tip; in this way, the analyte will be concentrated in the top 0.5–1 mm of the stationary phase.
5. Now, the probe tip can be introduced into the FAB-MS to acquire the spectra in the usual manner.

Mass spectrum was obtained by transferring an amount of a few micrograms of analyte from a separated spot on a TLC plate to a strip attached to the probe of an FAB MS. A Kratos MS-50 high-resolution mass spectrometer was used. The instrument was equipped with an FAB source and gun (supplied by M-Scan, Ltd., Ascot, Berkshire, England). The FAB probe tip was made of copper with a 1.5 mm × 7 mm cross-section and the sample surface was beveled at 70° to the probe axis. Xenon was used as the gas for FAB gun.

The TLC separation of coccidiostats was performed on silica-gel plated with ethyl acetate–dichloromethane (7:3, v/v) as mobile phase. The R_F values were: lasalocid, 0.89; septamycin, 0.79; and monensin, 0.27. Each fraction was readily identified based on the pseudomolecular ions such as: $[M + H]^+$, $[M + Na]^+$, $[M + K]^+$, etc., with $[M + Na]^+$ being the most prominent. These FAB spectra were comparable with those obtained with the individual product prior to mixing. Each fraction was readily identified based on the pseudomolecular ion: lasalocid, $[M + Na]^+$, *m/z* 613; septamycin, $[M + Na]^+$, *m/z* 937; and monensin, $[M + Na]^+$, *m/z*. 693. For monensin spotted on a TLC plate, the detection limit was established below 0.1 μg.

Oka et al. [174] developed a condensation technique including trapezoid technique for TLC-FAB-MS on both silica-gel and reversed phase on C18 plates. This technique prevented the diffusion of sample when a matrix was deposited on the sample spot on a TLC plate to obtain high sensitivity in MS. The trapezoid condensation technique was extended to analyze not only the samples, but also the nonvisible samples with or without a chromophore, that condenses the sample toward the center of the spot in a line, using methanol, and was found to be effective. The mass spectra obtained using this technique was able to improve the detection limits of molecular ion species in TLC-FAB-MS by 3–100 times with good reproducibility.

In a similar manner to that established earlier, a study examining the possibility of the identification of the 27 food dyes by TLC on RP-18 plates with methanol–acetonitrile–5% aqueous sodium sulfate (1:1:1, v/v) as mobile phase was carried out [175]. It described the technique for concentrating the diffuse sample spots on the plate. Identifications were performed by FAB-MS detection in a matrix of 1,4-dithiothreitol and 1,4-dithioerythritol (3:1, v/v).

A method for identification of surfactants and amine antioxidants by HPTLC separation using FAB-MS was described [176]. The HPTLC separation was performed on silica-gel F_{254} and nonfluorescent plates (Macherey–Nagel). The surfactants were developed with a mixture of toluene–methanol (80:20, v/v) and the amine antioxidants were developed with chloroform–ethanol (90:10, v/v) as mobile phase.

All FAB experiments were performed on a Kratos MS50 mass spectrometer. Ionization-sputtering was achieved using an Ion Tech FAB 11NF saddle field gun, supplied with argon and operated at 7.5 kV. All spectra were acquired via a Kratos DS50 data system. After development, the plate was submitted to FAB-MS and the following sample handling procedure was carried out: (1) FAB probe tip was covered with double-side adhesive tape; (2) the HPTLC spot was scraped loose from the plate using a melting-point tube; (3) the probe tip was placed over the loose material and 1 µL thioglycerol that acted as matrix material. The FAB-MS was carried out in the positive-ion mode in the conventional manner. By this method, identification of surfactants in mixed systems and the detection of amine antioxidants in gas oil at levels <20 ng/µL were possible.

A novel method was proposed by Zhongping et al. [177] for determining glycosyl sequence of glycosides by enzymatic hydrolysis followed by TLC-FAB-MS. The method involved three steps: (1) gradual hydrolysis of the glycoside, (2) separation of the hydrolyzate by TLC, and (3) direct analysis of TLC spot by FAB-MS utilizing a simple and rapid microtransfer technique.

Gypenoside A was isolated from *Gynostemma pentaphyllum,* Makino; on hydrolysis with sulfuric acid, its aglycone was identified as 2α-hydropanaxadiol. A VG ZAB-HS mass spectrometer equipped with an Ion Tech FAB gun was used to carry out FAB-MS data. The FAB with xenon was operated at 1 mA and 8 kV; the accelerated voltage was 7 kV, and the mass range was 100–1600.

The mixture of gypenoside A and cellulose in phosphate buffer was incubated. The hydrolyzate was applied in streak on a glass plate precoated with silica-gel G containing 0.6% carboxymethylcellulose as binder. A mixture of chloroform–ethyl acetate–methanol–water (15:40:22:10, v/v) was used as mobile phase. After development, one side of the thin-layer plate was covered, while the other side was sprayed with sulfuric acid solution to visualize the spots.

The TLC spot of interest was transferred with a small amount of glycerol and a droplet of a solution of sodium chloride was added to the FAB probe tip. In this way, the eight spots of enzymatic hydrolyzate were observed, with one of those being gypenoside A. All spectra of TLC spot gave obvious pseudomolecular ions: gypenoside A, [M + Na]$^+$, m/z 1131, R_F 0.16; A-rhamnose (Rham), m/z 985, R_F 0.28; A-glucose (Glu), m/z 969, R_F 0.37; A-Rham-Glu, m/z 823, R_F 0.52; A-Rham-Glu, m/z 823, R_F 0.56; A-2Glu, m/z 807, R_F 0.59; A-2Glu-Rham, m/z 661, R_F 0.67; and A-2Glu-Rham, m/z 661, R_F 0.71. This method can also be applied to the sequence determination of other compounds, e.g., polysaccharides, polypeptides, etc.

The combination of HPTLC with FAB-MS and MS-MS was used for the analysis of antipyrine and its metabolites in the extracts of human urine [178]. The chromatographic separation was performed on a glass-backed silica-gel HPTLC plates, with a fluorescent indicator (Merck). A mixture of chloroform–methanol–trifluoroacetic acid (95:5:1, v/v) was used as mobile phase. The standard samples were drug antipyrine (A) and three metabolites: norantipyrine (NORA), 4-hydroxyantipyrine (4OHA), dimethylaminoantipyrine (DMAA), and 3-hydroxymethylantipyrine (3OHA).

The urine sample was obtained from a normal healthy male volunteer who had received a single oral dose of antipyrine (1 g). The urine sample was treated with β-glucuronidase and incubated to hydrolyze the glucuronide conjugates. Sodium chloride was added and the sample was extracted in dichloromethane/hexane. The organic layer was evaporated and the residue was dissolved in methanol (0.2 mL) prior to application of 20 μL of the sample to the plate. A mixture of pure standards was also applied to the plate for a preliminary identification of the separated components on the basis of R_F values. The appropriate areas of the plate were then removed from the analysis.

The spectra data were acquired with a VG Analytical ZAB-HSQ tandem MS of BEqQ geometry. FAB-MS data were acquired using fast xenon atoms at an accelerated potential of 8000 V. Samples were dissolved (or suspended) in a liquid matrix (m-NBA) and introduced into the MS using a standard FAB probe. FAB-MS-MS data were acquired and processed by a VG Analytical 11–250 data system. The argon was used as collision gas at 40–60 eV.

The HPTLC separation of antipyrine and its metabolites was good, and the hR_F values were: 5, 15, 25, and 68 for 3OHA, A, 4OHA, and NORA, respectively. The molecular ions of antipyrine and its metabolites appeared at: m/z 189 [M + H]$^+$ (A); m/z 175 (NORA); both 3OHA and 4OHA compounds, therefore, produced molecular ions at m/z 205. The TLC-FAB-MS spectra were not conclusive to confirm unequivocal presence of antipyrine into urine extract. However, the TLC-FAB-MS-MS spectra data for antipyrin standard and a material extracted from urine unequivocally showed the presence of antipyrine. As was observed in the case of antipyrine, one part of the TLC-FAB-MS result compromised on the weakness of the molecular ion of the NORA from the extract, and the other part by a large interfering peak at m/z 176. The TLC-FAB-MS-MS experiment, in which a daughter-ion spectrum was obtained from the parent peak at m/z 174, showed a consistent spectrum, which confirmed the presence of NORA in the urine extract. The TLC-FAB spectra for 3OHA and 4OHA revealed the presence of a weak ion at m/z 205.

Furthermore, the TLC-FAB-MS-MS spectra showed a prominent ion at m/z 149, which resulted from dibutyl phthalate (278 Da), a common plasticizer, which also gave an ion at m/z 205 in the TLC-FAB-MS spectrum, in addition to the ions resulting from the analytes. By performing the appropriate background subtraction, this interference from the FAB-MS-MS spectra of 3OHA and 4OHA were removed, and thus, a characteristic spectrum was obtained. As such, this method may have wide applicability in the study of drug metabolites in the body fluids.

A study by Wilson and Morden [158] on the drugs metabolites used paracetamol (p-aminophenolacetanilide) and its sulfate, hippuric acid, phenolphthalein and its glucuronide, and paracetamol sulfate as model compounds representing acetyl and glucuronide. The HPTLC was carried out on silica-gel plates with chloroform–ethanol (3:2, v/v) as mobile phase. Mass spectra were obtained using a VG ZAB-HSQ MS-MS of BEqQ geometry. FAB-MS with xenon was then carried out on the appropriate zone of the plate using glycerol as matrix and removed areas of silica. The MS-MS data were acquired at collision energy of 40–60 eV with argon as the collision gas.

The compound chosen to represent glycine conjugates was hippuric (the glycine conjugate of benzoic acid). TLC-MS gave a mass spectrum dominated by the glycerol used as matrix, and glycerol–sodium adducts. For example, in the FAB-MS spectrum of hippuric acid, the protonated molecular ion $[M + H]^+$ (m/z 180) was very weak. Some improvements carried out with the glycerol background subtracted spectrum allowed the molecular ion for hippuric acid to be more easily observed; however, this was still relatively minor when compared with the glycerol–sodium adducts ions. MS-MS of the $[M + H]^+$ of the hippuric acid produced a clear and unequivocal spectrum containing a number of diagnostic ions. However, a major fragment at m/z 105 corresponded to the loss of glycine from a molecular ion. Also, minor ions were present at m/z: 162, 121, and 77.

Lafont et al. [179] reported the application of offline TLC-MS-MS in the identification of ecdysteroids in plant and arthropod samples. Initially, the extraction of the eggs of the desert locust, *Schistocerca gregaria*, was carried out with methanol. The solvent was removed with a stream of nitrogen. The sample was partitioned between the mixtures of methanol/water/hexane to remove the lipids. The aqueous layer, following the removal of methanol, was then partitioned with ethyl acetate, followed by SPE of the aqueous phase on to a C_{18} cartridge. The retained ecdysteroid conjugates were eluted from the cartridge with methanol/water. Half of this sample was reduced to dryness and redissolved in methanol. The other half was also reduced to dryness, and redissolved in acetate buffer and subjected to enzymatic hydrolysis (aryl sulfates and β-glucuronidase). The analytes were recovered from the incubation medium by SPE as described earlier. Sample of *Silene nutans* (roots) was air-dried and extracted in methanol, the solution was filtrated, and concentrated. Lipids were removed as described earlier, followed by precipitation with methanol/acetone to give an ecdysteroid-rich concentrate. Separations were performed on glass-backed silica-gel F_{254} HPTLC plates (Merck). Plant-derived samples were separated by double development with chloroform–ethanol (4:1, v/v). Samples of ecdysteroids obtained from locust eggs and their hydrolysates were also double developed once in the solvent system, followed by a single development with ethanol–ethyl acetate–water (80:20:5, v/v) (only to 2.3 cm from the origin).

Scanning densitometry was performed with a Desaga CD 60 scanning densitometer, in reflectance mode at 254 nm. TLC-MS and TLC-MS-MS were carried out with a VG Analytical Autospec-Q tandem double focusing spectrometer. Mass spectra and MS-MS spectra were obtained in negative mode.

The silica-gel from the zone with the analyte was removed from the plate and mixed with glycol as matrix and DMSO as cosolvent. This mixture was deposited on the LSI-MS probe and subjected to bombardment with 30 keV cesium ions. The ions generated were then accelerated to 8 keV for mass analysis. MS-MS data were acquired by setting the Autospec-Q double focusing mass spectrometer to transmit the precursor $[M - H]^-$ ion formed by the analytes. This ion was then retarded to 60 eV prior to transmission into a Q1 collision cell with xenon. The product ions were determined by scanning the voltages on a Q2 at unit mass resolution, prior to reacceleration to 8 keV, and for detection, a postacceleration conversion dynode (20 keV) and photomultiplier system were used.

The chromatogram of the HPTLC separation of a concentration extract of the roots of *Silene nutans* after densitometry revealed overloading, three UV-absorbing bands, and R_F values when compared with the corresponding standard sample of 20-hydroxyecdysone, polypodine B, and 2-deoxy-20-hydroxyecdysone. The appropriate area of silica-gel that contained the target analyte was removed from the HPTLC plate and mixed with matrix and cosolvent, and was subjected to LSIMS-MS-MS. The densitogram showed the separation of the concentrated extract of the roots of *Silene nutans*. The resulting daughter-ion spectra obtained by MS-MS of ecdysteroids in an extract of *Silene nutans* comprised: 20-hydroxyecdysone (*m/z* 461, 319, 159), polypodine B (*m/z* 477, 335), and 2-deoxy-20-hydroxyecdysone (*m/z* 445, 303, 159). This daughter-ion spectrum was similar to that of standard analyte, providing excellent confirmation of the identity of these compounds. Also, the daughter-ion spectra obtained from HPTLC of a hydrolyzed extract of the eggs of the desert locust by MS-MS included: 2-deoxy-ecdysone (*m/z* 429, 413, 355, 261) and ecdysone (*m/z* 445, 429, 373, 279, 183). The R_F values of these bands were consistent with the presence of ecdysone and 2-deoxy-ecdysone, and this was unambiguously confirmed through MS-MS.

A number of nonsteroidal anti-inflammatory and analgesic compounds, including salicylic acid and its glycine conjugate salicylhippuric acid, phenacetin, ibuprofen, diclofenac, indometacin, flurbiprofen, and naproxen were investigated using TLC-FAB-MS and MS-MS [158]. The separation was performed on HPTLC plates using chloroform–methanol system as mobile phase. To obtain the spectra, the silica-gel zone corresponding to each analyte was first moistened with *m*-nitrobenzyl alcohol as matrix, and then carefully removed from the plate, and placed on the tip of the FAB probe and mulled with a further $1–2\,\mu L$ of *m*-nitrobenzyl alcohol.

A VG ZAB-HSQ mass spectrometer of BEqQ geometry with fast xenon atoms with 1.2 mA and 8 kV was used for obtaining the spectra. The MS-MS data were acquired using argon as the collision gas at the energy of 10–20 eV. Two examples of the FAB-MS and FAB-MS-MS spectra obtained in this way for salicylic acid and phenacetin are provided. The negative ion TLC-FAB-MS spectrum of salicylic acid $[M - H]^{-1}$ (*m/z* 137) is the principal ion. This ion through MS-MS gave the product-ion spectrum, *m/z*: 137, 93, which was in agreement with that of the standard salicylic acid. The spectrum showed a major loss of 44 Da, corresponding to the loss of CO_2. Similarly, the positive ion of phenacetin $[M + H]^+$ (*m/z* 180) showed the MS-MS spectrum *m/z*: 180, by loss of 42 Da ($CH_2 = C = O$) to 138, and a further loss of 28 Da (C_2H_4) to 110. This spectrum was in excellent agreement with that of the standard.

Phenolphthalein glucuronide is a typical ether glucuronide, which revealed a TLC-MS spectra dominated by matrix-derived ions. The FAB-MS spectrum obtained directly from silica-gel was dominated by glycerol and the glycerol adduct ions. A peak for the phenolphthalein $[M + H]^+$ at m/z 319 was weak in comparison, and the spectrum was relatively uninformative. However, as observed for hippuric acid, the MS-MS together with the ion spectrum of the $[M + H]^+$ ion gave a clear spectra for phenolphthalein with a major fragment ion resulting from the loss of $C_6H_5OH^+$ from $[M + H]^+$ ion. A related compound, phenolphthalein glucuronide, a major metabolite of phenolphthalein, was also analyzed through this method. The FAB-MS phenolphthalein glucuronide from the silica-gel resulted in a weak $[M + H]^+$ ion observed at m/z 495 together with a weak sodium adduct ion $[M + H + Na]^+$ (m/z 517). The spectrum obtained by FAB-MS was dominated by background ions from matrix. However, the MS-MS spectrum was very clear with a major ion at m/z 319 as a result of the loss of glucuronic acid from the $[M + H]^+$ ion.

For aromatic amines, acetylation represents a major way of metabolism, and paracetamol was selected as a model to study for this class of conjugate. After TLC-MS-MS, specific classes of aromatic amides of 42 (CH_2CO) and 59 (CH_3CONH) were examined in the MS-MS spectra. On the other hand, the sulfate conjugate was demonstrated with ether sulfate of paracetamol using negative value in FAB. MS-MS spectrum of paracetamol sulfate showed m/z: 230 ion molecular $[M - H]^{-1}$, 150 principal ion resulting from loss of SO_3, and 107 from loss of CH_3CO, together with an ion at 80, specific for SO_3^- [180].

Morden and Wilson [181] studied the separation of nucleotides from their corresponding bases using HPTLC with offline identification by FAB-MS and FAB-MS-MS. For normal phase TLC, silica-gel, cyanopropyl-, aminopropyl-, and diol-bonded silica-gel HPTLC plates (Merck) were used. Mobile phase for the separation of nucleotides and their free bases on normal stationary phase were a mixture of methanol/dichloromethane with or without addition of aqueous ammonia at different rates. The samples used for separation were adenine, adenosine, cytidine, cytosine, thymidine, thymine, uridine, uracil, and methyluracil.

FAB-MS data from the bases and nucleotides were carried out on a VG ZAB-HSQ tandem mass spectrometer (VG Analytical Ltd., UK) with BEQQ geometry. Spectra were acquired using fast xenon atoms at an acceleration potential of 8 kV. The FAB-MS-MS data were performed at collision energy of 40 eV. Argon was used as collision gas. The FAB-MS and FAB-MS-MS data were acquired by a VG Analytical 11-250 data system.

The pure standards or zones of interest were removed from the HPTLC plate and mixed with a liquid matrix (glycerol or m-NBA), and then introduced into the ion source using a standard FAB probe fitted with a modified tip. The HPTLC results indicated that the possibility of the separation of the individual pairs of bases and nucleotides can be performed on all the phases with appropriate selection of mobile phase. This separation is important, because the nucleotides under FAB-MS conditions are readily decomposed and the major ions are produced by the free bases. In this condition, it is difficult to decide whether a given nucleotide is pure or contaminated with the free bases, using FAB-MS alone.

This work demonstrated that HPTLC separation is a simple and rapid method of separation of nucleotides and bases, with subsequent MS-MS identification. The spectrum of uridine performed by double focusing MS showed two major ions, which could be attributed to the analyte, one at m/z 245 $[M+H]^+$ as the molecular ion and other at m/z at 113. It is not possible to determine unequivocally if the ion observed at m/z 113 arises as a fragment ion from m/z 245, or whether it represents the $[M + H]^+$ uracil present as the contaminant. The MS-MS product ion spectrum of uridine, obtained by CID of the m/z 254 $[M+H]^+$ ion showed simple cleavage of the molecule fragment of m/z 113 $[Base+H]^+$ and $[Sugar + H]^+$ representing the sugar residue. In addition, in the typical FAB-MS spectrum of uridine on the silica-gel with glycerol as the matrix to product ion m/z 113 and m/z 133, the major product ions m/z 153 and m/z 185 resulted from the CID of the m/z 245 ion derived from glycerol. When the FAB-MS spectrum for uridine, using m-NBA as the matrix, was weaker when compared with glycerol, the product-ion spectrum obtained with the CID of the ion m/z 245 was free from major artifacts. The silica-gel together with all three bonded phases revealed good sensitivity and reproducible product-ion spectra.

10.7.3.2 TLC-SIMS, -LSI, and -ESI-MS

Several methods were developed for the incorporation of threitol into the chromatographic system [182]. A small amount of threitol (5%–10%) can be dissolved in the mobile phase. This approach has been used in TLC separation of phenothiazine-based drugs, but some tailing of the spots has been observed. Another procedure was to use threitol by spray deposition onto the surface of chromatogram after development. Methionine encephalin was used as the model compound to demonstrate the efficiency of the chromatographic extraction. Absolute sensitivity for discrete samples of this pentapeptide was eightfold better with the use of threitol when compared with glycerol probability, because of the differing surfactant properties. Threitol itself does contribute to a background spectrum below m/z 300, including abundant ions $[M + H]^+$ at m/z 123 and $[2M + H]^+$ at m/z 254. Currently, the relative contribution of matrix signals from threitol is quantitatively assessed. The positive ion SIMS spectrum of acepromazine separated by silica-gel TLC and extraction from the chromatogram by liquefied threitol showed $[M + H]^+$ at m/z 327 and m/z at 254, 245, respectively. The sample size of 5 μg revealed a persisted SIMS spectrum for 15 min.

Identification of food dyes by TLC-SIMS with a condensation technique was proposed by Masuda et al. [183]. A silica-gel on aluminum-backed thin-layer plate was impregnated with 10% liquid paraffin in petroleum ether as reversed-phase in TLC. A good separation was obtained for xanthene dyes (Acid Red, Erythrosine, Phloxine, and Rose Bengal) and triphenylmethane dyes (Brilliant Blue FCF and Fast Green FCF) using a methanol–methyl ethyl ketone–10% aqueous sodium sulfate solution (1:1:3.5, v/v), and a methanol–acetonitrile–aqueous sodium sulfate solution (3:3:10, v/v) as mobile phase, respectively. After development, thioglycerol matrix was applied around the spot on the TLC plate. After a few minutes, the dye was condensed to the center of a spot in line with the penetration of thioglycerol, and the magic bullet as a matrix was finally dropped on the condensed spots.

When the spot was linear, the thioglycerol would be rectangular-sampled. The plate was then set on to the TLC holder and an SIMS spectrum was measured. Two molecular ion species [M + H]+ and [M + Na]+ were clearly observed in spectra Acid Red (m/z 581 and 603) and Brilliant Blue (m/z 793 and 815).

A method for direct analysis of structures of glycolipids (accumulate in some lipidoses) directly on the TLC plates by matrix-assisted SIMS was developed [184]. For separation, HPTLC aluminum plates coated with silica gal 60 (Merck) were used. The development was performed with a mixture of chloroform–methanol–water (65:45:10, v/v) as mobile phase.

The area with the spot of interest on HPTLC plate was cut out and then the piece of plate was attached to a specially designed MS probe tip. A few microliters of methanol and about 2–3 µg of matrix liquid triethanolamine were added to the piece of plate with the sample lane mounted on the SIMS probe. Subsequently, the probe was introduced into Hitachi M-80 modified for SIMS with a TLC insertion instrument and an M-003 data processing system.

The analysis was carried out with detection in both the positive and negative ion modes. In case of manual manipulation for scanning pieces of an HPTLC plate, mass spectra were obtained at 0.25 mm intervals along the plate. The required amount of material to obtain an adequate spectrum was in the order of a few micrograms of lipids per band for both positive- and negative-ion detection. By scanning the plate, mass spectral for the qualitative identification and chromatographic information can be obtained simultaneously. Although the sensitivity of TLC-SIMS at microgram level is lesser than that of the TLC immunostaining method at pictogram level, these two methods are complementary for characterizing materials separated on TLC plate. Though the sensitivity and technical problems remain to be solved completely, TLC-SIMS would become a very powerful analytical method for the elucidation of glycolipids structures in conjunction with other analytical spectral methods.

A scrape and eluted method of TLC combined with LSI-MS was applied in the study of mixtures of organic sulfonium salts [185]. The sulfonium salts were performed on a silica-gel plate with methanol–water (1:1, v/v) containing 10% acetic acid as mobile phase.

The appropriated band of the analyte (ca. 1.5 cm²) was scraped from the plate and the analyte was eluted from the silica-gel using 1 mL of methanol. The methanol extract was concentrated to 0.5 mL and 5 µL, and was applied to the probe tip and mixed with m-NBA matrix. The positive LSI-MS spectrum was obtained on a VG 70SEQ hybrid mass spectrometer of EBqQ geometry. By direct injection of concentrated methanol extract into the ion source ESI of a VG-Quanttro II instrument, the positive ESI-MS spectrum was obtained.

In the first case, the positive LSI-MS spectrum data showed a range of ions, in addition (m/z: 461, 505, 550, 594, 638, 682) to those in the analyte (m/z 257). These ions correspond to m-NBA matrix and the polymeric binder that was used to make the plates. In this experiment, the analyte has been identified very well, but the problem occurs when the analyte is present in low concentration, because the background of matrix and polymeric binder peaks would cover it. In the second case, the ESI-MS spectrum data showed two ions at m/z 375 and 393, which were present in addition to the analyte ion at m/z 257, and exist as impurities owing to related structures.

10.7.3.3 TLC-MALDI-MS

10.7.3.3.1 Direct Matrix Deposed

Gusev et al. [153] considered that matrices and sample preparation is one of the most important problems to obtain a very good mass spectra. The matrix is believed to serve two major functions: (1) absorption of energy from the laser light and (2) isolation of the biopolymers from each other.

The useful matrix compounds in solid form include nicotinic acid, 2,5-dihydroxybenzoic acid (DHB), sinapinic acid (SA), feluric acid (FA), succinic acid, urea, sorbitol, threitol (1,2,3,4-butanetetrol), tris buffer (pH 7.3); and in liquid form include glycerol, thioglycerol, triethanolamine, α-cyano-4-hydroxycinamic acid (α-CHCA), and 3-nitrobenzyl alcohol.

In this method, the preparation of samples for analysis is very simple and fast. The molar ratio range of matrix to analyte is very large from 100:1 to 50,000:1. The ratio is thus relatively simple to determine whether a candidate matrix efficiently absorbs at the laser wavelength. However, it is not possible to predict a priori whether a compound would form a homogeneous solid or liquid solution with a given analyte molecule, in the presence of ionic contaminant. Some of the indications concerning the sample matrix are given for silica-gel and cellulose substrate evaluated for different MALDI matrices. Solutions of different matrices were pre-mixed with analyte in the optimum molar ratio before being deposited onto $5 \times 5\,mm$ of silica-gel and cellulose. The foremost observation between stainless steel and these substrates was that silica-gel and cellulose ($5\,mm \times 5\,mm$) in sample preparation required a larger volume of matrix solution to be deposited. The experiment established that a matrix/analyte mixture volume, $V_{silica,cellulose} = 20\text{--}40\,\mu L$, was required for both silica and cellulose to create a crystal structure sufficient for MALDI analysis, in contrast to a volume of $V_{stainless\ steel} = 0.5\text{--}2\,\mu L$. The matrix is characterized by performance coefficient:

$$K_{eff} = \frac{\dfrac{V_{stainless\ steel}}{S_1}}{\dfrac{V_{silica,cellulose}}{S_2}} \qquad (10.17)$$

where $V_{stainless\ steel}$ and $V_{silica,cellulose}$ are the matrix/analyte moisture volume deposited on stainless steel and silica-gel/cellulose substrates, and S_1 and S_2 are the corresponding sample areas, respectively. This coefficient has physical importance, because it reflects the efficiency of matrix crystal production on the silica-gel and cellulose substrates, compared with stainless steel. The matrix with the lower coefficient, K_{eff}, had higher crystal production on a TLC plate. Another criterion that was tested in the matrix search was that the matrix should generate a homogeneous crystal distribution into a surface of $10\,mm \times 10\,mm$ piece of silica-gel and cellulose thin-layer. The list of matrices from the best to worst according to this criterion was:

HABA > SA > FA/fructose $\approx$ FA > α-CHCA > 3-HPA > DHB

The experimental extraction sequences in the direct deposition of a matrix on a TLC plate are as follows:

1. Analyte was applied on ≈25 mm² surfaces of silica-gel or cellulose and dried with nitrogen for reduction at minimum diffusion of the analyte into thin-layer plate.
2. Then, the extraction solvent was added, and the experiment was carried out on a desiccators, to maintain the extraction volume on the substrate.
3. Experimental extraction time of small peptides was $t \approx 10\text{--}15\,\text{min}$.
4. Matrix to the extraction solvent was added bit by bit after the extraction was finished.
5. Time of crystallization was $<10\text{--}15\,\text{min}$.
6. TLC plate was thus prepared to be subjected to MS.

The time of analyte diffusion can be evaluated using the known simplified formula from a 1D diffusion equation

$$t_0 = \frac{d^2}{2\Delta} \qquad (10.18)$$

where

t_0 is the diffusion time
Δ is the diffusion coefficient
d is the substrate thickness

To evaluate the diffusion coefficient for peptide molecules, the Einstein–Stokes equation was used

$$\Delta = \frac{kT}{6\pi\eta R} \qquad (10.19)$$

where

η is the liquid viscosity
R is the peptide molecular radius

The molecular radius for a weight of 1000 amu and a denatured molecule was estimated to be ≈2 nm. Thus, the diffusion coefficient, Δ, calculated by Equation 10.19 was ≈10^{-6} cm²/s. Therefore, for a value of $d = 0.02$ cm, the stationary diffusion time, t_0, would be 200 s, because the diffusion process is more complicated and this time should increase at least by a factor of 2. Comparison of the experimental extraction time of small peptides, $t \approx 10\text{--}15\,\text{min}$ with the theoretical calculated and corrected time values, $t \approx 6\text{--}7\,\text{min}$ revealed that this is not a substantial difference in times.

In the last decade, the combination of MALDI with TLC was introduced [152,173,186–189]. The introduction of MALDI-MS as a soft ionization technique has greatly helped in introducing MS in the field of biomolecular science. MALDI-MS enables the mass analysis of large, nonvolatile, and thermally labile compounds, such as proteins, oligonucleotides, and synthetic polymers. During the last few years, MALDI has become a standard method for the MS analysis of biological macromolecules. The TLC-MALDI is carried out using a solvent to enrich the sample molecules onto the surface of the thin-layer plate, followed by a cocrystallization step of the analyte molecules with the matrix to form a structure-like conventional MALDI. Nevertheless, since an extraction solvent is used in TLC-MALDI, a certain loss of chromatographic resolution owing to lateral diffusion of separated compounds is inevitable. However, different optimization protocols have been developed to resolve this problem [173,186].

Direct quantitative analysis of cocaine by both normal and reversed-phase TLC using MALDI MS has also been reported [190]. A cocaine-D_3 was used as an IS for the cocaine sample. The mixture was examined using TLC on silica-gel and C-18 plates with chloroform–methanol–concentrated aqueous ammonia (100:20:1, v/v) and ethanol–water–acetic acid (50:50:1, v/v) as the respective mobile phase. A mixture of methanol–water–acetic acid (50:50:1, v/v) and aqueous 50% ethanol was used as extraction solvent, and 1.5% α-CHCA/3% L-(−)-fructose solution in acetone was used to provide the matrix for direct desorption from the plates. The detection limit of 60 pg was estimated for both the types of TLC plates, which was higher by a factor of six than that of the stainless steel.

The optimization of the protocol to analyze cationic pesticides by TLC-MALDI-MS was reported and tested on normal-phase, reversed-phase, and cellulose TLC plates [191]. Detection limits in the range of picogram were obtained for the analytes, phosphon (chlorphonium chloride), and Avenge (difenzoquat methylsulfate), but for glyodin, it was in the nanogram range. The applicability of this method on a variety of TLC plates was also demonstrated.

A combination of TLC with MALDI-MS for rapid identification of siderophores from microbial samples was also recently described [192]. The SPE was employed to recover the siderophores from the microbiological samples. After TLC was separated, the spots were visualized by spraying ferric chloride or chrome azurol sulfonate assay solution, and the MALDI matrix was applied to the gel surface. Several TLC-MALDI experimental parameters were optimized, such as the type and concentration of MALDI matrix, type and composition of solvent to facilitate analyte transport from the inside of the TLC gel to the surface. The impact of these parameters on sensitivity, precision, and ion formation of the various siderophores was investigated. The detection limits for the investigated siderophores were in the range 1–4 pmol. These values were about 4–24 times higher than the detection limits obtained directly from stainless steel MALDI targets. The differences were most likely owing to incomplete transport of the trapped analyte molecules from the deeper layers of the TLC gel to the surface and the matrix layer. In addition, chromatographic band broadening widened the analyte further in the TLC when compared with the steel plates, resulting in lesser analyte per surface area.

In a recent work by Nakamura et al. [193], rapid and convenient structural analysis of neutral GSLs was performed by direct coupling of TLC-MALDI-QIT-TOF MS-MS. Positions of unstained GLS spots on developed TLC plate were determined by comparing with the orcinol stained reference compounds.

A matrix solution DHB in acetonitrile–water (1:1, v/v) was then added directly to the unstained GSL spots, and the GSLs were directly analyzed by MALDI-QIT-TOF MS. The acetonitrile–water DHB solution proved to be suitable for MS-MS structural analysis with high sensitivity. MS-MS and MS-MS-MS of GSLs yielded simple and informative spectra that revealed the ceramide provided and the long-chain base structures, as well as the sugar sequences. Hydroxy fatty acids in ceramide provided characteristic MS-MS fragment ions. The GSLs were stained with primuline, a nondestructive dye, after TLC development, and were successfully analyzed by MALDI-QIT-TOF MS-MS with high sensitivity.

Immunostaining of GSLs after TLC development is a powerful method for the characterization of antibody-specific sugar, but not ceramides. However, by coupling TLC-immunostaining of GSLs with MALDI-QIT-TOF MS-MS, both the sugar and the ceramides structures could be identified. The detection limits of asialo GM1 (Galbeta1–3GalNAcbeta1–4Galbeta1–4Galbeta1–1'Cer) were 25 and 50 pmol in primuline staining and immunostaining, respectively.

A high-pressure matrix-assisted laser desorption ion source for Fourier transform MS (HP-MALDI-FT-MS) designed to accommodate large targets with diverse surfaces was presented by O'Connor et al. [194].

The instrument incorporates a large, 10 cm × 10 cm, sample 2D translation stage to accommodate and position the MALDI target. Combination with TLC was also presented. This instrument had a new nozzle design that allowed high-pressure collisional cooling, sufficient to stabilize gangliosides while minimizing the gas load on the system. The $X–Y$ stage (Parker-Daedal, series 800 stage, Irwin, PA) was mounted in a custom-built vacuum chamber. The stage was controlled by stepper motors. The stepper motor system was controlled from a computer. Software was written in-house to program the Parker 6 k controller. This software was designated to allow the user a lot of facilities: a digital photograph of the sample plate can be loaded; by a simple click of the mouse cursor, a photograph that points under the laser can be obtained. This capability added the flexibility of handling any material, which can be attached to the target, as a desorption surface. All of these made of it an interesting instrument.

High-pressure collisional cooling of gangliosides was previously reported [195], but owing to the difficulty in determining the pressure in the MALDI plume region, it was necessary to demonstrate similar collisional cooling with this new design. In this case, three ganglioside standards: GM2, GD1a, and GT1b (mono-, di-, and trisialylated, respectively) were ionized and detected with the new instrument. The three ganglioside samples were desorbed directly from the TLC plates using SA as the matrix. These gangliosides were deposited directly from 2 cm × 4 cm TLC plates. The ganglioside standards were individually deposited along the bottom of the TLC plate and developed vertically using a standard separation method. The plate was then attached to the MALDI sample target using double-sided tape. Matrix SA solution was deposited on top of the sample and allowed to dry. Two other labile species were tested, namely $P_{14}R$ peptide and fentomoles of a phosphopeptide. This prototype HP-MALDI-FT-MS instrument demonstrated detection limits in the low attomole to low femtomole range, which was impossible for the commercial MALDI-FT-MS to achieve.

Mehl and Hercules [188] reported a new direct TLC-MALDI method, which revealed a high recovery. Several TLC-MALDI direct coupling methodologies were evaluated by their group. This direct solution deposition method had a relatively low detection limits (2–4 ng) for small peptides.

The indirect solution deposit or pressing method was capable of providing picomole detection limits. An earlier study [187] of the indirect solution deposit methods outlined that the optimal analyte recovery was only approximately 22%. This low analyte recovery is owing to lateral spreading of analyte and loss of analyte to the intraparticle porosity of the stationary phase. However, the recovery degree can be improved

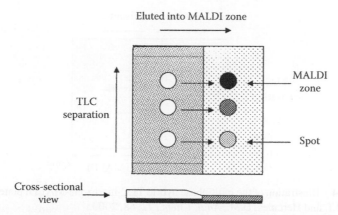

FIGURE 10.3 Schematic diagram of the hybrid TLC-MALDI plate. TLC separation is carried out first, and then the plate is rotated at 90° and spots are eluted into the MALDI zone. (Adapted from Mehl, J.T. and Hercules, D.M., *Anal. Chem.*, 72, 68, 2000.)

using smaller MALDI matrix transfer substance. A newer method that can recover approximately 100% of the analyte was recently developed [188]. This method was carried out on a hybrid TLC-MALDI plate in which two juxtaposed zones, a TLC zone and a MALDI zone, were formed on a common plate, as shown in Figure 10.3.

The TLC separation was carried out in the first direction, following the second elution, which is a perpendicular elution of the analyte from the TLC zone to the MALDI zone. The TLC separation was performed on aluminum-backed silica-gel plate (Merck) and polyester-backed, nonbinder, nonindicator silica-gel plates (Macherey-Nagel). The following mixtures were tested: valinomycin (Val), cyclosporine A (Cs A), and *N-t*-BOC-Phe-Lue-Phe-Lue-Phe (*N-t*-BOC-FLFLF). For separation, multiple developments with two solvents were used. The first solvent was $CHCl_3$, which had separated only *N-t*-BOC-FLFLF ($R_F = 0.8$) from Val and Cs A. The second solvent, ethyl ether–toluene (3:1, v/v), moved only Val ($R_F = 0.8$), but the distance from the eluent front was not as far as that in the first development, and the overlap of Val and *N-t*-BOC-FLFLF was prevented. A parallel zone using a higher 1.0 µg/spot of each analyte was used for iodine visualization.

The TLC-MALDI hybrid plate was carried out at home by scraping the thin-layer of silica-gel from the original 250 µm to zero over the distance of 1 cm, and then, the exposure of TLC aluminum-backing plate used for MALDI matrix zone was continued, as shown in Figure 10.3. Amount of matrix powder with the size distribution of adequate particles was swirled in toluene–ether mixture, and the dispersion/suspension on the MALDI zone was sprained. Subsequently, the TLC zone was covered to prevent the matrix from covering the silica-gel layer. This step was performed after the TLC separation of the test sample. A solvent mixture of chloroform–acetic acid–methanol–1% TFA (2:5:1:1, v/v) was used to elute the separated analyzed spots from the silica-gel zone into the MALDI zone. A mixture of 50% methanol water was deposited on the analyte spot following elution to the THA MALDI matrix to improve the analyte/matrix crystallization. This step is not necessary when DHB was used as the matrix or when Val is used as the test analyte with a THA matrix. There are some considerations that are to taken into account for realizing a good

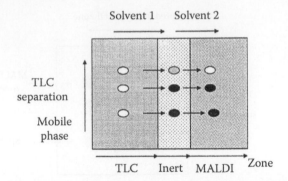

FIGURE 10.4 Illustration of the multiphase hybrid TLC-inert zone-MALDI plate. (Adapted from Mehl, J.T. and Hercules, D.M., *Anal. Chem.*, 72, 68, 2000.)

hybrid TLC-MALDI plate. A uniform tapered interface between the two zones is critical to avoid a distorted solvent front which would cause spreading of the analyte spot. If the interface is not tapered and the silica layer is thicker than the MALDI layer (≤25 μm thick), some analyte will remain in the thin-layer of silica-gel, and not be eluted into the MALDI layer. The elution solvent should have high-extraction efficiency, but it did not get dissolved, or was only partially dissolved, and the MALDI matrix layer was used.

This problem can be resolved by using a multiphase hybrid plate, as illustrated in Figure 10.4. A similar strategy could be used to prepare multiphase TLC-MALDI plates. Two solvents of different polarity can be used for eluting the analyte from the silica-gel to the MALDI layer. The first solvent would select the eluting analyte from the silica to the inert layer, followed by the second solvent, which would elute the analyte from the inert zone to the MALDI layer.

Comparison between the hybrid method and pressing TLC-MALDI coupling method revealed that the hybrid method provided significant improvement in the S/N ratio. This finding is consistent with the fact that the hybrid method recovers approximately 100% of the analyte, as determined by spectrofluorometry. For the cyclic peptides tested, the detection limits ranged from 11 to 116 fmol.

10.7.3.3.2 Direct Matrix-Graphite

In a recent work by Peng et al. [189], a combination from TLC-MS performed by a continuous wave near IR-diode laser with power lower than 4 W was presented. However, when this radiation was focused on a round spot with a diameter of 100 μm, the power density was at least one order of magnitude lower than the required power density for adsorption. To overcome this situation and enable efficient adsorption, a graphite suspension on the surface of the plate was applied to absorb the power of the diode-laser radiation. At the same time, a corona discharge was used at atmospheric pressure laser desorption/CI (AP-LD-CI) to desorb the molecules, before they entered the ion-trap MS. In the ion sources, a stainless steel target was placed with a diameter of 4 mm, which was fixed on an *X–Y* stage to support a small piece of a TLC plate. A target is placed in front of the heated capillary inlet (transfer capillary) of an LCQ Classic ion-trap mass spectrometer (ThermoFinigan, San Jose, California).

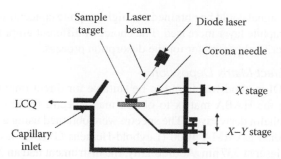

FIGURE 10.5 Schematic diagram of the ion source. (Adapted from Peng, S., Ahlmann, N., Kunze, K., Nigge, W., Edler, M., Hoffmann, T., and Franzke, J., *Rapid Commun. Mass Spectrom.*, 18, 1803, 2004.)

The LCQ Classic is a QIT LC mass spectrometer. A potential of 1.5 kV was applied to the target by an additional power supply to improve the ion-transmission efficiency. The corona needle was positioned on the axis of the transfer capillary and above the rear edge (in reference to the capillary inlet) of the target. The needle was positioned along this axis to obtain optimal APCI conditions (Figure 10.5).

The standard APCI power supply of the LCQ system was connected to the needle to generate the corona discharge. The diode laser with a wavelength of 985 and 807 nm and maximum power outputs of 1–4 W, respectively, were used to desorb the analyte molecules. Other parameters were as follows: voltage was set to 6 V capillary, tub lens voltage was 57 V, and capillary temperature was set at 150°C.

The model compound was reserpine and glycerol was the matrix. Graphite suspension was performed by diluting glycerol in methanol, followed by the addition of graphite powder and mechanical stirring. The sample of reserpine was deposited on TLC plate of $3 \times 3 \, mm^2$ without actual separation, followed by a suspension of graphite–methanol–glycerol. This piece was fixed onto the sample holder (target) using double-side adhesive tape. At this stage, the AP-LD-CI experiments were started and the protonated reserpine was detected by the LCQ Classic ion-trap mass spectrometer.

This study proposed to evaluate the applicability of desorption from a TLC plate by continuous diode-laser radiation. The discharge voltage was around 3 kV and the potential of 1.5 kV on the target was observed. No signal from reserpine sample was detected when only the corona discharge was in operation, owing to the fact that reserpine was not evaporated from the surface of the plate. Subsequently, the diode laser was also switched on and intensive signal appeared. When the corona discharge was switched off, the signal intensity decreased immediately to zero. Obviously, the reserpinol analyte ions can only be measured when both the laser and corona discharge are in operation. Variation in the measured signal for reserpine as the conditions changed was also noted. However, no optimized limit of detection for the reserpine test compound was reported.

Furthermore, the influence of the amount of glycerol matrix was also examined. The sample signal and spectral resolution were found to be improved with the increasing glycerol concentration. If the graphite concentration was low, it resulted in a surface that was not dark enough to absorb adequate power for desorption, and no stable sample signal could be obtained. At high-graphite concentration, the danger of igniting the graphite layer increased. The choice of different graphite and glycerol proportions is a critical parameter for the desorption process.

10.7.3.3.3   Indirect Matrix Deposition

The TLC-MALDI method was found to be suitable for direct imaging, which was performed using the HABA matrix to obtain mass spectra for bradykinin, angiotensin, and encephalin derivatives. The spectra were obtained using a TOF-laser MS, which was modified LAMMA 1000 (Leybold-Heraeus GMBH) laser microprobe, using a nitrogen laser at 337 nm. Additionally, the instrument had an $X–Y–Z$ manipulator and a microscope. All data acquisition and secondary processing software were written in-house (GOOGLY) [153].

In sample preparation, the following MALDI matrices were used for TLC coupling: FA, SA, FA/fucose, 3-HPA, HABA, DHB, and α-CHCA. The analyte, separated by TLC, was presumed to be situated up to 0.1–0.5 mm inside the bulk of the substrate. In this case, to produce a good analyte signal, it must be transferred onto the surface following matrix deposition and crystallization. The experimental extraction sequence of direct deposition of a matrix on a TLC plate has been presented earlier.

The experimental extraction efficiency and method were evaluated using small peptides: angiotensin I, Try-bradykinin, bradykinin, and a high cytochrome c. The small peptides (10–200 ng) or cytochrome c (300–600 ng) were deposited on ≈25 mm² of silica-gel or cellulose plates, dried, and extracted using protocol described earlier (Section 7.1). To decrease the extraction time, 15–20 μL of methanol–0.1% TFA (1:1, v/v) mixture was used onto a 25 mm² piece of plate, and for small peptide only methanol was used. Extraction with methanol was faster, although the ultimate extraction efficiency was similar to about 70%–90%. The experimental extraction time of small peptide was found to be between 10 and 15 min. The HABA matrix had the lowest coefficient of efficiency, $K_{eff} = 2$, when compared with the $K_{eff} = 5–6$ for FA and SA; however, the best detection limit was obtained with FA, FA/fucose, and SA matrices, with a value of 2–4 ng demonstrated for bradykinin, angiotensin, and encephalin derivatives.

Direct MALDI analysis of separated analytes was carried out using the following chromatographic conditions: the TLC separations were performed on either silica-gel 60 (with or without a fluorescent indicator) or cellulose plate (Merck). Aluminum-backed TLC plates were used, except for the direct experiment of TLC imaging, where glass-backed plates were used. The mobile phase used was selected in function of samples: for bradykinin derivatives, water–methanol–acetic acid (44:50:6, v/v) with either 3% KCl or 3% NaCl, or without salt; for encephalin derivatives, n-butanol–acetic acid–water (40:10:10, v/v); and for angiotensins, n-butanol–pyridine–acetic acid–water (15:15:3:12, v/v) for silica-gel, and (15:10:3:12, v/v) for cellulose plates.

Usually, two TLC plates were developed in parallel. The first plate was sprayed with ninhydrin reagent, and after the spots appeared, 5 mm × 5 mm was cut off. The spot positions on the second plate were determined using R_F values, but 7 mm × 7 mm were cut off. Analytes were extracted using a mixture of methanol–methanol 0.1% TFA (1:1, v/v) as extraction solvent. Both the samples with or without ninhydrin reagents were subjected to MS for analysis. The angiotensin derivatives without ninhydrin reagent were deposited on a silica-gel plate and were separated. The best sensitivity was again produced with FA, FA/fucose, and SA matrices. However, the spectral quality did not differ from those obtained without TLC separation. The absolute detection limit was ≈5 ng (S/N = 3) for the best FA/fucose matrix.

The influence of ninhydrin on spectral quality was investigated, and an increased signal was obtained for several analytes when ninhydrin was applied to the TLC plate. However, the mass spectra of peptides obtained from TLC plates, which had been treated with ninhydrin, were much more complicated (several groups of peak) than those from untreated plates. The TLC-MALDI method was found to be suitable for direct TLC 3D imaging, which was performed using the HABA matrix on silica-gel plate with bradykinin (1 μg), with a spatial resolution of ca. 1 mm.

A new approach for the direct coupling of TLC-MALDI-MS was introduced by Gusev et al. [186]. The protocol is presented in Figure 10.6. The MALDI is optimized from extraction efficiency (low detection limit) with minimal spread of the analyte. The instrument used in these studies is a TOF-laser MS, which was a modified LAMMA 1000 (Leybold-Heraeus GMBH) laser microprobe using a nitrogen laser emitted at 337 nm. The results were obtained with the laser beam defocused. Positive ions were postaccelerated to 16 keV onto a discrete dynode electron multiplier. The spectral acquisition system consisted of a 200 MHz transient recorder with a PC-based data processing system. The instrument incorporated a precise motorized X–Y–Z manipulator and a microscope, which allow control of plate position, visual selection, and examination of the point of laser interaction with the sample. Spatially resolved mass spectra (images) were obtained by moving the plate using X–Y–Z manipulator. All data acquisition and secondary processing software (except 3D images) were written in-house (GOOGLY). Standard processing included

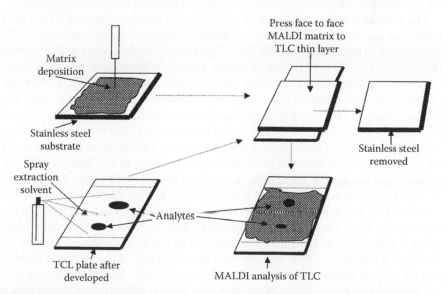

FIGURE 10.6 Schematic diagram of the TLC-MALDI coupling procedure. (Adapted from Gusev, A.I., Vasseur, O.J., Proctor, A., Sharkey, A.G., and Hercules, D.M., *Anal. Chem.*, 67, 4565, 1995.)

data acquisition, spectral smoothing, mass calibration, measurement of intensities, and plotting of contour or 3D images.

In this study, the TLC separation was performed on silica-gel 60 aluminum-backed plates having an organic binder, and silica-gel 60 plastic-backed with a gypsum binder or free binder. Both stationary phases had a fluorescent indicator. The mobile phases used for the development of bradykinin and angiotensin II were similar to that used in an earlier study [153], and for rodamine B and guinea green B, ethyl acetate–methanol–water (80:20:20, v/v) was used. A solution of ninhydrin was used for bradykinin and angiotensin sample visualization. The plates were cut to 20 mm × 20 mm, to 50 mm × 50 mm after separation.

The new technique MALDI involved the preparation of thin-layer matrix of ≈3 μm thick on a surface of 50 mm × 50 mm dimension, from α-CHCA and α-L-fucose (3:1) on a stainless-steel plate. This matrix layer was wetted with the extraction solvent, and was then transferred onto the developed TLC plate face to face via gentle pressure. The whole process of indirect matrix deposition is presented in Figure 10.6.

It must be emphasized that the critical point of the protocol is the selection of the best extraction solvent. For this matrix, a mixture of methanol–water (1:1, v/v) was selected. The extraction solvent had three principal functions: (1) matrix layer transfer, (2) analyte extraction, and (3) analyte incorporation into the matrix structure.

The homogeneity of the matrix crystal structure on a silica-gel plate was examined using an optical microscope and controlled during the analysis with the incorporated microscope in the MS. The surface coverage was found to be 100% over the 50 mm × 50 mm plate, and the matrix crystal layer on the silica-gel layer had a little heterogeneity of <5–10 μm. Other optimum parameters of TLC-MALDI coupling were: (1) 0.2–0.4 μL/mm^2 (4–8 μg/mm^2) of matrix solvent deposited on stainless steel, (2) 4–8 kg/mm^2 pressure for matrix transfer, and (3) 0.15–0.25 μL/mm^2 of extraction solvent.

The mass spectra of rhodamine B was separated on plastic-backed TLC plates (binder-free adsorbent) with different quantities of analyte. The main peak at 444 Da corresponded to the cation, formally [M – Cl]$^+$. Usually, matrix peaks were observed in the range up to 250–400 Da, which makes it possible for the analysis of low-mass compounds by TLC-MALDI. Using the S/N criterion, the detection limit for rhodamine B was found to be ≈40 pg. It is interesting to note that with the aluminum-backed TLC plates with organic binder, an increase in the polymer signal in the range of 800–1500 Da was observed. The laser energy is a very critical parameter, particularly for TLC plates with polymer binder. Excessive laser energy could lead to suppression of the analyte signal and an increase in the background. The optimum laser energy was found to be in the range of 40%–60% above the threshold for all analytes. Another experiment was performed in the same condition with the exception of matrix support. In the first case, stainless steel was used, while in the second case, a silica-gel aluminum-backed TLC plate (no separation) was used. Mass spectra for angiotensin II in both the cases were practically intense showing no significant differences, which demonstrated the efficiency of the proposed indirect methods of deposition of matrix and analyte extraction. The experimental data demonstrated that improvement of the detection limit is primarily correlated with a decrease in the matrix/analyte ratio. The proposed method of indirect matrix deposition/crystallization minimized the analyte spreading to 0.5 mm, and improved the detection to 50–200 pg, which was found to be at least 10 times better than that obtained with direct matrix deposition. The instrument was capable of obtaining a 3D image and contour plot 2D on the silica-gel plated by plotting the spatial dimensions X and Y versus absolute-on intensity. The intensity was found to be proportional to the analyte concentration. The 3D image is obtained for guinea green B (20 ng) and angiotensin II (100 ng) separated on silica-gel plastic-backed plate, and for rodamine B (5 ng) separated on silica-gel aluminum-backed TLC plate.

For optimization of the protocol, influence of factors, such as pressure and analyte R_F in the extraction solvent, extraction time on extraction efficiency and sensitivity, which were examined for analytes on silica-gel must be considered [187]. In these conditions, the following parameters were used: (1) extraction time ~2 min, (2) extraction solvent, which would give R_F values between 0.4 and 0.6, and (3) pressure of 4 kg/mm², for which the extraction efficiency was 22%.

Furthermore, separation and determination of carcinogen-adducted oligonucleotides by TLC-MALDI-MS was also described [196]. Plastic-backed polyethyleneimine cellulose-F TLC plates were predeveloped to remove contamination, and were refrigerated until further use. The TLC-MALDI-MS coupling protocol was carried out as described earlier [184], and involved the preparation of a MALDI matrix layer on the surface of a 20 mm × 20 mm polished stainless-steel plate. The TLC plate after separation was dried and then sprayed with an extraction solvent and pressed onto the matrix plate at 2.5–3 kg/mm² for 10 s. The entire TLC-MALDI matrix plate was then analyzed by MS. The detection limits for the oligonucleotides and their adducts were in the high-femtomole range.

The effect of local matrix crystal variation in the matrix-assisted ionization techniques for MS was reported by Luxemburg et al. [197]. Intense intact molecular ion signal was obtained from phosphatidylcholine, phosphatidylethanolamine, phosphatidylglycerol, and phosphatidylinositol using matrix-enhanced secondary ion MS (ME-SIMS). It was found that the high-mass (m/z >500) regions of the ME-SIMS spectra closely resembled those obtained using MALDI. The presented results demonstrated that the incorporation of an analyte in a matrix crystal softens the SIMS ionization process and thereby enhances the yield for larger pseudomolecular ions. The imaging SIMS was found to be a versatile tool to investigate the effect of the local physicochemical conditions on the detected molecular species.

The MS instrumentation included the development and use of ion microprobes and ion microscope. For the first time, macromolecular ion microscope images were recorded using MALDI-MS by Luxemburg et al. [198]. Single-shot, mass-resolved images of the spatial distributions of intact peptide and protein ions with 4 μm over an area of 200 μm in diameter were obtained in <1 ms at a repetition rate of 12 Hz. Thus, the ion microscopy approach offers an improvement with several orders of magnitude in speed of acquisition, when compared with the conventional (microprobe) approach of MALDI-MS imaging.

10.7.3.4 TLC-SALDI-MS

Earlier, Sunner et al. [199] suggested a new LDI method that used a carbon powder to couple the laser UV energy into a liquid solution. This method was called

SALDI [200]. Initially, 300 Å diameter cobalt particles were used to couple the laser energy into a glycerol solution for protein and polymer analyses [201]. A micrometer-sized activated carbon powder used in the SALDI process was found to be more reliable in yielding mass spectra than the graphite. Consequently, it was realized that SALDI had several advantages that were potentially useful for TLC interfacing. A single matrix system, such as the activated carbon and glycerol, was observed to work for most analytes. In addition, the analysis of organic compounds separated by TLC-SALDI-MS was also demonstrated [202]. SALDI has several advantages that makes it suitable as a TLC interface, and can be used for the analysis of a wide range of organic compounds [199,203]. In contrast to MALDI, only a few SALDI background ions appear at low mass.

The separations were performed on silica-gel 60 TLC plates, with a plastic backing and with a fluorescent indicator. In experiments for optimization, the procedures for preparation of the TLC plate for SALDI-MS were carried out without separation [202]. On the other hand, for peptide and amino acid separations, a developing solvent containing methanol–acetic acid–water (50:6:44, v/v) was used as mobile phase, and for diuretics separation, acetone–methanol (1:1, v/v) was employed as mobile phase. Each separation was run in two TLC lanes: one lane was used for visualization (ninhydrin, iodine vapor) and the other for MS. After development, the plates were dried and the lanes were separated by cutting.

The SALDI mass spectra were carried out on a Voyager RP TOF mass spectrometer (Framingham, Massachusetts). The nitrogen laser was used to desorb the analyte ions from the carbon particles. A 5.71 cm × 5.71 cm stainless-steel sample plate was inserted into the ion source by means of a pneumatic system. The incised piece of TLC plate was attached to the stainless-steel sample plate using double-sided sticky tape. Inside the source, the sample plate was attached to a computer-controller $X–Y$ stage that allowed precise manipulation, and by a TV camera and monitor, the desorption site was selected by visual inspection. The spectra were downloaded to a PC.

The size of the cut-out piece of different surface from 2 mm × 2 mm to 1 cm × 1 cm, and a volume of 0.5 μL of activated carbon suspension, which was sampled on the analyte TLC spot that covered about 50%, was found to give the best results. About 0.5 μL of pure glycerol was applied on top of the carbon-covered sample spot. Three activated carbon suspensions of different compositions were used in this study. Each suspension contained activated carbon, glycerol, sucrose, methanol, and water in different percentages.

For the first time, a typical TLC-SALDI background mass spectrum of a blank TLC plate was investigated. The mass spectrum obtained was found to be relatively clean and was dominated by a few mass peaks. The most abundant ions were: Na$^+$, K$^+$ adduct Na$^+$ (glycerol), m/z 115; Na$^+$ (sucrose), m/z 365, and a minor peak at m/z 203 was assigned to Na$^+$ (monosaccharide fragment). This was typical of the background mass spectrum in the positive-ion mode, but in the negative-ion mode, there were larger ion peaks than in the positive mode.

Furthermore, it is possible to choose between protonated (0.40:1, w/w activated carbon to glycerol) and cationized peptides (0.83:1, w/w activated carbon to glycerol) in TLC-SALDI, by varying the composition of activated carbon suspension.

For ideal TLC imaging, the distribution of activated carbon particles on the surface of silica-gel must be as uniform as possible.

Using this methodology, spectra were obtained for a variety of peptides (bradykinin and angiotensin II) and low-molecular mass organic compounds (herbicide like prometryn and diuretic hydrochlorothiazide [HCT]). The detection limit for bradykinin from a developed plate is approximately 25 ng (calculated for S/N = 3).

Techniques, such as MALDI, SALDI, and the ES interface, represent recent innovations and are still at an experimental stage. However, the SALDI method was developed recently. Here, particles or particle suspension were used to couple the laser energy and assist the desorption/ionization of sample molecule [202,204,205]. The UV laser desorption/ionization mechanism, the influence of preparative procedures, and the breadth of application of this methodology have been investigated in Dale et al. [203]. A simple and robust preparative procedure was presented for the analysis of proteins, oligosaccharides, and synthetic polymers. The graphite acted as an energy-transfer medium by absorbing the UV radiation, leading to thermal desorption of the liquid matrix and analyte. The liquid matrix was observed to fulfill several important roles. In some cases, it acted as a protonating agent (peptides and proteins) or enhanced the signal intensities of cationized species (e.g., polysaccharides and polar polymers) by assisting their desorption. An excess of liquid matrix served to cool the analyte during the desorption step and minimize decomposition. The presence of liquid matrices increased the sample lifetime at a particular desorption spot. This opened a wider choice of laser sources, including diode lasers, which emit invisible and near-IR wavelength range. Cercelius et al. [205] investigated a series of particle suspensions of different materials and sizes by applying them to TLC plates. The results from the graphite particle with a diameter of 2 μm showed the lowest background noise and the highest peak intensities. Graphite is advantageous as it not only produces relatively high absorption coefficient, but also heat of conductivity.

In a relatively recent review by Busch [206], some of the recent research results from TLC-MS and planar chromatography-MS were summarized. In the "Assessmentand Perspectives" section in the review by Busch, new results in relation with precedent and new instrumental developments were discussed. In the "Interconnections" section, synergies between MS and different methods of planar separations (PSs) were discussed. Finally, the uses of ESI and MALDI in conjunction with MS detection for PS were presented in contrast with the three areas of future innovative developments in PS, which include, spatial focusing, affinity focusing, and nanofluidic focusing.

10.7.4 TLC-MS Transfer Online Methods

Another important research direction was the improvement and development of instruments able to record mass spectra directly from a chromatographic plate. Ramaley et al. [207] developed a complex system for detection. The chromatographic plate (1 cm × 10 cm) can be moved (stepper motor) past a source of desorption energy. The desorption was made by small, pulsed CO_2 laser pulses. The CI reagent gas sweeps desorbed materials through a heated inlet system into the ion source of a Finnigan-MAT 4000 quadrupole mass spectrometer. The plate-scanning time was about 12 min. Good peak shapes and high resolutions were obtained for silica-gel layers using this system. However, the detection of polar compounds is more difficult.

Alcohols and carboxylic acids were less sensitive, and amino acids could not be detected. When the molecular mass of the compounds was higher, their volatility was lower and the detection became more difficult. Substances with molecular mass >300 cannot be detected. Furthermore, the chromatographic plate could be analyzed by other methods, because the MS is nondestructive.

Direct coupling of TLC with MS and scanning TLC-SIMS were also reported [208–210]. A strip of aluminum sheet coated with silica-gel was used for TLC separation. After development, this strip was subjected to SIMS, where it serves as the primary ion target (in place of the usual Ag plate). This method was demonstrated with some thermally unstable sample mixtures, such as peptides and drug metabolites [208]. In addition, direct TLC-MS using new sintered TLC plate was reported [209]. A TLC-MS system for separation and measurement of mass spectra of nonvolatile and thermally unstable mixture was described. A glass plate having many linear grooves engraved on the rear was used, which guided a cutter to cleave into strip, to be set on a usual holder for MS for qualitative identification. Separation and identification of nonvolatile and thermally unstable drug metabolites by scanning TLC-SIMS was also presented. Separation and identification was performed on a silica-gel and reversed-phase TLC on octyl-bonded silica with different solvent system. Scanning TLC-SIMS like GC-MS was carried out, which produced good quality of chromatograms and mass spectra [210].

The combination of TLC with laser mass spectrometry (LMS) was established as a potentially powerful analytical technique [211]. TLC-LMS has been applied for the detection and identification of aromatic hydrocarbons and purinic bases, on the polyamide layer. The R_F values could be accurately measured with TLC-LMS. The technique can determine whether the broad TLC spots are simply owing to the tailing or overlapping spots of the two components.

Pulsed laser desorption methodology has been shown to be a powerful method of volatilizing samples from TLC plates directly into the gas phase in the MS [212]. The resulting neutral plume of molecules was then entrained into a supersonic jet expansion of CO_2 and transported into a TOF-MS where resonant two-photon ionization was performed. Various thermally labile biological molecules, including indoleamines, catecholamines, peptides, and drugs have been detected from silica-gel layer with the production of molecular ion in a TOF-MS instrument. The technique combines the advantages of optic spectrometry and MS. In favorable cases, detection limits in the lower nanogram range can be obtained.

A CCD camera was fitted to the imaging SIMS instrument, replacing the MS detector [213,214]. A design for the conjunction of optical spectroscopy with TLC-MS was developed with a focus on integration of the measured spectroscopic data and online control of chromatogram movement and mass spectral measurement. Both the optical and mass spectrometric analytical data could be recorded, while the chromatogram remained in place on the manipulators and within the same coordinate system. Direct sputtering of most organic samples from Empore media was not as efficient as sputtering from aluminum-backed TLC plates, because the Empore sheets comprised 90% fibrillated polytetrafluoroethylene (PTFE) and 10% silica-gel or bonded C_8/C_{18}. This lowered efficiency in some parts is owing to the lowered sample densities, but it is also because of the fact that it is extremely difficult to moisten the surface of Empore, as the extraction process can only occur if the solvent is allowed to penetrate thoroughly within the thickness of the chromatographic medium. Concentration of samples separated in TLC into a wedge that fits directly onto the direct insertion FAB probe of commercial mass spectrometer (VG-70SEQ) was also described.

Separation of the alkaloid compounds was carried out on a 6 cm × 2.5 cm strip of the Empore silica-gel, or on a 6 cm × 4 cm strip of the aluminum-backed silica-gel HPTLC plate (Merck), and development was done with dioxane–hexane–ethanol–dimethylamine (44:48:4:4, v/v) as mobile phase. After development, the spot was removed from the TLC plate using a single whole-paper punch to produce a disk of 5 mm diameter. This disk was fitted to a standard probe tip on the VG instrument. Also, numerous solvent matrices were tested for extraction efficiency of the limit of detection for the model compounds. Comparison of the positive-ion FAB mass spectra for papaverine sputtered from Empore silica-gel and standard HPTLC silica-gel plates, obtained for the $[M + H]^+$ ion of papaverine at m/z 340 was relatively intense for HPTLC silica-gel, but was slightly intense for Empore, and similar in intensity for the glycerol cluster ion at m/z 369 $[4G + H]^+$. No background subtraction was used to enhance the appearance of this mass spectrum. The mass spectra recorded from 12 µg of noscapine, with the $[M + H]^+$ ion was expected at m/z 414. The protonated molecule of noscapine was clearly seen in the spectrum obtained for HPTLC silica-gel, but was barely visible above the background ions in the mass spectrum obtained from Empore silica-gel.

Furthermore, a simple method for sample concentration of the spot after separation by TLC was presented [213]. The first step was to carry out initial development ascendant of the TLC plate with an adequate mobile phase. The second step consisted of rotating the plate at 90° (as in 2D TLC), and the edge of the plate was cut in a regular pattern, similar to that of a saw. The third step included elution of the spots with one solvent in a descending concentration mode, with the sample spot only concentrated within the sample wedge, which is an equilateral triangle with a side length of 2 mm. Finally, after the sample concentration was completed, a cut was made along the indicated line, resulting in a number of small triangles, each of which were of proper size to fit on the probe tip for the direct insertion of FAB. This method was tested with two cationic dyes in positive ion FAB mass spectrum mode, which showed good results.

Stanley and Busch [215] described higher performance interface for the analysis of chromatographic plates by SIMS. The SIMS system adapted for TLC had a vacuum chamber of large volume ($h = 21.6$ cm and 66 cm × 66 cm in area) for chromatographic plates, which could be mounted on a 25-cm² heatable sample platform, mounted on piezoelectric manipulators that can be controlled by computer. The vacuum chamber had Plexiglas windows for sample insertion and removal. An Extrel C-50 quadrupole mass analyzer, with a mass range of 4700 Da, is an analyzer (model 616–1) with associated focusing lenses preceding the quadrupole. The primary particle beam was positioned at 45° relative to the sample plate, having the possibility to adjust the angle to the target surface, such that the point where the beam strikes the surface must be in alignment with the secondary-ion optics. This instrument has a flexible design, and several sources for primary beam can be used, including a Cs⁺ thermoionic gun or a saddlefield FAB (Ar°) gun, which produces quality mass spectra for discrete samples, while a focused liquid metal Ga-ion gun provides a small-diameter primary beam, needed for high-resolution imaging analysis.

The new, high-performance TOF-SIMS, which has become commercially available from Charles Evans and Associates was also used to explore the feasibility of high-spatial resolution TLC-MS [216]. An organic sulfonium salt $(C_{13}H_{17}OS \cdot AsF_6)$ and a common diuretic drug (amiloride hydrochloride) were tested. Each compound was chromatographed on the usual TLC plates, including silica-gel HPTLC plates and reversed-phase C_8 Empore TLC sheets. The common solvent matrices (sorbitol, glycerol) were used. The background mass spectrum measured by positive SIMS for the HPTLC silica-gel (methanol blank) contained predominant ions, such as Si^+, $SiOH^+$, and Na^+. On the other hand, for C_8 Empore sheet, the predominant ions were from the Teflon support, including C^+, CF^+, CF_3^+, $C_3F_3^+$, $C_2F_4^+$, $C_3F_5^+$, and Si^+. It is well known that the ion intensities measured in the background mass spectrum for the Empore sheet are about one-third of those from the silica-gel.

The positive-ion SIMS mass spectra of sulfonium salt and amiloride hydrochloride studied contained ions of much higher masses, and the interference from these background signals were not expected to be significant. The intact cation of the sulfonium salt at m/z 221 was the predominant ion in the spectrum. The next significant ion was $C_6H_5CO^+$ at m/z 105. There were several ions of lower relative intensity, which were distinct from the background mass spectrum, m/z: 55, 41, and 27. In the case of amiloride hydrochloride, the protonated molecular ion appeared at m/z 230, and the other ion that appeared to be characteristic of the compound included m/z: 171, 173, 116, 101, and 86.

For the sulfonium salt on the silica-gel plate, glycerol matrix excelled in providing a strong molecular-ion signal: the signal intensity observed for the cation m/z 221 was approximately 30 times more than that observed in the absence of matrix. However, for the cation signal, the Empore sheet and sorbitol matrix provided an increasing intensity by a factor of 5 over the observed value with no matrix. However, there exists a factor of 2 (sorbitol/glycerol) for the cation signal. Limit of detection for sulfonium salt was approximately 50 ng when silica-gel and glycerol matrix were used.

Busch et al. [217] described an instrument, particularly for the purpose of exploring the coupling of TLC-FAB-MS. This system was similar with that already described earlier [215]. The separation was performed on the aluminum-backed silica-gel 60 TLC plate containing a fluorescent indicator (Merck). A solvent mixture of methanol–chloroform (70:30, v/v) was used as mobile phase for the separation of distearoyl, dipalmitol, and dioctanoil fatty acids. The R_F values were measured as 0.80, 0.40, and 0.30, respectively, for the three compounds listed.

For the imaging experiments with the TLC-FAB instrument, the TLC plate was positioned on the sample platform such that the X-scan axis corresponded with the development direction. Sorbitol was used as the extraction and sputtering matrix. A period of 2 min was necessary to scan the entire 5 cm length of the chromatogram.

The positive-ion FAB mass spectra of the three phosphatidylcholines that formed the basis of an earlier study [215] contained, as expected, signals for the protonated molecules of each compound. The origins of these ions, also observed by other researchers who have recorded the FAB mass spectra of these compounds, were provided.

The ions of interest were those at m/z 86, m/z 184, and m/z 224, and the protonated molecule of distearoyl-phosphatidylcholine at m/z 790. In addition to the measurement of a complete mass spectrum at each indicated spot on TLC plate, the plate must be moved while monitoring the abundances of these class-specific ions. The traces of intensity were those recorded directly by an $X–Y$ recorder, and the location of the distearoyl-phosphatidylcholine was established, through the presence of a second related compound at another spot on the chromatographic plate. In this case, the other compound was in fact two overlapping spots from the protonated molecule, which were identified as dipalmitoyl- (m/z 734) and the dioctanoyl-phosphatidyl cholines (z/m 510).

Another study was dedicated to characterize the bioactive compounds from the bloodroot extract. The TLC was performed on the aluminum-backed silica-gel 60 plates containing a fluorescent indicator. The plates were developed with benzene–ethyl acetate–hexane (2:2:7, v/v) solvent mixture. Sanguinarine chloride was used as the standard sample. For TLC-FAB, sorbitol was selected as the extraction matrix. Extract-ion masses were measured on a ZAB-HF mass spectrometer with a resolution of 12,000 and 8 kV accelerated voltage. Daughter-ion MS-MS spectra were recorded on a Finnigan MAT TSQ-70 triple quadrupole mass spectrometer. For both the extract measurement and the daughter-ion MS-MS experiments, a glycerol matrix containing a small amount of thioglycerol and dimethylsulfoxide was used as the liquid solvent from which the sample was sputtered. FAB mass spectrum was recorded for the spot with the highest R_F values corresponding to the sanguinarine standard. The dominance of the ion at m/z 332 was consistent, with the suggestion that sanguinarine is a major component of the spot for the extract; however, higher-mass ions were not observed in the mass spectrum of the standard, suggesting that the separation was still incomplete. The two daughter-ion MS-MS spectra were virtually identical.

For TLC-FAB of diuretic samples, the six diuretic drugs were used. The separation of these drugs was carried out on aluminum-backed silica-gel HPTLC plates containing a fluorescent indicator and ethyl acetate–water (100:1.5, v/v) as mobile phase. After development, the R_F values determined for the diuretic drugs were: (a) amiloride hydrochloride (AMI) 0.00, (b) furosemide (FUR) 0.38, (c) HCT 0.63, (d) chlorthalidone (CTA) 0.71, (e) hydroflumethiazide (HFM) 0.80, and (f) trichorome-thiazide (TCM) 0.91. With the mass spectra of the diuretic measured and the characteristic ions identified, 1D or 2D imaging analysis of the developed thin-layer chromatogram can be easily performed by monitoring the protonated (positive-ion mode) or deprotonated (negative-ion mode) molecular ion of each diuretic compound, as a function of X and Y direction. The 2D image of the negative ions generated from these diuretic drugs was also obtained. Particularly, in this study, the ability of the TLC-FAB experiment to provide quantitative information about compound in the spot is quite significant. It was measured as the signal response curve for the protonated molecule of AMI, sputtered directly from a TLC plate. The calibration curve of urinary AMI concentration from about 10 to 1000 ng was linear (correlation coefficient of 0.997) in the dynamic range.

The analysis of phosphonium salt mixtures by TLC-SIMS was reported by Duffin and Busch [218]. The separation of five phosphonium salts were performed on aluminum-backed silica-gel TLC plate with fluorescent indicator, and eluted with acetonitrile–ammonium (9:1, v/v). The following are the compounds separated by TLC with their R_F values: (a) carbomethoxymethyltriphenylphosphonium bromide, 0.49, (b) 2-carboxyethyltriphenylphosphonium tribromide, 0.41, (c) phenacyltriphenylphosphonium bromide, 0.59, (d) 1,3-dithian-2-yltriphenylphosphonium bromide, 0.24, and (e) tetraphenylphosphonium bromide, 0.47. The instrument used in this study could only accommodate 2.5 cm × 2.5 cm plates, and hence, chromatograms must be cut into that dimension, under short-wave fluorescent detection. In these experiments, sorbitol was used as the matrix. The chromatogram sections were placed on a temperature-controlled platform that was translated in X–Y direction through the point of primary ion focused by a manually driven manipulator. The sample platform temperature was 5°C–10°C below the melting point of the matrix. The primary ion beam (Cs^+ at 8 kV) simultaneously melted the matrix and sputtered the sample compounds at the localized spot where it got impinged on the chromatogram. Ions derived from the sample spot were accelerated into a quadrupole mass analyzer. The axonometric plot of phosphonium salt a, b, and e was created for the two identified compounds a and b with similar mass using their different R_F values, whereas salts a and e with similar R_F values were identified by their different masses.

In another series of experiments, Brawn et al. [219] studied the use of optical spectroscopy in conjunction with TLC-MS. Two sector-based MS were used in these experiments. The first sector instrument used a VG-70SEQ mass spectrometer, which is a hybrid-configuration instrument—electric sector/magnetic sector with a collision cell and a quadrupole mass analyzer (EBqQ). The instrument was equipped with a cesium gun for LSIMS. Cesium ions were accelerated into the source with 35 keV ion energy, and the current to the gun was held in the range of 2–5 A. The second sector of the instrument consisted of a VG-70SE, and was an EB geometry, equipped with a xenon (8 kV and 1 mA) atom gun for FAB.

Several such specific interface devices for both TLC and electrophoresis have been described in this study. In general, these devices allow a large range of solvent systems to be used for the extraction of sample from the chromatographic matrix, while retaining the vacuum compatibility with the listed ionization methods. For TLC, a simple extraction and extraction devices were described. For the retention of the X–Y coordinate data of the sample compounds distributed on the surface of a thin-layer chromatogram, a surface-tracking capillary that can be moved across the X–Y surface in real time to generate spatially resolved mass spectra was reported.

Hsu et al. [220] described an interesting interface between TLC and an ESI mass spectrometer, working online with TLC separation process and mass spectrometric detection. The first system consisted of a TLC plate of 3 cm × 4 cm dimension and five parallel channels (2 mm wide, 1 mm deep, and 2.8 cm long, each) cut into a Teflon plate. These were then filled with C_{18}-bonded phase particles (TSG gel, ODS-80 T_M, 5 μm) packed at ambient pressure without inorganic binder. A single channel is presented in Figure 10.7.

The chromatographic process into a channel was carried out in a similar manner to that on a strip TLC plate. The sample spot (approximately 0.2 μL) was created near the reservoir of the developing solvent. After the sample spot was dried, the sample was developed continuously using the solution predeposited in the mobile-phase reservoir of methanol–dichloromethane (85–15, v/v). A syringe pump supplied

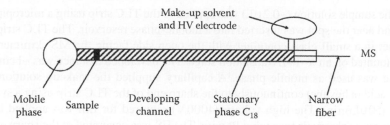

FIGURE 10.7 Schematic diagram of only one TLC channel. (Adapted from Hsu, F.-L., Chen, C.-H., Yuan, C.-H., and Shiea, J., *Anal. Chem.*, 75, 2493, 2003.)

the mobile phase (50 μL/min) to the makeup solution reservoir. For a single separation, a total of 200 μL of mobile phase was used. The developing distance was ~2 cm. One end of each TLC channel was connected to a small mobile-phase reservoir, and the TLC plate was then set in a small glass container mounted on an *X–Y–Z* translation stage. The glass container had an open end, namely, that facing the skimmer of a quadrupole mass analyzer. A direct ES probe was used as an interface to connect the small-size TLC channel to a quadruple mass analyzer. From this moment, the chromatographic separation began. The exit of each channel terminated in a narrow two optical fibers (125 μm in diameter, 1.2 cm long) into the packed C_{18} particles at the same TLC channel of 0.5 cm. The first 0.3 cm of the TLC channels was scarped off, and a makeup solution reservoir was connected to the TLC channel through a side arm. A suitable Pt wire was inserted into the makeup reservoir to conduct the high-electrical potential (3800 V) to create the condition necessary for ESI. The interface was operated as a single-channel separation device, with the MS used exactly as a detector for a single chromatographic column. The ions generated by direct ES probe were analyzed by a single-quadrupole mass analyzer.

The second TLC system was carried out by cutting a TLC silica-gel coated on an alumina plate (Merck, 5553) into small pieces (3 mm × 2 cm × 250 μm). One end of each strip was sharpened, and the silica-gel was scarped off at this sharp end (Figure 10.8).

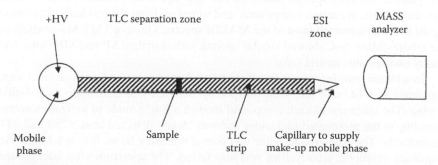

FIGURE 10.8 Schematic diagram of the TLC-ESI-MS system using small TLC strip with a sharp end. (Adapted from Hsu, F.-L., Chen, C.-H., Yuan, C.-H., and Shiea, J., *Anal. Chem.*, 75, 2493, 2003.)

The sample solution (~0.2 μL) was applied to the TLC strip using a micropipette. The end near the spot was inserted into a mobile-phase reservoir. The TLC strip was then set in a small glass container with the open side facing the MS skimmer, and was mounted on an $X–Y–Z$ translation stage. The developed distance was ~1 cm, and hexane was used as mobile phase. A capillary supplied the makeup solution (2% acetic acid in hexane) continuously to the sharp end of the TLC strip using a syringe pump (50 μL/min). The high voltage (4000 V) required for ESI was applied to the mobile-phase reservoir through a Pt wire. The ES was generated at the sharp end of the TLC stripe. The distance between the tip of the TLC strip and the sampling skimmer on MS interface was ~1.0–2.0 cm. The ions generated by ESI were then detected by the quadrupole mass analyzer.

In the first experimental system, clindamicyn and sildenafil (from a commercial pharmaceutical formulation in ~10^{-4} M concentration) were used as samples, and in the second system, a solution of ferrocene and biferrocene (10^{-4} M each) was used.

The organic mixtures were separated successfully and detected online using the TLC-ESI-MS techniques. In this work, no limit of detection for compounds, or dynamic range measurement, was reported.

Fast screening of low-molecular weight compounds by TLC and "on-spot" positive ion MALDI-TOF-MS (TMT-MS) using an UV-absorbing proton-donor ionic liquid as the matrix was described by Santos et al. [221]. Small commercial TLC aluminum sheets used (0.25 mm silica-gel with fluorescent indicator, Alugram SIL G/UV$_{254}$, Macherey-Nagel) were developed with chloroform–methanol (9:1, v/v), over a short distance. Three alkaloids, arborescidine A–C, were used as samples. The α-ciano-4-hydroxycinamic acid ionic liquid matrix (Et$_3$N-α-CHCA) was prepared by adding triethylamine to a solution of α-CHCA in acetonitrile. This ionic liquid matrix was added to the spot, which were well-separated and visually apparent on the TLC plate, and its spots were placed on top of a multiwell planar support. The laser was focused on these spots to cause desorption in the usual MALDI process, and the mass spectrum was acquired on a Micromass MALDI-TOF instrument in the reflection mode.

The positive-ion mass spectrum from the ionic matrix served as the internal mass calibration ion for the TOF mass analysis. It must be noted that the matrix upon positive-ion MALDI produced a nearly exclusive single ion of m/z 101, i.e., Et$_3$NH$^+$. The positive-ion mass spectra contained ions corresponding primarily to protonated molecules of the separated compounds, and were free from the residual background signal that characterized most of the MALDI spectra. On-spot TMT-MS analysis of the arborescidine A–C showed similar spectra with abundant M$^+$ and MH$^+$ ions, and nearly undetectable matrix ions.

In addition, two important anesthetics (levobupivacaine and mepivacaine) were separated by TLC and analyzed on-spot by MALDI-TOF-MS after ionic liquid doping. The spectrum of each compound showed a single mode of ionization corresponding to the protonated molecules and was clearly detected at m/z 289 and 247, respectively. The background noise was over a relatively large, low m/z range. An important antibiotic, tetracycline, was also tested. The spectrum after background subtraction predominantly revealed the protonated molecule of m/z 445.

A device with two possibilities of continuous extraction of substance from a spot, which was completely removed from the plate, was previously described [219]. The extracted sample was continuously introduced in the ionization source of an MS.

A small quantity of 5%–10% matrix liquid was introduced in the extraction solvent to increase the sputtering efficiency in the source.

Prosek et al. [222] constructed an original online TLC-SPE-APCI-MS interface in their laboratory. The computer-controlled extraction of substances from selected spots on a TLC or HPTLC plate, and its programmed injection into the MS are the advantages of this type of interface. Chromatographic separation was performed on 20 cm × 20 cm silica-gel HPTLC plates (Merck). The plates were developed with dichloromethane–methanol (90:10, v/v) as mobile phase. In the subsequent step, the selected spot was eluted from the sorbent (by SPE process) and injected in an LCQ MS^n ion-trap mass spectrometric detector equipped with APCI interface. The interface was tested and validated with a standard solution of caffeine as the test compound. The results were compared with those of the previously established and now routinely used offline TLC-SPE-APCI-MS extraction procedure.

The structural analysis of the small drug molecules by direct coupling of TLC with postsource-decay (PSD)-MALDI MS is reported in Cercelius et al. [223]. The PSD is a fragmentation technique used with MALDI-TOF MS to obtain structural information. In a typical MALDI analysis, the compound of interest is desorbed from the matrix by using a pulsed nitrogen laser at 337 nm. The resulting gas-phase ions of the matrix are determined by TOF mass analyzer. The decay of a mass-selected precursor ion occurs after the ion leaves the ion source in the flight tube, with the fragment ions separated in the reflectron.

The applicability of this technique was presented using two examples: the TLC-PSD-MALDI analysis of two representative nonsteroidal anti-inflammatory drugs (tenoxicam and piroxicam), and the analysis of the pharmaceutically active compound UK-137,457 and one of its related substances, UK-124,912. The separation was performed on the aluminum-backed TLC plate (10 cm ×10 cm) coated with 0.2 mm layer of silica-gel 60F$_{254}$ (Merck). For the development of tenoxicam and piroxicam, chloroform–methanol (9:1, v/v) was used, and for the separation of UK-137,457 from its related substance UK-124,912, chloroform–methanol–glacial acetic acid (60:5:1, v/v) as mobile phase in saturated chamber was used. Before the matrix was deposited on the silica-gel surface, a strip of developed TLC plate (60 mm × 2 mm) was attached to a modified MALDI target with the double-sided tape. The matrix α-CHCA was applied to the silica-gel surface of the TLC plate using an in-house modified commercial robotic X–Y–Z axis motion system. The graphite matrix of 30 μL was used to cover an area of 60 mm × 2 mm. An LT3 MS (SAI, Manchester, UK) equipped with a nitrogen laser of 337 nm was used for this study.

The generation of an impurity profile of UK-137,357 was also presented. The overlaid mass chromatograms of the parent compound UK-137,457 [M + H]$^+$ m/z 498 and its related substance UK-124,912 [M + H]$^+$ m/z 412 at 1% were constructed from a TLC-MALDI in a linear positive-ion mode. TLC-PSD-MALDI spectra of the drug analogs tenoxicam and piroxicam, and the pharmaceutically active compound UK-137,457 and its related substance UK-124,912 have been successfully obtained, and the effect of the precursor ion selection on this technique was evaluated.

The matrices α-CHCA and graphite were used to investigate the effect of the precursor-ion selection on the TLC-PSD-MALDI spectra of the drug molecules studied. The organic matrix α-CHCA enhanced the [M + H]$^+$ ion formation, because, in general, graphite produced only sodium adducts. Structural differentiation in the

PSD spectra of tenoxicam and piroxicam is possible only by selecting the sodium adducts of both drug molecules as the precursor ions. In the case of the TLC-PSD-MALDI analysis of UK-137,457 and related substance UK-124,912 at the level of 1% with α-CHCA matrix, structural differentiation in the PSD spectra was achieved when the protonated ion of both the molecules was selected.

Cercelius et al. [224–226] developed a technique for the direct determination of TLC plates by MALDI–MS. Data from the qualitative analysis of a range of pharmaceutical compounds and related substances were presented. Methods for the generation of TLC development solvent were described, and the use of PSD-MALDI experiments in conjunction with the TLC-MALDI-MS for compound identification was reported.

For the determination of piroxicam by TLC-MALDI-TOF-MS, three different approaches for the incorporation of the IS tenoxicam into the TLC plates were developed [226]. These were (1) adding the IS to the mobile phase and predeveloping the plate, (2) coating the plate with IS by electrospraying prior to matrix application, and finally, (3) mixing the IS into the matrix solution and electrospraying both. However, the first method was the most successful. In this case, linearity was observed over the range between 400 and 800 ng of piroxicam. The precision was found to be in the range of 1%–9% RSD from the average detected value ($n = 5$), dependent on the amount of analyte on the TLC plate, and was accurate with ±2% deviation from the known amount of piroxicam in the sample spot.

The direct coupling described previously was implemented on axial MALDI-TOF MS, showing the feasibility of this method [153,188,227,228]. However, this instrument requires a flat surface for sample desorption to obtain optimum resolving power and mass accuracy result. On the other hand, orthogonal TOF (oTOF) instruments do not have these limitations. An oTOF configuration implies decoupling of the desorption and detection step, resulting in an opportunity for desorbing TLC-separated gangliosides directly from a TLC plate without compromising on the mass spectral accuracy and resolution of the gangliosides analysis.

The goal of this study [229] was to evaluate the performance of the prOTOF 2000 Perkin-Elmer and MDS/Sciex (Boston, Massachusetts, and Concord, Ontario, Canada) instrument with respect to its cooling capabilities and the possibility of its direct coupling with the TLC-MALDI-oTOF-MS method for the analysis of gangliosides. The application of a declustering voltage provided a low level of excitation, allowed control of the matrix cluster and matrix adduct formation, and, thus, enhanced the detection of the gangliosides.

The optimized TLC separation procedure [230] was applied, using a solution of SA matrix in the methanol–acetonitrile–water (2:2:1, v/v), spotting on top of the TLC plate after separation. The two TLC plates were developed under similar conditions. The first plate was sprayed with a detecting reagent for determining the position of the spot, and the second plate was determined by laser scanning in the MS. For the analysis of biological mixtures, the standard ganglioside set alone was applied onto a reference TLC plate. The MALDI target was placed on a computer-controlled X–Y stage. All instrumental conditions were optimized and reported. In conclusion, coupling of the TLC methods with the prOTOF 2000 instrument yielded a simple and fast analysis for gangliosides.

A similar system was described for the analysis of gangliosides directly from TLC plates by IR-MALDI-oTOF MS with a glycerol matrix [231]. The oTOF MS is a modified Science prototype, similar to the one described in Ref. [232]. By default, the MS is equipped with a UV-MALDI ion source. In the modified version, an Er:YAG laser (Bioscope, BiOptics Laser System AG, Berlin, Germany) emitting pulses of ~100 ns, at a wavelength of 2.94 μm was used. The samples were observed with a CCD camera. The separation of GM3 (II^3-α-Neu5Ac-LacCer) gangliosides were performed by HPTLC on glass-backed silica-gel (No.5633; Merck) of 10 cm × 10 cm size. For development, a mixture of chloroform–methanol–water (120:85:20, v/v), supplemented with 2 mM $CaCl_2$ was used as mobile phase in a saturated chamber. On a single plate, up to eight separations were performed on parallel lanes. One of the lines was stained with orcinol. For MS analysis, the plate was cut into stripes with a width of 14 mm and a length of ~30 mm, and fixed on the sample probe with thin double-adhesive pads, and a liquid matrix, such as glycerol, was used for homogeneous wetting of the silica-gel. The analytical possibility of the method was demonstrated by the compositional mapping of a native GM3 ganglioside mixture from cultured Chinese hamster ovary cells. The analysis was characterized by a high relative sensitivity, allowing the simultaneous detection of various minor and major GM3 species, directly from separate HPTLC analyte bands. Also, resolution of the direct HPTLC-MALDI-MS analysis was defined by the laser focus diameter of ~200 μm.

A high-performance oTOF MS was developed specifically to be used in combination with a MALDI source [232]. The MALDI source features an ionization region containing a buffer gas with variable pressure. The pressure in the source influences the rate of cooling and allows control of the ion fragmentation. The instrument provides uniform resolution of up to 18,000 FWHM (full width at half maximum). Mass accuracy routinely achieved with a simple two-point calibration was sufficient for obtaining low part per million mass for peptides desorbed from a MALDI target, unlike the multipoint calibration that is required for MALDI-TOF instrument with delayed extraction. The instrument is also capable of recording spectra of sample containing compounds with a broad range of masses while using one set of experimental conditions and without compromising the resolution or mass accuracy.

Salo et al. [233] investigated the separation and identification of small drug molecules by ultra TLC-atmospheric pressure MALDI MS (UTLC-AP-MALDI-MS). The chromatographic separations were performed on silica-gel $60 F_{254}$ HPTLC plates of 10 cm × 10 cm (0.2 mm of thickness, and specific surface of about 500 m²/g) (Merck) and monolithic UTLC plates of 3.6 cm × 6 cm (10 μm and about 350 m²/g) (Merck), both on glass support. The mobile phase used after optimization was ethyl acetate–n-hexane (1:2, v/v) containing 2% acetic acid for triazoles, and ethyl acetate containing 0.5% ammonium hydroxide for drugs. The plates were developed in a saturated chamber to the distance of 5 cm for HPTLC and 2 cm for UTLC. The developed time was 5–8 min for HPTLC and 2–4 min for UTLC. In this experiment, two types of compounds were used: midazolam, verapamil, and metoprolol as reference standard, and five other compounds, which were all 1,2,3-triazoles. The α-CHCA was used as the matrix compound for MALDI-MS.

The AP-MALDI MS system consists of an AP-MALDI ion source (Agilent Technologies, Germany) combined with an Esquire 300plus ion trap instrument

(Bruker Daltonics, Germany). After adding the matrix, HPTLC and UTLC plates were attached to the face of an in-house-modified AP–MALDI target plate with double-sided conductive tape. The plate was cut to match the target plate, and the working parameter of MS system was optimized. The monolithic UTLC plates provided 10–100 times better sensitivity in MALDI analysis than the HPTLC plates. This is due to the thinner adsorbent layers. However, the UTLC-AP-MALDI-MS provides significantly improved sensitivity when compared with the HPTLC-AP-MALDI-MS. The applicability of UTLC-AP-MALDI-MS has been shown to be good enough for the identification of small drug molecules in relatively simple samples in MS mode.

A coupled TLC-ES-MS system for the analysis of TLC plates that makes use of a combined surface sampling probe-ES emitter as the coupling interface was reported [234–237]. This system exploits a sampling probe-to-TLC plate liquid microjunction and a self-aspirating ES emitter for the direct readout of the developed TLC plate by ES–MS. The illustration of the basic operation and analytical utility of this atmospheric pressure approach to TLC-MS have been demonstrated by using a variety of MS configurations, such as a single quadrupole MS, 3D QIT, and a hybrid triple quadrupole linear ion trap in the analysis of ES active compounds. TLC separation was performed in reversed-phase on C8 and C18 plates.

Van Berkel et al. [234] used the sampling-ES emitter probe design [238] with only a slight variation. The direct readout of TLC plates by ES-MS system was described. The performance of the system was illustrated by RP-TLC separation with a variety of dyes using positive- or negative-ion mode detection. The separation of the dyes was performed on glass-backed 10 cm × 10 cm HPTLC RP-C18 plates (EM Science, New Jersey). The positive-ion test mixture was separated in methanol–tetrahydrofuran (60:40, v/v) as the mobile phase containing 50–100 mM NH_4OAc. The negative-ion test mixture was separated using methanol–water (70:30, v/v) as the mobile phase. ES–MS experiments were carried out on a PE Sciex API165 single quadrupole mass spectrometer (MDS Sciex, Concord, Ontario, Canada). For the test, three-dye mixture of methylene blue, CV, and rhodamine 6G was used for the positive-ion mode detection, and a separate dye mixture containing fluorescein, naphthol blue black, and fast green FCF was used for the negative-ion mode detection. Figure 10.9 shows a schematic illustration of the sampling and probe detail.

One $X–Y–Z$ manipulator system was used to position the TLC surface relative to the sampling probe. Acquisition of mass spectra of the components of individual bends on the plate was carried out by manual stepping and sampling from specific locations within the band. Computer-controlled scanning of the development lanes on the plate was illustrated by using multiple ions, monitored in both positive- and negative-ion modes. The TLC condition, resolution of separation, the limits of scan speed, and detection limit were investigated and discussed.

Desorption ESI (DESI) was demonstrated as a method to couple TLC with MS. This method can be a new alternative to either the MALDI or the surface sampling probe ES approach to couple TLC-MS. Van Berkel et al. [235] considered that DESI provides many of the same advantages of the surface sampling probe ES approach to couple TLC-MS when compared with MALDI. Furthermore, since the surface to be analyzed is not touched with a continuous liquid stream, DESI

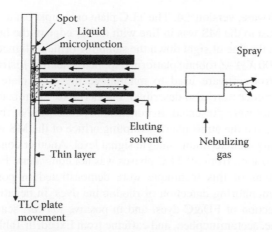

FIGURE 10.9 Schematic diagram of the sampling probe at the surface of a TLC with the formation of a liquid microjunction and sampling/emitter probe. (Adapted from Van Berkel, G. J., Sanchez, A.D., and Quirke, J.M.E., *Anal. Chem.*, 74, 6216, 2002.)

should circumvent surface wetting issues that have limited the current sampling probe ES approach [234,236,237].

The separation by RP-TLC of the rhodamine was performed using RP C8 plates (EM Science, Gibbstown, New York) and RP C2 plates (EM Science), and developed with methanol–water (80:20, v/v) containing ~200 mM ammonium acetate. For FD&C dyes, RP-TLC was carried out using wettable RP C18 plates (Whatman, UK) and developed in water–acetone (70:30, v/v) containing ~500 mM ammonium acetate. The TLC separation of aspirin, acetaminophen, and caffeine extracted from an Excedrin tablet was performed using silica-gel plates F_{254} (Alltech, IL), and developed in ethyl acetate–acetic acid. (99:1, v/v) [235].

The schematic illustration of the TLC–DESI–MS system is given in Figure 10.10 [235]. The MS was a 4000 QTrap (MSD SCIEX, Concord, Ontario, Canada) hybrid triple-quadrupole linear ion trap fitted with the nanospray interface orifice and operated

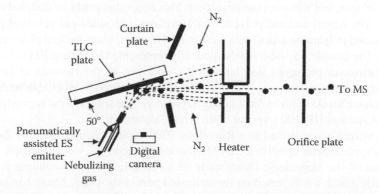

FIGURE 10.10 Schematic diagram of the TLC–DESI–MS experimental setup. (Adapted from Van Berkel, G.J., Ford, J.M., and Deibel, M.A., *Anal. Chem.*, 77, 1207, 2005.)

using Analyte Software, version 1.4. The TLC plate or sample band was mounted so that the edge nearest to the MS was in line with the far edge of the heater orifice and ~10° off-axis from the line of sight down the sampling orifice, aligned with the DESI plume. The MS2000 $X–Y–Z$ robotic platform (Applied Scientific Instrumentation Inc., Oregan) and control software used to manipulate the TLC plate relative to the stationary DESI emitter were also described. Furthermore, optimization and general performance metrics were presented. In conclusion, positioning of the DESI emitter, TLC plate surface, and the atmospheric sampling orifice of the MS were found to be crucial for obtaining the maximum analyte signal level. Another conclusion was that the desorption ionization from all TLC phases was not equivalent. Fundamental and practical applications of this technique were demonstrated in positive ion using selected reaction monitoring detection of rhodamine dyes, in negative-ion full scan mode using a selection of FD&C dyes, and in positive-ion full scan mode using a mixture of aspirine, acetaminophen, and caffeine from Excedrin Tablets. The separation results were also presented, such as pictures of TLC plates development, densitograms (relative abundance versus distance, [mm]), and mass spectra of the compounds.

Van Berkel et al. [239–241] presented an image analysis automation concept and associated software (HandsFree TLC/MS) to control the surface sampling probe-to-surface distance during operation of a surface sampling ES system. Sampling and imaging of rhodamine dyes separated on the reversed-phase TLC plates were used to illustrate some of the practical applications of this system. The image analysis concept and the practical implementation of the monitoring and automated adjustment of the sampling probe-to-surface distance (i.e., liquid microjunction thickness) were also presented. There is a possibility of the scan of a complete development lane or multiple lanes, or imaging of analyte bands in a development lane. Furthermore, the postdata acquisition processing and data display aspects of the software system were also presented.

The first device for the quantitative extraction with solvent of the spot from the thin-layer of the chromatographic plate was called the "Eluochrom" and was proposed by Esteban [159], and an alternative extractor was described by Anderson and Busch [242]. In addition, Luftmann [243] described an apparatus for the transfer of TLC separated compounds from thin-layer plate into a solvent, in short times, within a small volume, and without contamination. This apparatus could be coupled online to an ES MS. Aluminum and polyester-backed plates with silica-gel, reversed phase, cellulose, and polyamide adsorbents were used. However, glass-backed plates cannot be used. The complete system is schematically represented in Figure 10.11.

The extractor prototype was built in the workshop of the Institute of Organic Chemistry (Münster, Germany). The extractor is commercially available under the name ChromeXtrakt (ChromAn, Leipzig, Germany). The solvent flow is provided by the HPLC-pumps HP1100 (Agilent, Palo Alto, California).

The extractor was linked to a Rheodyne 7000 column switch valve. A Z-spray ES mass spectrometer Quattro LCZ (Micromass, Manchester, UK) was used. After separation of the oligomeric compounds on silica-gel, the corresponding regions were marked with a soft pencil on the untreated part of the plate. These bands were eluted from the thin-layer with an extractor (2 mm in diameter) and sent directly to the ES-MS by a 0.1 mL/min stream of chloroform–methanol (1:1, v/v). The ion signal

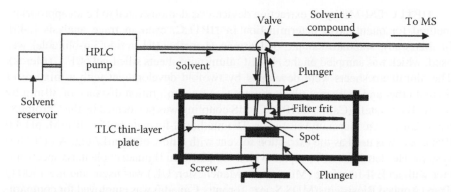

FIGURE 10.11 Schematic diagram of a simple device for the extraction of TLC spot and direct coupling with an ES–MS. (Adapted from Luftmann, H., *Anal. Bioanal. Chem.*, 378, 964, 2004.)

appeared ~20 s after the plunger was placed on a spot and the opening of the valve, and the extraction time was ~1.5 min/spot. The carbohydrate esters formed stable sodium adducts that can be detected easily. The ES mass spectrum from spot no. 5 corresponded to the pentanoic carbohydrate. The molecular weight (calculated 1626.5 Da) could be clearly identified by means of the $[M + Na]^+$ peak at m/z 1649.9 and the $[M + 2Na]^{2+}$ peak at m/z 836.7, and the peak m/z 413 corresponded to the plasticizer contaminant from the TLC plate. This extractor was successfully tested for the identification of certain complex lipids. Thus, it can be concluded that this extractor could be used online or offline.

A special device called ChromeXtractor for the direct extraction from TLC aluminum foils was developed by Luftmann [243]. Till date, glass-backed plates were not applicable for such devices, because it was not possible to reach sufficient contact pressure to seal the extraction area tightly enough without breaking the glass.

Alpmann and Morlock [244] described the possibility of extraction from glass plates through some modifications of the ChromeXtractor. In their work, two products of synthesis, namely, xanthyl ethyl carbamate (XEC) and dansyl ethylamide (DEA), capable of extraction of that modified new device, were demonstrated. The TLC separation of DEA standard and XEC standard were performed on HPTLC glass-backed (10 cm × 10 cm) with silica-gel $60F_{254}$ as 6 mm tracks on 9 tracks. Mobile phase for development was a mixture of acetone–n-hexane (1:4, v/v) and ethyl acetate, respectively. The extraction was performed with ChromeXtractor using a mixture of 95% methanol and 5% ammoniumformiate buffer (10 mM, pH 4), as the extraction solvent with a flow of 0.1 mL/min. Mass spectrometric measurement was carried out with VG Platform II quadrupole ES mass spectrometer (Micromass, Manchester, UK). The SIM and full-scan measurement were performed in ESI⁺ mode. Repeatability of the extraction from glass-backed plate, linearity of the signal obtained, and detection capability were shown to be comparable with the original device. Also, the influence of the elution solvent on the intensity of the MS signal that was demonstrated to be a compromise between the high elution power of the solvent and good solubility of the compound in the elution solvent was studied.

HPTLC-ESI-MS by an extraction device was demonstrated to be an appropriate method for quantitative determination in HPTLC, even in trace analysis [245]. In this study, standard sample of Harman (1-methyl-9H-pyrido[3,4-b]indole) was used, which was sampled on the HPTLC aluminum sheets silica-gel $60F_{254}$ (Merck). The aluminum sheets were developed by twofold development with a mixture of diethyl ether and methanol (98:2, v/v) up to a final migration distance of 30 mm by the AMD 2 system (Camag). HPTLC–MS coupling was performed by the ChromeXtrakt (ChromAn, Germany) extractor. The methanol-formate buffer, 10 mM, pH 4.0 (95:5, v/v) was used as the extraction solvent with a flow of 0.1 mL/min. An HP1100 system (Hewlett-Packard) coupled with a VG platform II quadrupole mass spectrometer with an ESI-interface (Micromass, Manchester, UK) was used, and the 4000 Q Trap (Applied Biosistems/MDS Sciex, Toronto, Canada) was employed for comparative HPTLC-ESI-MS measurement in the SIM mode. The experimental results showed a repeatability of the MS signal with a mean RDS of 12.5% for the Harman standard sample. Limit of detection by a single quadrupole was better than 40 pg, and the limit of quantitation by tandem MS was better than 20 pg.

In a very recent study [246], an analytical method capable of simultaneously detecting all relevant compounds contained in an energy or sport drink, i.e., vitamin B_2, B_3, B_6, caffeine, and taurine was presented. Owing to its flexibility regarding the detection, planar chromatographic method, i.e., HPTLC–UV/Vis-FLD followed by HPTLC-ESI-MS for mass confirmation, was taken into account. Chromatographic separation was performed using a mixture of chloroform–ethanol–acetic acid–acetone–water (54:27:10:2:2, v/v) as mobile phase. UV and Vis *in situ* were recorded using TLC Scanner 3 (Camag). All the instruments were controlled via the software platform viaCats 1.4.1. Planar Chromatography Manager (Camag). Statistical analysis was performed with GraphPad Prism 4.0 and QSM 2.1 software. After the plate was scanned, the position of each spot was marked. Using the interface ChromeXtrakt (ChromAn, Germany), the HPTLC plate was connected to the VG platform II signal-quadrupole mass spectrometer (Micromass, Manchester, UK). The analyte was eluted online from the thin-layer with a solvent mixture of methanol–formate buffer 10 mmol/L, pH 4.0 (95:5, v/v) at a flow rate of 0.1 mL/min. The operation parameters were optimized. The mass confirmation was realized by a single quadrupole MS in ESI$^+$ scan mode for all substances, except taurine in the ESI$^-$ mode. This method offers a good alternative for routine analysis owing to its simplicity and reliability.

REFERENCES

1. Szekely, G. Combination of PC and TLC with some spectroscopic methods. In *Pharmaceutical Applications of Thin-Layer and Paper Chromatography* (Ed. K. Macek) Elsevier, Amsterdam, the Netherlands, 1972, 102–111.
2. Siouffi, A. M., Mincsovics, E., and Tyihak, E. Planar chromatographic techniques in biomedicine: Current status. *J. Chromatogr.* 1989, 492, 471–538.
3. Touchstone, J. C. New developments in planar chromatography. *LC-GC* 1993, 11(6), 404–411.
4. Wilson, I. D. Thin-layer chromatography: A neglected technique. *Ther. Drug Monit.* 1996, 18, 484–492.
5. Somsen, G., Morden, W., and Wilson, I. D. Planar chromatography coupled with spectroscopic techniques. *J. Chromatogr. A.* 1995, 703, 613–665.

6. Gocan, S. and Cimpan, G. Compound identification in thin-layer chromatography using spectrometric methods. *Rev. Anal. Chem.* 1997, XVI, 1–26.
7. Wilson, I. D. The state of the art in thin-layer chromatography-mass spectrometry: A critical appraisal. *J. Chromatogr. A.* 1999, 856, 429–442.
8. Jork, H., Funk, W., Fischer, W., and Wimmer, H. *Thin-Layer Chromatography. Reagents and Detection Methods*, Vol. 1a, VCH, Weinheim, Germany, 1990.
9. Jork, H., Funk, W., Fischer, W., and Wimmer, H. *Thin-Layer Chromatography. Reagents and Detection Methods*, Vol. 1b, VCH, Weinheim, Germany, 1994.
10. Cimpan, G. Pre- and post-chromatographic derivatization. In *Planar Chromatography. A Retrospective View for the Third Millennium* (Ed. Sz. Nyiredy), Springer Scientific Publisher, Budapest, 2001, 410–445.
11. Kocjan, B. Detection and identification of quinines in TLC. *J. Planar Chromatogr.* 2000, 13, 396–397.
12. Bladek, J. New methods of visualizing thin-layer chromatograms by using liquid crystals. *J. Chromatogr.* 1987, 405, 203–211.
13. Bladek, J. Parameters of the liquid crystal method of visualizing thin-layer chromatograms. *J. Chromatogr.* 1988, 437, 131–137.
14. Bladek, J. The application of liquid crystals to the visualization of organic substances absorbed on thin carbon layers. *J. Planar Chromatogr.* 1990, 3, 307–310.
15. Bladek, J. The application of liquid crystals to the quantitative determination of organic compounds. *J. Planar Chromatogr.* 1993, 6, 487–491.
16. Nyiredy, Sz., Fater, Zs., and Szabady, B. Identification in planar chromatography by use of retention data measured using characterized mobile phase. *J. Planar Chromatogr.* 1994, 7, 406–409.
17. Snyder, L. R. Classification of the solvents properties of common liquids. *J. Chromatogr. Sci.* 1978, 16, 223–234.
18. Merck, E. *Merck Tox Screening System*, Darmstadt, Germany, 1991.
19. De Zeeuw, R. A., Witte D. T., and Franke J. -P. Identification power of thin-layer chromatographic color reaction and integration of color codes in a data base for computerized identification in systematic toxicological analysis. *J. Chromatogr.* 1990, 500, 661–671.
20. Moffat, A. C., Franke, J. P., Stead, A. H., Gill, G., Finkle, B. S., Moller, M. R., Muller, R. K., Wunsch, F., and de Zeeuw, R. A. *Thin-Layer Chromatography R_F Values of Toxicologically relevant substances on Standardized Systems*. VCH, Weinheim, New York, 1987.
21. Franke, J. P. and De Zeeuw, R. A., *Proc. 21st Int. Meet. TIAFT* (Dunned, N. and Kimber, K., Eds.), Brighton, September 13–17, 1984, 73–77.
22. Witting, B., Zeitler, M., Schmidt, W., De Zeeuw, R. A., and Franke J. P. *Tox Analysis Screening Program for TLC, GC, UV data for Toxicologically Relevant Substances*, VCH, Weinheim, New York, 1988.
23. Ebel, S. and Wuther, W. Photodiode array-scanning in TLC/HPTLC. *Proc. 3rd Int. Symp. Instrumental HPTLC* (R. E. Kaiser Ed.), Wurzburg, RFG, 1985, 381–386.
24. Frodyma, M. M. and Lieu, V. T. Analysis by means of spectral reflectance of substances resolved on thin-plates. in *Modern Aspects of Reflectance Spectroscopy* (Wendlandt W. W. Ed.), Plenum Publishing, New York, 1968.
25. Beroza, M., Hill, K. R., and Norris, K. H. Determination of reflectance of pesticide spots on thin-layer chromatograms using fiber optics. *Anal. Chem.* 1968, 40, 1608–1613.
26. Ebel, S., Alert, D., and Schaefer, U. Computer controller multiwavelength evaluation in TLC. *Proc. 3rd Int. Symp. Planar Chromatogr.*, Wurzburg RFG, 1985, 373–379.
27. Ebel, S. and Windmann, W. Development of a system for measuring the UV spectra of components separated by TLC. *J. Planar Chromatogr.* 1991, 4, 171–173.
28. Ebel, S. Datenverarbeitung in TLC/HPTLC and HPTLC. *Fresenius Z. Anal Chem.* 1984, 381, 201–205.
29. Diaz, A. N. Fiber optic remote sensor for *in situ* fluorometric quantification in thin-layer chromatography. *Anal. Chim. Acta.* 1991, 255, 297–303.

30. Aposte, R. L., Diaz, J. A., Perera, A. A., and Diaz, V. G. Simple thin-layer chromatography method with fiber optic remote sensor for fluorimetric quantification of tryptophan and related metabolites. *J. Liq. Chromatogr.* 1996, 19, 687–698.

31. Linares, R. M., Ayala, J. H., Alfonse, A. M., and Gonzales, V. Quantitative analysis of biogenic amines by high-performance thin-layer chromatography utilizing a fiber optic fluorescence detector. *Anal. Lett.* 1998, 31, 475–489.

32. Spangenberg, B. and Klein, K. -F. Fibre optical scanning with high resolution in thin-layer chromatography. *J. Chromatogr. A.* 2000, 898, 265–269.

33. Spangenberg, B. and Klein, K. -F. New evaluation algorithm in diode-array thin-layer chromatography. *J. Planar Chromatogr.* 2001, 14, 260–265.

34. Spangenberg, B., Post, P., and Ebel, S. Fiber optical scanning in TLC by use of a diode-array detector—Linearization models for absorption fluorescence evaluation. *J. Planar Chromatogr.* 2002, 15, 88–93.

35. Spangenberg, B., Klein, K. F., and Mannhardt, J. Proposal for error reduction in planar chromatography. *J. Planar Chromatogr.* 2002, 15, 204–209.

36. Ebel, S. and Kang, J. S. UV/Vis spectra and spectral libraries in TLC/HPTLC. 1. Background correction and normalization. *J. Planar Chromatogr.* 1990, 3, 42–46.

37. Allwohn, J. and Ebel, S. Testing and validation of TLC scanner. *J. Planar Chromatogr.* 1989, 2, 71–75.

38. Deveny, T. High-speed video-densitometry. I. Principles and instrument. *Hung. Sci. Instrum.* 1976, 1, 36–40.

39. Kerenyi, Gy., Pataki, T., Devenyi, J., and Hevesi, G. High-speed video-densitometry part IV. Telechrome S: The modified video-densitometer. *Hung. Sci. Instrum.* 1980, 47, 1–7.

40. Pongor, S. High-speed video-densitometry: Principles and applications. *J. Liq. Chromatogr.* 1982, 5, 1583–1595.

41. Ford-Holevinski, T. S. and Radin, N. Quantitation of thin-layer chromatograms with an Apple II computer-based videodensitometer. *Anal. Biochem.* 1985, 150, 359–363.

42. Touchstone, J. C. and Sherma, J. In *Densitometry in Thin-Layer Chromatography*, John Wiley & Sons, New York, 1979.

43. Ebel, S. and Post, P. Theory of reflectance measurements in quantitative evaluation in TLC and HPTLC. 1. Multilayer model of reflectance and transmittance. *J. HRC CC.* 1881, 4, 337–342.

44. Ebel, S. Quantitative analysis in TLC and HPTLC. *J. Planar Chromatogr.* 1996, 9, 4–15.

45. Spangenberg, B. Does the Kubelka-Munk theory describe TLC evaluations correctly? *J. Planar Chromatogr.* 2006, 19, 332–341.

46. Treiber, L. R. Standardization of the direct spectrophotometric quantitative analysis of chromatograms. IV. A comparative study on the most common mathematical treatments of thin-layer densitometric problems. *J. Chromatogr.* 1974, 100, 123–135.

47. Spangenberg, B. and Weyandt-Spangenberg, M. Fluorescence evaluation using the Kubelka–Munk formula. *J. Planar Chromatogr.* 2004, 17, 164–168.

48. Ferenczi-Fodor, K., Nagy-Turak, A., and Vegh, Z. Validation and monitoring of quantitative thin-layer chromatographic purity test for bulk drug substances. *J. Planar Chromatogr.* 1995, 8, 349–356.

49. Sane, R. T., Menon, S. N., Mote, M., Inamdar, S., and Menezes, A. High-performance thin-layer chromatographic determination of aceclofenac in bulk drug and in pharmaceutical preparations. *J. Planar Chromatogr.* 2004, 17, 238–240.

50. Kofman, I. Sh. and Linetsky, M. D. Identification of multicomponent mixtures by spectrodensitometric TLC. *Proc. 3rd Symp. Instr. HPTLC* (R. E. Kaiser, Ed.) Wurzburg, 1985, 277–287.

51. Poole, C. F. and Poole, S. K. Progress in densitometry for quantitation in planar chromatography. *J. Chromatogr.* 1989, 492, 539–584.

52. Jamshidi, A. A convenient and high throughput HPTLC method for determination of progesterone in release media of silicon-based controlled-release drug-delivery systems. *J. Planar Chromatogr.* 2004, 17, 229–232.

53. Olah, N. -K., Hanganu, D., Oprean, R., Mogosan, C., Dubai, N., and Gocan, S. Selective extraction of caffeic acid derivatives from *Orthosiphon stamineus* Benth. (Lamiaceae) leaves. *J. Planar Chromatogr.* 2004, 17, 18–21.

54. Poole, S. K. and Poole, C. F. Thin-layer chromatographic method for the determination of principal polar aromatic flavour compounds of cinnamons of commerce. *Analyst,* 1994, 119, 113–120.

55. Katay, Gy., Tyihak, E., and Szokan, Gy. Separation of ascorbigen, 1'-methylascorbigen, and their derivatives by liquid chromatographic techniques. Part 1. Thin-layer chromatography. *J. Planar Chromatogr.* 1996, 9, 98–102.

56. Ojanpera, I., and Vuori, E. Identification of drugs in autopsy liver by instrumental qualitative thin-layer chromatography. *J. Chromatogr. A.* 1994, 674, 147–152.

57. Ojanpera, I., Nokua, J., Vuori, E., Sunila, P., and Sippola E. Novel for combined dual-system R_F and UV library search software: Application to forensic drug analysis by TLC and RPTLC. *J. Planar Chromatogr.* 1997, 10, 281–285.

58. Spangenberg, B., Seigel, A., Kempf, J., and Weinmann, W. Forensic drug analysis by means of diode-array HPTLC using R_F and UV library search. *J. Planar Chromatogr.* 2005, 18, 336–343.

59. Brandt, C. and Kovar, K. -A. Determination of 11-nor-Δ^9-tetrahydrocannabinol-9-carboxilic acid in urine by use of HPTLC-UV/FTIR on-line coupling. *J. Planar Chromatogr.* 1997, 10, 348–352.

60. Butz, S. and Stan, H. J. Screening of 265 pesticides in water by thin-layer chromatography with automated multiple development. *Anal. Chem.* 1995, 67, 620–630.

61. Yucang, L., Wei, W., Mingzhi, W., Yichun, P., Leming, L., and Jun, Z. Simultaneous determination of the residues of TNT and its metabolites in human urine by thin-layer chromatography. *J. Planar Chromatogr.* 1991, 4, 146–149.

62. Stahlmann, S., Herkert, T., Roseler, C., Rager, I., and Kovar, K. A. New sorbent for HPTLC-FTIR in situ determination of impurities in flurazepam. *Eur. J. Pharm. Sci.* 2001, 12, 461–469.

63. Huf, F. A. In situ evaluation of thin-layer chromatograms. In *Quantitative Thin-layer chromatography and Its Industrial Applications,* Vol. 36 (Ed. L. R.Treiber) Marcel Dekker, New York and Basel. Chromatographic Science series, 1987, 17–66.

64. Brown, K. and Poole, C. F. Fluorescence stability for 1-aminopyrene on silica-gel HPTLC plates. *LC Magazine* 1984, 2, 526–530.

65. Poole, C. F., Poole, S. K., Dean, T. A., and Chirco, N. M. Sample requirement for quantification in thin-layer chromatography. *J. Planar Chromatogr.* 1989, 2, 180–189.

66. Baeyens, W. R. G. and Ling, B. L. Thin-layer chromatography applications with fluorodensitometric detection. *J. Planar Chromatogr.* 1988, 1, 198–213.

67. Poole, C. F., Coddens, M. N., Butler, H. T., Schuette, S. A., Ho, S. S. J., Khatib, S., Piet, L., and Brown, K. K. Some quantitative aspects of scanning densitometry in HPTLC. *J. Liq. Chromatogr.* 1985, 8, 2875–2926.

68. Kartnig, T. and Gobel, I. Effect of fluorescence intensifiers on the fluorodensitometric determination of flavones and flavonols after detection with diphenylboric acid 2-aminoethyl ester. *J. Chromatogr. A.* 1996, 740, 99–107.

69. Butler, H. T., Coddens, M. E., Khatib, S., and Poole, C. F. Determination of polycyclic aromatic hydrocarbons in environmental samples by high performance thin-layer chromatography and fluorescence scanning densitometry. *J. Chromatogr. Sci.* 1985, 23, 200–207.

70. Butler, H. T. and Poole, C. F. Optimization of a scanning densitometer for fluorescence detection in HPTLC. *J. High Resol. Chromatogr.* 1983, 6, 77–81.

71. Butler, H. T., Coddens, M. E., and Poole, C. F. Qualitative identification of polycyclic aromatic hydrocarbon by high-performance thin-layer chromatography and fluorescence scanning densitometry. *J. Chromatogr.* 1984, 290, 113–126.

72. Butler, H. T. and Poole, C. F. Two point calibration method applied to fluorescence scanning densitometry and HPTLC. *J. Chromatogr. Sci.* 1983, 21, 385–388.

73. Ma, Y., Koutny, L. B., and Yeung, E. S. Laser-based indirect fluorometric detection and quantification in thin-layer chromatography. *Anal. Chem.* 1989, 61, 1931–1933.

74. Ma, Y. and Yeung, E. S. Indirect fluorimetric detection of non-electrolytes in thin-layer chromatography. *J. Chromatogr.* 1988, 455, 382–390.

75. Ma, Y. and Yeung, E. S. Indirect fluorometric detection of anion in thin-layer chromatography. *Anal. Chem.* 1988, 60, 722–724.

76. Imasaka, T., Tanaka, K., and Ishibashi, N. Thin-layer chromatography with supersonic jet fluorometric detection. *Anal. Chem.* 1990, 62, 374–378.

77. Wagner, S. D., Kaufer, S. W., and Sherma, J. Quantification of creatine in nutrition supplements by thin-layer chromatography-densitometry with thermochemical activation of fluorescence quenching. *J. Liq. Chromatogr. Related Technol.* 2001, 24, 2525–2530.

78. Hofstrat, J. W., Engelsman, M., Van de Nesse, Q. J., Brinkman, U. A. Th., Gooijer, C., and Velthorst, N. H. Coupling of narrow-bore liquid chromatography to thin-layer chromatography. Part 2. Application of fluorescence-based spectroscopic techniques for off-line detection. *Anal. Chim. Acta* 1987, 193, 193–207.

79. Van de Nesse, R. J., Hoogland, G. J. M., De Moel, J. J. M., Gooijer, C., Brinkman, U. A. Th., and Velthorst, N. H. On-line coupling of liquid chromatography to thin-layer chromatography for the identification of polycyclic aromatic hydrocarbons in marine sediment by fluorescence excitation and emission spectroscopy. *J. Chromatogr.* 1991, 552, 613–623.

80. Komsta, L. Determination of fenofibrate and genofibrozil in pharmaceuticals by densitometric and videodensitometric thin-layer chromatography. *J. Assoc. Off. Anal. Chem.* 2005, 88, 1517–1524.

81. Komsta, L. and Misztal, G. Determination of benzofibrate and ciprofibrate in pharmaceutical formulation by densitometric and videodensitometric TLC. *J. Planar Chromatogr.* 2005, 18, 188–193.

82. Vovk, I., Golc-Wondra, A., and Prosek, M. Validation of TLC method for caffeine determination with CCD camera. *Proc. 9th Int. Symp. Instr. Chromatogr.*, Interlaken, April 9–11, 1997, 365–372.

83. Wieczorrek, C. Suitability of inexpensive image-generating systems for evaluation of thin-layer chromatography and gel electrophoresis. *J Planar Chromatogr.* 2005, 18, 181–187.

84. Brown, S. M. and Bush, K. L. A charge-coupled device for optical detection of sample bands in thin-layer chromatograms. *J. Planar Chromatogr.* 1992, 5, 338–342.

85. Sturm, P. A., Parkhurst, R. M., and Skinner, W. A. Quantitative determination of individual tocopherols by thin-layer chromatographic separation and spectrophotometry. *Anal. Chem.* 1966, 38, 1244–1247.

86. McCoy, R. N. and Fiebig, E. C. Technique for obtaining infrared spectra of microgram amounts of compounds separated by thin-layer chromatography. *Anal. Chem.* 1965, 37, 593–595.

87. Rice, D. D. A direct transfer technique for preparing micropellets from thin-layer chromatograms for infrared identification. *Anal. Chim.* 1967, 39, 1906–1907.

88. De Klein, W. J. Infrared spectra of compounds separated by thin-layer chromatography using a potassium bromide micropellet technique. *Anal. Chem.* 1969, 41, 667–668.

89. Issaq, H. J. A combined thin-layer chromatography/microinfrared. *J. Liq. Chromatogr.* 1983, 6, 1213–1220.

90. Amos, R. Improved procedures for handling thin-layer chromatographic spots. *J. Chromatogr.* 1970, 48, 343–352.

91. Goodman, G. W. In *Quantitative Paper and Thin-Layer Chromatography*, E. J. Shellard (Ed.), Academic Press. London and New York, 1968, p. 91.

92. Garner, H. R. and Packer H. New technique for the preparation of KBr pellets from microsamples. *Appl. Spectrosc.* 1968, 22, 122–123.
93. Alt, K. O. and Szekely, G. Combination of infrared spectroscopy and thin-layer chromatography for the identification of slightly soluble substances. *J. Chromatogr.* 1980, 202, 151–153.
94. Gurkan, R. and Savasci, S. Investigation of separation and identification possibilities of some metal-DEDTC complexes by sequential TLC-IR system. *J. Chromatogr. Sci.* 2005, 43, 324–328.
95. Chalmers, J. M., Mackenzie, M. W., and Sharp, J. L. Characterization of thin-layer chromatographically separated fractions by Fourier transform infrared reflectance spectrometry. *Anal. Chem.* 1987, 59, 415–418.
96. Iwaoka, T., Tsutsumi, S., Tada, K., and Suzuki, F. FTIR spectrometry for TLC adsorbate-KBr micropyramid technique. *Ann. Rep. Sankyo Res. Lab.* 1988, 40, 39–46.
97. Shafer, K. H., Griffiths, P. R., and Shu-Qin, W. Sample transfer accessory for thin-layer chromatography/Fourier transform infrared spectrometry. *Anal. Chem.* 1986, 58, 2708–2714.
98. Kiss, G. C., Forgacs, E., Cserhati, T., and Vizcaino, J. A. Color pigments of *Trichoderma harzianum* preliminary investigations with thin-layer chromatography-Fourier transform infrared spectroscopy and high-performance liquid chromatography with diode array and mass spectrometric detection. *J. Chromatogr. A.* 2000, 896, 61–68.
99. Cserhati, T., Forgacs, E., Morais, M. H., and Romos, A. C. TLC-FTIR of pigments of chestnut sawdust. *J. Liq. Chromatogr. Relat. Technol.* 2001, 24, 1435–1445.
100. Cserhati, T., Forgacs, E., Candeias, M., Vilas-Boas, L., Bronze, R., and Spranger, I. Separation and tentative identification of the main pigment fraction of raisins thin-layer chromatography-Fourier transform infrared and high performance liquid chromatography-ultraviolet detection. *J. Chromatogr. Sci.* 2000, 38, 145–150.
101. Cimpoiu, C. Quantitative and qualitative analysis by hyphenated (HP) TLC-FTIR technique. *J. Liq. Chromatogr. Related Technol.* 2005, 28, 1203–1213.
102. Fuller, M. P. and Griffiths, P. R. Diffuse reflectance measurements by infrared Fourier transform spectrometry. *Anal. Chem.* 1978, 50, 1906–1910.
103. Sthalmann, S. A. Ten-year report on HPTLC-FTIR on-line coupling. *J. Planar Chromatogr.* 1999, 12, 5–12.
104. Frey, O. R., and Kovar, K. -A. Possibilities and limitations of assays by on-line coupling of thin-layer chromatography and FTIR spectrometry. *J. Planar Chromatogr.* 1993, 6, 93–99.
105. Yamamoto, H., Yoshikawa, O., Nakatani, M., Tsuji, F., and Maeda, T. Densitometry on thin-layer chromatographic plates by Fourier transform near-infrared spectrometry. *Appl. Spectrosc.* 1991, 45, 1166–1170.
106. Percival, C. J. and Griffiths, P. R. Direct measurement of the infrared spectra of compounds separated by thin-layer chromatography. *Anal. Chem.* 1975, 47, 145–156.
107. Brimmer, P. J. and Griffiths, P. R. Effect of absorbing matrices on diffuse reflectance infrared spectra. *Anal. Chem.* 1986, 58, 2179–2184.
108. Otto, A., Bode, U., and Heise, H. M. Experiences with infrared microsampling for thin-layer and high-performance liquid chromatography. *Fresenius Z. Anal. Chem.* 1988, 331, 376–382.
109. Danielson, N. D., Katon, J. E., Bouffard, S. P., and Zhu, Z. Zirconium oxide stationary phase for thin-layer chromatography with diffuse reflectance Fourier transform infrared detection. *Anal. Chem.* 1992, 64, 2183–2186.
110. Bouffard, S. P., Katon, J. E., Sommer, A. J., and Danielson, N. D. Development of microchannel thin-layer chromatography with infrared microspectroscopic detection. *Anal. Chem.* 1994, 66, 1937–1940.
111. Bauer, G. K., Pfeifer, A. M., Hauck, H. E., and Kovar, K. -A. Development of an optimized sorbent for direct HPTLC-FTIR on-line coupling. *J. Planar Chromatogr.* 1998, 11, 84–89.

112. Zuber, G. E., Waren, R. J., Begosh, P. P., and O'Donnell, E. L. Direct analysis of thin-layer chromatography spots by diffuse reflectance Fourier transform infrared spectrometry. *Anal. Chem.* 1984, 56, 2935–2939.

113. Beauchemin Jr., B. T. and Brown, P. R. Quantitative analysis of diazonaphthoquinones by thin-layer chromatographic/diffuse reflectance infrared Fourier transform spectrometry. *Anal. Chem.* 1989, 61, 615–618.

114. Yamamoto, H., Wada, K., Tajima, T., and Ichimura, K. Simple transfer technique for thin-layer chromatography-Fourier transform infrared spectroscopy. *Appl. Spectrosc.* 1991, 45, 253–259.

115. Bauer, G. K. and Kovar, K. -A. Direct HPTLC-FTIR on-line coupling: Interaction of acid and bases with the binder in precoated HPTLC plates. *J. Planar Chromatogr.* 1998, 11, 30–33.

116. Roesch, C. and Kovar K. -A. Synthetic second-generation narcotics (the so-called "designer drugs"). II. Analysis of arylalkylamines (amphetamines). *Pharm. Unserer Zeit*, 1990, 19, 211–221.

117. Kovar, K. -A., Ensslin, H. K., Frey, O. R., Rienas, S., and Wolff, S. C. Applications of on-line coupling of thin-layer chromatography and FTIR spectroscopy. *J. Planar Chromatogr.* 1991, 4, 246–250.

118. Kovar, K. -A. and Hoffman, V. Possibilities and limits on-line TLC-FTIR coupling. *GIT Fachz. Lab.* 1991, 35, 1197–1201.

119. Kovar, K. -A. TLC-FTIR coupling for the identification of forensic sample. *Proc. 9th Int. Symp. Instr. Chromatogr.* Interlaken, April 9–11, 1997, 161–162.

120. Wolff, S. C. and Kovar, K. -A. Determination of edetic acid (EDTA) in water by on-line coupling of HPTLC and FTIR. *J. Planar Chromatogr.* 1994, 7, 286–290.

121. Wolff, S. C. and Kovar, K. -A. Direct HPTLC-FTIR measurement in combination with AMD. *J. Planar Chromatogr.* 1994, 7, 344–348.

122. Pfeifer, A. M. and Kovar, K. -A. Identification of LSD, MBDB, and atropine in real samples with on-line HPTLC-FTIR coupling. *J. Planar Chromatogr.* 1995, 8, 388–392.

123. Pfeifer, A. M., Tolimann G., Ammon, H. P. T., and Kovar, K. -A. Identification of adenosine in biological samples by HPTLC-FTIR on-line coupling. *J. Planar Chromatogr.* 1996, 9, 31–34.

124. Pisternic, W., Kovar, K. -A., and Ensslin, H. High-performance thin-layer chromatographic determination of *N*-ethyl-3,4-methylenedioxyamphetamine and its metabolites in urine and comparison with high-performance liquid chromatography. *J. Chromatogr. B. Biomed. Sci.* 1997, 688, 63–69.

125. Stahmann, S. and Kovar, K. -A. Analysis of impurities by high-performance thin-layer chromatography with Fourier transform infrared spectroscopy and UV absorbance detection in situ measurement: Chlordiazepoxide in bulk powder and its tablets. *J. Chromatogr.* 1998, 813, 145–152.

126. Cimpoiu, C., Miclaud, V., Damian, G., Puia, M., Casoni, D., Bele, C., and Hodisan, T. Identification of new phthalazine derivatives by HPTLC-FTIR and characterization of their separation using molecular properties. *J. Liq. Chromatogr. Related Technol.* 2003, 16, 2687–2696.

127. Rosencwaig, A. and Hall, A. Thin-layer chromatography and photoacoustic spectrometry. *Anal. Chem.* 1975, 47, 548–549.

128. Castieden, S. L., Kirkbright, G. F., and Spillane, D. E. M. Quantitative examination of thin-layer chromatography plates by photoacoustic spectrometry. *Anal. Chem.* 1979, 51, 2152–2153.

129. Gray, C. R., Fishman, V. A., and Bard, A. J. Simple sample cell for examination of solids and liquids by photoacoustic spectroscopy. *J. Anal. Chem.* 1977, 49, 697–700.

130. Fishman, V. A. and Bard, A. J. Open-ended photoacoustic spectroscopy cell for thin-layer chromatography and other applications. *Anal. Chem.* 1981, 53, 102–105.

131. Lloyd, L. B., Yeates, R. C., and Eyring, E. M. Fourier transform infrared photoacoustic spectrometry in thin-layer chromatography. *Anal. Chem.* 1982, 54, 549–552.

132. White, R. R. Analysis of thin-layer chromatographic adsorbates by Fourier transforms infrared photoacoustic spectroscopy. *Anal. Chem.* 1985, 57, 1819–1822.

133. Everall, N. J., Chalmers, J. M., and Newton, L. D., *In situ* identification of thin-layer chromatography fractions by FT Raman spectroscopy, *Appl. Spectrosc.* 1992, 46, 567–601.

134. Sollinger, S. and Sawatzki, J., TLC Raman for the R&D laboratory, *Bruker Report* 1998, 146, 16–17.

135. Sollinger, S. and Sawatzki, J. TLC-Raman for routine applications. *GIT Labor-Fachzeitschrift* 1999, 43, 14–18.

136. Amstrong, D. W., Spino, L. A., Ondrias, M. R., and Findsen, E. W. Characterization of nanogram level of metalloporphyrins with thin-layer chromatography-resonance Raman spectroscopy *J. Chromatogr.* 1986, 369, 227–230.

137. Cimpoiu, C., Casoni, D., Hosu, A., Miclaus, V., Hodisan T., and Damian, G. Separation and identification of eight hydrophilic vitamins using a new TLC method and Raman spectroscopy. *J. Liq. Chromatogr. Related Technol.* 2005, 28, 2551–2559.

138. Koglin, E. Combining HPTLC and micro-surface-enhanced Raman spectroscopy. *J. Planar Chromatogr.* 1989, 2, 194–197.

139. Creighton, J. A., Blatchford, C. G., and Albrecht, M. G. Plasma resonance enhancement of Raman scattering by pyridine adsorbed on silver or gold sol particles of size comparable to the excitation wavelength. *J. Chem. Soc. Faraday Trans.* 1979, 75, 790–798.

140. Sequaris, J. M. L. and Koglin, E. Direct analysis of high-performance thin-layer chromatography spot of nucleic purine derivatives by surface enhanced Raman scattering spectrometry. *Anal. Chem.* 1987, 59, 525–527.

141. Koglin, E. *In situ* identification of highly fluorescent molecules on HPTLC plates by surface-enhanced resonance Raman scattering. *J. Planar Chromatogr.* 1990, 3, 117–120.

142. Koglin, E. Characterization of cetylpyridinium chloride on HPTLC plates by surface-enhanced resonance Raman scattering. *J. Planar Chromatogr.* 1993, 6, 88–92.

143. Koglin, E. TLC (HPTLC)—SERS-kopplung. *GIT Labor-Fachzeitschrift* 1994, 38, 627–632.

144. Soper, S. A., Ratzlaff, K. L., and Kuwana, T. Surface-enhanced resonance Raman spectroscopy of liquid chromatographic analytes on thin-layer chromatographic plates. *Anal. Chem.* 1990, 62, 1438–1444.

145. Lee, P. C. and Meisel, D. Adsorption and surface-enhanced Raman of dyes on silver and gold sols. *J. Phys. Chem.* 1982, 86, 3391–3395.

146. Somsen, G. W., Patricia ter Riet, G. J. H., Gooijer, C., Velthorst, N. H., and Brinkman, U. A. Th. Characterization of aminotriphenylmethane dyes by TLC coupled with surface-enhanced resonance Raman spectroscopy, *J. Planar Chromatogr.* 1997, 10, 1017.

147. Matejka, P., Stavec, J., Volka, K., and Schrader, B. Near-infrared surface-enhanced Raman scattering spectra of heterocyclic and aromatic species adsorbed on thin-layer plates activates with silver. *Appl. Spectrosc.* 1996, 50, 409–414.

148. Burger, K. *Kopplung ADM-SERS/Raman.* InCom-Sonderband, 1997, 47–55

149. Wang, Y. *In situ* thin-layer chromatography-Fourier transform-surface-enhanced Raman scattering spectroscopy of amino acids. *Chin. Anal. Chem. (Fenxi Huaxue),* 1998, 26, 1047–1051. (From Camag Bibliogr. Service, 51–82.)

150. Wang, Y. Analysis of hyascine scopolamine by *in situ* thin-layer chromatography/surface enhanced near Fourier transform Raman spectroscopy. *Chin. Anal. Chem. (Fenxi Huaxue),* 1998, 26, 1406. (From Camag Bibliogr. Service, 51–82.)

151. Wang, Y., Wang, Y. F., and Ren, G. F. *In situ* identification of thin-layer chromatography-Fourier Raman spectroscopy of yohimbine in *Rauvolfia verticulata* Baill. *Yaowu Fenxi Zahi,* 2004, 24, 30–34 (*Anal. Abstr.* 67–33-G-10225).

152. Hillenkamp, F., Karas, M., Beavis, R., and Chait, B. T. Matrix-assisted laser desorption/ionization mass spectrometry of biopolymers. *Anal. Chem.* 1991, 63, 1193 A–1202 A.

153. Gusev, A. I., Proctor, A., Rabinovich, I., and Hercules, D. M. Thin-layer chromatography combined with matrix-assisted laser desorption/ionization mass spectrometry. *Anal. Chem.* 1995, 67, 1805–1814.

154. Kaiser, R. Thin-layer chromatography in direct coupling with gas chromatography and mass spectrometry. *Chem. Br.* 1969, 5, 54–61.

155. Flurer, R. A. and Busch, K. L. Data system for imaging analysis with a secondary-ion mass spectrometer. *Anal. Instrum. (NY)*, 1988, 17, 255–276.

156. Fiola, J. W., DiDonato, G. C., and Bush, K. L. Modular instrument for organic secondary ion mass spectrometry and direct chromatographic analysis. *Rev. Sci. Instrum.* 1986, 57, 2294–2302; Henion, J. and Maylin, G. A. Determination of drugs in biological samples by thin-layer chromatography-tandem mass spectrometry. *J. Chromatogr.* 1983, 271, 107–124.

157. Henion, J. and Maylin, G. A. Determination of drugs in biological samples by thin-layer chromatography-tandem mass spectrometry. *J. Chromatogr.* 1983, 271, 107–124.

158. Wilson I. D. and Morden, W. Advances and applications in the use of HPTLC-MS-MS. *J. Planar Chromatogr.* 1996, 9, 84–91.

159. Esteban, J. Eluocrom device. *Technicas Lab.* 1977, 5(59), 285–292.

160. Somogyi, G., Dinya, Z., Laczko, A., and Prokai, L. Mass spectrometric identification of thin-layer chromatographic spots: Implementation of a modified Wick-Stick technique. *J. Planar Chromatogr.* 1990, 3, 191–193.

161. Anderson, R. M. and Busch, K. Thin-layer chromatography coupled with mass spectrometry: Interfaces to electrospray ionization. *J. Planar Chromatogr.* 1998, 11, 336–341.

162. Fagarasan, M. and Fagarasan, E. Determination of non-polar metabolites of amitriptyline by mass spectrometry. *Rev. Chim.* (Bucharest), 1987, 38, 343–345.

163. Fagarasan, M. and Fagarasan, E. Determination of hydroxyl-metabolites of amitriptyline by mass spectrometry. *Rev. Chim.* (Bucarest), 1986, 37, 725–726.

164. Moise, M. -I., Marutoiu, C., Badea, D. N., Gavrila, C. -A., and Patroescu, C. Application of TLC and GC-MS to the detection of Capsaicin from hot peppers (*Capsicum annuum*). *J. Planar Chromatogr.* 2004, 17, 147–148.

165. Brzezinka, H. and Bertram, N. Combined thin-layer and mass spectrometry for the screening of pesticides in samples derived from biological origins. *J. Chromatogr. Sci.* 2002, 40, 609–613.

166. Simonovska, B., Srbinoska, M., and Vovk, I. Analysis of sucrose esters-insecticides from the surface of tobacco plant leaves. *J. Chromatogr. A* 2006, 1127, 273–277.

167. Liu, J. Q., Zhuang, H. Q., Mo, L. E., and Li, Q. N. TLC-MS identification of coumarins from extracts of *Cnidium monnieri* (L.) Cusson. *Fenxi Ceshi Xuebao*, 1999, 18, 26–28 (Anal. Abstr. 61–12-H-00219).

168. Danigel, H., Schmidt, L., Jungclas, H., and Pflueger, K. -H. Combined thin-layer chromatography-mass spectrometry: An application of californium-252 plasma desorption mass spectrometry for drug monitoring. *Biomed. Mass Spectrom.* 1985, 12, 542–544.

169. Walton, T. J., Mullins, C. J., Newton, R. P., Brenton, A. G., and Beynon, J. H. Tandem mass spectrometry in vitamin E analysis. *Biomed. Environ. Mass Spectrom.* 1988, 16, 289–298.

170. Toledo, M. S., Levery, S. B., Glushka, J., and Straus, A. Structure elucidation of sphingolipids from the mycopathogen *Sporothrix schenckii*: Identification of novel glycosylinositol phosphorylceramides with core Manα1 → 6Ins linkage. *Biochem. Biophys. Res. Commun.* 2001, 280, 19–24.

171. Meisen, I., Peter-Katalinic, J., and Muthing, J. Direct analysis of silica-gel extract from immunostained glycosphingolipids by nanoelectrospray ionization quadrupole time-of-flight mass spectrometry. *Anal. Chem.* 2004, 76, 2248–2255.

172. Prosek, M., Golc-Wondra, A., Vovk, I., and Andrensek, S. Quantification of caffeine by off-line TLC-MS. *J. Planar Chromatogr.* 2000, 13, 452–456.

173. Chang, T., Lay, J., and Francel, R. Direct analysis of thin-layer chromatography spots by fast atom bombardment mass spectrometry. *Anal. Chem.* 1984, 56, 109–111.

174. Oka, H., Ikai, Y., Konodo, F., Hayakawa, N., Masuda, K., Harada, K. -I., and Suzuki, M. Development of a condensation technique for thin-layer chromatography/fast-atom bombardment mass spectrometry of non-visible compounds. *Rapid Commun. Mass Spectrom.* 1992, 6, 89–94.

175. Oka, H., Ikai, Y., Ohno, T., Kawamura, N., Hayakawa, J., Harada, K., and Suzuki, M. Identification of unlawful food dyes by thin-layer chromatography-fast atom bombardment mass spectrometry. *J. Chromatogr.* 1994, 674, 301–307.

176. Bare, K. J. and Red, H. Use of fast atom bombardment mass spectrometry to identify materials separated on high-performance thin-layer chromatography plates. *Analyst,* 1987, 112, 433–436.

177. Zhongping, Y., Guoqiao, L., Hanhui, W., Weide, H., Shankai, Z., and Wenmei, H. Determination of the glycosyl sequence of glycosides by enzymatic hydrolysis followed by TLC-FABMS. *J. Planar Chromatogr.* 1994, 7, 410–412.

178. Martin, P., Morden, W., Wall, P., and Wilson, I. D. TLC combined with tandem mass spectrometry: Application to the analysis of antipyrine and its metabolites in extracts of human urine. *J. Planar Chromatogr.* 1992, 5, 255–258.

179. Lafont, R., Porter, C. J., Williams, E., Read, H., Morgan, E. D., and Wilson, I. D. The application of off-line HPTLC-MS-MS to the identification of ecdysteroids in plant and arthropod samples. *J. Planar Chromatogr.* 1993, 6, 421–424.

180. Wilson, I. D. and Morden, W. Application of thin-layer chromatography-mass spectrometry to drugs and their metabolites. Advantages of tandem MS-MS. *J. Planar Chromatogr.* 1991, 4, 226–229.

181. Morden, W. and Wilson, I. D. Separation of nucleotides and bases by high performance thin-layer chromatography with identification by tandem mass spectrometry. *J. Planar Chromatogr.* 1995, 8, 98–102.

182. DiDonato, G. C. and Busch, K. L. Phase-transition matrix for chromatography/secondary ion mass spectrometry. *Anal. Chem.* 1986, 58, 3231–3232.

183. Masuda, K., Harada, K. -I., Suzuki, M., Oka, H., Kawamura, N., and Yamada, M. Identification of food dyes by TLC/SIMS with a condensation technique. *Org. Mass Spectrom.* 1989, 24, 74–75.

184. Kushi, Y., Rokukawa, C., and Handa, S. Direct analysis on thin-layer plates by matrix-assisted secondary ion mass spectrometry: Application for glycolipid storage disorders. *Anal. Biochem.* 1988, 175, 167–176.

185. Xia, Y. and Busch, K. L. Analysis of organic sulfonium salt mixtures by capillary electrophoresis and thin-layer chromatography-mass spectrometry. *J. Planar Chromatogr.* 1998, 11, 186–190.

186. Gusev, A. I., Vasseur, O. J., Proctor, A., Sharkey, A. G., and Hercules, D. M. Imaging of thin-layer chromatograms using matrix assisted laser desorption/ionization mass spectrometry. *Anal. Chem.* 1995, 67, 4565–4570.

187. Mehl, J. T., Gusev, A. I., and Hercules, D. M. Coupling protocol for thin-layer chromatography matrix-assisted laser desorption ionization. *Chromatographia* 1997, 46, 358–364.

188. Mehl, J. T. and Hercules, D. M. Direct TLC-MALDI coupling using a hybrid plate. *Anal. Chem.* 2000, 72, 68–73.

189. Peng, S., Ahlmann, N., Kunze, K., Nigge, W., Edler, M., Hoffmann, T., and Franzke, J. Thin-layer chromatography combined with diode laser desorption/atmospheric pressure chemical ionization mass spectrometry. *Rapid Commun. Mass Spectrom.* 2004, 18, 1803–1808.

190. Nicola A. J., Gusev, A. I., and Hercules, D. M. Direct quantitative analysis from thin-layer chromatography plates using matrix-assisted laser desorption-ionization mass spectrometry. *Appl. Spectrom.* 1996, 50, 1479–1482.

191. Vermilion-Salsburi, R. L., Hoops, A. A., Gusev, A. I., and Hercules, D. H. Analysis of cationic pesticides by thin-layer chromatography-matrix-assisted laser desorption mass spectrometry. *Int. J. Environ. Anal. Chem.* 1999, 73, 179–190.

192. Hayen, H. and Volmer, D. A. Rapid identification of siderophores by combined thin-layer chromatography/matrix-assisted laser desorption/ionization mass spectrometry. *Rapid Commun. Mass Spectrom.* 2005, 19, 711–720.

193. Nakamura, K., Suzuki, Y., Gato-Inoue, N., Yoshida-Noro, C., and Suzuki, A. Structural characterization of neutral glycosphingolipids by thin-layer chromatography coupled to matrix-assisted laser desorption/ionization quadrupole ion trap time-of-flight MS/MS. *Anal. Chem.* 2006, 78, 5736–5743.

194. O'Connor, P. B., Budnik, B. A., Ivleva, V. B., Kaur, P., Moyer, S. C., Pittman, J. L., and Costello, C. E. A high pressure matrix-assisted laser desorption ion source for Fourier transform mass spectrometry designed to accommodate large targets with diverse surfaces. *J. Am. Soc. Mass Spectrom.* 2004, 15, 128–132.

195. O'Connor, P. B., Mirgorodskaya, E., and Costello, C. E. High pressure matrix-assisted laser desorption/ionization Fourier transform mass spectrometry for minimization of ganglioside fragmentation. *J. Am. Soc. Mass Spectrom.* 2002, 13, 402–407.

196. Isbell, D. T., Gusev, A. I., Taranenko, N. I., Chen, C. H., and Hercules, D. M. Separation and detection of carcinogen-adducted oligonucleotides. *J. Mass Spectrom.* 1999, 34, 1076–5174.

197. Luxemburg, S. L., McDonnell, L. A., Duursma, M. C., Gou, X., and Heeren, R.M.A. Effect of local matrix crystal variations in matrix-assisted ionization techniques for mass spectrometry. *Anal. Chem.* 2003, 75, 2333–2341.

198. Luxemburg, S. L., Mize, T. H., McDonnell, L. A., and Heeren, R. M. A. High-spatial resolution mass spectrometric imaging of peptide and protein distributions on a surface. *Anal. Chem.* 204, 76, 5339–5344.

199. Sunner, J., Dratz, E., and Chen, Y. -C. Graphite surface-assisted laser desorption/ionization time-of-flight mass spectrometry of peptides and proteins from liquid solutions. *Anal. Chem.* 1995, 67, 4355–4342.

200. Chen, Y. -C., Shiea, J., and Sunner, J. Thin-layer chromatography-mass spectrometry using activated carbon, surface-assisted laser desorption/ionization. *J. Chromatogr. A.* 1998, 826, 77–86.

201. Tanaka, K., Waki, H., Ido, Y., Akita, S., Yoshida, Y., Yoshida, T., and Matsuo, T. Protein and polymer analyses up to *m/z* 100 000 by laser ionization time-of-flight mass spectrometry. *Rapid Commun. Mass Spectrom.* 1988, 2, 151–153.

202. Chen, Y. -C., Shiea, J., and Sunner, J. In *Proc. 44th ASMS Conf. Mass Spectrom. Allied Topics*, May 12–16, 1996, Portland, OR, p. 642.

203. Dale, M. J., Knochenmuss, R., and Zenobi, R. Graphite/liquid mixed matrices for laser desorption/ionization mass spectrometry. *Anal. Chem.* 1996, 68, 3321–3329.

204. Therisod, H., Labas, V., and Caroff, M. Direct microextraction and analysis of rough-type lipopolysaccharides by combined thin-layer chromatography and MALDI mass spectrometry. *Anal. Chem.* 2001, 73, 3804–3807.

205. Cercelius, A., Clench, M. R., Richard, D. S., and Parr, V. Thin-layer chromatography-matrix-assisted laser desorption ionization-time-of-flight mass spectrometry using particle suspension matrices. *J. Chromatogr. A* 2002, 958, 249–260.

206. Busch, K. L. Planar separation and mass spectrometric detection. *J. Planar Chromatogr.* 2004, 17, 398–403.

207. Ramaley, L., Vaughan, M. -A., and Jamieson, W. Characteristic of a thin-layer chromatogram scanner-mass spectrometer system. *Anal. Chem.* 1985, 57, 353–358.

208. Nakagawa, Y. and Iwatani, K. Scanning thin-layer chromatography-liquid secondary ion mass spectrometry and its application for investigation of drug metabolites. *J. Chromatogr.* 1991, 562, 99–110.

209. Iwatani, K., Kadono, T., and Nakagawa, Y. Direct coupling of thin-layer chromatography [TLC] with mass spectrometry [MS]. Direct TLC/MS using new sintered TLC plate. *Jpn. Mass Spectrosc.* 1986, 34, 181–187.

210. Iwatani, K. and Nakagawa, Y. Direct coupling of thin-layer chromatography with mass spectrometry. Direct TLC/SIMS. *Jpn. J. Mass Spectrom.* 1986, 34, 169–196.

211. Kubis, A. J., Somayajula, K. V., Sharkey, A. G., and Hercules, D. M. Laser mass spectrometric analysis of compounds separated by thin-layer chromatography. *Anal. Chem.* 1989, 61, 2516–2523.

212. Li, L. and Lubman, D. M. Resonant two-photon ionization spectroscopic analysis of thin-layer chromatography using pulsed laser desorption/volatilization into supersonic jet expansions. *Anal. Chem.* 1989, 61, 1911–1915.

213. Busch, K. L., Mullis, J. O., and Carlson, R. E. Planar chromatography coupled to mass spectrometry. *J. Liq. Chromatogr.* 1993, 16, 1695–1713.

214. Doherty, S. J. and Busch, K. L. An optical CCD camera integrated with a secondary ion mass spectrometer for analysis of planar chromatograms. *J. Planar Chromatogr.* 1089, 2, 149–151.

215. Stanley, M. S. and Busch, K. L. Primary beam and ion extraction optics optimization for an organic secondary ion mass spectrometer. *Anal. Instrum.* 1989, 18, 243–264.

216. Busch, K. L., Mullis, J. O., and Chakel, J. A. High resolution imaging of samples in thin-layer chromatograms using a time-of-flight secondary ion mass spectrometer. *J. Planar Chromatogr.* 1992, 5, 9–15.

217. Busch, K. L., Brown, S. M., Doherty, S. J., Dunphy, J. K., and Buchanan, M. V. *J. Liq. Chromatogr.* 1990, 13, 2841–2869.

218. Duffin, K. L. and Busch, K. L. Analysis of phosphonium salt mixtures by thin-layer chromatography/secondary ion mass spectrometry. *J. Planar Chromatogr.* 1988, 1, 249–251.

219. Brown, S. M., Schurz, H., and Busch, K. L. Interface devices for combination of planar chromatography with mass spectrometry. *J. Planar Chromatogr.* 1990, 3, 222–227.

220. Hsu, F. -L., Chen, C. -H., Yuan, C. -H., and Shiea, J. Interface to connect thin-layer chromatography with electrospray ionization mass spectrometry. *Anal. Chem.* 2003, 75, 2493–2496.

221. Santos, L. S., Haddad, R., Hoehler, N. F., Pilli R. A., and Eberlin, M. N. *Anal. Chem.* 2004, 76, 2144–2147.

222. Prosek, M., Milivojevic, L., Krizman, M., and Fir, M. On-line TLC-MS, *J. Planar Chromatogr.* 2004, 17, 420–423.

223. Crecelius, A., Clench, M. R., Richards, D. S., Evason, D., and Parr, V. Thin-layer chromatography-postsource-decay matrix-assisted laser desorption/ionization time-of-flight mass spectrometry of small drug molecules. *J. Chromatogr. Sci.* 2002, 40, 614–620.

224. Crecelius, A., Clench, M. R., and Richards, D. S. TLC-MALDI in pharmaceutical analysis. *LC-GC Europe*, 2003, 16(4), 225–229.

225. Crecelius, A., Clench, M. R., and Richards, D. S., TLC-MALDI in pharmaceutical analysis. *Curr. Trends Mass Spec.* 2004, 19(55), 28–34.

226. Crecelius, A., Clench, M. R., Richards, D. S., and Parr, V. Quantitative determination of piroxicam by TLC-MALDI TOF MS. *J. Pharm. Biomed. Anal.* 2004, 35, 31–39.

227. Guitard, J., Hronowski, X. L., and Costello, C. E. Direct matrix-assisted laser desorption/ionization mass spectrometric analysis of glycosphingolipids on thin-layer chromatographic plates and transfer membranes. *Rapid Commun. Mass Spectrom.* 1999, 13, 1838–1849.

228. Crecelius, A., Clench, M. R., Richards, D. S., and Parr, T. Thin-layer chromatography-matrix-assisted laser desorption ionization-time-of-flight mass spectrometry using particle suspension matrices. *J. Chromatogr. A* 2002, 958, 249–260.

229. Ivleva, V. B., Sapp, L. M., O'Connor, P. B., and Costello, C. E. Ganglioside analysis by thin-layer chromatography matrix-assisted laser desorption/ionization orthogonal time-of-flight mass spectrometry. *J. Am. Soc. Mass Spectrom.* 2005, 16, 1552–1560.

230. Ivleva, V. B., Elkin, Y. N., Budnik, B. A., Moyer, S. C., O'Connor, P. B., and Costello, C. E. Coupling thin-layer chromatography with vibrational cooling matrix-assisted laser desorption/ionization Fourier transform mass spectrometry for the analysis of ganglioside mixtures. *Anal. Chem.* 2004, 76, 6484–6512.

231. Dreisewerd, K., Muthing, J., Rohlfing, A., Meisen, I., Vukelic, Z., Peter-Katalinic, J., Hillenkamp, F., and Berkenkamp, S. Analysis of gangliosides directly from thin-layer chromatography plates by infrared matrix-assisted laser desorption/ionization orthogonal time-of-flight mass spectrometry with a glycerol matrix. *Anal. Chem.* 2005, 77, 4098–4107.

232. Loboda, A. V., Ackloo, S., and Chernushevich, I. V. A high-performance matrix assisted lased desorption orthogonal time-of-flight mass spectrometer with collisional cooling. *Rapid Commun. Mass Spectrom.* 2003, 17, 2508–2516.

233. Salo, P. K., Salomies, H., Harju, K., Ketola, R. A., Kotiaho, T., Yli-Kauhaluoma, J., and Kostiainen, R. Analysis of small molecules by ultra thin-layer chromatography-atmospheric pressure matrix-assisted laser desorption/ionization mass spectrometry. *J. Am. Soc. Mass Spectrom.* 2005, 16, 906–915.

234. Van Berkel, G. J., Sanchez, A. D., and Quirke, J. M. E. Thin-layer chromatography and electrospray mass spectrometry coupled using a surface sampling probe. *Anal. Chem.* 2002, 74, 6216–6223.

235. Van Berkel, G. J., Ford, J. M., and Deibel, M. A. Thin-layer chromatography and mass spectrometry coupled using desorption electrospray ionization. *Anal. Chem.* 2005, 77, 1207–1215.

236. Ford, M. J. and Van Berkel, G. J. An improved thin-layer chromatography/mass spectrometry coupling using a surface sampling probe electrospray ion trap system. *Rapid Commun. Mass Spectrom.* 2004, 18, 1303–1309.

237. Ford, M. J., Kertesz, V., and Van Berkel, G. J. Thin-layer chromatography/electrospray ionization triple quadrupole ion trap mass spectrometry system: Analysis of rhodamine dyes separated on reversed-phase C8 plate. *J. Mass Spectrom.* 2005, 40, 866–875.

238. Wachs, T. and Henion, J. Electrospray device for coupling microscale separations and other miniaturized devices with electrospray mass spectrometry. *Anal. Chem.* 2001, 73, 632–636.

239. Van Berkel, G. J. and Kertesz, V. Automated sampling and imaging of analytes separated on thin-layer chromatography plates using desorption electrospray ionization mass spectrometry. *Anal. Chem.* 2006, 78, 4938–4944.

240. Kertesz, V., Ford, M. J., and Van Berkel, G. J. Automation of a surface sampling probe/electrospray mass spectrometry system. *Anal. Chem.* 2005, 77, 7183–7189.

241. Ford, M. J., Deibel, M. A., Tomkins, B. A., and Van Berkel, G. J. Quantitative thin-layer chromatography/mass spectrometry analysis of caffeine using a surface sampling probe electrospray ionization tandem mass spectrometry system. *Anal. Chem.* 2005, 77, 4385–4389.

242. Anderson, R. M. and Busch, K. L. Thin-layer chromatography coupled with mass spectrometry: Interface to electrospray. *J. Planar Chromatogr.* 1998, 11, 336–341.

243. Luftmann, H. A simple device for the extraction of TLC spots: Direct coupling with an electrospray mass spectrometer. *Anal. Bioanal. Chem.* 2004, 378, 964–968.

244. Alpmann, A. and Morlock, G. Improved online coupling of planar chromatography with electrospray mass spectrometry: Extraction of zones from glass plates. *Anal. Bioanal. Chem.* 2006, 386, 1543–1551.

245. Jautz, U. and Morlock, G. Efficacy of planar chromatography coupled to (tandem) mass spectrometry for employment in trace analysis. *J. Chromatogr. A* 2006, 1128, 244–250.

246. Aranda, M. and Marlock, G. Simultaneous determination of riboflavin, pyridoxine, nicotinamide, caffeine and taurine in energy drinks by planar chromatography-multiple detection with confirmation by electrospray ionization mass spectrometry. *J. Chromatogr. A* 2006, 1131, 253–260.

Index

A

Acrylamide-based monoliths, 136–137
Adenovirus proteins characterization, 21–24
   reverse phase-high-performance liquid
      chromatography (RP-HPLC), 22–24
   structure, 21–22
Aminopropyl silica (APS) reactions, 129–131
Arrhenius' reaction rate equation, 288–289
Automated multiple development (AMD), 369

B

Back-extrusion CCC method (BECCC), 344
Batch-gradient chromatography, 173–174
Binding strength, ion-exchange purification
   adsorption isotherms, 216
   anion–cation exchangers, ranking
      strength, 205
   anion-exchange resins, BSA, 211
   biopharmaceutical industry, 217
   cation-exchange resins, lysozyme, 212
   conductivity data, 204–205
   fractogel EMD resins, 211
   isocratic runs, 202
   Langmuir fits, 214
   pure test proteins, gradient mode, 230
   static capacities and absorption
      isotherms, 207
Bioproducts fingerprinting
   carbohydrate-based bioprocesses
      alternate energy source, 268
      lignocellulosic material, 267
      spirizyme fuel, 268
   enzyme characterization
      extraction efficiency values, 265
      Hanes plots, 265
      Michaelis–Menten enzyme kinetic
         parameters, 264
      salicin, 266
   glucuronides, toxicological applications
      alcohol, 271
      ethyl glucuronide, 271
      PED chromatogram, 273
   high-performance anion-exchange
      chromatography
      bacterial polysaccharides, carbohydrate
         analysis, 261
      Bureau of Alcohol, Tobacco, and
         Firearms agency, 259
      flue-curing process, 258

      peptones characterization, 260–261
      tobaccco classification, 258–260
   in vitro microdialysis
      MD sampling system, 261
      molecular weight cutoff, 263
Bovine serum albumin (BSA)
   adsorption isotherms, 213
   anion-exchange resins, 211
   in binding capacity strength, 208
   common test proteins, 195
   in pH range, 197
   properties, 196

C

Capillary electrochromatography (CEC)
   monolithic columns
      acrylamide-based monoliths, 136–137
      C_{17} monolithic stationary phase, 142–144
      ionexchange chromatography, 144–146
      RP chromatographic retention, 146–147
      SAX–RP single step preparation, 138–142
      SCX–RP mixed-mode, 138–142
      sol-gel silica backbone, 147–154
      2-(sulfooxy)ethyl methacrylate (SEMA)
         copolymerization, 137–138
   octadecyl silica columns, 127–128
   open-tubular columns, 154
   packed columns
      aminopropyl silica (APS) reactions,
         129–131
      porous silica particles, 133–135
      SCX–RP mixed-mode, 128–129
   separation mechanism
      charged analytes, 154–155
      protein separation, 156–157
Capillary liquid chromatography (CLC), 275
Capto MMC resin, 215–216
Charge-coupled device (CCD), 361
Chip-based online nanospray mass
      spectrometry, 288
ChromeXtractor
   HPTLC–MS coupling, 432
   TLC aluminum foils, 431
C_{17} monolithic stationary phase, 142–144
Collision-induced dissociation (CID), 391
Column capacity, chromatography, 210
Column configuration modulation, 180–182
Conventional SMB process
   advantages and disadvantages, 168–169
   basic design, complete separation, 170–172

Printed and bound by CPI Group (UK) Ltd, Croydon, CR0 4YY

22/10/2024

01777815-0001